AF868429

Astrophysics and Space Science Library

Volume 461

Series Editor

Steven N. Shore, Dipartimento di Fisica "Enrico Fermi", Università di Pisa, Pisa, Italy

The Astrophysics and Space Science Library is a series of high-level monographs and edited volumes covering a broad range of subjects in Astrophysics, Astronomy, Cosmology, and Space Science. The authors are distinguished specialists with international reputations in their fields of expertise. Each title is carefully supervised and aims to provide an in-depth understanding by offering detailed background and the results of state-of-the-art research. The subjects are placed in the broader context of related disciplines such as Engineering, Computer Science, Environmental Science, and Nuclear and Particle Physics. The ASSL series offers a reliable resource for scientific professional researchers and advanced graduate students.

Series Editor:
STEVEN N. SHORE, Dipartimento di Fisica "Enrico Fermi", Università di Pisa, Pisa, Italy

Advisory Board:
F. BERTOLA, University of Padua, Italy
C. J. CESARSKY, Commission for Atomic Energy, Saclay, France
P. EHRENFREUND, Leiden University, The Netherlands
O. ENGVOLD, University of Oslo, Norway
E. P. J. VAN DEN HEUVEL, University of Amsterdam, The Netherlands
V. M. KASPI, McGill University, Montreal, Canada
J. M. E. KUIJPERS, University of Nijmegen, The Netherlands
H. VAN DER LAAN, University of Utrecht, The Netherlands
P. G. MURDIN, Institute of Astronomy, Cambridge, UK
B. V. SOMOV, Astronomical Institute, Moscow State University, Russia
R. A. SUNYAEV, Max Planck Institute for Astrophysics, Garching, Germany

More information about this series at http://www.springer.com/series/5664

Tomaso M. Belloni • Mariano Méndez •
Chengmin Zhang
Editors

Timing Neutron Stars: Pulsations, Oscillations and Explosions

Springer

Editors
Tomaso M. Belloni
INAF Osservatorio Astronomico di Brera
Merate, Italy

Mariano Méndez
Kapteyn Astronomical Institute
University of Groningen
Groningen, The Netherlands

Chengmin Zhang
National Astronomical Observatories
and
The University of Chinese Academy
of Sciences
Chinese Academy of Sciences
Beijing, China

ISSN 0067-0057 ISSN 2214-7985 (electronic)
Astrophysics and Space Science Library
ISBN 978-3-662-62108-0 ISBN 978-3-662-62110-3 (eBook)
https://doi.org/10.1007/978-3-662-62110-3

© Springer-Verlag GmbH Germany, part of Springer Nature 2021
This work is subject to copyright. All rights are reserved by the Publisher, whether the whole or part of the material is concerned, specifically the rights of translation, reprinting, reuse of illustrations, recitation, broadcasting, reproduction on microfilms or in any other physical way, and transmission or information storage and retrieval, electronic adaptation, computer software, or by similar or dissimilar methodology now known or hereafter developed.
The use of general descriptive names, registered names, trademarks, service marks, etc. in this publication does not imply, even in the absence of a specific statement, that such names are exempt from the relevant protective laws and regulations and therefore free for general use.
The publisher, the authors, and the editors are safe to assume that the advice and information in this book are believed to be true and accurate at the date of publication. Neither the publisher nor the authors or the editors give a warranty, expressed or implied, with respect to the material contained herein or for any errors or omissions that may have been made. The publisher remains neutral with regard to jurisdictional claims in published maps and institutional affiliations.

Cover illustration: Composite image showing an artist's impression of the SGR 1806-20 magnetar during its giant flare in 2004. The magnetic field lines are shown, together with the presence of hot spots. These are responsible for the oscillations that can be seen in the overlaid light curve shown in yellow. The time axis spans about six minutes. Artist's Image credit: NASA Light curve credit: T. Belloni

This Springer imprint is published by the registered company Springer-Verlag GmbH, DE part of Springer Nature.
The registered company address is: Heidelberger Platz 3, 14197 Berlin, Germany

Preface

The study of neutron stars is important in several aspects. On the astrophysical side, from their birth from supernova events to their evolution when isolated or in binary systems, they constitute an important end point of stellar evolution. In accreting binaries, they provide an important class of objects for our understanding of accretion onto compact objects, together with black-hole binaries. We now know transitional objects between accreting neutron stars and radio pulsars that can help us understand better both classes of sources. On the physical side, they constitute formidable and unique laboratories to study fundamental processes. First, they are the only objects in the Universe where matter is at extremely high density and we still do not know the associated equation of state. Second, they offer us the possibility to study their internal structure, affected by high angular momentum and large magnetic field. Third, but not least, they are being used as test objects for experimental measurements of the effects of general relativity (GR). Binary pulsars provide very precise clocks to monitor the orbit of the object and detect modifications due to GR, while in accreting neutron stars the strength of the gravitational field close to the compact object modifies the properties of accretion giving us the possibility to study GR in the strong-field regime.

All these studies are based on the time variability of the flux, from the radio band to X-rays. This book is focused on timing aspects. As the title says, this variability is in three forms. Pulsations, observed historically from radio pulsars, but now detected also in accreting low-mass X-ray binaries, the progenitors of millisecond pulsars. Oscillations, from the noisy variability of accreting systems, where strong quasi-periodicities are observed at a period of a few milliseconds, comparable to the dynamical time close to the surface of the neutron star. Explosions, powerful nuclear-powered X-ray bursts that take place on the surface of accreting neutron stars when the accretion level is in the favorable range.

Merate, Italy
Groningen, The Netherlands
Beijing, China
June 2020

Tomaso M. Belloni
Mariano Méndez
Chengmin Zhang

Contents

Contributors

Tomaso M. Belloni INAF - Osservatorio Astronomico di Brera, Merate, Italy

Marta Burgay INAF-Osservatorio Astronomico di Cagliari, Selargius, CA, Italy

Paolo Esposito Scuola Universitaria Superiore IUSS Pavia, Pavia, Italy
INAF - Istituto di Astrofisica Spaziale e Fisica Cosmica di Milano, Milano, Italy

Duncan K. Galloway School of Physics & Astronomy, Monash University, Clayton, VIC, Australia

Gian Luca Israel Osservatorio Astronomico di Roma, INAF, Monteporzio Catone, Rome, Italy

Laurens Keek CRESST and X-ray Astrophysics Laboratory NASA/GSFC, Greenbelt, MD, USA
Department of Astronomy, University of Maryland, College Park, MD, USA

Mariano Méndez Kapteyn Astronomical Institute, University of Groningen, Groningen, The Netherlands

M. Coleman Miller Department of Astronomy and Joint Space-Science Institute, University of Maryland, College Park, MD, USA

Alessandro Patruno Institute of Space Sciences, CSIC (Consejo Superior de Investigaciones Cientificas), Barcelona, Spain

Delphine Perrodin INAF-Osservatorio Astronomico di Cagliari, Selargius, CA, Italy

Andrea Possenti INAF-Osservatorio Astronomico di Cagliari, Selargius, CA, Italy

Nanda Rea Institute of Space Sciences (ICE, CSIC), Barcelona, Spain
Institut d'Estudis Espacials de Catalunya (IEEC), Barcelona, Spain
Astronomical Institute "Anton Pannekoek", University of Amsterdam, Amsterdam, The Netherlands

Anna L. Watts Astronomical Institute "Anton Pannekoek", University of Amsterdam, Amsterdam, The Netherlands

Chapter 1
Astrophysical Constraints on Dense Matter in Neutron Stars

M. Coleman Miller

Contents

M. C. Miller (✉)
Department of Astronomy and Joint Space-Science Institute, University of Maryland, College Park, MD, USA
e-mail: miller@astro.umd.edu

© Springer-Verlag GmbH Germany, part of Springer Nature 2021

T. M. Belloni et al. (eds.), *Timing Neutron Stars: Pulsations, Oscillations and Explosions*, Astrophysics and Space Science Library 461,
https://doi.org/10.1007/978-3-662-62110-3_1

Abstract Ever since the discovery of neutron stars it has been realized that they serve as probes of a physical regime that cannot be accessed in laboratories: strongly degenerate matter at several times nuclear saturation density. Existing nuclear theories diverge widely in their predictions about such matter. It could be that the matter is primarily nucleons, but it is also possible that exotic species such as hyperons, free quarks, condensates, or strange matter may dominate this regime. Astronomical observations of cold high-density matter are necessarily indirect, which means that we must rely on measurements of quantities such as the masses and radii of neutron stars and their surface effective temperatures as a function of age. Here we review the current status of constraints from various methods and the prospects for future improvements.

1.1 Introduction

The nature of the matter in the cores of neutron stars is of great interest to nuclear physicists and astrophysicists alike, but its properties are difficult to establish in terrestrial laboratories. This is because neutron star cores reach a few times the density of matter in terrestrial nuclei and yet they are strongly degenerate and they have far more neutrons than protons. The core matter thus occupies a different phase than is accessible in laboratories. Within current theoretical uncertainties there are many possibilities for the state of this matter: it could be primarily nucleonic, or dominated by deconfined quark matter, or mainly hyperons, or even mostly in a condensate.

Only astrophysical observations of neutron stars can constrain the properties of the cold supranuclear matter in their cores. Because we cannot sample the matter directly, we need to infer its state by measurements of neutron star masses, radii, and cooling rates. For the last two of these, the method of measurement is highly indirect and thus subject to systematic errors. Note, to be precise, that throughout this review we mean by mass the gravitational mass (which would be measured by using Kepler's laws for a satellite in a distant orbit around the star) rather than the baryonic mass (which is the sum of the rest masses of the individual particles in the star); for a neutron star, the gravitational mass is typically less than the baryonic mass by $\sim$20%. We also mean by radius the circumferential radius, i.e., the circumference at the equator divided by 2π, rather than other measures such as the proper distance between the stellar center and a point on the surface. Again, for objects as compact as neutron stars, the difference can amount to tens of percent.

In this review we discuss current attempts to measure the relevant stellar properties. We also discuss future prospects for constraints including those that will come from analysis of gravitational waves. For each of the constraint methods, we discuss the current uncertainties and assess the prospects for lowering systematic errors in the future. In Sect. 1.2 we set the stage by discussing current expectations from nuclear theory and laboratory measurements. In Sect. 1.3 we examine mass measurements in binaries. In Sect. 1.4 we discuss current attempts to measure the radii of neutron stars and show that most of them suffer severely from systematic errors. In Sect. 1.5 we explore what can be learned from cooling of neutron stars, and the difficulties in getting clean measurements of temperatures. In Sect. 1.6 we investigate the highly promising constraints that could be obtained from the detection of gravitational waves from neutron star–neutron star or neutron star–black hole systems. We summarize our conclusions in Sect. 1.7. For other recent reviews of equation of state constraints from neutron star observations, see [27, 101, 116, 117, 129, 165, 171, 199, 206, 234].

1.2 Expectations from Nuclear Theory

Any observations of neutron stars bearing on the properties of high-density matter must be put into the context of existing nuclear theory. This theory, which relies primarily on laboratory measurements of matter at nuclear density that has approximately equal numbers of protons and neutrons, must be extrapolated significantly to the asymmetric matter at far higher density in the cores of neutron stars. We also note that the inferred macroscopic properties of neutron stars depend on the nature of strong gravity as well as on the properties of dense matter (e.g., [183]), but for this review we will assume the correctness of general relativity.

In this section we give a brief overview of current thinking about dense matter. We begin with simple arguments motivating the zero-temperature approximation for the core matter and giving the basics of degenerate matter. We then address a commonly-asked question: given that the fundamental theory of quantum chromodynamics (QCD) exists, why can we not simply employ computer calculations (e.g., using lattice gauge theory) to determine the state of matter at high densities? Given that in fact such calculations are not practical, we explore the freedom that exists in principle to construct models of high-density matter; the fundamental point is that because the densities are well above what is measurable in the laboratory, one could always imagine, in the context of a model, adding terms that are negligible at nuclear density or for symmetric matter but important when the matter is a few times denser and significantly asymmetric. After discussing some example classes of models, we survey current constraints from laboratory experiments and future prospects. We conclude with a discussion of how one would map an idealized future data set of masses, radii, temperatures, etc. of neutron stars onto the equation of state of cold dense matter.

1.2.1 The Basics: Dense Matter and Neutron Stars

Consider a set of identical fermions (e.g., electrons or neutrons) of mass m and number density n. The linear space available to each fermion is thus $\Delta x \sim n^{-1/3}$, and the uncertainty principle states that the uncertainty in momentum (and hence the minimum momentum) is given by $\Delta p \Delta x \sim \hbar$ and thus $p_{\rm min} \sim \hbar n^{1/3}$, where

$\hbar = 1.05457 \times 10^{-27}$ erg s is the reduced Planck constant. Done more precisely and assuming isotropic matter we find that this minimum is the Fermi momentum $p_F = (3\pi^2\hbar^3 n)^{1/3}$. The Fermi energy adds to the rest-mass energy via $E_{\rm tot} = (m^2c^4 + p_F^2c^2)^{1/2} = mc^2 + E_F$, and is $E_F \approx p_F^2/2m$ for $p_F \ll mc$ and $E_F \approx p_F c$ for $p_F \gg mc$, where $c = 2.99792458 \times 10^{10}$ cm s^{-1} is the speed of light.

Matter is degenerate when $E_F > kT$, and strongly degenerate when $E_F \gg kT$, where $k = 1.38065 \times 10^{-16}$ erg K^{-1} is the Boltzmann constant and T is the temperature. As a result, electrons (with their low masses $m_e = 9.109382 \times 10^{-28}$ g $= 0.510999$ MeV$/c^2$) become degenerate at much lower densities than do neutrons or protons. Matter dominated by nuclei heavier than hydrogen has ~2 baryons per electron, and hence electrons become relativistically degenerate ($p_F = m_e c$) at a density of $\approx 2 \times 10^6$ g cm^{-3}. In addition, above $\approx 2.5 \times 10^7$ g cm^{-3} the total energy of electrons becomes larger than $m_n c^2 - m_p c^2 = 1.294$ MeV, where $m_n = 1.674927 \times 10^{-24}$ g $= 939.566$ MeV$/c^2$ and $m_p = 1.672622 \times 10^{-24}$ g $= 938.272$ MeV$/c^2$ are respectively the rest masses of neutrons and protons. As a result, at these and greater densities electrons and protons can undergo inverse beta decay $e^- + p \to n + \nu_e$. At higher densities the ratio of neutrons to protons in nuclei increases, and then at the "neutron drip" density $\rho_{nd} \approx 4.3 \times 10^{11}$ g cm^{-3} neutrons are stable outside nuclei. The neutron drip density is derived in detail in, e.g., [22], but a good estimate can be obtained by a simple argument. Neutrons can drip out of the nucleus when the total electron energy per nucleon equals the nuclear binding energy per nucleon (as described in [22] there are small corrections due to lattice energy and the nonzero energy of neutron continuum states). The binding energy per nucleon is ~8 MeV (see, e.g., Table 3 of [22]). At high densities, $Z/A \sim 0.3$ in contrast to the $Z/A \sim 0.5$ common at lower densities. Thus the condition on the electron Fermi energy for neutron drip is $E_{F,e} \approx 8(A/Z)$ MeV. The density at which this happens is therefore approximately 2×10^6 g cm$^{-3}(0.5/0.3)[(8\ {\rm MeV}/0.3)/0.5\ {\rm MeV}]^3 \approx 4 \times 10^{11}$ g cm^{-3}, where we extrapolate from the density at which the electron Fermi energy becomes relativistic.

At infinite density, equilibrium matter consisting of just neutrons, protons, and electrons would have eight times as many neutrons as protons (and electrons, because charge balance has to be maintained). To see this, note that at infinite density all species are ultrarelativistic and their chemical potentials are thus dominated by their Fermi energies. Charge balance means that $n_p = n_e$, so equilibrium implies

$$\begin{aligned} E_{F,n} &= E_{F,p} + E_{F,e^-} \\ &= 2E_{F,p} \\ n_n^{1/3} &= 2n_p^{1/3} \\ n_n &= 8n_p\ . \end{aligned} \tag{1.1}$$

For a neutron star with a canonical mass $M = 1.4\ M_\odot$ (where $M_\odot = 1.989 \times 10^{33}$ g is the mass of the Sun) and radius $R = 10$ km that for simplicity we will treat as made entirely of free neutrons, the average number density is $n = (M/m_n)/(4\pi R^3/3) = 4 \times 10^{38}$ cm^{-3}. This implies $p_F = 2.4 \times 10^{-14}$ g cm s^{-1} and thus $E_F \approx 2 \times 10^{-4}$ erg$=$ 100 MeV. This corresponds to a temperature of

$T_F = E_F/k \approx 10^{12}$ K, which is much hotter than the expected interior temperatures $T < 10^{10}$ K typical of neutron stars more than a few years old [172]. Neutron stars are strongly degenerate.

Note, however, that the high Fermi energy of neutrons suggests the possibility of additional particles at high densities. For example, the lambda particle has a rest mass of $m_\Lambda = 1115.6\,\mathrm{MeV/c^2}$, so if the neutron Fermi energy exceeds 176 MeV then the lambda is in principle stable because 176 MeV plus the neutron rest-mass energy 939.6 MeV exceeds 1115.6 MeV. Several other particles are within 300 MeV/c^2 of the neutron. In addition, because the density at the center of a neutron star is a few times nuclear saturation density $\rho_{\mathrm{nuc}} \approx 2.6 \times 10^{14}\,\mathrm{g\,cm^{-3}}$, quarks may become deconfined or matter might transition to a state that is lower-energy than nucleonic matter even at zero pressure (strange matter; see, e.g., [77]). Various density thresholds are summarized in Fig. 1.1.

Stars supported by nonrelativistic degeneracy pressure (a reasonable approximation for neutron stars, because $E_F < m_n c^2$) have radii that decrease with increasing mass in contrast to most other objects. To see this, consider a star with a mass M and radius R supported by nonrelativistic fermions of mass m. The Fermi energy per particle is $E_F \sim p_F^2/2m \sim \hbar^2 n^{2/3} \sim (M/R^3)^{2/3} \sim M^{2/3}/R^2$. The gravitational

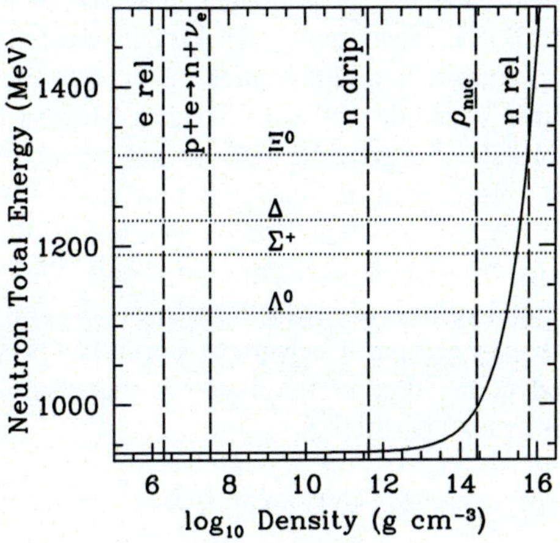

Fig. 1.1 Total energy per free neutron versus mass density (solid line). Above $\sim 10^{13}\,\mathrm{g\,cm^{-3}}$ the Fermi energy starts to contribute palpably to the total, and above $\sim 10^{15}\,\mathrm{g\,cm^{-3}}$ the total energy can exceed the rest mass energy of particles such as Λ^0, Σ^+, Δ, and Ξ^0 (marked by horizontal dotted lines). Interactions between these particles can change the threshold density. The central densities of realistic neutron stars range from $\sim 5 \times 10^{14}\,\mathrm{g\,cm^{-3}}$ to $\sim$ few $\times\, 10^{15}\,\mathrm{g\,cm^{-3}}$, so some of these exotic particles may indeed be energetically favorable. Also marked are the densities at which free electrons become relativistic; where those electrons have enough total energy to make $p + e^- \to n + \nu_e$ possible; where free neutrons can exist stably (i.e., at neutron drip); nuclear saturation density ρ_{nuc}; and where free neutrons have a Fermi energy equal to their rest-mass energy. To calculate the neutron Fermi energy we assume that all the mass is in free neutrons; in reality at least a few percent of the mass is in protons and other particles, and below ρ_{nuc} a significant fraction of mass is in nuclei

energy per particle is $E_G \sim -GMm/R$, where $G = 6.67 \times 10^{-8}\,\mathrm{g}^{-1}\,\mathrm{cm}^3\,\mathrm{s}^{-2}$ is Newton's gravitational constant, so the total energy per particle is $E_{\rm tot} = C_1 M^{2/3}/R^2 - C_2 M/R$ where C_1 and C_2 are constants. Minimizing with respect to R gives $R \sim M^{-1/3}$. Effects associated with interactions can change this slightly, but in practice most equations of state produce a radius that either decreases with increasing mass or is nearly constant over a broad range in mass. This led [130] to note that even for a star of unknown mass a measurement of the radius to within ~10% would provide meaningful constraints on the equation of state.

1.2.2 Models of Matter at High Densities

There are several classes of matter beyond nuclear density: ones in which neutrons and protons are the only baryons, ones in which other baryons enter (especially those with strange quarks), ones involving deconfined quark matter, and so on. Within each class there are a number of adjustable parameters, some of which are constrained by laboratory measurements at nuclear density or below but many of which can be changed to accommodate observations of neutron stars.

When confronted by this complexity a common question is: why is there uncertainty about dense matter? The fundamental theory, QCD, is well-established. Asymptotic freedom is not reached at neutron star densities, so the coupling constant is large enough that expansions similar to those in quantum electrodynamics are not straightforward, but in principle one could imagine Monte Carlo calculations that establish the ground state of degenerate high-density matter.

This approach is unfortunately not currently practical, due to the lack of a viable algorithm for high baryon densities. This is because of the so-called "fermion sign problem", which has been known for many years. We start by considering a representative but small volume of matter at some density and chemical potential μ that can exchange energy and particles with its surroundings but has a fixed volume [109]. The thermodynamic state of the matter is therefore described by a grand canonical ensemble using a partition function

$$Z = Z(T, \mu) = \mathrm{Tr}\,\{\exp[-(\mathcal{H} - \mu N)/kT]\} \tag{1.2}$$

where $\mathcal{H}$ is the Hamiltonian and N is the particle number operator. It is common to use $\beta \equiv 1/(kT)$. The trace is evaluated over Fock space, which makes this formulation inconvenient. One can instead rewrite Z as

$$Z = \int \mathcal{D}A\, \det M(A) e^{-S_G(A)}\,, \tag{1.3}$$

that is, as a Euclidean functional integral over classical field configurations. Here A represents the degrees of freedom (quarks, gluons, ...), $S_G(A;\beta) = \int_0^\beta dx^4 \int d^3x \mathcal{L}_G^E(A)$ is the thermal Euclidean gauge action, and the quark propagator matrix is $M(A) = \not{D}(A) - m - \mu\gamma_4$ where $\not{D} = \gamma_\mu(\partial_\mu - igA_\mu^a t^a)$, t^a are the Hermitian group generators, g is the strong coupling constant, and γ are the Euclidean gamma matrices.

For a vanishing chemical potential $\mu = 0$, det$M(A)$ is positive definite, meaning that all contributions add in the same direction and Z is comparatively straightforward to compute. If instead μ is real and nonzero, then det$M(A)$ is complex in general. Thus although Z is still real and strictly positive, the integrands have various phases and the integral takes the form of a cancellation between large quantities. The bad news is that the general fermion sign problem is NP-hard [216], but work is proceeding on better approximation methods. If a first-principles evaluation of Z yields, without ambiguity, the equilibrium state of dense matter at low temperatures, then measurements of the properties of neutron stars would serve as important tests of QCD itself and would thus be probes of very fundamental physics indeed. Until that point, however, it is necessary to use phenomenological models.

It is very difficult to rule out an entire class of models (e.g., those with only nucleonic degrees of freedom or those with significant contributions from hyperons). This is because neutron star core densities and the asymmetry in the number densities of neutrons and protons are significantly greater than those that can be probed in laboratories. As a result, one could always imagine adding contributions that involve high powers of the density or asymmetry. These contributions would have a negligible impact on laboratory matter but would have important effects in the cores of neutron stars. One can make some general statements: for instance, if non-nucleonic components become important above some density the equilibrium radius at a given mass and the maximum mass both tend to be smaller than if only nucleonic degrees of freedom contribute (because the presence of a new energetically favorable composition softens the equation of state). Unfortunately, it is difficult to establish a particular mass or radius that would eliminate such exotic models. For example, hyperonic and hybrid quark models of neutron stars have been constructed with maximum masses $>2.0\ M_\odot$ [117, 122]. Nonetheless, although neutron star observations cannot entirely rule out model classes in principle, their role is important because they probe a different realm of matter than what is accessible in laboratories. Ockham's razor should then be used to judge between different model classes: if one class fits all data using a small number of parameters that have reasonable values and other classes require great complexity or unreasonable values, the first class would be preferred.

One basic category of models, relativistic mean field theories, is quite phenomenological in nature. In these models the degrees of freedom are nucleons and mesons (which couple minimally to the nucleons but the coupling could have some density dependence). The coupling strengths can be adjusted to laboratory data and/or neutron star observations. In a more microscopically oriented approach, one starts instead from some given nucleon-nucleon interaction (which can be extended to more than two nucleons) that is fitted to data including the binding energy of light nuclei and scattering data (for a recent effort in the context of chiral effective field theory, see [99]). In both approaches there is considerable freedom about the types of particles considered, e.g., the particles could be nucleons or the particles could

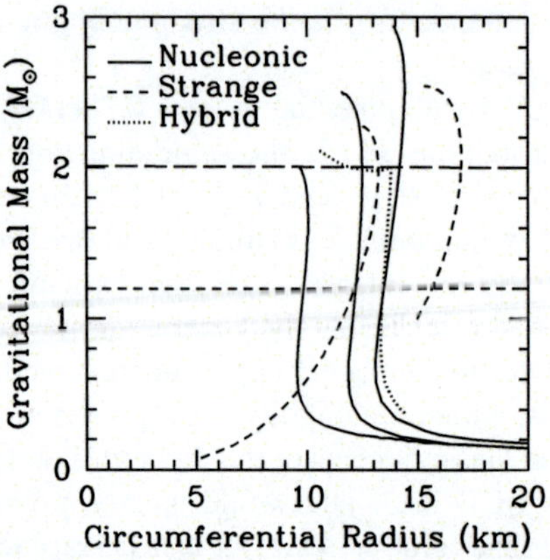

Fig. 1.2 Mass versus radius for nonrotating stars constructed using several different high-density equations of state. Rotation changes the radius to second order in the spin rate, but the corrections are minor for known neutron stars. The solid curves include only nucleonic degrees of freedom (these are the mass-radius relations for the soft, medium, and hard equations of state from [100]), the short dashed lines assume bare strange matter [121], and the dotted curve uses a hybrid quark equation of state with a phase transition [34]. The horizontal dashed line at 1.2 $M_\odot$ represents approximately the minimum gravitational mass for a neutron star in current formation scenarios, whereas the horizontal dashed line at 2.01 $M_\odot$ shows the highest precisely measured gravitational mass for a neutron star

include hyperons or deconfined quarks. We plot some mass-radius relations from representative equations of state in Fig. 1.2. It is clear that models can be constrained tightly if more massive neutron stars are discovered, or if neutron star radii can be measured with accuracy and precision (especially for stars of known mass).

Constraints on the equation of state of cold dense matter can be obtained from astronomical observations or laboratory experiments. Some of the more useful experimental data come from relativistic heavy-ion collisions, which can reach 2 to 4.5 times nuclear saturation density [70] but which have relativistic temperatures and are therefore not degenerate. In addition, in such collisions the time for weak interactions to occur is short, in contrast to the effectively infinite time available in neutron stars. Laboratory data also include the binding energies of light nuclei and recent measurements of the neutron skin thickness of heavy nuclei such as ^{208}Pb ($0.33^{+0.16}_{-0.18}$ fm according to the PREX team [3]; see [178] for some of the implications of the expected more precise future measurements), which provide a rare glimpse of neutron-rich matter because the neutron wavefunctions extend slightly beyond the proton wavefunctions. These experiments thus measure the microphysics semi-directly, whereas all astrophysical observations place indirect constraints. In order to make explicit contact between microphysics and astrophysics we now discuss briefly how to construct models of neutron stars given a high-density equation of state.

1.2.3 Construction of Neutron Star Models from Microphysics

We argued earlier that the Fermi energy in the cores of neutron stars is much greater than the thermal energy. If we also assume that the matter is in its ground state at a given density, this implies that the pressure is only a function of density: $P = P(\rho)$ (this is called a *barotropic* equation of state). If we have such an equation of state we can compute the structure of a nonrotating and hence spherically symmetric star using the Tolman-Oppenheimer-Volkoff (TOV) equation [164]:

$$\frac{dP(r)}{dr} = -\frac{G}{r^2}\left[\rho(r) + \frac{P(r)}{c^2}\right]\left[M(r) + 4\pi r^3 \frac{P(r)}{c^2}\right]\left[1 - \frac{2GM(r)}{c^2 r}\right]^{-1} \tag{1.4}$$

where $M(r) = \int_0^r 4\pi\rho(r)r^2 dr$ is the gravitational mass integrated from the center to a circumferential radius r. Note that for $P/c^2 \ll \rho$ and $GM/c^2 \ll r$ this reduces to the Newtonian equation of hydrostatic equilibrium $dP/dr = -GM\rho/r^2$. Thus one can construct a model star by choosing a central density or pressure and integrating to the surface, which is defined by $P = \rho = 0$. This gives the radius and gravitational mass of the star. Similar constructions are possible for stars that rotate uniformly or differentially in some specified manner (see [63, 207]), but it is conceptually clearer to focus on the nonrotating case.

One can therefore determine the neutron star mass-radius relation from a given equation of state. If we suppose that in the future we will have precise measurements of the radii and gravitational masses of a large number of neutron stars, say from the minimum possible to the maximum possible mass, then comparison of the observed $M - R$ curve with predicted curves will strongly constrain the parameters of a given class of models. But is it possible to go in the other direction, that is, could one take observed (M, R) pairs and infer $P(\rho)$ directly while remaining agnostic about the microphysics that produces the equation of state?

This is not trivial. One might imagine that the construction of $P(\rho)$ would proceed as follows. First, we assume that we know the equation of state up to nuclear saturation density $\rho_{\rm nuc}$. Even in this first step we therefore make an extrapolation from the nearly symmetric nuclear matter in nuclei to the highly asymmetric matter in neutron stars. We use this equation of state to compute the mass M_s and radius R_s of a star with a central density of $\rho_{\rm nuc}$. We then observe a star with a slightly larger mass than M_s. The microscopic unknowns would be the central density (slightly larger than nuclear saturation) and the pressure at that density, which our two measurements (of M and R) are sufficient to constrain. We then bootstrap $P(\rho)$ by measuring the mass and radius of successively more massive neutron stars.

The difficulty with this procedure is evident from Figure 11 of [4], which shows that a star whose central density is exactly nuclear saturation density has a total mass of only $\sim$0.1 $M_\odot$. To get to the $M \approx 1.2\ M_\odot$ minimum for neutron star masses [163, 237] requires densities that are more than twice nuclear saturation. We will thus be required to extrapolate well beyond known matter in density and nuclear asymmetry to fit neutron star data. This is not fatal, because it means that we are simply comparing model predictions with data, but it does mean that $P(\rho)$ cannot be inferred blindly without models.

If additional assumptions are made, for example that between fiducial densities the equation of state is a polytrope $P \propto \rho^\gamma$, then [166] have shown that precise mass and radius measurements of as few as three neutron stars could suffice to give an empirically determined $P(\rho)$. The inferred $P(\rho)$ would then be compared with the predictions from microphysical equations of state. As discussed by [131], there are additional relatively model-independent constraints on the equation of state that one can infer from observations. For example, the typical radius of a neutron star scales as the quarter power of the pressure near nuclear saturation density, and the maximum density that can be reached in a neutron star is $\rho_{\max} \approx 1.5 \times 10^{16}\ \mathrm{g\ cm^{-3}}(M_\odot/M_{\max})^2$ if the maximum mass is $M_{\max}$.

Our final note about the theoretical predictions is that there are some phenomena that have little effect on the mass-radius relation but are important for other observables. For example, the existence of a proton superconducting gap can modify core cooling dramatically [172], but the predicted gap energies of $\sim 0.1 - 1$ MeV [172] are so small compared to the Fermi energy that the overall structure of neutron stars will be affected minimally. Thus if neutron star temperatures and ages, particularly those of isolated neutron stars, can be inferred reliably, then they will provide a beautiful complement to the mass and radius measurements that are emphasized more in this review.

1.3 Constraints on Mass from Binary Observations

Mass measurements of neutron stars in binaries provide the most certain of all constraints on the properties of cold high-density matter, particularly when the companion to the neutron star is also a neutron star and thus the system approaches the ideal of two point masses. In this section we discuss such measurements, beginning with what can be learned from purely Newtonian observations and moving on to the greater precision and breaking of degeneracies that are enabled by measurements of post-Keplerian parameters from systems involving pulsars.

1.3.1 Newtonian Observations of Binaries

The classical approach to mass measurements in binaries assumes that one sees periodic variation in the energy of spectral lines from one of the stars in the binary, which we will call star 1. The period of variation is the orbital period $P_{\rm orb}$, the shape of the variation gives the eccentricity e of the orbit, and the magnitude K_1 (which has dimensions of speed) of the variation indicates the line-of-sight component of the orbital speed of star 1. Using Kepler's laws these observed quantities can be combined to form the "mass function", which is

$$f_1(M_1, M_2) = \frac{K_1^3 P_{\rm orb}}{2\pi G}(1 - e^2)^{3/2} = \frac{M_2^3 \sin^3 i}{(M_1 + M_2)^2} \,. \qquad (1.5)$$

Here M_1 is the mass of the star being observed, M_2 is the mass of the other star, and i is the inclination of the binary orbit to our line of sight ($i = 0$ means a face-on orbit, $i = \pi/2$ means an edge-on orbit). From this expression, f_1 is the minimum possible mass for M_2; if $M_1 > 0$ or $i < \pi/2$ then $M_2 > f_1$. If periodically shifting spectral lines are also observed from the second star (and thus the binary is a so-called double-line spectroscopic binary), then the mass ratio is known and only i is uncertain.

The inclination can be constrained for eclipsing systems. Particular precision is possible in some extrasolar planet observations because of the Rossiter-McLaughlin effect (in which the apparent color of the star varies in a way dependent on inclination as a planet transits across its disk; see [147, 194]). This effect also produces a velocity offset, which might have been seen in 2S 0921–63 [110]. The inclination can also be constrained in systems for which the companion to a compact object just fills its Roche lobe. This is because as the companion orbits it presents different aspects to us, and the amplitude of variation depends on the inclination; for example, a star in a face-on orbit looks the same to us at all phases, whereas star in an edge-on orbit varies maximally in its aspect [13, 145]. In practice this analysis is limited to systems that have low-mass companions (because Roche lobe overflow from a high-mass companion to a lower-mass compact object is usually unstable; see [81]) and that have transient accretion phases and hence have long intervals in which there is effectively no accretion disk (because an active accretion disk easily outshines a low-mass star and thus the binary periodicity is very difficult to observe). Neutron star X-ray binaries might be less likely to be transient than black hole X-ray binaries, although the data are ambiguous on this point, and their companions tend to be much less massive and hence dimmer than the companions to black holes [81]. Thus despite the great success of this method for black hole binaries it has found limited application for neutron star binaries.

1.3.2 *Post-Keplerian Measurements of Pulsar Binaries*

The most precise measurements of the masses of neutron stars in binaries are made for systems in which additional parameters can be measured. The extreme timing precision of pulsars makes pulsar binaries especially good candidates for such measurements. The new effects that can be measured are:

- Precession of the pericenter of the system, $\dot{\omega}$.
- Einstein delay γ. At pericenter, the gravitational redshift from the system is maximized.

- Binary orbital decay $\dot{P}_b$. Gravitational waves are emitted by anything that has a time-variable quadrupole (or higher-order) mass moment. This shrinks and circularizes binary orbits.
- Shapiro delay, r and s. When the signal from the pulsar passes near its companion, time dilation in the enhanced potential delays the signal compared to the arrival time of a photon in flat spacetime. The magnitude of the delay over the orbit is characterized by the range r and shape s of the delay as a function of phase. See [83] for a recent reparameterization of the Shapiro delay that may represent the error region better for some orbital geometries.

Using the notation of [82], the dependences of these post-Keplerian parameters on the properties of the binary are

$$
\begin{aligned}
\dot{\omega} &= 3\left(\frac{P_b}{2\pi}\right)^{-5/3}(T_\odot M)^{2/3}(1-e^2)^{-1} \\
\gamma &= e\left(\frac{P_b}{2\pi}\right)^{1/3} T_\odot^{2/3} M^{-4/3} m_c(m_p+2m_c) \\
\dot{P}_b &= -\frac{192\pi}{5}\left(\frac{P_b}{2\pi}\right)^{-5/3} f(e) T_\odot^{5/3} m_p m_c M^{-1/3} \\
r &= T_\odot m_2 \\
s &= \sin i
\end{aligned}
\tag{1.6}
$$

where m_p is the pulsar mass, m_c is the companion mass, $M = m_p + m_c$ is the total mass (all masses are in units of a solar mass), $T_\odot = GM_\odot/c^3 = 4.925590947\,\mu\text{s}$, and $f(e) = (1 + 73e^2/24 + 37e^4/96)(1-e^2)^{-7/2}$. For a given system, there are thus three Keplerian parameters that can be measured (binary period, radial velocity, and eccentricity) along with the five post-Keplerian parameters. For a system such as the double pulsar J0737–3039A/B [47] additional quantities can be measured. Hence double neutron star systems in which at least one is visible as a pulsar are superb probes of general relativity and yield by far the most precise masses ever obtained for any extrasolar objects.

As discussed by Freire [82], the neutron stars with the greatest timing precision are the millisecond pulsars. These, however, are spun up by accretion in Roche lobe overflow systems, and that accretion also circularizes the system to high precision. As a result, precession of the pericenter and the Einstein delay cannot be measured. The Shapiro delay, however, can be measured even for circular binaries, and because the Shapiro delay does not have classical contributions from tides (unlike pericenter precession, for example), r and s can yield unbiased mass estimates. As pointed out by Scott Ransom, Shapiro delay measurements are likely to become more common due to the development of very high-precision timing for gravitational wave detection via pulsar timing arrays. The consequence is that currently the most constrained systems are field NS-NS binaries, in which little mass transfer has taken place in the system and the stars are thus close to their birth masses. In contrast, recycled millisecond pulsars have had an opportunity to acquire an additional several tenths of a solar mass via accretion.

Another possibility, which is described clearly by Freire [82], is that in high stellar density environments such as globular clusters binary-single interactions could play a major role. For example, a pulsar could be recycled to millisecond periods and then an exchange interaction could leave it in an eccentric binary with a white dwarf or another neutron star. Such a system would have a measurable $\dot{\omega}$ and γ, and the neutron star could be high-mass and have excellent timing precision.

We do note that there are two drawbacks to NS-WD systems in globulars compared to field NS-NS systems. First, although white dwarfs are small they are not point masses to the degree that neutron stars are. As a result, there is a small contribution to the precession from the finite structure of the white dwarfs. Second, even at the high stellar densities of globulars a comparatively large orbit is required for there to be a significant probability of interaction. To see this, note that the rate of interactions for a binary of interaction cross section σ is $\tau^{-1} = n\sigma v$, where n is the number density of stars (typically in the core $n = 10^5 - 10^6\,\mathrm{pc}^{-3}$) and $v \sim 10\,\mathrm{km\,s}^{-1}$ is the velocity dispersion. For a system of mass M, the interaction cross section for a closest approach of a, roughly equal to the semimajor axis of the binary, is $\sigma \approx \pi a(2GM/v^2 + a)$. If $M \approx 2\,M_\odot$ and $n = 10^5\,\mathrm{pc}^{-3}$, this implies $\tau = 10^{10}\,\mathrm{yr}$ when $a \approx 0.04\,\mathrm{AU}$. This implies orbital periods greater than a day, so dynamically formed NS-WD binaries are systematically larger than NS-NS binaries formed in situ. Thus longer observation times are required to achieve a given precision.

1.3.3 Dynamically Estimated Neutron Star Masses and Future Prospects

For a recent compilation of dynamically estimated neutron star masses and uncertainties, see [115]. From the standpoint of constraints on dense matter, the most important development over the last few years has been the discovery of neutron stars with masses $M \sim 2\,M_\odot$, and possibly more. The first such established mass was for PSR J1614–2230. Demorest et al. [72] determined that its mass is $M = 1.97 \pm 0.04\,M_\odot$, which they obtained via a precise measurement of the Shapiro delay. This measurement was aided by the nearly edge-on orientation of the system (inclination angle 89.17°), which increases the maximum magnitude of the delay and produces a cuspy timing residual that is easily distinguished from any effects of an eccentric orbit.

The second large mass that has been robustly established belongs to PSR J0348+0432. Antoniadis et al. [11] observed gravitationally redshifted optical lines from the companion white dwarf. The observed Doppler modulation of the energy of these lines yields a mass ratio when combined with the modulation of the observed spin frequency of the pulsar. In addition, interpretation of the Balmer lines from the white dwarf in the context of white dwarf models gives a precise mass for the white dwarf, and indicates that the neutron star has a mass of $M = 2.01 \pm 0.04\,M_\odot$.

In addition to these well-established high masses, there are hints that some black widow pulsars (those that are currently evaporating their companions) might have even higher masses. This was first reported for the original black widow pulsar PSR B1957+20 [223]. For this star the best-fit mass is $M = 2.40 \pm 0.12\ M_\odot$, but at the highest allowed inclination and lowest allowed center of mass motion the mass could be as low as 1.66 $M_\odot$. More recently, [188] analyzed the gamma-ray black widow pulsar PSR J1311–3430 and found a mass of $M = 2.7\ M_\odot$ for simple heated light curves (but with significant residuals in the light curve), and no viable solutions with a mass less than 2.1 $M_\odot$. They conclude that better modeling and more observation is needed to establish a reliable mass, but it is an intriguing possibility that black widow pulsars have particularly large masses.

From the astrophysical standpoint, it has been proposed that neutron star birth masses are bimodal, depending on whether the core collapse occurs due to electron capture or iron core collapse [197]. There is also mounting evidence for systematically higher masses in systems that are expected to have had substantial accretion [241]. From the standpoint of nuclear physics, [132] point out that 2.0 $M_\odot$ neutron stars place interesting upper limits on the physically realizable energy density, pressure, and chemical potential. Higher masses would present even stronger constraints.

To a far greater extent than with the other constraints described in this review, we can be confident that the mere passage of time will greatly improve the mass measurements, and indeed all of the timing parameters. Table II of [68] shows that as a function of the total observation time T (assuming a constant rate of sampling), the fractional uncertainties in the post-Keplerian parameters scale as $\Delta\dot{\omega} \propto T^{-3/2}$, $\Delta\gamma \propto T^{-3/2}$, $\Delta\dot{P}_b \propto T^{-5/2}$, $\Delta r \propto T^{-1/2}$, and $\Delta s \propto T^{-1/2}$; for the r and s parameters the improvements are simply due to having more measurements, whereas the others improve faster with time because the effects accumulate. Particularly good improvement is expected for the NS-WD systems because as we describe above they have larger orbits and thus slower precession than NS-NS systems. There is thus reason to hope that additional high-mass systems will be discovered.

There are also planned observatories and surveys that will dramatically increase the number of known pulsars of all types, which will likely include additional NS-NS and NS-WD systems. An example of such a planned observatory is the Square Kilometer Array, which has been projected to increase our known sample of pulsars by a factor of $\sim$10. In addition, as [215] pointed out recently, future high-precision astrometry will be able to deconvolve the parallactic, proper, and orbital motion of the two components of a high-mass X-ray binary. They estimate that for parameters appropriate to the Space Interferometry Mission [201] this will yield a neutron star mass accurate to 2.5% in X Per, to 6.5% in Vela X-1, and to $\sim$10% in V725 Tau and GX 301–2.

It is thus probable that in the next $\sim$10 years we will have far more, and far better, estimates of the masses of individual neutron stars. We do not, however, have a guarantee that any of those masses will be close to the maximum allowed. We thus need additional ways to access the properties of high-density matter. In particular, many equations of state imply similar maximum masses but widely different radii. We therefore turn to constraints on the radius.

1.4 Constraints on Radius, and Other Mass Constraints

As discussed in Sect. 1.2, accurate measurements of the radii of neutron stars would strongly constrain the properties of neutron star core matter. Unfortunately, all current inferences of neutron star radii are dominated by systematic errors, and hence no radius estimates are reliable enough for such constraints. However, future measurements using the approved missions NICER [90] and Athena+ [161] and the proposed missions LOFT [78] and AXTAR [185], and hold great promise for precise radius measurements if the effects of systematic errors can be shown to be small. In this section we discuss various proposed methods for measuring radii and the diverse results obtained by applying these methods. Some of the methods also result in mass estimates, so we discuss those implications along the way.

1.4.1 Thermonuclear X-ray Bursts

More than 30 years ago it was proposed that the masses and radii of neutron stars could be obtained via measurement of thermonuclear X-ray bursts [224]. These bursts occur when enough hydrogen or helium (or carbon for the long-lasting "superbursts") accumulates on a neutron star in a binary. Nuclear fusion at the base of the layer becomes unstable, leading to a burst that lasts for seconds to hours. For a selection of observational and theoretical papers on thermonuclear bursts, see [23, 64, 85, 86, 91, 112, 125, 137, 204, 212, 213, 236]. In some bursts, fits of a Planck function to the spectra reveal a temperature that initially increases, then decreases, then increases again before finally decreasing [107, 136, 203]. These are called photospheric radius expansion (PRE) bursts [168]. The usual assumption is that PREs occur because the radiative luminosity exceeds the Eddington luminosity

$$L_E = \frac{4\pi GMc}{\kappa} \tag{1.7}$$

where M is the mass of the star and κ is the radiative opacity. At luminosities greater than L_E, an optically thick wind can be driven a potentially significant distance from the star. This leads to an increase in the radiating area and a consequent decrease in the temperature [75, 167]. For Thomson scattering in fully ionized matter with a hydrogen mass fraction X, $\kappa = 0.2(1 + X)\,\text{cm}^2\,\text{g}^{-1}$ and thus

$$L_E = 2.6 \times 10^{38}\,\text{erg s}^{-1}(1 + X)^{-1}(M/M_\odot)\,. \tag{1.8}$$

The basic method of [224] involves several assumptions. These are:

1. The full surface radiates uniformly after the photosphere has retreated to the radius of the star.
2. The stellar luminosity is the Eddington luminosity at the point of "touchdown", which is defined as the time after the peak inferred photospheric radius is reached when the color temperature, derived from a Planck fit to the spectrum, is maximal. The luminosity can then be determined via measurement of the flux at Earth and the distance to the star, assuming that the flux is emitted isotropically.
3. The spectral model for the atmosphere is correct. Thus the model must have been verified against data good enough to distinguish between models, and the atmospheric composition must be known. It is typically assumed that the color factor $f_c \equiv T_{\rm col}/T_{\rm eff}$, which is the ratio between the fitted Planck temperature and the effective temperature, is not only known but is constant throughout the cooling phase.
4. All other sources of emission from the system are negligible.

Using these assumptions, and using the notation of [206], if we have measured the distance D to the star and know κ, we can measure the touchdown flux

$$F_{\rm TD,\infty} = \frac{GMc}{\kappa D^2}\sqrt{1-2\beta(r_{\rm ph})} \tag{1.9}$$

where $\beta(r) \equiv GM/rc^2$, the factor before the square root is the Eddington flux diluted by distance, and $r_{\rm ph}$ is the radius of the photosphere. We can also use the cooling phase of the burst to define a normalized angular surface area

$$A = \frac{F_\infty}{\sigma_{\rm SB} T^4_{\rm col,\infty}} = f_c^{-4}\left(\frac{R}{D}\right)^2 (1-2\beta)^{-1} \ . \tag{1.10}$$

Here $\sigma_{\rm SB} = 5.6704 \times 10^{-5}\,{\rm erg\,cm^{-2}\,s^{-1}\,K^{-4}}$ is the Stefan-Boltzmann constant and F_∞ and $T_{\rm col,\infty}$ are the flux and fitted Planck temperature that we measure in the cooling phase. Then the combinations of observed quantities

$$\begin{aligned} \alpha &\equiv \frac{F_{\rm TD,\infty}}{\sqrt{A}}\frac{\kappa D}{c^3 f_c^2} \\ \gamma &\equiv \frac{Ac^3 f_c^4}{F_{\rm TD,\infty}\kappa} \end{aligned} \tag{1.11}$$

can be related to β and R by $\alpha = \beta(1-2\beta)$ and $\gamma = R[\beta(1-2\beta)^{3/2}]^{-1}$ and solved to yield

$$\begin{aligned} \beta &= \tfrac{1}{4} \pm \tfrac{1}{4}\sqrt{1-8\alpha} \ , \\ R &= \alpha\gamma\sqrt{1-2\beta} \ , \\ M &= \beta R c^2/G \ . \end{aligned} \tag{1.12}$$

The problem is that for several bursters, the most probable values of the observationally inferred quantities $F_{TD,\infty}$, A, and D, combined with the model parameter f_c, yield $\alpha > 1/8$. This would imply that the mass and radius are complex numbers. For example, in their analysis of 4U 1820–30, [95] used a Gaussian prior probability distribution for $F_{TD,\infty}$, which had a mean $F_0 = 5.39 \times 10^{-8}\,\mathrm{erg\,cm^{-2}\,s^{-1}}$ and a standard deviation $\sigma_F = 0.12 \times 10^{-8}\,\mathrm{erg\,cm^{-2}\,s^{-1}}$. They also used a Gaussian prior probability distribution for A, with $A_0 = 91.98\ (\mathrm{km}/10\ \mathrm{kpc})^2$ and $\sigma_A = 1.86\ (\mathrm{km}/10\ \mathrm{kpc})^2$. Their prior probability distribution for D was a boxcar distribution with a midpoint $D_0 = 8.2\,\mathrm{kpc}$ and a half-width of $\Delta D = 1.4\,\mathrm{kpc}$. Finally, they assumed a boxcar prior probability distribution for the color factor, with $f_{c0} = 1.35$ and a half-width $\Delta f_c = 0.05$. If we take the midpoint of each distribution and also follow [95] by assuming that the opacity is dominated by Thomson scattering and thus $\kappa = 0.2\,\mathrm{cm^2\,g^{-1}}$ for the pure helium composition appropriate to 4U 1820–30, we find $\alpha = 0.179 > 1/8$.

Güver et al. [95] note that the probability of obtaining a viable M and R for 4U 1820–30 from these equations drops with increasing distance, but if we reduce D to the 6.8 kpc, which is the smallest value allowed in the priors of [95], and keep the other input parameters fixed, we find $\alpha = 0.148$. If we also increase f_c to its maximum value of 1.4 and take the $+2\sigma$ value of A and the -2σ value of $F_{TD,\infty}$, α is still 0.129. In fact [206] showed that if we consider the prior probability distribution of $F_{TD,\infty}$, A, D, and f_c used by Güver et al. [95], only a fraction 1.5×10^{-8} of that distribution yields real numbers for M and R. This demonstrates that the 4% fractional uncertainties on the mass and radius of this neutron star obtained by Güver et al. [95] emerge from the theoretical assumptions rather than from the data. Thus such apparent precision is actually a red flag that one or more of the model assumptions is incorrect.

The first suggestion for which assumption is in error came from [206], who proposed that although the entire surface still emits uniformly throughout the cooling phase, the photospheric radius might be larger than the radius of the star, i.e., $r_{ph} > R$. However, analysis of the cooling phase of the superburst from 4U 1820–30 [158] demonstrates that such a solution is disallowed for at least the superburst emission from this star, because any detectable change in photospheric radius would require a flux very close to Eddington, and such fluxes give extremely poor fits to the data. The work of [158] used and verified the fully relativistic Comptonized spectral models of [211], and also showed that the *fraction* of the surface that emits changes systematically throughout the superburst (the emitting area drops by ~20% during the ~1600 s analyzed). Moreover, there is no guarantee that the whole surface was emitting at *any* time. Thus the star does not emit uniformly over its entire surface during the superburst, and hence it cannot be assumed that it has uniform emission during shorter bursts when the data quality is insufficient to check this assumption. Indeed, the presence of burst oscillations (see [230] for a recent review) demonstrates that there is nonuniformity in burning during many bursts. Additional concerns are that the color factor is likely to evolve during the burst, and that some of the bursts are not fit well using existing spectral models [43, 55, 87, 89, 210, 242].

Suleimanov et al. [210] find a radius of more than 14 km from a long PRE burst from 4U 1724–307 based on fits of their spectral models to the decaying phase of the burst. As part of their fits, they find that the Eddington flux occurs not at touchdown, but at a 15% lower luminosity; this, therefore, calls into question another of the assumptions in the standard approach to obtaining mass and radius from bursts. Their radius value is based on the good fit they get to the bright portion of the burst, when according to their fits the local surface flux exceeds ∼50% of Eddington. This is an intriguing method that should be considered carefully when data are available from the next generation of X-ray instruments. However, a potential concern is that the spectra does not agree with their models below ∼50% of Eddington. This suggests that there is other emission in the system at least at those lower luminosities, and hence that some of this emission might be present at higher luminosities as well. Against this is the excellent fit without extra components that [158] obtained for the superburst from 4U 1820–30 using the models of [211]. More and better data are the key.

If truly excellent spectra can be obtained, then as pointed out by [143], inference of the surface gravity g and surface redshift z leads uniquely to determination of the stellar mass and radius. However, even the $\sim 2 \times 10^7$ counts observed using *RXTE* from the 4U 1820–30 superburst are insufficient to determine both g and z uniquely [158], so this appears to require much larger collection areas. The combination $(1+z)/g^{2/9}$ can be measured precisely using sufficiently good continuum spectra, and then combined with other measurements to, possibly, constrain M and R [138, 158], so this is promising for the future. The net result is that currently inferred masses and radii from spectral fits to thermonuclear X-ray bursts must be treated with caution; none are reliable enough to factor into equation of state constraints.

1.4.2 Fits of Thermal Spectra to Cooling Neutron Stars

In principle, observations of cooling neutron stars with known distances allow us to measure the radii of those stars, modulo an unknown redshift. In practice, as with radius estimates from bursts, systematic errors dominate and thus current radius determinations are not reliable enough to help constrain the properties of dense cold matter.

To understand the basic principles, suppose that the star is at a distance d and that we measure a detector bolometric flux $F_{\rm det,bol}$ from the star that is fit by a spectrum with an effective temperature $T_{\rm eff,\infty}$ at infinity. Suppose that we also assume that the entire surface radiates uniformly. If the surface redshift is z then the luminosity at the surface is $L_{\rm surf} = (1+z)^2 L_\infty = (1+z)^2 F_{\rm det,bol} 4\pi d^2 = 4\pi R^2 \sigma_{\rm SB} T_{\rm surf}^4 = 4\pi R^2 (1+z)^4 \sigma_{\rm SB} T_{\rm eff,\infty}^4$. This implies

$$R = (1+z)^{-1} d [F_{\rm det,bol}/(\sigma_{\rm SB} T_{\rm eff,\infty}^4)]^{1/2} . \qquad (1.13)$$

The key questions are therefore (1) how well can the distance be determined, (2) how well can $T_{\rm eff,\infty}$ be established, given that it depends on a spectral fit, and (3) how well is the flux known, since only the thermal component is relevant?

There are two categories of sources that have been studied carefully in this manner to estimate neutron star radii. The first is the so-called quiescent low-mass X-ray binaries (qLMXBs). These are transiently accreting neutron stars that, in the ideal case, do not accrete at all when they are not accreting actively. Some qLMXBs are in globular clusters, so for those sources the distance to the qLMXBs can be determined by measuring the distance to the cluster. If a qLMXB has a phase of active accretion, then it has a steady supply of hydrogen or helium, which should rise to the top and dominate the surface composition after a few seconds [5]. In addition, the dipolar magnetic field strengths of LMXBs are weak, typically $\sim 10^8 - 10^9$ G averaged over the surface [7, 42]. Thus magnetic effects will be minimal (note that magnetic fields can affect energy spectra significantly at photon energies comparable to or less than the electron cyclotron energy $\hbar\omega_B = \hbar eB/m_e c = 11.6\,\mathrm{keV}(B/10^{12}\,\mathrm{G})$). Therefore, the spectra might be well-modeled by nonmagnetic atmospheres. These characteristics are all good for estimating radii. Guillot et al. [94] recently applied this method to five qLMXBs in globular clusters. They assumed that all the stars have the same radius, that the atmospheres are nonmagnetic and composed purely of hydrogen (an assumption that they supported on the basis of the companion type in two cases), and that the surface emission was uniform. With these assumptions, they found $R = 9.1^{+1.3}_{-1.5}$ km.

However, the atmospheric composition matters greatly. For example, [200] find that whereas fits of hydrogen atmospheres to a qLMXB in the globular cluster M28 yield a radius of $R = 9 \pm 3$ km at 90% significance, a helium atmosphere gives $R = 14^{+3}_{-8}$ km with an equally good fit (χ_ν/dof $= 0.87/141$ for the H atmosphere, 0.88/142 for the He atmosphere). Similarly, [52] fit data from a qLMXB in the globular cluster M13 and find $R = 9.0^{+3.0}_{-1.5}$ km for an H atmosphere and $R = 14.6^{+3.5}_{-3.1}$ km for a He atmosphere, again with a comparable quality of fit. Although one might argue [94] that if the companion is hydrogen rich the neutron star atmosphere will be as well, it has been proposed that after $\sim 10^{2-4}$ years diffusive burning of hydrogen will leave helium as the main surface composition ([56, 58, 193]; see [26] for the first step in a re-evaluation of this work in the light of a better treatment of Coulomb separation of ions). Given that $\sim$15 years of RXTE observations have not revealed any accretion-powered outbursts from any of the qLMXBs included in the analysis of [94], the recurrence time could be significantly greater than a century and thus the surface composition might be helium rather than hydrogen (see [134] for a discussion of what the larger implied radii would mean for dense matter).

Moreover, it is not necessarily valid to assume that the entire surface radiates uniformly. Nonaxisymmetric emission could theoretically show up as detectable pulses, but if the magnetic pole is close to the rotational pole (as was suggested to explain phenomena including the lack of pulsations in most LMXBs; see [126, 127]) the pulsations could be undetectable [181, 182, 226] while leading to inferred radii that are too low.

In addition, there are several puzzling aspects to the quiescent emission. For stars that have undergone several outbursts we have a decent estimate of their overall accretion rate. We can thus estimate the luminosity produced in the crust by electron capture reactions (e.g., [46]), which should be the minimum emergent luminosity. We also expect that the flux we see from the star should decline smoothly after cessation of accretion, and thus not have short timescale increases and decreases. In addition, if cooling is the only process, the spectrum should be purely thermal. All these expectations are violated in several stars. The quiescent thermal luminosity of 1H 1905+00 is $L < 2 \times 10^{30}\,\mathrm{erg\,s^{-1}}$, which is significantly below standard predictions [111]. Several stars vary in intensity by a factor of a few on timescales of days to years during their decline (for a discussion see [217]). In addition, many of the field low-luminosity stars have significant, even dominant, nonthermal tails (e.g., Cen X-4; [54, 195]).

It is possible that these discrepancies can be understood within the basic picture. For example, enhanced neutrino emission in the crust would not be observed, so this could explain the underluminous stars. Short timescale variability might be caused by the motion of obscuring matter in these binaries. Nonthermal spectral tails could be caused by coronal emission from the companion star [32, 51], magnetospheric activity, or some small residual accretion. Thus although the simple model is contradicted, plausible additions could rescue it. This nonetheless contributes an additional note of caution to radius inferences from these stars, and means that although the small radii reported from current qLMXB fits are similar to the small radii reported from some fits to X-ray bursts, both methods are dominated by systematic errors.

The second category of sources for spectral modeling and radius inferences is isolated neutron stars. We will encounter these again in Sect. 1.5 when we discuss cooling processes, but here we focus on what can be learned about radii. Exactly the same principles apply as for the qLMXBs, except that these stars do not accrete (the accretion rate from the interstellar medium is negligible; see [33] for a discussion). The same questions apply for these stars as they do for qLMXBs. For example, the spectra of young isolated neutron stars (such as the Crab pulsar) have significant nonthermal components probably caused by magnetospheric emission. Distances are also often difficult to establish, and the spectra can be fit using various spectra that give significantly different answers for the radius.

A case in point is RX J1856.5–3754, which is the brightest of the eight thermally emitting isolated neutron stars discovered using ROSAT. Its spectrum is featureless and can be fitted using a Planck spectrum, a heavy element spectrum, or a spectrum appropriate for condensed matter with a thin hydrogen envelope [206]. An interesting puzzle that constrains the atmospheric model is that the best-fit Planck spectrum to the X-ray data underpredicts the optical flux by a factor of $\sim$6 [48]. The distance has been estimated to be 117 ± 12 pc [228] or 161 ± 12 pc [222], although the data for the latter were re-analyzed to find a distance of 123^{+11}_{-15} pc [229]. It has been argued that this star has a radius of >14 km, which if true would make a strong case for hard equations of state. However, better atmospheric modeling suggests a radius of 11.5 ± 1.2 km [105, 106] assuming a distance of 120 pc, which makes the radius consistent with standard nucleonic and quark matter equations of state.

Other stars have been fit in the same manner; see for example [92, 93, 233]. Table 2 of [233] gives fitted radii for three neutron stars in globular clusters, and Table 4 of [92] gives fits for $R_\infty \equiv R(1 - 2GM/Rc^2)^{-1/2}$ for a number of qLMXBs and neutron stars in globulars. The uncertainties are large in many cases and thus these measurements tend not to be good discriminants between equations of state. In addition, it must be kept in mind that, as with the qLMXBs, the stars are so dim that their observed spectra cannot easily discriminate between quite different models, whether these be hydrogen atmospheres, helium atmospheres, Planck spectra, or heavy elements (see Fig. 1.3). Finally, it must be recalled that we know of nonaccreting neutron stars that pulse in the X-rays. If the isolated neutron stars we see that do not appear to pulse simply have their magnetic axes nearly aligned with their rotational axes, then single-temperature fits are likely to be misleading. Large-area instruments or long observations such as those planned for NICER [90] or LOFT [78] would help greatly in distinguishing between models.

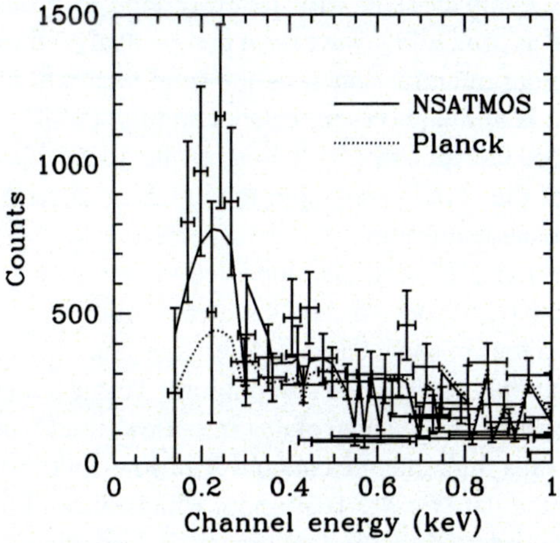

Fig. 1.3 XMM spectrum (error bars) and fits to a neutron star atmosphere model (solid line) and a blackbody (dotted line) for a neutron star in the globular cluster M13 (see [233]). The total chi squared for the blackbody fit is $\chi^2 = 88$ for 59 degrees of freedom, compared with $\chi^2 = 64$ in the NSATMOS model, so for the three extra parameters the difference is slightly greater than 4σ. The fitted radius of emission in the blackbody model is only $\sim$3 km, which is unphysically low unless only a small portion of the polar caps is emitting. The blackbody model can thus be tentatively ruled out on physical grounds and by a goodness of fit measure. However, a Planck function with an efficiency much less than the 100% efficiency of a blackbody is viable. Note also that the differences between a nonmagnetic hydrogen atmosphere and other candidates (e.g., pure helium, heavy atmospheres, or condensed surfaces; see [104–106]) are much less than their differences from a blackbody, and as these give significantly different inferred stellar radii, caution is essential in these inferences. Data and model kindly provided by Natalie Webb

1.4.3 Modeling of Waveforms

Thermonuclear bursts that display brightness oscillations, and isolated millisecond pulsars that are detectable in X-rays, have had their periodic waveforms analyzed in attempts to constrain their masses and radii. The basic principle is simple: if one makes the standard assumption that the oscillations are rotational modulations of the flux produced by a hot spot fixed on the star, then the shape of the waveform encodes information about the mass and radius (see [182], in which the relevant formulae were derived and which was the first paper to use millisecond pulsar waveforms to constrain the neutron star mass and radius). For example, a star of a given rotation frequency that has a large radius will have a higher linear surface rotational speed at a given latitude than a star with a small radius. Thus the waveform will have greater asymmetries produced by Doppler effects if the star is large than if it is small. The mass to radius ratio affects the fraction of the cycle when the spot is visible; in the Newtonian limit $R \gg GM/c^2$ exactly half the surface is visible, but for radii appropriate to neutron stars light-bending effects make more of the surface visible, and for a slowly rotating star with $R < 3.5GM/c^2$ the entire surface can be seen. More compact stars will therefore tend to produce lower amplitude waveforms. Hence in principle a waveform can be analyzed to infer the radius and mass, as well as other quantities such as the rotational latitude of the spot center, the rotational latitude of the observer, the spot angular radius (which turns out to be unimportant, as does the spot shape, as long as the spot radius is significantly less than the latitude of the spot center), and the emission pattern from the surface as seen in the spot's local rest frame.

Work by Muno et al. [160] initially appeared to cast doubt on this model of burst oscillations, because when they stacked data from multiple bursts from a given star they seemed to find that in some cases high energy photons arrive after low energy photons, i.e., the hard photons lag the soft photons. This is unexpected in the rotating hot spot model because as the spot rotates into view, the Doppler effect blueshifts the spectrum and thus high energies should lead low energies. However, a careful re-examination of the data for 4U 1636–536, which showed the strongest hard-lag trend in [160], demonstrated that in fact the oscillation phase versus photon energy is entirely consistent with the rotating hot spot model [12]. Statistical fluctuations, and probably the consequences of stacking burst data, appear to have led to the opposite conclusion in [160]. Thus currently it does seem that rotating hot spots are consistent with the data on burst oscillations.

The most common assumption in such modeling is the "Schwarzschild plus Doppler" approximation (e.g., [29, 154, 162, 235]), in which the star is assumed to be spherical and the spacetime external to the star is assumed to be Schwarzschild (nonrotating), but all the special relativistic effects associated with the rotation of the surface are treated exactly. Full simulations that trace rays in the numerical spacetimes appropriate for rotating objects have demonstrated that the waveforms generated using the Schwarzschild + Doppler approximation are indistinguishable from the waveforms in the full simulation when there are $\lesssim 10^5$ total counts and the stars have radii $R < 15\,\text{km}$ and spin frequencies $\nu < 600\,\text{Hz}$ [44, 50]. The current best analyses of isolated millisecond pulsars and bursting stars yield radii that are consistent with expectations but not very constraining; [38] find that $J0030+0451$ has a radius $>10.4\,\text{km}$ at 99.9% confidence, and [39] find that J2124–3358 has a radius $>7.8\,\text{km}$ at 68% confidence (both of these assume $M = 1.4\,M_\odot$), and [29] find $Rc^2/GM > 4.2$ at 90% confidence for the burster XTE J1814–338. Analyses using the "oblate Schwarzschild" approximation (in which the star is allowed to be oblate due to rotation but the spacetime is still assumed to be Schwarzschild) for SAX J1808.4–3658 [159] and XTE J1807–294 [135] are similarly unconstraining. The strongest current constraints from this method come from a recent analysis of PSR J0437–4715 assuming a hydrogen atmosphere, for which the result is $R > 11.1\,\text{km}$ at 3σ confidence [37].

A key assumption in the analysis of [37] is that the angular distribution of radiation from a point on the surface can be described by the pattern that emerges when the energy is deposited deep and propagates through a pure nonmagnetic hydrogen atmosphere. This could be a correct assumption, but in addition to our previous comments that hydrogen might not be the dominant surface composition, we note that the assumption of deep deposition of energy (which for this source comes from the return current of relativistic pairs from the magnetosphere) is based on the idea that the current gives up its energy via Coulomb collisions and nothing else. Given that plasma instabilities can shorten by orders of magnitude the column depth of energy deposition (see [45] for a recent example in the context of how AGN jet energy is injected in the intergalactic medium), this might not be a safe assumption.

Similar caution is appropriate for analyses of the waveforms from accretion-powered pulsars. For example, the oblate Schwarzschild analyses of the accretion-powered pulsars SAX J1808.4–3658 [159] and XTE J1807–294 [135] use a model with a blackbody component (assumed to be isotropic) and a Comptonized component (assumed to have an angular variation $\propto 1 - a\cos\alpha$, where α is the angle from the surface normal and a is a free parameter). These authors also included a scattered light component and assumed an infinitesimal spot. None of these assumptions or models can be perfect, and it is not known how serious an effect they have on the derived radii of the stars.

There are two reasons for the large credible regions that currently arise from analyses of waveforms: (1) the total number of counts is small, and (2) there are significant degeneracies between the parameters that produce the waveform. The effects of both factors are expected to be addressed using the next generation of large-area X-ray timing satellites. As discussed by Lo et al. [138], if a million counts are received from the spot (comparable to the total number expected from combining several bursts observed using LOFT, or to the integrated counts from non-accreting neutron stars or bursting sources using NICER) and the center of the hot spot and the observer are both within 10° of the rotational equator, then M and R can both be obtained to 10% precision.

As with the other methods we discuss, a key question is the role of systematic errors: if some aspect of the real system differs from what we assume in our model fits, how badly will our inferred mass and radius be skewed? There are clearly an unlimited number of possible sources of systematic error, but an encouraging conclusion from the work done by Lo et al. [138] with synthetic data is that even if the assumed surface beaming pattern, spot shape, or spectrum differ significantly from the actual ones, fits using the standard model do not *simultaneously* produce (1) a statistically good fit, (2) apparently strong constraints on M and R, and (3) significant bias in M and R. Thus, at least for the systematic differences from the model explored by Lo et al. [138], if the fit is good and the constraints are strong, the inferred values of the mass and radius are reliable.

Valuable extra information could be obtained from the identification of atomic lines from the surfaces of rotating neutron stars. No such line has been confirmed, and indeed even if a line-like feature is seen in a spectrum it is not trivial to identify the $z = 0$ atomic transition corresponding to the line. One such identification was claimed from an analysis of stacked bursts from EXO 0748−676 [65], but an additional long look at the star found it in another state that had no lines at all, whether zero redshift or from the surface, and thus was unable to confirm the lines [66]. The spin frequency of this star is 552 Hz [88] rather than the originally claimed 45 Hz [227], and hence one might expect that Doppler smearing would make a sharp line undetectable (although note that [20] suggest that sharp lines would still be visible; if this result is confirmed, it means that there are better prospects for sharp lines than previously thought). If future large-area instruments are able to not only detect such features but also measure them precisely, then both the redshift from the surface and the linear speed of the surface at the spot, as well as possibly even frame-dragging effects, could be inferred [30]. This would allow many degeneracies to be broken and would lead to much more precise constraints on neutron star masses and radii (and moments of inertia from frame-dragging). Note that such measurements will only be possible from actively accreting stars, because heavy elements sink in the atmospheres of isolated neutron stars within seconds [5]. It has also been proposed that the equivalent width of the line will allow a measurement of the surface gravity, and hence that M/R (from the redshift) and M/R^2 (from the surface gravity) can be measured independently (e.g., [57]). In principle this is also possible using a non-thermal continuum spectrum, but this would require exceptional data.

1.4.4 Maximum Spin Rate

Another method that has been suggested to constrain the radius (or more properly, the average density) is measurements of spin frequencies: a high enough spin frequency from any star would rule out the hardest equations of state. Unfortunately, no confirmed spin frequency is high enough to place significant limits on dense matter. The maximum spin frequency for a star of mass M and radius R is roughly [63]

$$\nu_{\rm max} = 1250\,{\rm Hz}\,(M/M_\odot)^{1/2}\,(R/10\,{\rm km})^{-3/2}\,. \tag{1.14}$$

This is $\approx$800 Hz for $M = 1.4\, M_\odot$ and $R = 15$ km. In comparison, the highest frequency ever established is 716 Hz, for PSR J1748–2446ad in the globular cluster Terzan 5 [102]. Weak evidence for an 1122 Hz signal during thermonuclear bursts from XTE J1739–285 has been claimed [113] but not confirmed. Even if it were confirmed there would have to be strong evidence that this was the fundamental frequency instead of the first overtone; the overtone can be dominant, as was shown for IGR J17511–3057 [8].

Given that neutron stars could spin faster than they do, and that X-ray observations using the *Rossi* X-ray Timing Explorer are not biased against signals with $\nu > 1000$ Hz [53], why do the stars not spin faster? A conservative answer that is in agreement with all data is that magnetic torques during accretion and spinup limit the frequency. The required dipolar field strengths $\sim 10^8$ G agree with the fields inferred from the spindown of their descendants (the rotation-powered millisecond pulsars), and in particular with the adherence of those pulsars to the spin-up line [144]. Fields of this strength are also consistent with the spin behavior of stars between transient outbursts [97, 176]. Another possibility is that in many cases not enough matter has been accreted to reach maximum spin. An exciting longshot that has received much attention due to the rapid improvements in ground-based gravitational wave detectors is that nonaxisymmetries in neutron stars, e.g., Rossby waves [9, 10, 40, 59, 84] (see [98, 142] for recent observational constraints) or perhaps accretion-induced lumpiness in the stars [31, 218], might counteract accretion spinup via emission of gravitational radiation. One way to test this hypothesis is to observe systems that have had multiple transient episodes, because the spindown between active phases could indicate whether gravitational radiation (which would depend only on the long-term average accretion rate rather than the instantaneous rate) emits angular momentum at the required rate. Two such systems (SAX J1808–3658 and IGR J00291+5934) have been observed with the required precision. In both cases there is no evidence that gravitational radiation induced spindown is occurring, but within the observational uncertainties there is room for contributions at the tens of percent level [96, 175]. Whatever the reason for the spin ceiling, at this stage there are no known neutron stars with spin frequencies high enough to rule out any plausible equation of state.

1.4.5 Kilohertz QPOs

Kilohertz quasi-periodic brightness oscillations (kHz QPOs) from accreting neutron stars have been proposed to constrain the masses and radii of the stars. The basic phenomenology of kHz QPOs is that there are commonly two relatively narrow ($Q \equiv \nu/\mathrm{FWHM} \sim 20-200$) QPOs that appear in the power density spectra of more than 25 neutron star low-mass X-ray binaries. Both frequencies vary by hundreds of Hertz between and during observations. The higher-frequency of the two often reaches $\nu > 1000$ Hz, and the lower-frequency peak (which is often the sharper one, and also commonly has a larger fractional root mean square amplitude) has a frequency less than that of the upper peak by a characteristic but not exactly constant amount that may be related to the spin frequency of the star (but see [149] for a dissenting opinion).

Most, but not all, modelers identify the upper peak frequency with an orbital frequency at some characteristic radius around the star. If this is true, it means that the star must fit inside that radius, as must the radius of the innermost stable circular orbit (ISCO) predicted by general relativity; the latter condition follows because matter inside the ISCO will fall rapidly towards the star and thus prevent it from forming high quality factor oscillations. As derived by [157], for a neutron star with a dimensionless rotation parameter $j \equiv cJ/GM^2$ these conditions limit the mass and radius to $M_{\rm max} = 2.2\,M_\odot(1+0.75j)(1000\,{\rm Hz}/\nu_{\rm QPO})$ and $R_{\rm max} = 19.5\,{\rm km}(1+0.2j)(1000\ {\rm Hz}/\nu_{\rm QPO})$ for an upper peak frequency $\nu_{\rm QPO}$. The highest confirmed QPO frequencies are all less than 1300 Hz (see, e.g., [41] for a discussion of the 1330 Hz QPO once suggested for 4U 0614+09), so at this stage the constraints are not restrictive. If broad iron lines from the inner disk are discovered simultaneously with kilohertz QPOs, this will provide another measure of the mass because the line breadth gives $\sqrt{M/r}$ whereas the QPO frequency gives $\sqrt{M/r^3}$ at the orbital radius r (see [49] for current data and [28] for future prospects; note that the Kerr spacetime is not an adequate approximation for sufficiently rapidly rotating neutron stars [156]). Similarly, if reverberation mapping can establish an absolute time scale for the system, this might yield masses and radii [15].

If the orbital frequency of the ISCO is established for a star, then the mass of the star is known to within a small uncertainty related to the star's dimensionless angular momentum parameter. After doubt was cast on initial claims of ISCO signatures [240] because of the complex relation between count rate and QPO frequency in these stars [150], recent analysis of the RXTE database has suggested that in many stars the predicted sharp drop in quality factor and gradual drop in amplitude [157] are seen at a frequency that is consistent across a wide range in count rate and X-ray colors [16–19]. The independence of this behavior from proxies of mass accretion rate such as count rate and colors led these authors to suggest that a spacetime marker such as the ISCO was the most likely reason for the observed phenomena. If so, this represents a confirmation of a key prediction of strong-gravity general relativity (the ISCO), and implies masses greater than 2.0 $M_\odot$ for some neutron stars, which would be highly constraining on equations of state.

Such important implications demand careful examination. For example, [148] notes that the maximum quality factor achieved by neutron star LMXBs, versus their average luminosity, has a shape similar to the quality factor versus radius in individual stars, and uses this to conclude that other factors operate and that the drop in quality factor might not be caused by approach to the ISCO. Indeed, other factors do influence the quality factor; for example, the high-luminosity stars plotted by Méndez [148] have geometrically thick disks that cannot produce high-Q oscillations. This is thus not directly relevant to the arguments made by Barret et al. [16–19] because the stars they examined are all low-luminosity, and hence the results of [148] do not address whether the ISCO causes the behavior observed in those stars. Nonetheless, the complex phenomenology of kHz QPOs and the lack of first-principles numerical simulations that display them means that claims that the ISCO has been detected must be treated with care.

1.4.6 Other Methods to Determine the Radius and Future Prospects

There are two other noteworthy ways to measure the radius that have been suggested with particular applications to the double pulsar J0737–3039. The first involves the binding energy of the lower-mass Pulsar B in that system. Based on an idea originally proposed by Nomoto [163], Podsiadlowski et al. [179] suggested that this neutron star formed via an electron-capture supernova (in which electron captures onto Mg and then Ne in a core cause a loss of pressure support) rather than the usually considered collapse of an iron core when it goes above the Chandrasekhar mass. As this electron capture happens at a very specific central density of $4.5 \times 10^9\,\mathrm{g\,cm^{-3}}$ that corresponds to a well-defined baryonic mass of $M_{\rm bary} = 1.366 - 1.375\,M_\odot$ [179], if one could identify this baryonic mass with the gravitational mass $M_{\rm grav} = 1.2489 \pm 0.0007\,M_\odot$ of Pulsar B then one would have an extremely precise constraint that would suggest a fairly hard equation of state. The degree to which this constraint is useful depends on our theoretical certainty that there is not significant subsequent fallback or expulsion of matter. Current models do suggest that fallback or expulsion only introduce a spread of $\sim 0.01\,M_\odot$ in the final baryonic mass (K. Nomoto, personal communication). If this tight range is confirmed by further work and if Pulsar B is indeed produced by this mechanism, this could be a useful constraint.

The second method stems from the observation that spin-orbit coupling, which is dominated in this system by the 23 ms period Pulsar A instead of the 2.8 s period Pulsar B, leads to precession of the orbital plane and additional pericenter precession [14, 67, 118, 133, 141]. Orbital plane precession will be difficult to measure, but the required precision for the extra pericenter precession could be reached in the next few years [133]. Such a measurement would effectively determine the moment of inertia I of Pulsar A (potentially to 10%; [133]), and given that $I \sim MR^2$ and M is known well, this would amount to a $\sim$5% measurement of the radius. It is unclear whether this will be reachable in practice, because of the confusing effect of the unknown acceleration of the center of mass of the binary within the Galactic potential [118].

A final method to mention for completeness (see also the introduction to Sect. 1.6) is neutron star seismology. Asteroseismology has greatly improved our understanding of normal stars, and it would do so for neutron stars as well if particular modes could be identified with confidence. Indeed, fast oscillations that might be torsional modes have been seen in the tails of giant flares from the soft gamma-ray repeaters SGR 1900+14 [208] and SGR 1806–20 [209, 231] (see [232] for some perspectives on these oscillations). Unfortunately, the data are insufficient to make clear identifications. For example, the candidate torsional modes for these two soft gamma-ray repeaters skip many modes that would have been expected to be seen. In addition, the theory for these oscillations is not settled. Nonetheless, as data and models progress mode identifications may become clearer, and if they do then they will provide a powerful and complementary set of constraints.

In general, the methods discussed in this section have uncertainties that are dominated by systematics. As a result, future progress depends on both better observations and better models. For example, there is a recent set of magnetohydrodynamic simulations that aim to reproduce kHz QPO phenomenology in accreting weakly magnetic neutron stars [36, 119, 120, 139, 140, 189–192]. These simulations may ultimately result in solid understanding of kHz QPOs that could be used to interpret observations. The simulations are extremely challenging, however, and may take many years to get to the point of full reliability. Some of the apparently disparate constraints may eventually be coupled through the recently discovered "I-Love-Q" relations between the moment of inertia I, Love number, and rotational quadrupole moment Q [238], although the important issues of systematics must be solved first.

1.5 Cooling of Neutron Stars

As noted in Sect. 1.2, cooling processes are sensitive to different aspects of the equation of state than are the maximum mass and the mass-radius relation. This in principle means that observations of cooling neutron stars can give us a complementary tool with which to constrain the properties of dense matter. In fact, current data are broadly consistent with the dense matter in neutron stars not having significant contributions from exotic phases (modulo some complications we shall discuss), but unfortunately as we will see this is a very blunt tool and there is plenty of room for exotic matter.

X-rays from cooling neutron stars were originally proposed in the mid-1960s [62] as one of the few ways that these stars could be detected. In this section we will focus on cooling theory and observations that bear on the matter at the cores of neutron stars. We will thus not discuss, e.g., the cooling of transiently accreting neutron stars (see [172, 239] for recent reviews) that return to quiescence after a years-long outburst has raised the crust out of thermal equilibrium with the core, because their cooling curves depend primarily on processes in the crust.

Broadly speaking, after a neutron star forms in a supernova (where at birth its temperature is roughly the virial temperature $T_{\rm vir} \sim GMm_n/(Rk) \sim 10^{12}$ K), the star goes through a phase of duration $\sim 10^{4-6}$ yr in which its cooling is dominated by neutrino losses from the core. After this point the star cools mainly by photon luminosity from the surface, where the energy from the core is transported conductively until the density is low enough that radiative processes take over (this typically occurs in the outer crust). The temperature of a neutron star at a given age therefore depends strongly on how long neutrino emission dominates. The temperature tool is blunt because there are many processes that can enhance neutrino production, and many processes that can suppress it, and thus mere measurement of the temperature with age of neutron stars would not allow us to distinguish easily between the multiple candidate effects.

1.5.1 The URCA Processes

To start, we note that as the star cools down from its high-temperature birth the processes $p+e^- \rightarrow n+\nu_e$ and $n \rightarrow p+e^- +\bar{\nu}_e$ can produce neutrinos efficiently. In this context these are known as the URCA processes, named thus by George Gamow after the Urca casino in Rio de Janeiro because the URCA processes are a perfect sink for energy just as the casino is a perfect sink for money! Phase space considerations indicate that the URCA emissivity (energy per volume per time) scales as T^6.

As the temperature drops below the Fermi temperature, however, we can see that these processes become impossible unless the proton to neutron ratio is sufficiently high. The derivation of the critical ratio is similar to what we presented in Sect. 1.2: the momenta of the particles, which are dominated by Fermi momenta, must satisfy the triangle inequality $p_n \leq p_p + p_e$. The Fermi momenta of all three particles are determined by their respective number densities, so this means $n_n^{1/3} \leq n_p^{1/3} + n_e^{1/3}$. Charge neutrality means $n_e = n_p$, so $n_n^{1/3} \leq 2n_p^{1/3}$ and thus the criterion for the URCA process to be possible is $n_n \leq 8n_p$. This is on the low side for many traditional equations of state but can be achieved in some cases at high density. Note that if muons are also present (these are higher-mass analogs to electrons with the same electric charge), and thus electrons have a number density $n_e = xn_p$ with $x \leq 1$, the criterion becomes $n_n \leq (1 + x^{1/3})^3 n_p$, so even higher proton fractions would be required.

If the URCA process is suppressed, bystander particles can soak up the extra momentum, e.g., $n+n \rightarrow n+p+e^- +\bar{\nu}_e$. This is usually called the modified URCA process to distinguish it from the direct URCA (sometimes DURCA) processes described above. Given that the neutrons are degenerate, only a fraction T/T_F of them can interact. Thus the modified URCA process is suppressed by a factor $(T/T_F)^2$ (one factor is for the initial neutron and one is for the final) compared to the direct URCA process. Given that in a neutron star core $T_F \sim 10^{12}$ K and T can be $\sim 10^9$ K, the suppression factor can easily be a million.

The huge difference between the direct and modified URCA rates means that if enough protons are present, or if there are any other channels for neutrino production, then cooling can be enhanced dramatically. We now consider such channels.

1.5.2 Additional Neutrino Production Channels and Suppression

As described in the lucid review of [172], hyperons can produce neutrinos via, e.g., $\Sigma^- \to \Lambda + e^- + \bar{\nu}_e$, $\Lambda + e^- \to \Sigma^- + \nu_e$, and processes that involve both hyperons and nucleons. If condensates form (the leading candidates are of pions or kaons) then the condensate acts as an effectively infinite reservoir of momentum. This produces channels such as $n + \langle \pi^- \rangle \to n + e^- + \bar{\nu}_e$ and $n + \langle K^- \rangle \to n + e^- + \bar{\nu}_e$, where the angle brackets indicate the condensate. The kaon process involves a strangeness change and is thus less efficient than the pion process, modulo effects related to the medium [172]. If deconfined quarks are a significant degree of freedom then other direct URCA-like processes can emerge, such as $u + e^- \to d + \nu_e$ and its inverse (e^+ could be present in negatively charged quark matter, but the processes are the same), and $u + e^- \to s + \nu_e$ and its inverse (suppressed by an order of magnitude because of the strangeness change). All of these processes are orders of magnitude more efficient than modified URCA, and all scale as T^6.

A separate channel has the interesting property that it can both increase and suppress cooling, depending on the details. This is the transition to a superfluid state. This transition occurs via Cooper pairing of the neutrons, and the immediate effect is the emission of a neutrino-antineutrino pair: $n + n \to [nn] + \nu + \bar{\nu}$. Given that one fewer effective particle is involved than in modified URCA the emissivity scales as T^7 instead of T^8, so at $T \sim 10^9$ K the emissivity of Cooper pairing can be comparable to or greater than modified URCA. As the shell where the temperature is less than the superfluid critical temperature moves inwards, this can therefore enhance neutrino emission.

In the long term, however, the pairing produces an energy gap $\Delta \sim kT_c$ (where T_c is the superfluid critical temperature) at the Fermi surface that suppresses processes by a factor of order $e^{-\Delta/kT}$ for $T \ll T_c$, modulo details of the phase space and the temperature dependence of Δ. The suppression can thus be extremely dramatic. It occurs for both the neutrino emissivity and for the specific heat (with different factors). See, e.g., Figure 5 of [172] for plots of some of the suppression factors (called control functions there). The critical temperature and energy gap are extremely difficult to calculate from first principles. Current estimates of the ^{3}P-F$_2$ gap suggest $\Delta \sim 0.05 - 0.1$ MeV [74, 169, 198], which corresponds to $T \sim 5 \times 10^8 - 10^9$ K. This is comparable to the expected core temperatures and thus could make a significant difference.

Cooper pairing can also occur in deconfined quark matter [6]. In that context it is much more complicated than in ordinary matter because quarks have different colors, flavors, and masses. Multiple types of condensation are therefore possible. The color gap is estimated to be $\sim$50 – 100 MeV [172], which is huge compared to internal temperatures and is thus potentially quite important.

There are also neutrino emission processes in the crust of the star, such as plasmon decay and neutrino pair bremsstrahlung from electron-ion and electron-electron interactions. These could contribute for middle-aged stars, when the core processes are suppressed by the superfluidity gap and the crust is warm enough to produce neutrinos.

1.5.3 Photon Luminosity and the Minimal Cooling Model

When neutrino emission has tapered off, further cooling of the star is controlled by conductive transport in the dense layers of the star (where electrons are degenerate and thus have large mean free paths) and by radiative transport nearer the surface. There is a minimum in the overall efficiency of energy transport at the "sensitivity layer" where conduction hands off to radiation (where electrons are only partially degenerate). This thus acts as a bottleneck and largely determines the overall cooling rate; the density at the sensitivity layer increases with increasing temperature. For temperatures in the observed range $\sim 10^{5-6}$ K the sensitivity layer is at a density $< 10^9\ \mathrm{g\,cm^{-3}}$, where pycnonuclear fusion is not guaranteed to convert light elements to heavy ones. The electron thermal conductivity in liquid ions depends on their charge Z as $\sim 1/Z$ [172], so differences in the composition of the upper layers (due, e.g., to different fall-back after core collapse) could have an influence on the thermal evolution of neutron stars.

Conduction dominates at high enough densities that magnetic fields are likely to be unimportant: at typical densities $\rho \gtrsim 10^6$ g cm^{-3} the Fermi energy is $E_F \gtrsim m_e c^2$, implying that the magnetic field needs to be $B \gtrsim B_c = m_e c^3/(\hbar e) = 4.414 \times 10^{13}$ G to have a significant influence. However, more moderate magnetic fields can affect radiative transport near the surface. More specifically, suppression of electron motion across field lines yields anisotropic conduction. This can produce strong anisotropies in the emergent radiation, but the overall effect on cooling is relatively small [172]. Ultimately, the radiation emerges with some spectrum and an effective temperature $T_{\rm eff}$ that can be defined as

$$T_{\rm eff}^4 \equiv \frac{1}{4\pi} \int\int T_s^4(\theta,\phi) \sin\theta d\theta\, d\phi\ . \tag{1.15}$$

Here, the local effective temperature $T_s(\theta,\phi)$ at each location (θ,ϕ) on the surface is defined via $\sigma_{\rm SB} T_s^4(\theta,\phi) = F(\theta,\phi)$ where F is the emergent photon flux and these quantities are appropriately redshifted. If the emergent spectrum is close to a blackbody then one can estimate the temperature without knowing the distance to the star (see Sect. 1.4), but as we discussed earlier atmospheric effects cause the spectrum to deviate from a blackbody form, which complicates inferences.

With this physics in place one can construct a surface temperature versus age curve by including a specific choice for the uncertain core processes as well as choosing the composition of the surface layers. A particularly useful choice is the "minimal cooling model" (see [173] for a recent treatment), in which one assumes no exotic components or direct URCA but does include Cooper pairing, crustal bremsstrahlung, and all relevant photon processes. This represents the smallest amount of cooling that is realistic. As a result, if there is evidence that at least some neutron stars are significantly cooler than they are predicted to be in this model, that might suggest exotic components (although it does not work the other way around; conformity with minimal cooling might mean the suppression effects are important). To see how this cooling model fares we now turn to observations.

1.5.4 Observations and Systematic Errors

Before discussing the observations we must issue a series of caveats. There are many reasons why it is difficult to generate reliable points on a temperature-age curve, including the interpretation of the spectrum, other possible heating sources, and challenges with age estimation. We now discuss these in order before finally evaluating the best current data on cooling neutron stars.

A neutron star with a surface effective temperature of 10^6 K and a radius of 10 km has a luminosity of $L \sim 10^{32}\,\mathrm{erg\,s^{-1}}$, which at a distance of 3 kpc gives a detector flux of $F \approx 7 \times 10^{-13}\,\mathrm{erg\,cm^{-2}\,s^{-1}}$ and corresponds to ~2 counts per second for a 1000 cm^2 detector. It is therefore possible to get a reasonable number of photons over a long observation, although the thermal peak of $\sim 0.2\,\mathrm{keV}(T/10^6\,\mathrm{K})$ is strongly susceptible to interstellar absorption.

There are also complications with the atmospheric model, as we discussed when we examined radius estimates for cooling neutron stars. For example, compared to a blackbody with the same effective temperature, an unmagnetized hydrogen atmosphere has a strong excess at higher energies when absorption dominates the opacity, because these opacities scale as ν^{-3} (e.g., [196]). Magnetized hydrogen has less of an excess because strong fields increase the binding energy of atoms (e.g., Problem 3 in §112 of [128]), and heavier elements also have more bound electrons and thus less of an opacity deficit at high energies compared to lighter elements [152, 155, 184, 187]. Observational support for the diffusive burning of hydrogen or helium [56, 58] may have been obtained from the evidence that the atmosphere of Cas A is dominated by carbon [104]. As a result, unless there is a clear statistical preference for one atmospheric model versus another (which is not currently the case for any star), the temperature will be uncertain.

We must also be cautious because in addition to simple cooling there are various heat sources that could contribute. These include magnetic dissipation (likely only important for highly magnetic neutron stars) and magnetospheric emission. The latter is likely to produce a nonthermal spectrum, which is indeed seen in some stars. One could argue with some justice that if we are looking for cases where the temperature is *less* than predicted by the minimal cooling model, other heat sources will only mean that such evidence is even stronger. The tricky part comes when one subtracts off nonthermal components to estimate the underlying thermal emission;

if the other components were oversubtracted this would lead to an underestimate of the cooling temperature.

The final caveat relates to the age. If the star was born with a much more rapid spin than it has now and it has slowed down exclusively by magnetic dipole radiation then its current age is $P/(2\dot{P})$, where P is the current spin period and $\dot{P}$ is the current spin derivative. In some cases this age estimate can be checked using a kinematic age, either from a supernova remnant or from the angular distance above the Galactic plane (where massive stars are born) divided by the angular proper motion away from the plane. Unfortunately there is often a discrepancy between these estimates of a factor of three or more, so all of these numbers must be treated with caution.

With the preceding in mind, Figure 8 in [173] shows the comparison between the minimal cooling model (with light or heavy element envelopes) and the data. Within the significant uncertainties we note that all of the data are consistent with the minimal cooling model, although there is some evidence that variation in the envelope composition is needed to explain the data. As [170] note, however, there is a significant selection effect at work: if there are stars that have cooled rapidly, they are obviously more difficult to see. It is thus possible that using our current satellites we can only observe comparatively hot stars.

Recently it was suggested that the neutron star in the supernova remnant Cas A has cooled very rapidly over the past decade or so [104]. Further observational analyses, especially taking into account the complexity of the surrounding supernova remnant emission [76] and the possibility of changes in the calibration, emitting region size, or absorbing column [180] have made it much less clear that there actually is anomalously fast cooling. If the evidence strengthens for such cooling, it has potentially exciting implications for the physics of the interior of this neutron star, with the leading idea being Cooper pair creation in the superfluid [174, 202]. One-pion exchange and polarization effects could also play a role [35].

1.5.5 Current Status and Future Prospects

Currently there is no evidence that exotic components are *necessary*, although given the large uncertainties in data they could certainly be *accommodated*. The possible lack of exotic components is consistent with the tentative evidence presented in Sect. 1.4 that some neutron stars have masses $\gtrsim 2.0\ M_\odot$, and could mean that nucleonic degrees of freedom dominate the internal structure of neutron stars. However, the data are not clear.

To improve the observational situation it will be necessary to have much larger-area future X-ray observatories, such as the approved mission Athena+ [161]. The resulting high-signal observations would play two important roles: (1) they would reduce the bias against rapidly cooled stars, and would thus possibly reveal evidence for exotic components, and (2) they would allow us to distinguish empirically between different candidate atmospheric spectra (nonmagnetic hydrogen or helium, magnetic atmospheres, and even the straw man blackbody spectrum), and thus have greater confidence in the inferred temperatures. It will be more challenging to come up with much more precise and accurate ages, but it could be that future radio arrays such as the Square Kilometer Array will allow us to determine the proper motion and kinetic ages more accurately.

1.6 Gravitational Waves from Coalescing Binaries

In the next few years it is expected that the worldwide gravitational wave detector network will achieve sufficient sensitivity to detect ∼0.4–400 NS-NS mergers per year, and an uncertain number of NS-BH coalescences (see [2] for a recent discussion of the predicted rates and their substantial uncertainties). In this section we summarize what can be learned from such observations. It has also been proposed that observation of neutron star oscillation modes stimulated by mergers or (at much lower amplitude) glitches would yield important information about the stars (see, e.g., [21, 60, 61, 73, 221]); this is true, but the amplitudes are likely to be significantly below the amplitudes of the coalescence signal, so we will not focus on oscillations here.

In brief, for a binary of masses m_1 and m_2 and thus total mass $M = m_1 + m_2$ and symmetric mass ratio $\eta = m_1 m_2/M^2$ (note that $\eta \leq 0.25$, with the maximum occurring for $m_1 = m_2$), the combination $\eta M^{5/3}$ will be determined with high precision and could lead to significant constraints if a high enough mass binary is detected. The individual masses and the average density of the neutron stars will be more challenging to measure, but they seem within reach for the strongest events.

In more detail, we note that to lowest order gravitational radiation changes the binary orbital frequency at a rate (see [177] for the rate of change of the semimajor axis)

$$\frac{df}{dt} = \frac{96}{5}(4\pi^2)^{4/3} G^{5/3} \eta M^{5/3} f^{11/3} c^{-5} (1-e^2)^{-7/2} (1 + 73e^2/24 + 37e^4/96) \,. \tag{1.16}$$

Now consider a NS-NS binary. When the orbital separation is much greater than the radii of the neutron stars, the inspiral of the binary proceeds almost as it would if the stars were point masses. From Eq. 1.16 we see that the mass combination $M_{\rm ch}^{5/3} \equiv \eta M^{5/3}$ (where $M_{\rm ch}$ is called the "chirp mass") determines the frequency and amplitude evolution. It can thus be measured with precisions better than 0.1% in many cases [220]. Even with no additional information we can obtain a strong lower limit to the total mass M by setting $\eta = 0.25$, and thus can obtain a strong lower limit $M/2$ to the greater of the two masses. It could be that if tens to hundreds of NS-NS mergers are observed per year, and if these can be distinguished clearly from NS-BH and BH-BH mergers, then a small number of them will have $M > 4.0\ M_\odot$ and thus it will be possible to establish rigorously that $M_{\rm max} > 2.0\ M_\odot$.

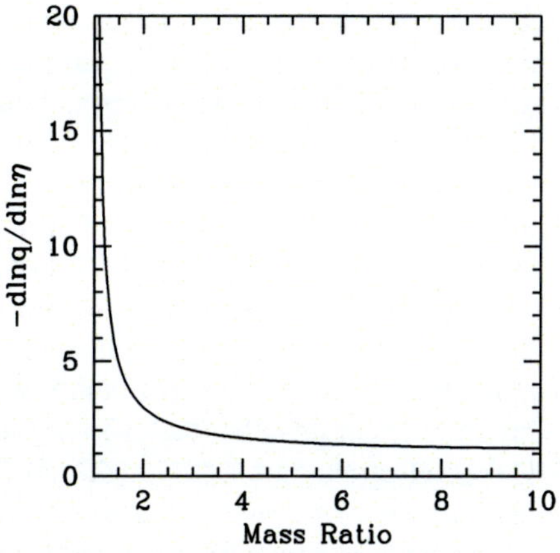

Fig. 1.4 Logarithmic derivative of the mass ratio $q \equiv m_1/m_2 \geq 1$ with respect to the symmetric mass ratio η. For double neutron star binaries, in which $q \sim 1$, even a small error in η leads to significant uncertainty in q. As a result, gravitational waves from double neutron star systems are not ideal for precise mass measurements of individual neutron stars

We might not be this lucky. If the NS-NS detection rates are as low as $\sim$1 yr^{-1} then high-mass binaries might not be sampled. If the rate is tens per year but the chirp masses do not imply a high minimum mass it could be that the upper limit to neutron star masses is close to the $\sim$2 $M_\odot$ that we have already established from electromagnetic observations, but it could also be that there is essentially only one way to form NS-NS systems and thus that those systems will tend to have similar chirp masses.

In these cases more information is needed. When the inspiral is followed to higher post-Newtonian order the additional terms, which involve tidal effects, have different dependences on η and M than does the lowest-order expression, so this can be used to break the degeneracy and infer the two masses separately. For systems in which the two masses are comparable (as in a NS-NS system) this requires very high precision measurements of η. Note, for instance, that whereas $\eta = 0.25$ implies a 1:1 mass ratio, $\eta = 0.24$ implies 1.5:1. Figure 1.4 shows that for nearly equal masses a very small fractional error in η can still imply a large fractional error in the mass ratio $q \equiv m_1/m_2 \geq 1$. The sensitivity is naturally less away from the maximum, which might lead one to suppose that NS-BH binaries, which are asymmetric in mass, would provide greater prospects for NS mass measurements. For example, suppose that $M_{\rm ch} = 2.994\ M_\odot$ is measured with effectively zero uncertainty, and that η is constrained to be between 0.1 and 0.11. Then the lighter component of the binary has a mass between 1.34 $M_\odot$ and 1.42 $M_\odot$ and is thus well known despite the significant fractional uncertainty in η.

The tradeoff is that black holes may typically have large enough masses that neutron stars spiral into them without significant tidal effects [153], although higher harmonics can still be important and as partial compensation they will appear at frequencies of greater sensitivity in ground-based detectors than will the corresponding harmonics in NS-NS systems. In addition black holes, unlike the neutron stars in NS-NS binaries, may well have large enough spins to affect the last part of the inspiral and thus compromise parameter estimation due to the greater complexity of the waveforms. No pulsar in a NS-NS binary has a spin period shorter than 23 ms [205], so the spin parameters of these neutron stars are $j \equiv cJ/GM^2 \lesssim 0.02$. In contrast, although inferences of the spins of stellar-mass black holes are not yet fully vetted, there is growing evidence that many of them have spin parameters of several tenths, with some possibly approaching $j = 1$, the mathematical maximum for black holes [146, 151]. The likely evolutionary difference is that in the supernova that creates a neutron star or black hole, much greater amounts of mass fall back to create a black hole than a neutron star, and this mass will come from significant radii that thus carry considerable angular momentum. Accretion from a binary subsequent to the production of a black hole has little effect on either the mass or the angular momentum of the hole (see, e.g., [114]). Neutron stars in the high-mass X-ray binaries that could create a double neutron star pair accrete little mass in the active phase and, empirically, seem to have relatively strong magnetic fields, so accretion does not spin them up to high rotation rates.

These statements are not absolute. It could be that there are slowly-spinning black holes in BH-NS binaries, or rapidly spinning neutron stars in NS-NS binaries (especially if they are produced dynamically in globular clusters). However, the expectation at this time is that until tidal effects become important NS-NS waveforms will be easier to interpret than those from BH-NS coalescences.

This leads us to what those tidal effects might tell us about neutron stars. At large separations compared to the neutron star radii, nonlinear mode couplings due to tides will not affect the inspiral phase significantly [225]. Most of the information that can be obtained from tidal phase deviations from point-mass inspiral exists at high frequencies [103], and recent comparisons of analytical theory with numerical simulations suggest that with some calibration the theory does extremely well [25, 69, 79, 108, 124]. An explicit comparison of the expected signals from piecewise-polytropic equations of state suggests that a difference of only 1.3 km in radius could be distinguished for NS-NS systems out to 300 Mpc with optimal direction and binary orientation (this corresponds roughly to a direction- and orientation-averaged distance of 140 Mpc) given a full hybrid post-Newtonian plus numerical relativity waveform [186]. See Fig. 1.5 for an indication of the different frequencies of tidal disruptions implied by two candidate equations of state.

Inspirals of neutron stars into black holes have not yet been examined using as much care with respect to phase deviations. Recent work suggests that although if the mass ratio is as high as 6:1 a NS-BH coalescence will be indistinguishable from a BH-BH coalescence with the same masses [80], if the mass ratio is as small as 2:1 or 3:1, single events at a distance of 100 Mpc could reveal the neutron star radius to within 10–50% [123, 124], depending on details. This is a case in which a combination of gravitational wave and electromagnetic observations, along with simulations, would work very well: the gravitational wave observation identifies the masses of the black hole and neutron star and the spin of the black hole, and the electromagnetic observation plus simulations derives information about the remaining mass of the disk (particularly if the recent promising observations related to kilonovae hold up; see [24, 214]). Simulations are still very much in their infancy,

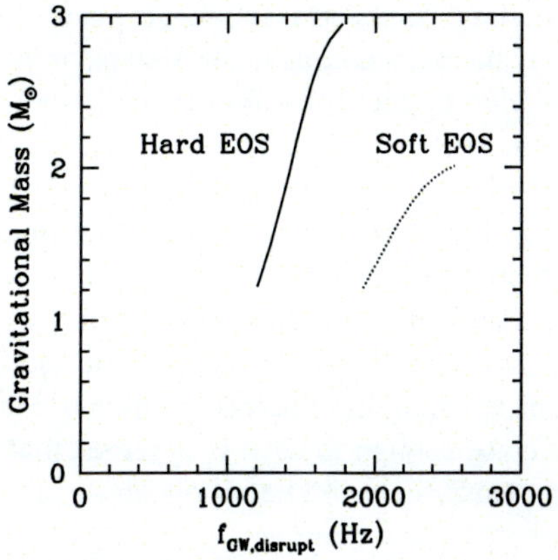

Fig. 1.5 Gravitational wave frequency at the point of tidal disruption as a function of neutron star mass. The solid and dotted lines refer, respectively, to the hard and soft equations of state from [100]. In both cases we assume that the companion is a neutron star of equal mass. Tidal effects will affect the phasing of the inspiral below $f_{\rm GW,disrupt}$, but the effects fall off rapidly with decreasing frequency. Good sensitivity at high frequencies is essential for these effects to be detected

but this is a promising avenue to pursue. It has also been noted that the precision of constraints will be improved significantly by combining the analyses of >10 bursts, if systematic errors are under control [71], and that observations with the planned third-generation gravitational wave detector the Einstein Telescope will improve precision by an order of magnitude [219].

In summary, observations of gravitational waves from compact object coalescences will yield promising new constraints on the properties of neutron stars. This will occur because of new mass measurements and radius constraints. Most of the information exists at high frequencies, so in order to maximize this information it may be necessary to explore techniques beyond the second generation of detectors, such as squeezing of light [1].

1.7 Summary

Significant progress has been made in the last decade on mass estimates of neutron stars. The maximum mass is $\gtrsim 2\ M_\odot$, which along with the lack of evidence of rapid cooling suggests that non-nucleonic degrees of freedom appear unnecessary to explain current data. There is, however, considerable freedom that would allow exotic phases. The main limitation of existing observations is that radius estimates are shrouded in systematic errors. No current method is both precise and reliable enough to pose significant constraints on the structure of neutron stars. This is unfortunate, because good radii would do more than any other single measurement to inform us about the matter in the cores of neutron stars [130].

Where, then, do we stand? In the near future it is plausible that our understanding of thermonuclear X-ray bursts, accreting neutron stars, or the emission from cooling neutron stars may evolve to the point that the complex phenomena associated with them can be interpreted with confidence, and reliable radii and masses will emerge. In principle the needed data have already been collected, but our evolving understanding that, e.g., bursts do not settle quickly to a constant area of emission and that cooling neutron stars display nonthermal emission has led to caution. At the same time, numerical simulations are improving rapidly in sophistication. It is possible, although far from guaranteed, that within a few years magnetohydrodynamic simulations of accretion disks or bursting atmospheres will yield results that are close enough to what is observed that our understanding will solidify and radius measurements will become a powerful tool for constraining the state of matter in the cores of neutron stars.

If ground-based gravitational wave facilities make detections as expected within 5 years, then as we discussed in the previous section this will open up a new way to measure masses and radii, with systematic errors that are at a minimum different from those that bedevil current attempts, and more optimistically might be less significant than statistical uncertainties. However, most of the information resides at high frequencies where, at least for standard configurations, second-generation detectors will be insensitive enough that it may require a rare high-signal event to derive restrictive constraints. A third-generation detector such as the Einstein Telescope should be able to obtain all the required information, and even before such detectors exist it may be possible to use configurations optimized for high frequencies or 2.5 generation technology such as squeezed light to obtain the information.

Electromagnetic observatories are also improving, and some of the old reliable methods may improve our understanding substantially. For example, the Jansky Very Large Array or (in roughly a decade) the Square Kilometer Array might have enough sensitivity, bandwidth, and computer power to detect a much larger population of double neutron star binaries, among which we might by chance have some with stars of $M > 2.0\ M_{\odot}$. Even without such serendipitous discoveries, the mere accumulation of time and data on NS-WD binaries in globular clusters seems likely to yield high-mass objects that will provide firm lower limits to the maximum mass that are much stronger than currently exist. As we have discussed, many other improvements are expected using the large area and excellent spectral resolution of Athena+ [161], and the high area and timing resolution of NICER [90] and LOFT [78]. This is especially true of radius estimates from fits to X-ray waveforms.

Overall, although we expect that eventually radius measurements will play a major role, in the next several years it appears that mass measurements of neutron stars in binaries will continue to dominate the discussion of the cold high-density equation of state. The expected improvements in data and models will allow us to provide nuclear physicists with more certain constraints, and we await the theoretical developments that result.

Acknowledgments This work was supported in part by NSF grant AST0708424, by NASA ATP grants NNX08AH29G and NNX12AG29G, and by grant number 230349 from the Simons Foundation. We appreciate helpful suggestions from Didier Barret, Paulo Bedaque, Sudip Bhattacharyya, David Blaschke, Tom Cohen, Peter Jonker, Fred Lamb, Jim Lattimer, Simin Mahmoodifar, Ilya Mandel, Dany Page, Bettina Posselt, Scott Ransom, Ingrid Stairs, and Natalie Webb. We also thank the referee for a highly constructive report.

References

1. J. Aasi, J. Abadie, B.P. Abbott, R. Abbott, T.D. Abbott, M.R. Abernathy, C. Adams, T. Adams, P. Addesso, R.X. Adhikari et al. Enhanced sensitivity of the LIGO gravitational wave detector by using squeezed states of light. Nat. Photonics **7**, 613–619 (2013). https://doi.org/10.1038/nphoton.2013.177
2. J. Abadie, B.P. Abbott, R. Abbott, M. Abernathy, T. Accadia, F. Acernese, C. Adams, R. Adhikari, P. Ajith, B. Allen et al. TOPICAL REVIEW: predictions for the rates of compact binary coalescences observable by ground-based gravitational-wave detectors. Classical Quantum Gravity **27**(17), 173001 (2010). https://doi.org/10.1088/0264-9381/27/17/173001
3. S. Abrahamyan et al. Measurement of the Neutron Radius of Pb208 through parity violation in electron scattering. Phys. Rev. Lett. **108**(11), 112502 (2012). https://doi.org/10.1103/PhysRevLett.108.112502
4. A. Akmal, V.R. Pandharipande, D.G. Ravenhall, Equation of state of nucleon matter and neutron star structure. Phys. Rev. C **58**, 1804–1828 (1998). https://doi.org/10.1103/PhysRevC.58.1804
5. C. Alcock, A. Illarionov, The surface chemistry of stars. I – Diffusion of heavy ions in white dwarf envelopes. II – Fractionated accretion of interstellar matter. Astrophys. J. **235**, 534–553 (1980). https://doi.org/10.1086/157656
6. M.G. Alford, A. Schmitt, K. Rajagopal, T. Schäfer, Color superconductivity in dense quark matter. Rev. Mod. Phys. **80**, 1455–1515 (2008). https://doi.org/10.1103/RevModPhys.80.1455
7. M.A. Alpar, A.F. Cheng, M.A. Ruderman, J. Shaham, A new class of radio pulsars. Nature **300**, 728–730 (1982). https://doi.org/10.1038/300728a0
8. D. Altamirano, A. Watts, M. Linares, C.B. Markwardt, T. Strohmayer, A. Patruno, Type I X-ray bursts and burst oscillations in the accreting millisecond X-ray pulsar IGR J17511-3057. Mon. Not. R. Astron. Soc. **409**, 1136–1145 (2010). https://doi.org/10.1111/j.1365-2966.2010.17369.x
9. N. Andersson, A new class of unstable modes of rotating relativistic stars. Astrophys. J. **502**, 708 (1998). https://doi.org/10.1086/305919
10. N. Andersson, K. Kokkotas, B.F. Schutz, Gravitational radiation limit on the spin of young neutron stars. Astrophys. J., **510**, 846–853 (1999). https://doi.org/10.1086/306625
11. J. Antoniadis, P.C.C. Freire, N. Wex, T.M. Tauris, R.S. Lynch, M.H. van Kerkwijk, M. Kramer, C. Bassa, V.S. Dhillon, T. Driebe, J.W.T. Hessels, V.M. Kaspi, V.I. Kondratiev, N. Langer, T.R. Marsh, M.A. McLaughlin, T.T. Pennucci, S.M. Ransom, I.H. Stairs, J. van Leeuwen, J.P.W. Verbiest, D.G. Whelan, A massive pulsar in a compact relativistic binary. Science **340**, 448 (2013). https://doi.org/10.1126/science.1233232
12. R. Artigue, D. Barret, F.K. Lamb, K.H. Lo, M.C. Miller, Testing the rotating hotspot model using X-ray burst oscillations from 4U 1636-536. Mon. Not. R. Astron. Soc. Lett. **433**, L64–L68 (2013). https://doi.org/10.1093/mnrasl/slt059
13. Y. Avni, Mass estimates from optical-light curves for binary X-ray sources, in *Physics and Astrophysics of Neutron Stars and Black Holes*, ed. by R. Giacconi, R. Ruffini (North Holland Publishing Co., Amsterdam, 1978), pp. 43–62

14. B.M. Barker, R.F. O'Connell, Gravitational two-body problem with arbitrary masses, spins, and quadrupole moments. Phys. Rev. D **12**, 329–335 (1975). https://doi.org/10.1103/PhysRevD.12.329
15. D. Barret, Soft lags in neutron star kHz quasi-periodic oscillations: evidence for reverberation? Astrophys. J. **770**, 9 (2013). https://doi.org/10.1088/0004-637X/770/1/9
16. D. Barret, J.-F. Olive, M.C. Miller, Drop of coherence of the lower kilo-Hz QPO in neutron stars: is there a link with the innermost stable circular orbit? Astron. Nachr. **326**, 808–811 (2005). https://doi.org/10.1002/asna.200510417
17. D. Barret, J.-F. Olive, M.C. Miller, An abrupt drop in the coherence of the lower kHz quasi-periodic oscillations in 4U 1636-536. Mon. Not. R. Astron. Soc. **361**, 855–860 (2005). https://doi.org/10.1111/j.1365-2966.2005.09214.x
18. D. Barret, J.-F. Olive, M.C. Miller, The coherence of kilohertz quasi-periodic oscillations in the X-rays from accreting neutron stars. Mon. Not. R. Astron. Soc., **370**, 1140–1146 (2006). https://doi.org/10.1111/j.1365-2966.2006.10571.x
19. D. Barret, J.-F. Olive, M.C. Miller, Supporting evidence for the signature of the innermost stable circular orbit in Rossi X-ray data from 4U 1636-536. Mon. Not. R. Astron. Soc. **376**, 1139–1144 (2007). https://doi.org/10.1111/j.1365-2966.2007.11491.x
20. M. Bauböck, D. Psaltis, F. Özel, Narrow atomic features from rapidly spinning neutron stars. Astrophys. J. **766**, 87 (2013). https://doi.org/10.1088/0004-637X/766/2/87
21. A. Bauswein, H.-T. Janka, K. Hebeler, A. Schwenk, Equation-of-state dependence of the gravitational-wave signal from the ring-down phase of neutron-star mergers. Phys. Rev. D: Part. Fields **86**(6), 063001 (2012). https://doi.org/10.1103/PhysRevD.86.063001
22. G. Baym, C. Pethick, P. Sutherland, The ground state of matter at high densities: equation of state and stellar models. Astrophys. J. **170**, 299 (1971). https://doi.org/10.1086/151216
23. R.D. Belian, J.P. Conner, W.D. Evans, The discovery of X-ray bursts from a region in the constellation Norma. Astrophys. J. **206**, L135–L138 (1976). https://doi.org/10.1086/182151
24. E. Berger, W. Fong, R. Chornock, An r-process kilonova associated with the short-hard GRB 130603B. Astrophys. J. **774**, L23 (2013). https://doi.org/10.1088/2041-8205/774/2/L23
25. S. Bernuzzi, A. Nagar, M. Thierfelder, B. Brügmann, Tidal effects in binary neutron star coalescence. Phys. Rev. D: Part. Fields **86**(4), 044030 (2012). https://doi.org/10.1103/PhysRevD.86.044030
26. M.V. Beznogov, D.G. Yakovlev, Diffusion and coulomb separation of ions in dense matter. Phys. Rev. Lett. **111**(16), 161101 (2013). https://doi.org/10.1103/PhysRevLett.111.161101
27. S. Bhattacharyya, Measurement of neutron star parameters: a review of methods for low-mass X-ray binaries. Adv. Space Res. **45**, 949–978 (2010). https://doi.org/10.1016/j.asr.2010.01.010
28. S. Bhattacharyya, Ways to constrain neutron star equation of state models using relativistic disc lines. Mon. Not. R. Astron. Soc. **415**, 3247–3252 (2011). https://doi.org/10.1111/j.1365-2966.2011.18936.x
29. S. Bhattacharyya, T.E. Strohmayer, M.C. Miller, C.B. Markwardt, Constraints on neutron star parameters from burst oscillation light curves of the accreting millisecond pulsar XTE J1814-338. Astrophys. J. **619**, 483–491 (2005). https://doi.org/10.1086/426383
30. S. Bhattacharyya, M.C. Miller, F.K. Lamb, The shapes of atomic lines from the surfaces of weakly magnetic rotating neutron stars and their implications. Astrophys. J. **644**, 1085–1089 (2006). https://doi.org/10.1086/503860
31. L. Bildsten, Gravitational radiation and rotation of accreting neutron stars. Astrophys. J. **501**, L89 (1998). https://doi.org/10.1086/311440
32. L. Bildsten, R.E. Rutledge, Coronal X-ray emission from the stellar companions to transiently accreting black holes. Astrophys. J. **541**, 908–917 (2000). https://doi.org/10.1086/309464
33. O. Blaes, O. Warren, P. Madau, Accreting, isolated neutron stars. III. Preheating of infalling gas and cometary H II regions. Astrophys. J. **454**, 370 (1995). https://doi.org/10.1086/176488
34. D. Blaschke, D.E. Alvarez-Castillo, S. Benic, Mass-radius constraints for compact stars and a critical endpoint. ArXiv e-prints (2013)

35. D. Blaschke, H. Grigorian, D.N. Voskresensky, Nuclear medium cooling scenario in light of new Cas A cooling data and the $2M_{\odot}$ pulsar mass measurements. Phys. Rev. C **88**(6), 065805 (2013). https://doi.org/10.1103/PhysRevC.88.065805
36. A.A. Blinova, M. Bachetti, M.M. Romanova, Oscillations of the boundary layer and high-frequency QPOs, in *European Physical Journal Web of Conferences*, vol. 64 (2014), p. 5009. https://doi.org/10.1051/epjconf/20136405009
37. S. Bogdanov, The nearest millisecond pulsar revisited with XMM-Newton: improved mass-radius constraints for PSR J0437-4715. Astrophys. J. **762**, 96 (2013). https://doi.org/10.1088/0004-637X/762/2/96
38. S. Bogdanov, J.E. Grindlay, Deep XMM-Newton spectroscopic and timing observations of the isolated radio millisecond pulsar PSR J0030+0451. Astrophys. J. **703**, 1557–1564 (2009). https://doi.org/10.1088/0004-637X/703/2/1557
39. S. Bogdanov, J.E. Grindlay, G.B. Rybicki, Thermal X-Rays from millisecond pulsars: constraining the fundamental properties of neutron stars. Astrophys. J. **689**, 407–415 (2008). https://doi.org/10.1086/592341
40. R. Bondarescu, I. Wasserman, Nonlinear development of the R-mode instability and the maximum rotation rate of neutron stars. Astrophys. J. **778**, 9 (2013). https://doi.org/10.1088/0004-637X/778/1/9
41. M. Boutelier, D. Barret, M.C. Miller, kHz quasi-periodic oscillations in the low-mass X-ray binary 4U 0614+09. Mon. Not. R. Astron. Soc. **399**, 1901–1906 (2009). https://doi.org/10.1111/j.1365-2966.2009.15430.x
42. S. Boutloukos, F.K. Lamb, Implications of kHz QPOs for the spin frequencies and magnetic fields of neutron stars: new results from circinus X-1, in *40 Years of Pulsars: Millisecond Pulsars, Magnetars and More*, ed. by C. Bassa, Z. Wang, A. Cumming, V.M. Kaspi, vol. 983. American Institute of Physics Conference Series (2008), pp. 533–535. https://doi.org/10.1063/1.2900291
43. S. Boutloukos, M.C. Miller, F.K. Lamb, Super-eddington fluxes during thermonuclear X-ray bursts. Astrophys. J. **720**, L15–L19 (2010). https://doi.org/10.1088/2041-8205/720/1/L15
44. T.M. Braje, R.W. Romani, K.P. Rauch, Light curves of rapidly rotating neutron stars. Astrophys. J. **531**, 447–452 (2000). https://doi.org/10.1086/308448
45. A.E. Broderick, P. Chang, C. Pfrommer, The cosmological impact of luminous TeV blazars. I. Implications of plasma instabilities for the intergalactic magnetic field and extragalactic gamma-ray background. Astrophys. J. **752**, 22 (2012). https://doi.org/10.1088/0004-637X/752/1/22
46. E. Brown, S.S. Gupta, H. Schatz, P. Möller, K.-L. Kratz, Electron capture reactions in neutron star crusts: deep heating and observational constraints, in *International Symposium on Nuclear Astrophysics - Nuclei in the Cosmos* (2006)
47. M. Burgay, N. D'Amico, A. Possenti, R.N. Manchester, A.G. Lyne, B.C. Joshi, M.A. McLaughlin, M. Kramer, J.M. Sarkissian, F. Camilo, V. Kalogera, C. Kim, D.R. Lorimer, An increased estimate of the merger rate of double neutron stars from observations of a highly relativistic system. Nature **426**, 531–533 (2003). https://doi.org/10.1038/nature02124
48. V. Burwitz, V.E. Zavlin, R. Neuhäuser, P. Predehl, J. Trümper, A.C. Brinkman, The chandra LETGS high resolution X-ray spectrum of the isolated neutron star RX J1856.5-3754. Astron. Astrophys. **379**, L35–L38 (2001). https://doi.org/10.1051/0004-6361:20011304
49. E.M. Cackett, J.M. Miller, S. Bhattacharyya, J.E. Grindlay, J. Homan, M. van der Klis, M.C. Miller, T.E. Strohmayer, R. Wijnands, Relativistic iron emission lines in neutron star low-mass X-ray binaries as probes of neutron star radii. Astrophys. J. **674**, 415–420 (2008). https://doi.org/10.1086/524936
50. C. Cadeau, S.M. Morsink, D. Leahy, S.S. Campbell, Light curves for rapidly rotating nNeutron stars. Astrophys. J. **654**, 458–469 (2007). https://doi.org/10.1086/509103
51. S. Campana, L. Stella, On the bolometric quiescent luminosity and luminosity swing of black hole candidate and neutron star low-mass X-ray rransients. Astrophys. J. **541**, 849–859 (2000). https://doi.org/10.1086/309493

52. A. Catuneanu, C.O. Heinke, G.R. Sivakoff, W.C.G. Ho, M. Servillat, Mass/radius constraints on the quiescent neutron star in M13 using hydrogen and helium atmospheres. Astrophys. J. **764**, 145 (2013). https://doi.org/10.1088/0004-637X/764/2/145
53. D. Chakrabarty, The spin distribution of millisecond X-ray pulsars, in *American Institute of Physics Conference Series*, ed. by R. Wijnands, D. Altamirano, P. Soleri, N. Degenaar, N. Rea, P. Casella, A. Patruno, M. Linares, vol. 1068 (2008), pp. 67–74. https://doi.org/10.1063/1.3031208
54. D. Chakrabarty, J.A. Tomsick, B.W. Grefenstette, D. Psaltis, M. Bachetti, D. Barret, S.E. Boggs, F.E. Christensen, W.W. Craig, F. Fuerst, C.J. Hailey, F.A. Harrison, V.A. Kaspi, J.M. Miller, M.A. Nowak, V. Rana, D. Stern, D.R. Wik, J. Wilms, W.W. Zhang, A hard X-ray power-law spectral cutoff in centaurus X-4. Astrophys. J. **797**, 92–106 (2014). https://doi.org/10.1088/0004-637X/797/2/92
55. M. Chakraborty, S. Bhattacharyya, Thermonuclear X-ray bursts from the 401-Hz accreting pulsar IGR J17498-2921: indication of burning in confined regions. Mon. Not. R. Astron. Soc. **422**, 2351–2356 (2012). https://doi.org/10.1111/j.1365-2966.2012.20786.x
56. P. Chang, L. Bildsten, Diffusive nuclear burning in neutron star envelopes. Astrophys. J. **585**, 464–474 (2003). https://doi.org/10.1086/345551
57. P. Chang, L. Bildsten, I. Wasserman, Formation of resonant atomic lines during thermonuclear flashes on neutron stars. Astrophys. J. **629**, 998–1007 (2005). https://doi.org/10.1086/431730
58. P. Chang, L. Bildsten, P. Arras, Diffusive nuclear burning of helium on neutron stars. Astrophys. J. **723**, 719–728 (2010). https://doi.org/10.1088/0004-637X/723/1/719
59. C. Chirenti, J. Skákala, Effect of magnetic fields on the r-modes of slowly rotating relativistic neutron stars. Phys. Rev. D: Part. Fields **88**(10), 104018 (2013). https://doi.org/10.1103/PhysRevD.88.104018
60. C. Chirenti, P.R. Silveira, O.D. Aguiar, Non-radial oscillations of neutron stars and the detection of gravitational waves. Int. J. Mod. Phys. Conf. Ser. **18**, 48–52 (2012). https://doi.org/10.1142/S2010194512008185
61. C. Chirenti, J. Skákala, S. Yoshida, Slowly rotating neutron stars with small differential rotation: equilibrium models and oscillations in the Cowling approximation. Phys. Rev. D: Part. Fields **87**(4), 044043 (2013). https://doi.org/10.1103/PhysRevD.87.044043
62. H.-Y. Chiu, E.E. Salpeter, Surface X-ray emission from neutron stars. Phys. Rev. Lett. **12**, 413–415 (1964). https://doi.org/10.1103/PhysRevLett.12.413
63. G.B. Cook, S.L. Shapiro, S.A. Teukolsky, Rapidly rotating neutron stars in general relativity: realistic equations of state. Astrophys. J. **424**, 823–845 (1994). https://doi.org/10.1086/173934
64. R.L. Cooper, R. Narayan, Theoretical models of superbursts on accreting neutron stars. Astrophys. J. **629**, 422–437 (2005). https://doi.org/10.1086/431273
65. J. Cottam, F. Paerels, M. Mendez, Gravitationally redshifted absorption lines in the X-ray burst spectra of a neutron star. Nature **420**, 51–54 (2002). https://doi.org/10.1038/nature01159
66. J. Cottam, F. Paerels, M. Méndez, L. Boirin, W.H.G. Lewin, E. Kuulkers, J.M. Miller, The burst spectra of EXO 0748-676 during a long 2003 XMM-newton observation. Astrophys. J. **672**, 504–509 (2008). https://doi.org/10.1086/524186
67. T. Damour, G. Schäfer, Higher-order relativistic periastron advances and binary pulsars. Nuovo Cimento B Serie **101**, 127–176 (1988). https://doi.org/10.1007/BF02828697
68. T. Damour, J.H. Taylor, Strong-field tests of relativistic gravity and binary pulsars. Phys. Rev. D **45**, 1840–1868 (1992). https://doi.org/10.1103/PhysRevD.45.1840
69. T. Damour, A. Nagar, L. Villain, Measurability of the tidal polarizability of neutron stars in late-inspiral gravitational-wave signals. Phys. Rev. D: Part. Fields **85**(12), 123007 (2012). https://doi.org/10.1103/PhysRevD.85.123007
70. P. Danielewicz, R. Lacey, W.G. Lynch, Determination of the equation of state of dense matter. Science **298**, 1592–1596 (2002). https://doi.org/10.1126/science.1078070

71. W. Del Pozzo, T.G.F. Li, M. Agathos, C. Van Den Broeck, S. Vitale, Demonstrating the feasibility of probing the neutron-star equation of state with second-generation gravitational-wave detectors. Phys. Rev. Lett. **111**(7), 071101 (2013). https://doi.org/10.1103/PhysRevLett.111.071101
72. P.B. Demorest, T. Pennucci, S.M. Ransom, M.S.E. Roberts, J.W.T. Hessels, A two-solar-mass neutron star measured using Shapiro delay. Nature **467**, 1081–1083 (2010). https://doi.org/10.1038/nature09466
73. D.D. Doneva, E. Gaertig, K.D. Kokkotas, C. Krüger, Gravitational wave asteroseismology of fast rotating neutron stars with realistic equations of state. Phys. Rev. D: Part. Fields **88**(4), 044052 (2013). https://doi.org/10.1103/PhysRevD.88.044052
74. J.M. Dong, U. Lombardo, W. Zuo, $^3PF_2$ pairing in high-density neutron matter. Phys. Rev. C **87**(6), 062801 (2013). https://doi.org/10.1103/PhysRevC.87.062801
75. T. Ebisuzaki, D. Sugimoto, T. Hanawa, Are X-ray bursts really of super-Eddington luminosities? Pub. Astr. Soc. Jap. **36**, 551–566 (1984)
76. K.G. Elshamouty, C.O. Heinke, G.R. Sivakoff, W.C.G. Ho, P.S. Shternin, D.G. Yakovlev, D.J. Patnaude, L. David, Measuring the cooling of the neutron star in cassiopeia a with all chandra X-ray observatory detectors. Astrophys. J. **777**, 22 (2013). https://doi.org/10.1088/0004-637X/777/1/22
77. E. Farhi, R.L. Jaffe, Strange matter. Phys. Rev. D **30**, 2379–2390 (1984). https://doi.org/10.1103/PhysRevD.30.2379
78. M. Feroci et al.: The large observatory for X-ray timing (LOFT). Exp. Astron. **34**, 415–444 (2012). https://doi.org/10.1007/s10686-011-9237-2
79. V. Ferrari, L. Gualtieri, A. Maselli, Tidal interaction in compact binaries: a post-Newtonian affine framework. Phys. Rev. D: Part. Fields **85**(4), 044045 (2012). https://doi.org/10.1103/PhysRevD.85.044045
80. F. Foucart, L. Buchman, M.D. Duez, M. Grudich, L.E. Kidder, I. MacDonald, A. Mroue, H.P. Pfeiffer, M.A. Scheel, B. Szilagyi, First direct comparison of nondisrupting neutron star-black hole and binary black hole merger simulations. Phys. Rev. D: Part. Fields **88**(6), 064017 (2013). https://doi.org/10.1103/PhysRevD.88.064017
81. J. Frank, A. King, D.J. Raine, *Accretion Power in Astrophysics*, 3rd edn. (Cambridge University Press, Cambridge, 2002)
82. P.C.C. Freire, Eccentric binary millisecond pulsars. ArXiv e-prints (2009)
83. P.C.C. Freire, N. Wex, The orthometric parametrization of the Shapiro delay and an improved test of general relativity with binary pulsars. Mon. Not. R. Astron. Soc. **409**, 199–212 (2010). https://doi.org/10.1111/j.1365-2966.2010.17319.x
84. J.L. Friedman, S.M. Morsink, Axial instability of rotating relativistic stars. Astrophys. J. **502**, 714 (1998). https://doi.org/10.1086/305920
85. B.A. Fryxell, S.E. Woosley, Finite propagation time in multidimensional thermonuclear runaways. Astrophys. J. **261**, 332–336 (1982). https://doi.org/10.1086/160344
86. I. Fushiki, D.Q. Lamb, New insights from a global view of X-ray bursts. Astrophys. J. **323**, L55–L60 (1987). https://doi.org/10.1086/185056
87. D.K. Galloway, N. Lampe, On the consistency of neutron-star radius measurements from thermonuclear bursts. Astrophys. J. **747**, 75 (2012). https://doi.org/10.1088/0004-637X/747/1/75
88. D.K. Galloway, J. Lin, D. Chakrabarty, J.M. Hartman, Discovery of a 552 Hz burst oscillation in the low-mass X-ray binary EXO 0748-676. Astrophys. J. **711**, L148–L151 (2010). https://doi.org/10.1088/2041-8205/711/2/L148
89. F. García, G. Zhang, M. Méndez, The cooling phase of Type I X-ray bursts observed with RXTE in 4U 1820-30 does not follow the canonical $F \propto T^4$ relation. Mon. Not. R. Astron. Soc. **429**, 3266–3271 (2013). https://doi.org/10.1093/mnras/sts583
90. K.C. Gendreau, Z. Arzoumanian, T. Okajima, The neutron star interior composition exploreR (NICER): an explorer mission of opportunity for soft x-ray timing spectroscopy, in *Society of Photo-Optical Instrumentation Engineers (SPIE) Conference Series*, vol. 8443 (2012) https://doi.org/10.1117/12.926396

91. J. Grindlay, H. Gursky, H. Schnopper, D.R. Parsignault, J. Heise, A.C. Brinkman, J. Schrijver, Discovery of intense X-ray bursts from the globular cluster NGC 6624. Astrophys. J. **205**, L127–L130 (1976). https://doi.org/10.1086/182105
92. S. Guillot, R.E. Rutledge, L. Bildsten, E.F. Brown, G.G. Pavlov, V.E. Zavlin, X-ray spectral identification of three candidate quiescent low-mass X-ray binaries in the globular cluster NGC 6304. Mon. Not. R. Astron. Soc. **392**, 665–681 (2009). https://doi.org/10.1111/j.1365-2966.2008.14076.x
93. S. Guillot, R.E. Rutledge, E.F. Brown, Neutron star radius measurement with the quiescent low-mass X-ray binary U24 in NGC 6397. Astrophys. J. **732**, 88 (2011). https://doi.org/10.1088/0004-637X/732/2/88
94. S. Guillot, M. Servillat, N.A. Webb, R.E. Rutledge, Measurement of the radius of neutron stars with high signal-to-noise quiescent low-mass X-ray binaries in globular clusters. Astrophys. J. **772**, 7 (2013). https://doi.org/10.1088/0004-637X/772/1/7
95. T. Güver, P. Wroblewski, L. Camarota, F. Özel, The mass and radius of the neutron star in 4U 1820–30. Astrophys. J. **719**, 1807–1812 (2010). https://doi.org/10.1088/0004-637X/719/2/1807
96. J.M. Hartman, A. Patruno, D. Chakrabarty, C.B. Markwardt, E.H. Morgan, M. van der Klis, R. Wijnands, A decade of timing an accretion-powered millisecond pulsar: the continuing spin down and orbital evolution of SAX J1808.4-3658. Astrophys. J. **702**, 1673–1678 (2009). https://doi.org/10.1088/0004-637X/702/2/1673
97. B. Haskell, A. Patruno, Spin equilibrium with or without gravitational wave emission: the case of XTE J1814-338 and SAX J1808.4-3658. Astrophys. J. **738**, L14 (2011). https://doi.org/10.1088/2041-8205/738/1/L14
98. B. Haskell, N. Degenaar, W.C.G. Ho, Constraining the physics of the r-mode instability in neutron stars with X-ray and ultraviolet observations. Mon. Not. R. Astron. Soc. **424**, 93–103 (2012). https://doi.org/10.1111/j.1365-2966.2012.21171.x
99. K. Hebeler, J.M. Lattimer, C.J. Pethick, A. Schwenk, Constraints on neutron star radii based on chiral effective field theory interactions. Phys. Rev. Lett. **105**(16), 161102 (2010). https://doi.org/10.1103/PhysRevLett.105.161102
100. K. Hebeler, J.M. Lattimer, C.J. Pethick, A. Schwenk, Equation of state and neutron star properties constrained by nuclear physics and observation. Astrophys. J. **773**, 11 (2013). https://doi.org/10.1088/0004-637X/773/1/11
101. C.O. Heinke, Constraints on physics of neutron stars from X-ray observations. J. Phys. Conf. Ser. **432**(1), 012001 (2013). https://doi.org/10.1088/1742-6596/432/1/012001
102. J.W.T. Hessels, S.M. Ransom, I.H. Stairs, P.C.C. Freire, V.M. Kaspi, F. Camilo, A radio pulsar spinning at 716 Hz. Science **311**, 1901–1904 (2006). https://doi.org/10.1126/science.1123430
103. T. Hinderer, B.D. Lackey, R.N. Lang, J.S. Read, Tidal deformability of neutron stars with realistic equations of state and their gravitational wave signatures in binary inspiral. Phys. Rev. D: Part. Fields **81**(12), 123016 (2010). https://doi.org/10.1103/PhysRevD.81.123016
104. W.C.G. Ho, C.O. Heinke, A neutron star with a carbon atmosphere in the Cassiopeia A supernova remnant. Nature **462**, 71–73 (2009). https://doi.org/10.1038/nature08525
105. W.C.G. Ho, D.L. Kaplan, P. Chang, M. van Adelsberg, A.Y. Potekhin, Thin magnetic hydrogen atmospheres and the neutron star RX J1856.5 3754. Astrophys. Space Sci. **308**, 279–286 (2007). https://doi.org/10.1007/s10509-007-9366-2
106. W.C.G. Ho, D.L. Kaplan, P. Chang, M. van Adelsberg, A.Y. Potekhin, Magnetic hydrogen atmosphere models and the neutron star RX J1856.5-3754. Mon. Not. R. Astron. Soc. **375**, 821–830 (2007). https://doi.org/10.1111/j.1365-2966.2006.11376.x
107. J.A. Hoffman, L. Cominsky, W.H.G. Lewin, Repeatable, multiple-peaked structure in Type I X-ray bursts. Astrophys. J. **240**, L27–L31 (1980). https://doi.org/10.1086/183317
108. K. Hotokezaka, K. Kyutoku, M. Shibata, Exploring tidal effects of coalescing binary neutron stars in numerical relativity. Phys. Rev. D: Part. Fields **87**(4), 044001 (2013). https://doi.org/10.1103/PhysRevD.87.044001
109. S.D.H. Hsu, D. Reeb, On the sign problem in dense QCD. Int. J. Mod. Phys. A **25**, 53–67 (2010). https://doi.org/10.1142/S0217751X10047968

110. P.G. Jonker, D. Steeghs, G. Nelemans, M. van der Klis, The radial velocity of the companion star in the low-mass X-ray binary 2S 0921-630: limits on the mass of the compact object. Mon. Not. R. Astron. Soc. **356**, 621–626 (2005). https://doi.org/10.1111/j.1365-2966.2004.08476.x
111. P.G. Jonker, D. Steeghs, D. Chakrabarty, A.M. Juett, The cold neutron star in the soft X-ray transient 1H 1905+000. Astrophys. J. **665**, L147–L150 (2007). https://doi.org/10.1086/521079.
112. P. C. Joss, X-ray bursts and neutron-star thermonuclear flashes. Nature **270**, 310–314 (1977). https://doi.org/10.1038/270310a0
113. P. Kaaret, Z. Prieskorn, J.J.M. in 't Zand, S. Brandt, N. Lund, S. Mereghetti, D. Götz, E. Kuulkers, J.A. Tomsick, Evidence of 1122 Hz X-ray burst oscillations from the neutron star X-ray transient XTE J1739-285. Astrophys. J. Lett. **657**, L97–L100 (2007). https://doi.org/10.1086/513270
114. A.R. King, U. Kolb, The evolution of black hole mass and angular momentum. Mon. Not. R. Astron. Soc. **305**, 654–660 (1999). https://doi.org/10.1046/j.1365-8711.1999.02482.x
115. B. Kiziltan, A. Kottas, M. De Yoreo, S.E. Thorsett, The neutron star mass distribution. Astrophys. J. **778**, 66 (2013). https://doi.org/10.1088/0004-637X/778/1/66
116. T. Klähn, D. Blaschke, S. Typel, E.N.E. van Dalen, A. Faessler, C. Fuchs, T. Gaitanos, H. Grigorian, A. Ho, E.E. Kolomeitsev, M.C. Miller, G. Röpke, J. Trümper, D.N. Voskresensky, F. Weber, H.H. Wolter, Constraints on the high-density nuclear equation of state from the phenomenology of compact stars and heavy-ion collisions. Phys. Rev. C **74**(3), 035802 (2006). https://doi.org/10.1103/PhysRevC.74.035802
117. T. Klähn, R. Łastowiecki, D. Blaschke, Implications of the measurement of pulsars with two solar masses for quark matter in compact stars and heavy-ion collisions: a Nambu-Jona-Lasinio model case study. Phys. Rev. D **88**(8), 085001 (2013). https://doi.org/10.1103/PhysRevD.88.085001
118. M. Kramer, I.H. Stairs, The double pulsar. Annu. Rev. Astron. Astrophys. **46**, 541–572 (2008). https://doi.org/10.1146/annurev.astro.46.060407.145247
119. A.K. Kulkarni, M.M. Romanova, Variability profiles of millisecond X-ray pulsars: results of pseudo-newtonian three-dimensional magnetohydrodynamic simulations. Astrophys. J. **633**, 349–357 (2005). https://doi.org/10.1086/444489
120. A.K. Kulkarni, M.M. Romanova, Possible quasi-periodic oscillations from unstable accretion: 3D magnetohydrodynamic simulations. Mon. Not. R. Astron. Soc. **398**, 701–714 (2009). https://doi.org/10.1111/j.1365-2966.2009.15186.x
121. A. Kurkela, P. Romatschke, A. Vuorinen, Cold quark matter. Phys. Rev. D **81**(10), 105021 (2010). https://doi.org/10.1103/PhysRevD.81.105021
122. B.D. Lackey, M. Nayyar, B.J. Owen, Observational constraints on hyperons in neutron stars. Phys. Rev. D **73**(2), 024021 (2006). https://doi.org/10.1103/PhysRevD.73.024021
123. B.D. Lackey, K. Kyutoku, M. Shibata, P.R. Brady, J.L. Friedman, Extracting equation of state parameters from black hole-neutron star mergers: nonspinning black holes. Phys. Rev. D **85**(4), 044061 (2012). https://doi.org/10.1103/PhysRevD.85.044061
124. B.D. Lackey, K. Kyutoku, M. Shibata, P.R. Brady, J. L. Friedman, Extracting equation of state parameters from black hole-neutron star mergers: aligned-spin black holes and a preliminary waveform model. Phys. Rev. D **89**(4), 043009 (2014). https://doi.org/10.1103/PhysRevD.89.043009
125. D.Q. Lamb, F.K. Lamb, Nuclear burning in accreting neutron stars and X-ray bursts. Astrophys. J. **220**, 291–302 (1978). https://doi.org/10.1086/155905
126. F.K. Lamb, S. Boutloukos, S. Van Wassenhove, R.T. Chamberlain, K.H. Lo, A. Clare, W. Yu, M.C. Miller, A model for the waveform behavior of accreting millisecond X-ray pulsars: nearly aligned magnetic fields and moving emission regions. Astrophys. J. **706**, 417–435 (2009). https://doi.org/10.1088/0004-637X/706/1/417

127. F.K. Lamb, S. Boutloukos, S. Van Wassenhove, R.T. Chamberlain, K.H. Lo, M.C. Miller, Origin of intermittent accretion-powered X-ray oscillations in neutron stars with millisecond spin periods. Astrophys. J. Lett. **705**, L36–L39 (2009). https://doi.org/10.1088/0004-637X/705/1/L36
128. L.D. Landau, E.M. Lifshitz, *Quantum Mechanics* (Pergamon Press, Oxford, 1965)
129. J.M. Lattimer, The nuclear equation of state and neutron star masses. Annu. Rev. Nucl. Part. Sci. **62**, 485–515 (2012). https://doi.org/10.1146/annurev-nucl-102711-095018
130. J.M. Lattimer, M. Prakash, Neutron star structure and the equation of state. Astrophys. J. **550**, 426–442 (2001). https://doi.org/10.1086/319702
131. J.M. Lattimer, M. Prakash, Equation of state, neutron stars and exotic phases. Nucl. Phys. A **777**, 479–496 (2006). https://doi.org/10.1016/j.nuclphysa.2005.01.014
132. J.M. Lattimer, M. Prakash, What a two solar mass neutron star really means. ArXiv e-prints (2010)
133. J.M. Lattimer, B.F. Schutz, Constraining the equation of state with moment of inertia measurements. Astrophys. J. **629**, 979–984 (2005). https://doi.org/10.1086/431543
134. J.M. Lattimer, A.W. Steiner, Neutron star masses and radii from quiescent low-mass X-ray binaries. Astrophys. J. **784**, 123 (2014). https://doi.org/10.1088/0004-637X/784/2/123
135. D.A. Leahy, S.M. Morsink, Y. Chou, Constraints on the mass and radius of the neutron star XTE J1807-294. Astrophys. J. **742**, 17 (2011). https://doi.org/10.1088/0004-637X/742/1/17
136. W.H.G. Lewin, J.A. Hoffman, J. Doty, D.R. Hearn, G.W. Clark, J.G. Jernigan, F.K. Li, J.E. McClintock, J. Richardson, Discovery of X-ray bursts from several sources near the galactic centre. Mon. Not. R. Astron. Soc. **177**, 83P–92P (1976)
137. W.H.G. Lewin, J. van Paradijs, R.E. Taam, X-ray bursts. Space Sci. Rev. **62**, 223–389 (1993). https://doi.org/10.1007/BF00196124
138. K.H. Lo, M.C. Miller, S. Bhattacharyya, F.K. Lamb, Determining neutron star masses and radii using energy-resolved waveforms of X-ray burst oscillations. Astrophys. J. **776**, 19 (2013). https://doi.org/10.1088/0004-637X/776/1/19
139. M. Long, M.M. Romanova, R.V.E. Lovelace, Accretion to stars with non-dipole magnetic fields. Mon. Not. R. Astron. Soc. **374**, 436–444 (2007). https://doi.org/10.1111/j.1365-2966.2006.11192.x
140. R.V.E. Lovelace, M.M. Romanova, Three disk oscillation modes of rotating magnetized neutron stars. Astrophys. J. **670**, L13–L16 (2007). https://doi.org/10.1086/524034
141. A.G. Lyne, M. Burgay, M. Kramer, A. Possenti, R.N. Manchester, F. Camilo, M.A. McLaughlin, D.R. Lorimer, N. D'Amico, B.C. Joshi, J. Reynolds, P.C.C. Freire, A double-pulsar system: a rare laboratory for relativistic gravity and plasma physics. Science **303**, 1153–1157 (2004). https://doi.org/10.1126/science.1094645
142. S. Mahmoodifar, T. Strohmayer, Upper bounds on r-mode amplitudes from observations of low-mass X-ray binary neutron stars. Astrophys. J. **773**, 140 (2013). https://doi.org/10.1088/0004-637X/773/2/140
143. A. Majczyna, J. Madej, Mass and radius determination for the neutron star in X-ray burst source 4U/MXB 1728-34. Acta. Astronom. **55**, 349–366 (2005)
144. R.N. Manchester, Radio emission properties of pulsars, in *Astrophysics and Space Science Library*, ed. by W. Becker, vol. 357, pp. 19–39 (2009). https://doi.org/10.1007/978-3-540-76965-1_2
145. J.E. McClintock, R.A. Remillard, The X-ray nova Centaurus X-4 - comparisons with A0620-00. Astrophys. J. **350**, 386–394 (1990). https://doi.org/10.1086/168392
146. J.E. McClintock, R. Narayan, L. Gou, J. Liu, R.F. Penna, J.F. Steiner, Measuring the spins of stellar black holes: a progress report, in *American Institute of Physics Conference Series*, ed. by A. Comastri, L. Angelini, M. Cappi, vol. 1248, pp. 101–106 (2010). https://doi.org/10.1063/1.3475156
147. D.B. McLaughlin, The spectrum of Nova Herculis. Publ. Michigan Observatory **6**, 107–214 (1937)

148. M. Méndez, On the maximum amplitude and coherence of the kilohertz quasi-periodic oscillations in low-mass X-ray binaries. Mon. Not. R. Astron. Soc. **371**, 1925–1938 (2006). https://doi.org/10.1111/j.1365-2966.2006.10830.x
149. M. Méndez, T. Belloni, Is there a link between the neutron-star spin and the frequency of the kilohertz quasi-periodic oscillations? Mon. Not. R. Astron. Soc. **381**, 790–796 (2007). https://doi.org/10.1111/j.1365-2966.2007.12306.x
150. M. Méndez, M. van der Klis, E.C. Ford, R. Wijnands, J. van Paradijs, Dependence of the frequency of the kilohertz quasi-periodic oscillationson X-ray count rate and colors in 4U 1608-52. Astrophys. J. **511**, L49–L52 (1999). https://doi.org/10.1086/311836
151. J.M. Miller, Relativistic X-ray lines from the inner accretion disks around black holes. Annu. Rev. Astron. Astrophys. **45**, 441–479 (2007). https://doi.org/10.1146/annurev.astro.45.051806.110555
152. M.C. Miller, Model atmospheres for neutron stars. Mon. Not. R. Astron. Soc. **255**, 129–145 (1992)
153. M.C. Miller, Prompt mergers of neutron stars with black holes. Astrophys. J. Lett. **626**, L41–L44 (2005). https://doi.org/10.1086/431583
154. M.C. Miller, F.K. Lamb, Bounds on the compactness of neutron stars from brightness oscillations during X-ray bursts. Astrophys. J. Lett. **499**, L37+ (1998). https://doi.org/10.1086/311335
155. M.C. Miller, D. Neuhauser, Atoms in very strong magnetic fields. Mon. Not. R. Astron. Soc. **253**, 107–122 (1991)
156. M.C. Miller, F.K. Lamb, G.B. Cook, Effects of rapid stellar rotation on equation-of-state constraints derived from quasi-periodic brightness oscillations. Astrophys. J. **509**, 793–801 (1998). https://doi.org/10.1086/306533
157. M.C. Miller, F.K. Lamb, D. Psaltis, Sonic-point model of kilohertz quasi-periodic brightness oscillations in low-mass X-ray binaries. Astrophys. J. **508**, 791–830 (1998). https://doi.org/10.1086/306408
158. M.C. Miller, S. Boutloukos, K.H. Lo, F.K. Lamb, Implications of high-precision spectra of thermonuclear X-ray bursts for determining neutron star masses and radii, in *Fast X-ray Timing and Spectroscopy at Extreme Count Rates (HTRS 2011)* (2011)
159. S.M. Morsink, D.A. Leahy, Multi-epoch analysis of pulse shapes from the neutron star SAX J1808.4-3658. Astrophys. J. **726**, 56 (2011). https://doi.org/10.1088/0004-637X/726/1/56
160. M.P. Muno, F. Özel, D. Chakrabarty, The energy dependence of millisecond oscillations in thermonuclear X-ray bursts. Astrophys. J. **595**, 1066–1076 (2003). https://doi.org/10.1086/377447
161. K. Nandra, D. Barret, X. Barcons, A. Fabian, J.-W. den Herder, L. Piro, M. Watson, C. Adami, J. Aird, J.M. Afonso, et al. The hot and energetic universe: a white paper presenting the science theme motivating the Athena+ mission. ArXiv e-prints (2013)
162. N.R. Nath, T.E. Strohmayer, J.H. Swank, Bounds on compactness for low-mass X-ray binary neutron stars from X-ray burst oscillations. Astrophys. J. **564**, 353–360 (2002). https://doi.org/10.1086/324132
163. K. Nomoto, Evolution of 8-10 solar mass stars toward electron capture supernovae. I - Formation of electron-degenerate O + NE + MG cores. Astrophys. J. **277**, 791–805 (1984). https://doi.org/10.1086/161749
164. J.R. Oppenheimer, G.M. Volkoff, On massive neutron cores. Phys. Rev. **55**, 374–381 (1939). https://doi.org/10.1103/PhysRev.55.374
165. F. Özel, Surface emission from neutron stars and implications for the physics of their interiors. Rep. Prog. Phys. **76**(1), 016901 (2013). https://doi.org/10.1088/0034-4885/76/1/016901
166. F. Özel, D. Psaltis, Reconstructing the neutron-star equation of state from astrophysical measurements. Phys. Rev. D **80**(10), 103003 (2009). https://doi.org/10.1103/PhysRevD.80.103003
167. B. Paczynski, Models of X-ray bursters with radius expansion. Astrophys. J. **267**, 315–321 (1983). https://doi.org/10.1086/160870

168. B. Paczynski, N. Anderson, Models of extended relativistic envelopes of neutron stars. Astrophys. J. **302**, 1–10 (1986). https://doi.org/10.1086/163966
169. D. Page, *Pairing and the Cooling of Neutron Stars* (World Scientific Publishing Co., Singapore, 2013), pp. 324–337. https://doi.org/10.1142/9789814412490_0024
170. D. Page, J.H. Applegate, The cooling of neutron stars by the direct URCA process. Astrophys. J. Lett. **394**, L17–L20 (1992). https://doi.org/10.1086/186462
171. D. Page, S. Reddy, Dense matter in compact stars: theoretical developments and observational constraints. Annu. Rev. Nucl. Part. Sci. **56**, 327–374 (2006). https://doi.org/10.1146/annurev.nucl.56.080805.140600
172. D. Page, U. Geppert, F. Weber, The cooling of compact stars. Nucl. Phys. A **777**, 497–530 (2006). https://doi.org/10.1016/j.nuclphysa.2005.09.019
173. D. Page, J.M. Lattimer, M. Prakash, A.W. Steiner, Neutrino emission from cooper pairs and minimal cooling of neutron stars. Astrophys. J. **707**, 1131–1140 (2009). https://doi.org/10.1088/0004-637X/707/2/1131
174. D. Page, M. Prakash, J.M. Lattimer, A.W. Steiner, Rapid cooling of the neutron star in Cassiopeia A triggered by neutron superfluidity in dense matter. Phys. Rev. Lett. **106**(8), 081101 (2011). https://doi.org/10.1103/PhysRevLett.106.081101
175. A. Papitto, A. Riggio, L. Burderi, T. di Salvo, A. D'Aí, R. Iaria, Spin down during quiescence of the fastest known accretion-powered pulsar. Astron. Astrophys. **528**, A55 (2011). https://doi.org/10.1051/0004-6361/201014837
176. A. Patruno, B. Haskell, C. D'Angelo, Gravitational waves and the maximum spin frequency of neutron stars. Astrophys. J. **746**, 9 (2012). https://doi.org/10.1088/0004-637X/746/1/9
177. P.C. Peters, Gravitational radiation and the motion of two point masses. Phys. Rev. **136**, 1224–1232 (1964). https://doi.org/10.1103/PhysRev.136.B1224
178. J. Piekarewicz, Neutron skins and neutron stars, in *American Institute of Physics Conference Series*, ed. by R. Milner, R. Carlini, F. Maas, vol. 1563 (2013), pp. 227–230. https://doi.org/10.1063/1.4829416
179. P. Podsiadlowski, J.D.M. Dewi, P. Lesaffre, J.C. Miller, W.G. Newton, J.R. Stone, The double pulsar J0737-3039: testing the neutron star equation of state. Mon. Not. R. Astron. Soc. **361**, 1243–1249 (2005). https://doi.org/10.1111/j.1365-2966.2005.09253.x
180. B. Posselt, G.G. Pavlov, V. Suleimanov, O. Kargaltsev, New constraints on the cooling of the central compact object in Cas A. Astrophys. J. **779**, 186 (2013). https://doi.org/10.1088/0004-637X/779/2/186
181. J. Poutanen, A.M. Beloborodov, Pulse profiles of millisecond pulsars and their Fourier amplitudes. Mon. Not. R. Astron. Soc. **373**, 836–844 (2006). https://doi.org/10.1111/j.1365-2966.2006.11088.x
182. J. Poutanen, M. Gierliński, On the nature of the X-ray emission from the accreting millisecond pulsar SAX J1808.4-3658. Mon. Not. R. Astron. Soc. **343**, 1301–1311 (2003). https://doi.org/10.1046/j.1365-8711.2003.06773.x
183. D. Psaltis, Testing general metric theories of gravity with bursting neutron stars. Phys. Rev. D **77**(6), 064006 (2008). https://doi.org/10.1103/PhysRevD.77.064006
184. M. Rajagopal, R.W. Romani, M.C. Miller, Magnetized iron atmospheres for neutron stars. Astrophys. J. **479**, 347–356 (1997). https://doi.org/10.1086/303865
185. P.S. Ray, B.F. Phlips, K.S. Wood, D. Chakrabarty, R.A. Remillard, C.A. Wilson-Hodge, The advanced X-ray timing array (AXTAR): a US MIDEX mission concept. ArXiv e-prints (2011)
186. J.S. Read, L. Baiotti, J.D.E. Creighton, J.L. Friedman, B. Giacomazzo, K. Kyutoku, C. Markakis, L. Rezzolla, M. Shibata, K. Taniguchi, Matter effects on binary neutron star waveforms. Phys. Rev. D **88**(4), 044042 (2013). https://doi.org/10.1103/PhysRevD.88.044042
187. R.W. Romani, Model atmospheres for cooling neutron stars. Astrophys. J. **313**, 718–726 (1987). https://doi.org/10.1086/165010
188. R.W. Romani, A.V. Filippenko, J.M. Silverman, S.B. Cenko, J. Greiner, A. Rau, J. Elliott, H.J. Pletsch, PSR J1311-3430: a heavyweight neutron star with a flyweight helium companion. Astrophys. J. Lett. **760**, L36 (2012). https://doi.org/10.1088/2041-8205/760/2/L36

189. M.M. Romanova, G.V. Ustyugova, A.V. Koldoba, J.V. Wick, R.V.E. Lovelace, Three-dimensional simulations of disk accretion to an inclined dipole. I. Magnetospheric flows at different Θ. Astrophys. J. **595**, 1009–1031 (2003). https://doi.org/10.1086/377514
190. M.M. Romanova, G.V. Ustyugova, A.V. Koldoba, R.V.E. Lovelace, Three-dimensional simulations of disk accretion to an inclined dipole. II. Hot spots and variability. Astrophys. J. **610**, 920–932 (2004). https://doi.org/10.1086/421867
191. M.M. Romanova, A. Kulkarni, M. Long, R.V.E. Lovelace, J.V. Wick, G.V. Ustyugova, A.V. Koldoba, 2D and 3D MHD simulations of disk accretion by rotating magnetized stars: search for variability. Adv. Space Res. **38**, 2887–2892 (2006). https://doi.org/10.1016/j.asr.2006.04.019
192. M.M. Romanova, A.K. Kulkarni, R.V.E. Lovelace, Unstable disk accretion onto magnetized stars: first global three-dimensional magnetohydrodynamic simulations. Astrophys. J. **673**, L171–L174 (2008). https://doi.org/10.1086/527298
193. L.C. Rosen, Hydrogen and helium abundances in neutron-star atmospheres. Astrophys. Space Sci. **1**, 372–387 (1968). https://doi.org/10.1007/BF00656008
194. R.A. Rossiter, On the detection of an effect of rotation during eclipse in the velocity of the brigher component of beta Lyrae, and on the constancy of velocity of this system. Astrophys. J. **60**, 15–21 (1924). https://doi.org/10.1086/142825
195. R.E. Rutledge, L. Bildsten, E.F. Brown, G.G. Pavlov, V. E. Zavlin, The quiescent X-ray spectrum of the neutron star in centaurus X-4 observed with chandra/ACIS-S. Astrophys. J. **551**, 921–928 (2001). https://doi.org/10.1086/320247
196. G.B. Rybicki, A.P. Lightman, *Radiative Processes in Astrophysics* (Wiley-VCH, Weinheim, 1986)
197. J. Schwab, P. Podsiadlowski, S. Rappaport, Further evidence for the bimodal distribution of neutron-star masses. Astrophys. J. **719**, 722–727 (2010). https://doi.org/10.1088/0004-637X/719/1/722
198. A. Schwenk, B. Friman, Polarization contributions to the spin dependence of the effective interaction in neutron matter. Phys. Rev. Lett. **92**(8), 082501 (2004). https://doi.org/10.1103/PhysRevLett.92.082501
199. A. Sedrakian, The physics of dense hadronic matter and compact stars. Prog. Part. Nucl. Phys. **58**, 168–246 (2007). https://doi.org/10.1016/j.ppnp.2006.02.002
200. M. Servillat, C.O. Heinke, W.C.G. Ho, J.E. Grindlay, J. Hong, M. van den Berg, S. Bogdanov, Neutron star atmosphere composition: the quiescent, low-mass X-ray binary in the globular cluster M28. Mon. Not. R. Astron. Soc. **423**, 1556–1561 (2012). https://doi.org/10.1111/j.1365-2966.2012.20976.x
201. M. Shao, B. Nemati, C. Zhai, Space interferometry mission (SIM)-lite status, in *EAS Publications Series*, vol. 45 (2011), pp. 141–142. https://doi.org/10.1051/eas/1045024
202. P.S. Shternin, D.G. Yakovlev, C.O. Heinke, W.C.G. Ho, D.J. Patnaude, Cooling neutron star in the Cassiopeia A supernova remnant: evidence for superfluidity in the core. Mon. Not. R. Astron. Soc. **412**, L108–L112 (2011). https://doi.org/10.1111/j.1745-3933.2011.01015.x
203. A.P. Smale, A second intense burst with photospheric radius expansion from X2127+119 in M15. Astrophys. J. **562**, 957–962 (2001). https://doi.org/10.1086/323517
204. A. Spitkovsky, Y. Levin, G. Ushomirsky, Propagation of thermonuclear flames on rapidly rotating neutron stars: extreme weather during type I X-ray bursts. Astrophys. J. **566**, 1018–1038 (2002). https://doi.org/10.1086/338040
205. I.H. Stairs, Testing general relativity with pulsar timing. Living Rev. Relativ. **6**, 5 (2003)
206. A.W. Steiner, J.M. Lattimer, E.F. Brown, The equation of state from observed masses and radii of neutron stars. Astrophys. J. **722**, 33–54 (2010). https://doi.org/10.1088/0004-637X/722/1/33
207. N. Stergioulas, Rotating stars in relativity. Living Rev. Relativ. **6**, 3 (2003)
208. T.E. Strohmayer, A.L. Watts, Discovery of fast X-ray oscillations during the 1998 giant flare from SGR 1900+14. Astrophys. J. Lett. **632**, L111–L114 (2005). https://doi.org/10.1086/497911

209. T.E. Strohmayer, A.L. Watts, The 2004 hyperflare from SGR 1806-20: further evidence for global torsional vibrations. Astrophys. J. **653**, 593–601 (2006). https://doi.org/10.1086/508703
210. V. Suleimanov, J. Poutanen, M. Revnivtsev, K. Werner, A neutron star stiff equation of state derived from cooling phases of the X-ray burster 4U 1724-307. Astrophys. J. **742**, 122 (2011). https://doi.org/10.1088/0004-637X/742/2/122
211. V. Suleimanov, J. Poutanen, K. Werner, X-ray bursting neutron star atmosphere models using an exact relativistic kinetic equation for Compton scattering. Astron. Astrophys. **545**, A120 (2012). https://doi.org/10.1051/0004-6361/201219480
212. R.E. Taam, R.E. Picklum, Nuclear fusion and carbon flashes on neutron stars. Astrophys. J. **224**, 210–216 (1978). https://doi.org/10.1086/156367
213. R.E. Taam, S.E. Woosley, T.A. Weaver, D.Q. Lamb, Successive X-ray bursts from accreting neutron stars. Astrophys. J. **413**, 324–332 (1993). https://doi.org/10.1086/173000
214. N.R. Tanvir, A.J. Levan, A.S. Fruchter, J. Hjorth, R.A. Hounsell, K. Wiersema, R.L. Tunnicliffe, A 'kilonova' associated with the short-duration γ-ray burst GRB130603B. Nature **500**, 547–549 (2013). https://doi.org/10.1038/nature12505
215. J.A. Tomsick, M.W. Muterspaugh, Masses of neutron stars in high-mass X-ray binaries with optical astrometry. Astrophys. J. **719**, 958–965 (2010). https://doi.org/10.1088/0004-637X/719/1/958
216. M. Troyer, U.-J. Wiese, Computational complexity and fundamental limitations to fermionic Quantum Monte Carlo simulations. Phys. Rev. Lett. **94**(17), 170201–+ (2005). https://doi.org/10.1103/PhysRevLett.94.170201
217. G. Ushomirsky, R.E. Rutledge, Time-variable emission from transiently accreting neutron stars in quiescence due to deep crustal heating. Mon. Not. R. Astron. Soc. **325**, 1157–1166 (2001). https://doi.org/10.1046/j.1365-8711.2001.04515.x
218. G. Ushomirsky, C. Cutler, L. Bildsten, Deformations of accreting neutron star crusts and gravitational wave emission. Mon. Not. R. Astron. Soc. **319**, 902–932 (2000). https://doi.org/10.1046/j.1365-8711.2000.03938.x
219. C. Van Den Broeck, Astrophysics, cosmology, and fundamental physics with compact binary coalescence and the Einstein Telescope. J. Phys. Conf. Ser. **484**(1), 012008 (2014). https://doi.org/10.1088/1742-6596/484/1/012008
220. M. van der Sluys, I. Mandel, V. Raymond, V. Kalogera, C. Röver, N. Christensen, Parameter estimation for signals from compact binary inspirals injected into LIGO data. Classical Quantum Gravity **26**(20), 204010 (2009). https://doi.org/10.1088/0264-9381/26/20/204010
221. C.A. van Eysden, A. Melatos, Gravitational radiation from pulsar glitches. Classical Quantum Gravity **25**(22), 225020–+ (2008). https://doi.org/10.1088/0264-9381/25/22/225020
222. M.H. van Kerkwijk, D.L. Kaplan, Isolated neutron stars: magnetic fields, distances,and spectra. Astrophys. Space Sci. **308**, 191–201 (2007). https://doi.org/10.1007/s10509-007-9343-9
223. M.H. van Kerkwijk, R.P. Breton, S.R. Kulkarni, Evidence for a massive neutron star from a radial-velocity study of the companion to the black-widow pulsar PSR B1957+20. Astrophys. J. **728**, 95 (2011). https://doi.org/10.1088/0004-637X/728/2/95
224. J. van Paradijs, Possible observational constraints on the mass-radius relation of neutron stars. Astrophys. J. **234**, 609–611 (1979). https://doi.org/10.1086/157535
225. T. Venumadhav, A. Zimmerman, C.M. Hirata, The stability of tidally deformed neutron stars to three- and four-mode coupling. Astrophys. J. **781**, 23 (2014). https://doi.org/10.1088/0004-637X/781/1/23
226. K. Viironen, J. Poutanen, Light curves and polarization of accretion- and nuclear-powered millisecond pulsars. Astron. Astrophys. **426**, 985–997 (2004). https://doi.org/10.1051/0004-6361:20041084
227. A.R. Villarreal, T.E. Strohmayer, Discovery of the neutron star spin frequency in EXO 0748-676. Astrophys. J. Lett. **614**, L121–L124 (2004). https://doi.org/10.1086/425737
228. F.M. Walter, J.M. Lattimer, A revised parallax and its implications for RX J185635-3754. Astrophys. J. Lett. **576**, L145–L148 (2002). https://doi.org/10.1086/343850

229. F.M. Walter, T. Eisenbeiß, J.M. Lattimer, B. Kim, V. Hambaryan, R. Neuhäuser, Revisiting the parallax of the isolated neutron star RX J185635-3754 using HST/ACS imaging. Astrophys. J. **724**, 669–677 (2010). https://doi.org/10.1088/0004-637X/724/1/669
230. A.L. Watts, Thermonuclear burst oscillations. Annu. Rev. Astron. Astrophys. **50**, 609–640 (2012). https://doi.org/10.1146/annurev-astro-040312-132617
231. A.L. Watts, T.E. Strohmayer, Detection with RHESSI of high-frequency X-ray oscillations in the tailof the 2004 hyperflare from SGR 1806-20. Astrophys. J. Lett. **637**, L117–L120 (2006). https://doi.org/10.1086/500735
232. A.L. Watts, T.E. Strohmayer, Neutron star oscillations and QPOs during magnetar flares. Adv. Space Res. **40**, 1446–1452 (2007). https://doi.org/10.1016/j.asr.2006.12.021
233. N.A. Webb, D. Barret, Constraining the equation of state of supranuclear dense matter from XMM-newton observations of neutron stars in globular clusters. Astrophys. J. **671**, 727–733 (2007). https://doi.org/10.1086/522877
234. F. Weber, R. Negreiros, P. Rosenfield, Neutron star interiors and the equation of state of superdense matter. in *Astrophysics and Space Science Library*, ed. by W. Becker, vol. 357 (2009), p. 213. https://doi.org/10.1007/978-3-540-76965-1_0
235. N. Weinberg, M.C. Miller, D.Q. Lamb, Oscillation waveforms and amplitudes from hot spots on neutron stars. Astrophys. J. **546**, 1098–1106 (2001). https://doi.org/10.1086/318279
236. S.E. Woosley, R.E. Taam, Gamma-ray bursts from thermonuclear explosions on neutron stars. Nature **263**, 101–103 (1976). https://doi.org/10.1038/263101a0
237. S.E. Woosley, A. Heger, T.A. Weaver, The evolution and explosion of massive stars. Rev. Mod. Phys. **74**, 1015–1071 (2002). https://doi.org/10.1103/RevModPhys.74.1015
238. K. Yagi, N. Yunes, I-Love-Q relations in neutron stars and their applications to astrophysics, gravitational waves, and fundamental physics. Phys. Rev. D **88**(2), 023009 (2013). https://doi.org/10.1103/PhysRevD.88.023009
239. D.G. Yakovlev, C.J. Pethick, Neutron star cooling. Annu. Rev. Astron. Astrophys. **42**, 169–210 (2004). https://doi.org/10.1146/annurev.astro.42.053102.134013
240. W. Zhang, A.P. Smale, T.E. Strohmayer, J.H. Swank, Correlation between energy spectral states and fast time variability and further evidence for the marginally stable orbit in 4U 1820-30. Astrophys. J. **500**, L171–174 (1998). https://doi.org/10.1086/311418
241. C.M. Zhang, J. Wang, Y.H. Zhao, H.X. Yin, L.M. Song, D.P. Menezes, D.T. Wickramasinghe, L. Ferrario, P. Chardonnet, Study of measured pulsar masses and their possible conclusions. Astron. Astrophys. **527**, A83 (2011). https://doi.org/10.1051/0004-6361/201015532
242. G. Zhang, M. Méndez, T.M. Belloni, J. Homan, Coherent oscillations and the evolution of the apparent emission area in the decaying phase of radius-expansion bursts from 4U 1636-53. Mon. Not. R. Astron. Soc. **436**, 2276–2285 (2013). https://doi.org/10.1093/mnras/stt1728

Chapter 2
General Relativity Measurements from Pulsars

Marta Burgay, Delphine Perrodin, and Andrea Possenti

Contents

Abstract In this chapter, after a brief description of radio pulsars and their different manifestations, we illustrate the case of the so-called relativistic binary pulsars, which are fast-spinning neutron stars in orbit with a compact companion star, either a white dwarf or a second neutron star. Thanks to the clock-like nature of their radio signals, some of these binary pulsars can be used as test beds for General Relativity and other theories of gravity. We describe the methodology used to exploit these sources as laboratories for relativistic gravity and review some of the most important recent results in this field of research.

M. Burgay (✉) · D. Perrodin · A. Possenti
INAF-Osservatorio Astronomico di Cagliari, Selargius, CA, Italy
e-mail: burgay@oa-cagliari.inaf.it; delphine@oa-cagliari.inaf.it; possenti@oa-cagliari.inaf.it

© Springer-Verlag GmbH Germany, part of Springer Nature 2021

T. M. Belloni et al. (eds.), *Timing Neutron Stars: Pulsations, Oscillations and Explosions*, Astrophysics and Space Science Library 461,
https://doi.org/10.1007/978-3-662-62110-3_2

2.1 Why Radio Pulsars

Since their discovery in 1967, the number of known radio pulsars has been almost exponentially growing (Fig. 2.1). Even though major steps forward have been made in recent years in the understanding of pulsar electrodynamics (particularly thanks to new gamma-ray observations; e.g. [2, 3]), the exact physical mechanism responsible for their broadband emission is still under debate. However, the lack of an accurate model for their emission has not prevented the in-depth study of a great number of physical and astrophysical phenomena using these objects in the past decades. In fact, the very stable and accurately measurable ticking of some radio pulsars (see Sects. 2.3 and 2.4 for details) makes them superb time keepers, opening the possibility of using these sources as excellent tools to investigate a wide range of topics: for instance (1) the dispersion of their pulsed signal in the interstellar medium is a primary way to measure the distribution of free electrons in our Galaxy (e.g. [4, 5]); (2) the Faraday rotation of their polarised signals helps determine the large-scale structure of the Galactic magnetic field (e.g. [6]); (3) if found in globular clusters, they allow us to study cluster dynamics and potential wells, in some cases uncovering the presence of non-luminous matter (e.g. black holes [7]); (4) in recent years, arrays of pulsars with particularly stable and accurately measurable signals, have been used in so-called Pulsar Timing Array experiments aimed at detecting gravitational waves, thanks to the fact that gravitational waves can affect—in a correlated way—the measured times of arrival at Earth of the radio signals emitted by different pulsars (e.g. [8]); (5) when found orbiting a companion star, pulsars provide the possibility of studying binary evolution and in particular (6) when the companion star is a compact object (such as a white dwarf, a second neutron star or even a black hole), the unique possibility of precisely measuring post-Keplerian

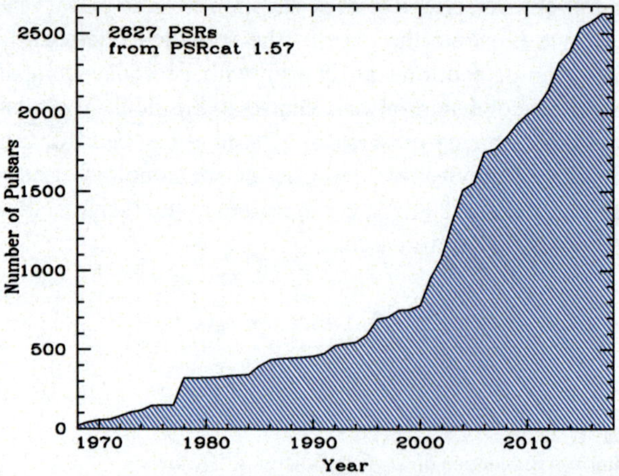

Fig. 2.1 Total number of pulsar discoveries published from 1968 to 2017. Data extracted from PSRcat v. 1.57 http://www.atnf.csiro.au/research/pulsar/psrcat/ [1]

effects on the orbital motions, leading to the best tests of relativistic gravity in the strong-field regime.

In this chapter, after briefly summarising the 'zoology' of radio pulsars (Sect. 2.2), we will concentrate on the description of the so-called *relativistic binary pulsars* (Sect. 2.3), which are neutron stars orbiting compact objects at relativistic speed, and whose radio signals, studied through pulsar timing techniques (Sect. 2.4), are the best probes of the deformation of the space-time in the vicinity of a compact star, hence making these pulsars unique laboratories for testing gravitational theories with steadily increasing precision.

2.2 The Many Faces of the Radio Pulsar Zoo

Non-accreting neutron stars (NSs) are, for the vast majority, found in the form of radio pulsars, which are sources of steady pulsations in the radio band. In recent years, however, more and more different manifestations of slowing-down NSs have been discovered, either emitting solely in other energy bands (such as the Central Compact Objects, CCOs [9], or the X-ray Dim Isolated Neutron Stars, XDINSs [10]) or as more or less extreme and irregular transients in the radio band, like the sporadically emitting Rotational RAdio Transients, RRATs [11], or the puzzling intermittent pulsars [12]. For most of the aforementioned families of NSs, the energy budget is coming from their rotational energy (i.e. they are rotational-powered NSs), but this does not hold true for the so-called 'magnetars' [13] and it is still under debate for CCOs.

The diagram of Fig. 2.2 collects all of the non-accreting NSs for which the spin period P and its derivative $\dot{P}$ are known (but for a subsample of the RRATs, which do not have yet a measured $\dot{P}$). In the following sections, we will describe radio-emitting pulsars.

2.2.1 *Radio Pulsars*

When a radio pulsar is discovered, the first parameters measured are its approximate position in the sky, its period and its dispersion measure (DM). The latter is computed by measuring the delay in the arrival times of a pulse at different frequencies. Radio pulsar observations are in fact usually done over a wide band (ideally up to about 30% of the central observing frequency) and, because of the free electrons in the interstellar medium (ISM), waves emitted at higher frequencies arrive at the telescope before those emitted at lower ones, following the law:

$$\Delta t_{\rm DM} = t(\nu_2) - t(\nu_1) = \frac{e^2}{2\pi m_e c}\left(\frac{1}{\nu_1^2} - \frac{1}{\nu_2^2}\right) {\rm DM} \qquad (2.1)$$

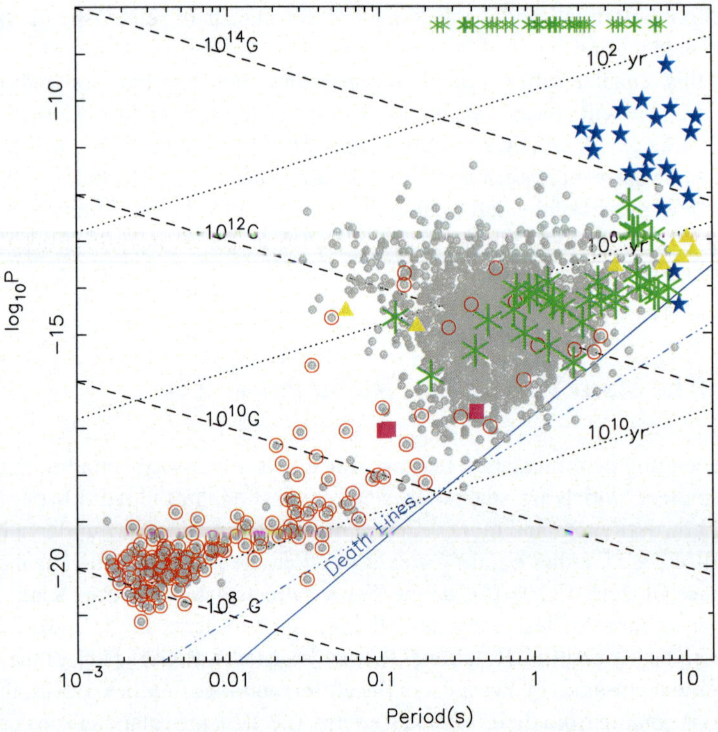

Fig. 2.2 Period-period derivative diagram. Updated from [14]. "Grey dots are Galactic field radio pulsars (those surrounded by a red circle are in a binary system), purple squares are XDINSs, yellow triangles are CCOs, green asterisks are RRATs (the top smaller ones do not yet have a measured $\dot{P}$), blue stars are magnetars. Dashed lines denote equal dipolar magnetic field, calculated as in Eq. (2.3), while dotted ones are equal spin-down age (Eq. (2.4)) lines. The violet line is the so-called *death line* (in particular the *death-line C* of [15]) for values of P and $\dot{P}$ below which, the mechanism responsible for radio emission is not efficient anymore and the pulsar switches off. We point out that the use of a specific line is only for the sake of simplicity; a death valley [15], across which the pulsar signal slowly fades out with time, would better replace a single line. Data taken from [1] http://www.atnf.csiro.au/research/pulsar/psrcat/ and http://www.physics.mcgill.ca/~pulsar/magnetar/main.html"

where e is the electron charge, m_e is the electron mass, c is the speed of light, ν_1 and ν_2 are two observing frequencies and DM is the dispersion measure, i.e. the free electron column density along the line of sight. Because the time delay over the observing frequency band is often larger than the period of repetition of the pulses, the pulse is broadened so much that it would be impossible to catch it. To overcome this problem, the observing band is split into several frequency channels, over each of which the effects of the ISM are proportionally smaller; in each channel, the pulse is almost unaffected by dispersion and its time of arrival is delayed with respect to the following channel according to Eq. (2.1), where, in this case, ν_1 and ν_2 are the central frequencies of adjacent channels. The DM is hence simply measured by correcting this delay (de-dispersion; Fig. 2.3) and this operation allows one to

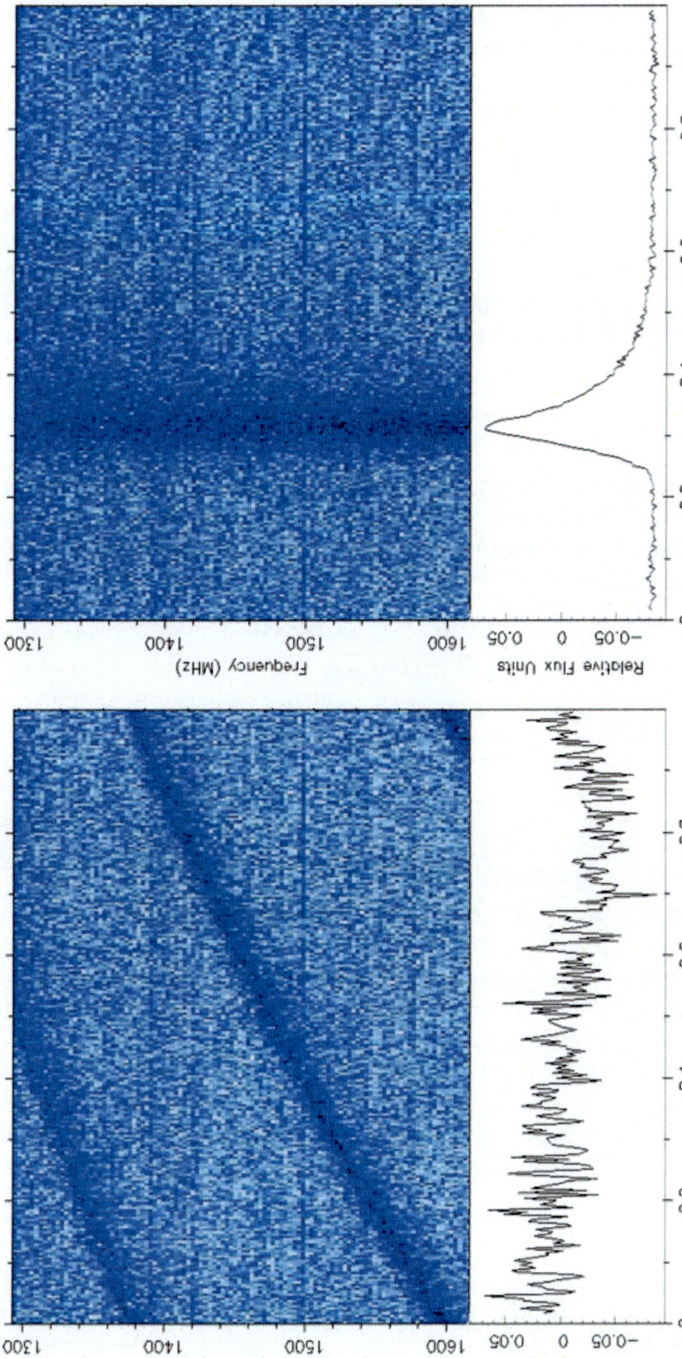

Fig. 2.3 Frequency vs. pulse phase plots for PSR J1721−3532: the observing band is split into 96 3 MHz-wide channels in each of which the pulse is clearly visible (darker parts of the gray-scale). *On the left panel* the signal is dispersed and the resulting pulse profile (reported at the *bottom* of the panel) is broadened so much over the entire bandwidth that its impulsive nature is completely washed out. *On the right panel*, the various frequency channels are delayed one with respect to the other, taking into account the effects of the dispersion in the ISM. The integrated pulse profile is thus neatly recovered

recover the pulsed signal over the whole band, obtaining a high signal-to-noise ratio (S/N) pulse profile (Fig. 2.3, *right panel*).

After the discovery of a new pulsar, in order to fully exploit its potential as an astrophysical laboratory, it is necessary to start a follow-up campaign of observations to precisely measure its spin, astrometric and, in the case of a binary system, orbital parameters (see Sect. 2.4). One of the basic parameters is the first derivative of the spin period: $\dot{P}$. This is a crucial parameter because, assuming, as the standard electrodynamics model [16] states,[1] that the emission is due to a spinning magneto-dipole losing rotational energy (see Eq. (2.2)), its value is directly related to the dipolar surface magnetic field B_s and the age (or an estimate of the age) of the pulsar. The equation describing the dipole spin-down emission, obtained by equalling the rotational energy loss to the dipole radiation power, can be written as:

$$-I_{NS}\omega\dot{\omega} = \frac{2}{3}\frac{1}{c^3}\omega^4 B_s^2 R_{NS}^6 \sin^2\alpha \tag{2.2}$$

where I_{NS} is the moment of inertia of the NS, $\omega = 2\pi/\mathrm{P}$ its angular velocity, R_{NS} its radius and α the angle between the magnetic and the rotational axes. Rewriting Eq. (2.2) as a function of the spin period and its derivative and using $\alpha = 90°$, $I_{NS} = 10^{45}$ g cm^2 and $R_{NS} = 10^6$ cm, we can derive an estimate of the magnetic field as:

$$B_s = 3.2 \times 10^{19}\sqrt{P\dot{P}} \qquad \mathrm{G} \tag{2.3}$$

and, integrating in time, we can derive the so-called spin-down age of the pulsar as:

$$\tau_c = \frac{P}{2\dot{P}}\left(1 - \frac{P_0^2}{P^2}\right) \sim \frac{P}{2\dot{P}} \tag{2.4}$$

where P_0 is the spin period of the pulsar at birth (assumed to be negligible with respect to the current one).

The measurement of $\dot{P}$ and all of the other parameters (e.g. position, proper motion, Keplerian and post-Keplerian parameters) is done through the *timing* technique that is fully described in Sect. 2.4.

According to the ATNF pulsar catalogue PSRcat (www.atnf.csiro.au/research/pulsar/psrcat/ [1]), as of the end of 2017, more than 2600 pulsars have been discovered (and published, see Fig. 2.4) in our Galaxy, including Galactic globular clusters and in the Magellanic Clouds. The sub-set of *millisecond*, or *recycled*,

[1]This is a very crude model not suitable for explaining the nature of the processes involved in the broad band radio emission from the pulsars. However, barring the detailed physics, the energy budget resulting from this model and the related Eqs. (2.2), (2.3), (2.4), are used as a reference for classification of the rotational powered NSs and for comparison with more advanced and physically sound modelling.

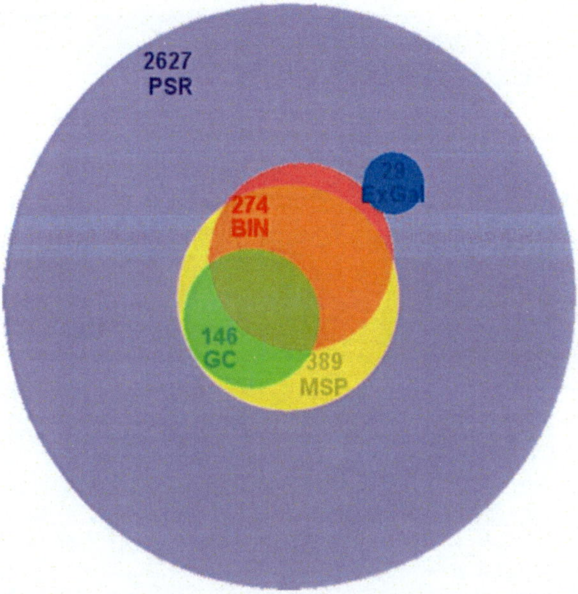

Fig. 2.4 Graphical representation of the number of members of various classes of known pulsars. The purple circle labelled PSR includes the entire pulsar population (as from the ATNF pulsar catalogue v1.57). The blue circle labelled ExGal refers to pulsars discovered outside the Milky Way (in the Magellanic Clouds). Binary pulsars are represented by the red circle labelled BIN. Pulsars discovered in globular clusters are enclosed in the green circle labelled GC, whereas the recycled pulsars are reported in the yellow MSP circle. Data are extracted from http://www.atnf.csiro.au/research/pulsar/psrcat/ [1]

pulsars (see Sect. 2.3), located in the bottom-left corner of the $P - \dot{P}$ diagram of Fig. 2.2, are of special interest: the short duration of their signals and the extreme regularity of their pulsations, make them particularly useful for studying the fundamental topics mentioned in Sect. 2.1.

2.2.2 *Intermittent Pulsars*

Among the NSs with pulsating radio emission, there is a well-known subclass: the so-called *nulling* pulsars (first noticed by Backer [17]) in which the emission appears to suddenly switch off for several pulse periods (e.g. the case displayed in Fig. 2.5). The nulling fraction is very variable from pulsar to pulsar and the mechanism beyond the disappearance of pulses is still a matter of debate.

In recent years, a handful of objects, dubbed *intermittent pulsars*, in a way similar to nulling pulsars but more extreme, were discovered. These objects show a quite peculiar behaviour: their pulsed radio emission, regular for intervals of time variable

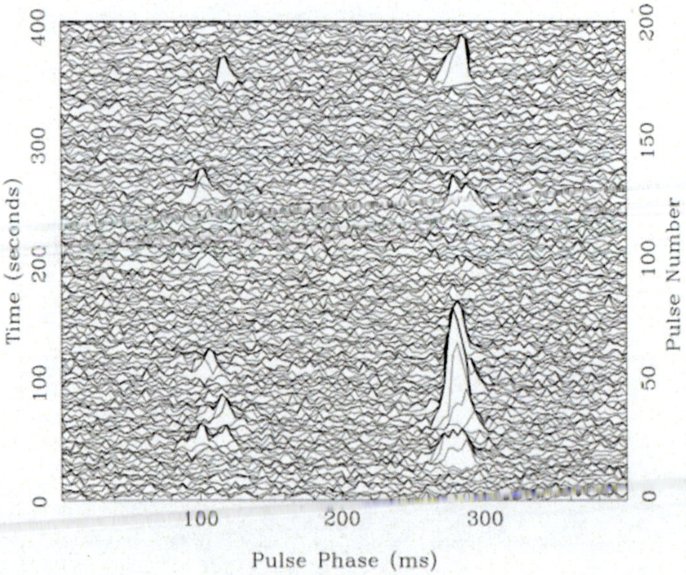

Fig. 2.5 Sequence of 200 consecutive single pulses from the 2 s pulsar J1944+1745. For clarity, only the central 400 ms of pulse phase of each pulse is shown. It appears that phases of emission of the radio pulsations alternate to phases where no pulsed signal emerges above the noise (figure courtesy of D. Lorimer 2020) [18]

from days to years, stops in a quasi-periodic way for longer intervals in which they show no emission at all. A careful study of one such object, PSR B1931+24 [12], shows that the rate of spin-down during the "on" and "off" phases is different, almost doubling during the phases in which the pulsar emission is "on" (Fig. 2.6, left). The same behaviour has been seen, e. g., in the intermittent pulsar J1832+0029 (Fig. 2.6, right).

Kramer et al. [12] interpret this peculiar emission pattern as due to a failure of charged particles in the magnetosphere. This both explains the sudden switch off of the pulsations and the large changes in slow-down rate linked to the changes in radio emission: the sudden absence of plasma in the magnetosphere can in fact decrease the braking torque on the NS, which will hence, at the same time, turn off as a radio pulsar, because of the lack of particles, and slow down less. From the difference in loss of rotational energy during the "on" and "off" phases, Kramer et al. [12] estimated the charge density of the current, and, interestingly, they found that the plasma current associated with radio emission carries a charge density very close to that computed in the Goldreich-Julian model for the pulsar magnetosphere [16].

Given their transient nature, many more such objects may exist in our Galaxy, but have been overlooked so far because they were "off" during either the search or the confirmation observations.

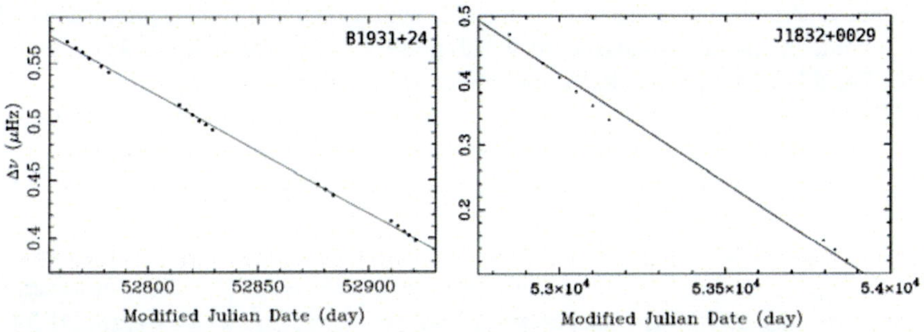

Fig. 2.6 The variation of the rotational frequency for PSR B1931+24 (*left*) and PSR J1832+0029 (*right*). The *solid* almost diagonal *line* represents the average spin frequency derivative over the plotted time-span. We note, in both cases, the increasing magnitude of the spin frequency derivative in the time intervals during which the pulsar was "on" (corresponding to the location of the black dots). Figure courtesy of A. G. Lyne 2020

2.2.3 *Rotating RAdio Transients*

An even more extreme family of transient radio pulsars is that of the so-called *Rotating RAdio Transients (RRATs)*. These sporadically-emitting NSs were first discovered [11] during the re-analysis of the data of the Parkes Multibeam Survey (e.g. [19]) by using an algorithm [20] to search for single dispersed pulses, and are now steadily increasing in number [21]. RRATs emit short (2–30 ms), infrequent (from 1 every few minutes to 1 every few hours) and relatively bright (0.1–10 Jy at 1.4 GHz) dispersed pulses (see Fig. 2.7). Their repetition periods, found by computing the minimum common denominator between the differences in the pulse

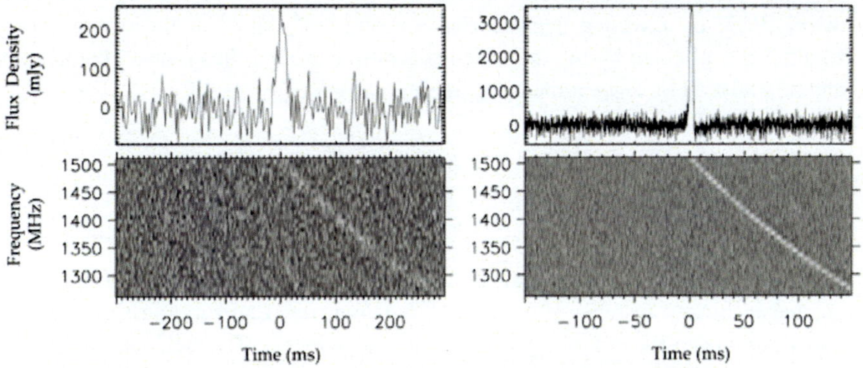

Fig. 2.7 Single dedispersed pulses from two RRATs. Top panels: the dedispersed pulse integrated over the entire observing band. Bottom panels: the observing band subdivided into 96 frequency channels over which the dispersion effect is clearly visible (figure courtesy of M. McLaughlin 2020)

times of arrival, range from 0.1 s to 7.7 s, with a median of 1.5 s, which is much higher than that of normal (non recycled) pulsars (0.6 s) (from PSRcat [1])

For 20 RRATs, it has been possible to also measure the first derivative of the spin period [22] and it was found to be positive. This, in addition to the detection of RRAT J1819−1458 [23] in the X-rays, where it behaves like a quite normal pulsar, allowed us to establish that RRATs are indeed a subclass of the rotational-powered NSs.

The mechanism responsible for the transient behaviour of RRATs is not established yet. Several hypotheses have been proposed, none of which, however, is able to satisfactorily account for all of RRATs' characteristics (see e.g. [24] for a brief overview). Whatever they are, whether they are really only emitting occasional bursts or their steady pulsed emission is just too faint to be seen, RRATs possibly represent a very significant fraction of the entire NS population: the sporadicity of their detectable emission, in fact, results in a substantially increased estimate of the total number of Galactic active radio-emitting neutron stars, with RRATs being probably as many as 4 times the number of normal radio pulsars (but see [25] for a thorough discussion of RRAT and NS populations)!

2.3 Relativistic Binary Pulsars

As mentioned in Sect. 2.1, some radio pulsars—most notably some of the *recycled* radio pulsars—have a high potential for studying many physical and astrophysical phenomena. Thanks to their periodic signals, they can be used as precision cosmic clocks, with a stability that is only slightly worse than that of the last-generation laboratory clocks [26]; they have the capability of providing that stability over time intervals longer than those tested for the best atomic clocks so far. When such a precision clock is placed in a tight orbit around a compact object, such as a white dwarf (WD) or a second neutron star, the times of arrival of the pulses can be significantly affected by the deformed space-time around the compact companion, and relativistic effects become measurable.

2.3.1 Basic Evolution

Recycled pulsars, of which relativistic binaries are a sub-class, are thought to be created in binary systems in which a companion star transfers mass and angular momentum onto the NS surface (*recycling model*) [27]. Through stellar winds or Roche lobe overflow, the NS can hence be accelerated to spin periods of few or few tens of milliseconds, depending on the amount of mass accreted, hence on the initial mass of the companion star.

During the mass-transfer phase, the NS is seen as a bright, pulsating X-ray source, if the magnetic field is high enough to channel the accreted matter onto the polar caps. During the recycling process, the radio emission is switched off, both because, in some stages, the combination of spin and magnetic field is not suitable for sustaining the mechanism responsible for pair production and particle acceleration (hence the system lies below the death-line in the $P - \dot{P}$ diagram of Fig. 2.2), and because the matter engulfing the system quenches the radio emission.

In the mass accretion phase, not only is the star spun up, moving right to left in the $P - \dot{P}$ diagram (or, analogously, in the $P - B_s$ diagram of Fig. 2.8), but its magnetic field is also believed to decrease, either because of accretion itself

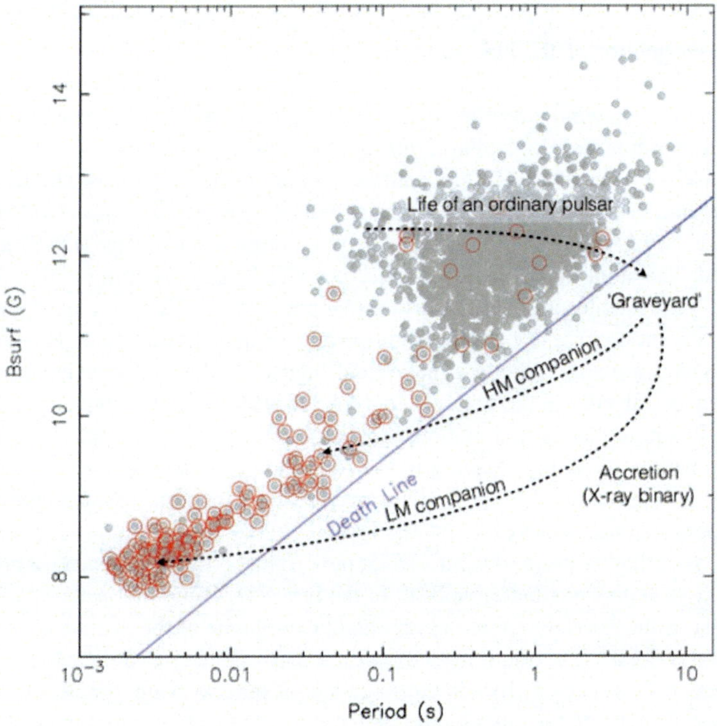

Fig. 2.8 Evolution of a neutron star into a recycled pulsar, in the period—magnetic field diagram (analogous to the $P - \dot{P}$ diagram and obtained using Eq. (2.3)). At birth, after the supernova explosion, the NS spins fast (tens of ms) and has a high magnetic field ($10^{11 \div 13}$ G), hence a high $\dot{P}$. Given the high $\dot{P}$, the pulsar slows down on relatively short time-scales moving left in the diagram; it is still under discussion whether a significant spontaneous decay of the surface magnetic field does [33] or does not [34] occur during this stage. When the death-line is crossed, the emission mechanism is not efficient anymore and the pulsar switches off. If the NS is isolated, it ends its electromagnetic life in the so-called pulsar graveyard. If, on the other hand, it belongs to a binary system with suitable orbital parameters, the pulsar, according to the *recycling model*, can be re-accelerated to millisecond periods by mass accretion from the companion star. During the spin-up process, the pulsar surface magnetic field is also lowered by $\sim 3-4$ orders of magnitude. When the accretion stops, the NS is again above the death line and shines again as a radio pulsar, spinning at a rate of hundreds of Hertz and slowing down at a much smaller rate than that of ordinary pulsars

(*accretion-driven decay*; [28]), or because of the variation in the rotational regime (*spin-driven decay*; [29]), hence making the NS move downwards in the diagram. As a result, when the accretion stops, the neutron star, spinning at a period of only several milliseconds and with a magnetic field of 10^8–10^{10} G, is again above the death-line (or, more precisely, the death-valley) and its radio emission switches on.[2] The typical evolutionary path of a NS, leading to a recycled pulsar, is reported in the period-magnetic field diagram of Fig. 2.8.

2.3.1.1 Pulsars with a Neutron Star Companion

If the progenitor of the NS companion is a massive star, i.e. more massive than 8-10 $M_{\odot}$, the transfer of its swollen external layers towards the compact object is done mainly via wind; accretion can occur also in a short Roche lobe overflow phase or in a common envelope phase, during which the orbit shrinks. The evolution of such a big star is relatively fast and the amount of accreted mass, hence angular momentum, is small. The companion star ends its life in a supernova explosion and, if the binary is not disrupted by the event, a double neutron star (DNS) system is created. The spun-up NS is now a *mildly recycled* pulsar, rotating with a period of tens of milliseconds and with a surface magnetic field, which decreased, during the short accretion phase, by only a couple orders of magnitude ($B_s \sim 10^{10}$ G). Since it possesses a much higher magnetic field (hence $\dot{P}$) than its recycled companion, the new pulsar born in the second supernova explosion slows down and reaches the death-line much faster than the other pulsar in the system. A graphical description of the evolutionary steps leading to the formation of a *double neutron star system* (DNS) is reported in Fig. 2.9 for the specific case of a binary system starting with two main sequence stars of about 13 $M_{\odot}$ and 10 $M_{\odot}$.

We note that, in order to form a DNS, the original binary must survive both of the supernova explosions that give birth to the two NSs. If the first supernova disrupts the system, no accretion phase occurs and the outcome of the evolution will be two isolated ordinary (i.e. non recycled) pulsars that will later end their life, on a time scale of $\sim 10^8$ yrs since birth, in the so-called pulsar graveyard, below the death line (see Fig. 2.8). If, on the other hand, the system is disrupted by the second supernova explosion, one of the two stars will end as an isolated mildly-recycled pulsar (known as a disrupted recycled pulsar, DRP [36]), and the other star as an isolated ordinary pulsar.

---

[2] In recent years a small number of *transitional millisecond pulsars*, binary systems switching from an accretion powered X-ray pulsar state to a rotation-powered radio pulsar state, have been discovered [30–32]. Their existence is a direct observational proof that, at least for a sub-class of binaries, the recycling model is correct.

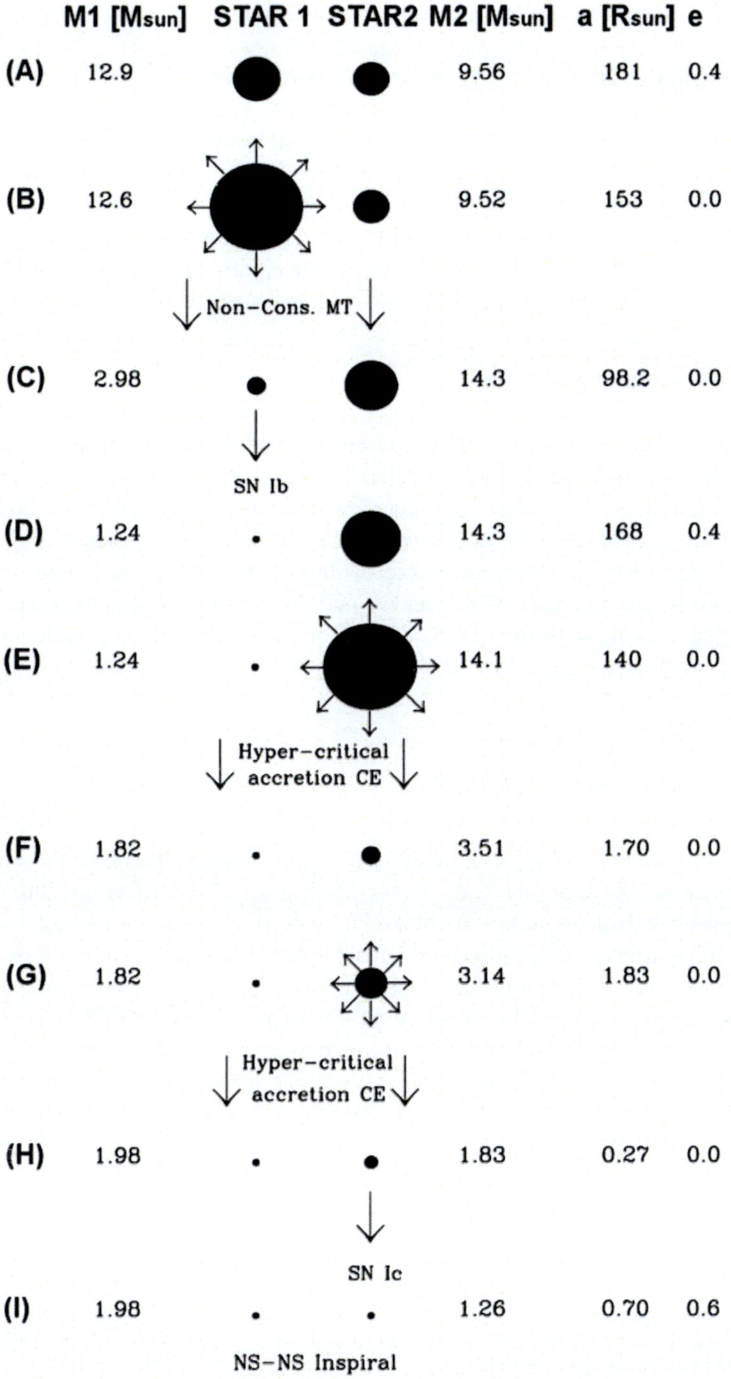

Fig. 2.9 Steps of the evolution of a binary system containing a $12.9\,M_{\odot}$ and a $9.5\,M_{\odot}$ stars (courtesy of K. Belczynski 2020, adapted from [35]). From top to bottom: main sequence—

2.3.1.2 Pulsars with a White Dwarf Companion

If the companion star was born with a mass below ~8 $M_\odot$, its evolution is slower. When the star enters the Red Giant phase, it fills its Roche lobe and starts transferring mass towards the NS. This phase can last up to $\sim 10^8$ yrs during which the matter accreted onto the NS surface at a rate of $\lesssim 10^{-8}\ M_\odot yr^{-1}$ is able to spin it up to millisecond periods. At the same time, its magnetic field gets down to $\sim 10^8$ or 10^9 G. At the end of the accretion phase, the NS parameters are such that its radio emission can turn on again. The NS turns, hence, into a *fully recycled* millisecond pulsar, while its companion evolves into a white dwarf (see Fig. 2.10).

2.3.1.3 Additional Reading

In the sections above, we have only depicted general features of the evolutionary paths leading to the formation of relativistic binary pulsars. Actually, at every step, the evolution can take various alternate branches, according to the initial masses of both stars, to their relative distances and to the details of the supernova explosion(s). Therefore, the final outcomes of the evolution can also cover a wide range of possibilities, whose discussion goes beyond the scope of this chapter. For a more complete picture of (some of) the different possible evolutionary paths, we refer, e.g., to the following reviews: [37–39].

2.3.2 The Current Sample

Here we present the sample of binary pulsars for which at least one post-Keplerian parameter of the orbit has been measured (see Sect. 2.5). Of course this is not a completely exhaustive definition of a relativistic binary, since, prolonging the time-span of the timing observations or increasing the precision of single measurements can lead to the inclusion of already known but not yet completely studied pulsars in the sample; on the other hand, some of the measured parameters may not be related (or not exclusively related) to relativistic gravity but, for instance, to tidal effects (if the size of the companion is not negligible with respect to the size of the orbit, as in the case of non-compact objects) or to the effects of the gravitational potential well of a globular cluster (for a pulsar inside the cluster) or of the Galaxy

---

Fig. 2.9 (continued) main sequence; non conservative mass transfer from primary to secondary star; helium core of the primary—accreted secondary; primary supernova explosion and formation of an ordinary pulsar; non conservative mass transfer from the secondary towards the NS and possible common envelope (CE) phase; further evolution of helium core of the secondary and second mass transfer phase with CE; second supernova explosion; formation of a double NS system with a young and a mildly recycled pulsar

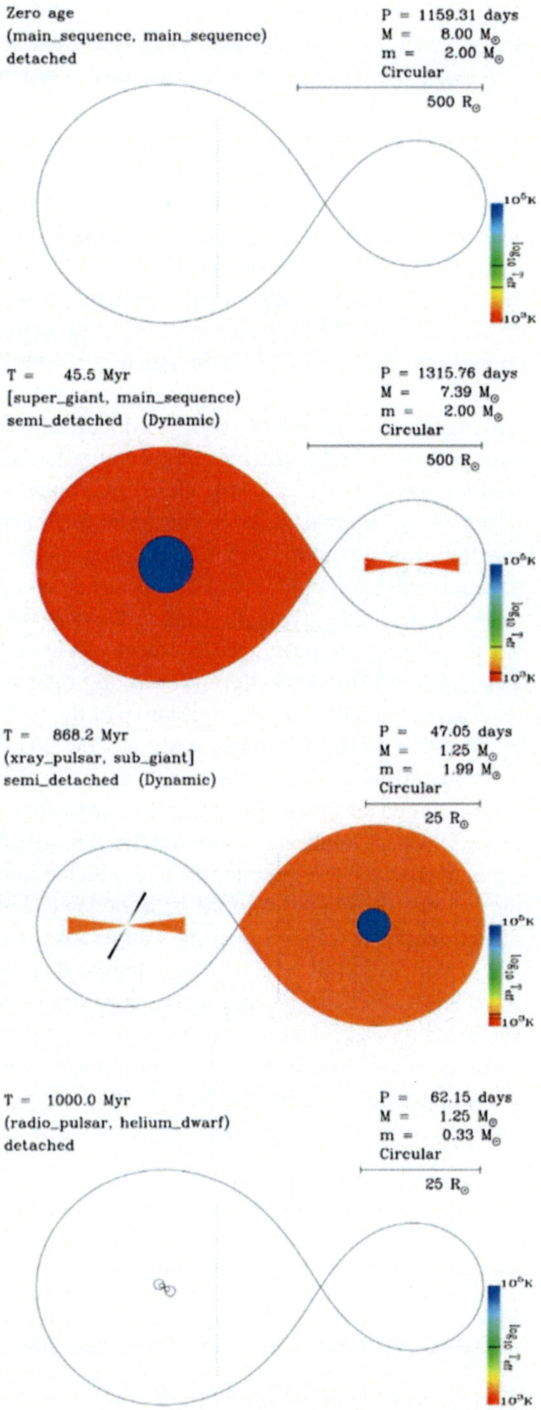

Fig. 2.10 Basic steps of the standard evolution of a binary system containing a 9 $M_\odot$ and a 2 $M_\odot$ stars (from http://www.manybody.org/cgi-bin/starlab/binary_demo.pl). From top to

itself. Following this simple definition, the total number of published binaries with measured PK parameters to date is 72 (see Table 2.1).

Among these, 12 have a neutron star companion: J0453+1559 [40], a DNS system with a large mass asymmetry, J0737−3039A [41] and J0737−3039B [42], composing the first (and so far the only) double pulsar system (see Sect. 2.5.2.2); J1518+4904 [43]; B1534+12 [44, 45] and J1756−2251 [46], the only other DNSs besides the double pulsar for which all 5 post-Keplerian parameters have been measured; J1811−1736 [47, 48]; J1829+2456 [49]; J1906+0746 [50], whose undetected companion is likely the recycled pulsar in the system; B1913+16, the first known binary pulsar, whose discoverers were awarded the Nobel prize for physics in 1993 [51, 52]; J1930−1852 [53], belonging to the widest DNS system known, and J2129+1210C in the globular cluster M15 [54, 55].

The rest of the sample is mostly composed of NS–WD systems, with the remarkable case of J0348−0432 [56], for which timing measurements, coupled with time-resolved optical spectroscopy, allowed astronomers to derive that the neutron star has a mass very close to 2 solar masses ($2.01 \pm 0.04\ M_{\odot}$) [57], which is the highest neutron star mass yet measured with such accuracy. This constrains the equation of state for nuclear matter, effectively ruling out the softest equations of state. As for the masses of the companion star in NS–WD systems, they range from $\sim$1 $M_{\odot}$ (typically Carbon-Oxygen WDs) to 0.02 $M_{\odot}$ (typically Helium WDs). Among the pulsars with a heavy WD companion, two, J1141−6545 [58] and J2305+4707 [59] (believed for a long time to be part of a DNS system) have relatively long spin periods, suggestive of the fact that they have not been recycled by accretion from the companion star. The NS–WD binary J1738+0333 is so far the best pulsar for constraining tensor-scalar theories because of a very precise determination of its orbital decay PK parameter, as will be seen in Sect. 2.5.2.3 [60, 61]. The evolutionary path for these peculiar systems starts from two almost equal mass stars, the more massive of which evolves first and transfers matter onto the companion. The mass of the latter becomes higher than that of the primary, and the subsequent evolution leads to a supernova explosion and to the formation of a NS, which will emit as an ordinary pulsar. The initially more massive star, on the other hand, after losing mass to the companion, has not enough mass anymore to become a second NS and evolves into a heavy WD.

In the sample of Table 2.1, there are also a few binary systems with massive main sequence companions, (e.g. J0045−7319 [62] and J1740−3052 [63]). The post-Keplerian parameters measured for these pulsars, however, are not (or not

Fig. 2.10 (continued) bottom: main sequence—main sequence; first Roche lobe overflow (followed by a supernova explosion that leads to the formation of a NS, possibly shining as an ordinary pulsar); second Roche lobe overflow (the NS shines as an X-ray pulsar and is spun up by the accreted mass and angular momentum); end of the mass transfer: the NS shines as a fully recycled millisecond pulsar and the secondary star is a white dwarf

Table 2.1 Known binary pulsars for which at least one post-Keplerian parameter is published, listed with their orbital (Keplerian and post-Keplerian) parameters

PSR	P_b (days)	$a \sin i$ (lt-s)	T0 (MJD)	Ecc	ω (deg)	$\dot{\omega}(deg/yr)$	$\dot{P}_b$	γ (sec)	$\sin i$	M2 ($M_\odot$)
B0021−72E[a]	2.2568483(9)	1.9818427(4)	51001.7900(8)	3.159E−4(4)	218.6(1)	0.090(16)	4.8E−12(2)	–	–	–
B0021−72H[a]	2.357696895(10)	2.152813(2)	51602.186289(7)	7.0558E−2(1)	110.603(1)	0.06725(19)	−0.7E−12(6)	–	–	–
B0021−72I[a]	0.2297922489(4)	3.8446E−2(1)	–	–	–	–	−0.8E−12(2)	–	–	–
B0021−72J[a]	0.12066493772(13)	4.040	51600.1084250(6)	0	0	–	3.06E−23(5)	–	–	–
J0024−7204O[a]	0.13597430589(10)	58E−2(6)	51600.0757563(3)	0	0	–	1.350E−21(19)	–	–	–
J0024−7204Q[a]	1.1890840496(4)	1.4622043(9)	–	8.03E−5(15)	129.4(12)	–	−1.0E−12(2)	–	–	0.187(13)
J0024−7204R[a]	6.623147751E−2(6)	4.51533E−2(3)	–	0.000104(3)	253(4)	–	0.19E−12(4)	–	–	–
J0024−7204S[a]	1.2017242354(6)	0.7662686(8)	–	0.000398(3)	13.2(5)	0.331(75)	−4.9E−12(4)	–	–	0.175(25)
J0024−7204T[a]	1.126176771(1)	1.338501(5)	–	0.000400(7)	62.5(11)	–	2.5E−12(11)	–	–	0.168(18)
J0024−7204U[a]	0.42910568324(8)	0.5269494(7)	–	0.000146(3)	348.5(16)	1.17(32)	0.66E−12(5)	–	–	0.1705(25)
J0024−7204V[a]	0.2121312098(16)	0.74191(2)	51803.775137(2)	–	–	–	1.053E−19(18)	–	–	–
J0024−7204W[a]	0.1329444433(5)	0.243443(2)	51585.3327393(2)	–	–	–	2.23E−20(8)	–	–	–
J0024−7204X[a]	10.921183545(1)	11.9170570(9)	–	4.5E−7(9)	117(7)	–	6E−12(2)	–	–	–
J0024−7204Y[a]	0.5219386107(1)	0.6685965(7)	–	3.0E−6(3)	270(39)	–	−0.82E−12(7)	–	–	0.17(2)
J0045−7319	51.169451(5)	174.2576(7)	49169.21361(3)	0.807949(3)	115.2540(5)	0.0259(5)	−3.03E−7(9)	–	–	–
J0337+1715	1.629401788(5)	1.21752844(4)	–	0.00069178(2)	97.6182(17)	–	–	–	0.6327(2)	0.19751(15)
J0348+0432	0.102424062722(7)	0.14097938(7)	–	2.4E−6(10)	54(25)	–	−2.73E−12(45)	–	–	0.172(3)
J0437−4715	5.7410459(4)	3.36671444(5)	54501.4671(3)	1.91811E−5(15)	1.363(17)	0.0138(13)	3.728E−12(6)	–	–	0.224(7)
J0453+1559	4.072468649(4)	14.4667896(42)	56344.0031965(9)	0.11251844(8)	223.06965(8)	0.03793(3)	–	–	0.969(3)	1.172(4)
J0514−4002A[a]	18.78517915(4)	36.296588(9)	53623.1550879(4)	0.8879773(3)	82.266550(18)	0.01289(4)	–	–	–	–
J0613−0200[b]	1.198512575184(13)	1.0914409(6)	–	5.40E−6(11)	47.2(12)	–	4.8E−14(11)	–	–	–
J0614−3329	53.5846127(8)	27.638787(2)	55146.821(7)	0.0001801(1)	15.92(4)	–	–	–	–	0.24(4)
J0621+1002	8.3186812(3)	12.0320732(4)	49746.86675(19)	0.00245724(7)	188.774(9)	0.0113(6)	–	–	–	–
J0737−3039A	0.10225156248(5)	1.415032(1)	53155.9074280(2)	0.0877775(9)	87.0331(8)	16.89947(68)	−1.252E−12(17)	0.0003856(26)	0.99974(39)	1.2489(7)
J0737−3039B	0.10225156248(5)	1.5161(16)	53155.9074280(2)	0.0877775(9)	267.0331(8)	16.89947(68)	−1.252E−12(17)	0.00038(5)	–	–
J0751+1807	0.263144270792(7)	0.3966158(3)	–	3.3E−6(5)	83(9)	–	−3.50E−14(25)	–	–	–
B0820+02	1232.404(11)	162.14564(6)	44286.49(5)	0.0118689(8)	332.022(15)	0.0008(9)	–	–	–	–
J1012+5307[b]	0.604672722901(13)	0.58181703(12)	50700.229(13)	1.30E−6(17)	88(8)	–	6.1E−14(4)	–	–	0.16(2)

(continued)

Table 2.1 (continued)

PSR	P_b (days)	$a \sin i$ (lt-s)	T0 (MJD)	Ecc	ω (deg)	$\dot{\omega}(deg/yr)$	$\dot{P}_b$	γ (sec)	$\sin i$	M2 ($M_\odot$)
J1022+1001	7.8051348(11)	16.7654104(5)	50246.7166(7)	9.7229E−5(14)	97.68(3)	0.0097(23)	–	–	0.69(18)	–
J1023+0038[b]	0.1980962019(6)	0.3433494(3)	–	–	–	–	2.5E−10(4)	–	–	–
J1048+2339[b]	0.250519045(6)	0.836122(3)	–	–	–	–	5.209E−19(19)	–	–	–
J1141−6545	0.1976509593(1)	1.858922(6)	51369.8545515(9)	0.171884(2)	42.4561(16)	5.3096(4)	−0.403E−12(25)	0.000773(11)	–	–
J1227−4853	0.287887519(1)	0.668468(4)	–	–	–	–	−8.7E−10(1)	–	0.8(1)	–
J1231−1411	1.860143882(9)	2.042633(3)	55016.8(2)	4E−6(3)	320(40)	–	–	–	–	0.28(10)
B1257+12	25.262(3)	0.0000030(1)	49765.1(2)	0.0	0.0	–	–	–	–	0.060E−6(6)
B1259−63[b]	1236.724526(6)	1296.27448(14)	53071.2447290(7)	0.86987970(6)	138.665013(11)	7.81E−5(3)	1.4E−8(7)	–	–	–
J1417−4402	5.37372(3)	4.876(9)	–	–	–	–	–	–	–	0.33(3)
J1518+0204C[a]	0.08682882865(3)	0.05732042(57)	–	–	–	–	−0.914E−12(23)	–	–	–
J1518+4904[b]	8.6340050964(11)	20.0440029(4)	52857.71084163(17)	0.24948451(3)	342.554394(7)	0.0113725(19)	2.4E−13(22)	–	–	–
B1534+12	0.420737298879(2)	3.7294636(6)	52076.827113263(11)	0.27367752(7)	283.306012(12)	1.7557950(19)	−0.1366E−12(3)	2.0708E−03(5	0.9772(16)	1.35(5)
J1603−7202[b]	6.3086296691(5)	6.8806577(6)	–	9.34E−6(5)	170.1(4)	–	3.10E−13(15)	–	–	–
J1614−2230[b]	8.68661942171(9)	11.29119744(7)	–	1.333E−6(8)	175.9(4)	–	1.3E−12(7)	–	–	–
B1620−26[a,b]	191.44281(2)	64.809460(4)	48728.26242(12)	0.02531545(12)	117.1291(2)	−5E−5(8)	4E−10(6)	–	–	–
J1701−3006B[a]	0.14454541718(58)	0.2527565(28)	55038.62576448(54)	0.0E−5(45)	–	–	−5.12E−12(62)	–	–	–
J1713+0747[b]	67.8251383185(17)	32.34242187(13)	53761.0328(3)	0.0000749402(6)	176.1966(14)	–	0.36E−12(17)	–	0.951(4)	0.286(12)
J1723−2837	0.615436473(8)	1.225807(9)	–	–	–	–	−3.50E−9(12)	–	–	–
J1738+0333	0.3547907398724(13)	0.343429130(17)	–	3.4E−7(11)	336(19)	–	−17.0E−15(31)	–	–	0.181(7)
J1740−3052	231.029630(2)	756.90794(14)	52970.719801(11)	0.57887011(19)	178.646811(17)	0.000112(5)	–	–	–	–
J1741+1351	16.3353478266(7)	11.0033168(5)	–	9.98E−6(3)	204.00(17)	–	–	–	0.99(1)	0.14(3)
J1748−2021B[a]	20.5500072(6)	4.466994(6)	54005.480292(7)	0.5701606(15)	314.31935(13)	0.00391(18)	–	–	–	–
J1750−3703A[a]	17.3342759(7)	24.39312(8)	54003.127812(11)	0.712431(2)	131.3547(2)	0.00548(30)	–	–	–	–
J1750−3703B[a]	3.60511446(5)	2.865858(13)	54002.7705(11)	0.004046(9)	323.07(11)	0.00391(18)	–	–	–	–
J1756−2251	0.31963390143(3)	2.756457(9)	53562.7809359(2)	0.1805694(2)	327.8245(3)	2.58240(4)	−2.29E−13(5)	0.001148(9)	0.93(4)	1.6(6)
J1802−2124	0.698889243381(5)	3.7188533(5)	–	2.47E−6(5)	20.3(20)	–	–	–	0.984(2)	0.78(4)
B1802−07[a]	2.61676335(10)	3.92055(6)	48354.48538(7)	0.21204(3)	164.752(10)	0.0595(20)	–	–	–	–
J1807−2459B[a]	9.9566681588(27)	28.920391(44)	54881.34735776(22)	0.747033198(40)	11.334600(18)	0.018319(12)	–	0.026(14)	0.99715(20)	1.02(17)
J1811−1736	18.7791691(4)	34.7827(5)	50875.02452(3)	0.828011(9)	127.6577(11)	0.0090(2)	–	–	–	–

(continued)

Table 2.1 (continued)

B1820−11	357.76199(5)	200.6720(12)	47260.5438(4)	0.794608(7)	99.1719(11)	7.4E−5(130)	–	–	–	–
J1829+2456	1.176027941(16)	7.238(2)	52848.579775(3)	0.1391412(19)	229.92(2)	0.2919(16)	–	–	–	–
B1855+09	12.3271713831(3)	9.2307819(9)	46432.781(3)	2.170E−5(4)	276.47(7)	–	–	–	0.9987(6)	0.27(3)
J1903+0327	95.174118753(14)	105.5934643(5)	55015.58158859(4)	0.436678409(3)	141.6524786(6)	0.0002400(2)	–	–	0.9760(15)	1.03(3)
J1906+0746	0.16599304683(11)	1.4199620(18)	54288.9298810(2)	0.0853028(6)	76.3320(6)	7.5841(5)	−0.56E−12(3)	0.000470(5)	–	–
J1909−3744[b]	1.533449474406(13)	1.89799118(4)	53631.39(4)	1.14E−7(10)	156(8)	–	5.03E−13(6)	–	–	0.2067(19)
J1910−5959A[a]	0.83711347691(3)	1.2060418(7)	51919.206480057(55)	–	–	–	–	–	0.9997(3)	0.180(18)
J1913+1102	0.206252330(6)	1.754623(8)	56241.029660(5)	0.08954(1)	264.279(9)	5.632(18)	–	–	–	–
B1913+16	0.322997448918(3)	2.341782(3)	52144.90097849(3)	0.6171340(4)	292.54450(8)	4.226585(4)	−2.423E−12(1)	4.307E−06(4)	–	–
J1930−1852	45.0600007(5)	86.890277(7)	56526.642330(3)	0.39886340(17)	292.07706(2)	0.00078(4)	–	–	–	–
J1946+3417	27.01994783(5)	13.8690718(5)	56819.4529124(10)	0.134495389(17)	223.364186(14)	0.001363(9)	–	–	–	–
J1950+2414	22.1913727(10)	14.2199738(11)	55846.0226219(15)	0.07981158(12)	274.4155(3)	0.0020(3)	–	–	–	–
J1957+2516	0.2381447210(7)	0.283349(6)	–	2.8E−5(26)	45(52)	–	12E−12(2)	–	–	–
B1957+20[b]	0.3819666069(8)	0.0892253(6)	48196.0635242(6)	0.00000(4)	–	–	1.47E−11(8)	–	–	–
J2019+2425	76.51163479(2)	38.7676297(8)	50054.6439021(40000)	0.00011109(4)	159.03(2)	–	−3E−11(6)	–	–	–
J2032+4127	16835(835)	13580(6490)	58061(9)	0.964(30)	34(18)	–	–	–	–	15.0(28)
J2051−0827	0.09911025490(4)	0.0450720(3)	–	5.1E−5(10)	36(11)	–	−5.9E−12(3)	–	–	–
J2129−0429[b]	0.6352274131(3)	1.855(42)	–	–	–	–	–	–	–	0.44(4)
B2127+11C[a]	0.33528204828(5)	2.51845(6)	50000.0643452(3)	0.681395(2)	345.3069(5)	4.4644(1)	−3.96E−12(5)	0.00478(4)	–	–
J2129−5721	6.6254930923(13)	3.50056678(14)	–	1.219E−5(8)	197.1(4)	–	7.9E−13(36)	–	–	–
J2215+5135	0.172502105(8)	0.468141(13)	–	1.1E−5(29)	68(92)	–	−439E−12(2)	–	–	0.23
J2222−0137	2.44576469(13)	10.8480239(6)	56001.38381(8)	3.80967E−4(30)	119.900(11)	0.1033(29)	0.20E−12(9)	–	0.99659(30)	1.293(25)
B2303+46	12.33954454(17)	32.6878(3)	47452.560747(17)	0.658369(9)	35.0776(7)	0.0099(2)	–	–	–	–
J2339−0533	0.19309840181(4)	0.611656(4)	–	0.0002102(5)	177(3)	–	−2.226E−20(19)	–	0.87	0.32

Left to right: pulsar name, orbital period, projected semimajor axis, epoch of periastron, eccentricity, longitude of periastron, periastron advance, orbital decay, time-dilation and gravitational red-shift parameter, sine of the inclination of the orbit (equal, in general relativity, to the shape parameter of the Shapiro delay) and mass of the companion star (equal, in general relativity, to the range parameter of the Shapiro delay). Note that for the globular cluster pulsars (denoted by[a]), some of the parameters can be severely affected by the cluster potential well and not be directly related to relativistic gravity. More in general pulsars with positive derivative of the orbital period are certainly affected by non-relativistic effects (mass losses, or kinematic effects due to the influence of the Galactic potential and/or proper motion of the pulsar); pulsars with positive $\dot{P}_b$ are hence marked with[b]. Data extracted from PSRcat v1.57. http://www.atnf.csiro.au/research/pulsar/psrcat/ [1]

exclusively) related to relativistic gravity, hence, strictly speaking, they are not relativistic binaries: the measured $\dot{\omega}$ of J1740−3052, for instance, is likely due to a mixture of relativistic orbital precession and precession due to spin quadrupole of the massive companion [64], while the orbital decay measured in J0045−7319 is likely caused by tidal effects of a retrograde rotation of the companion [65].

For the pulsars for which the measured $\dot{P}_b$ is positive, marked with[b] in Table 2.1, the variation of the orbital period is certainly not relativistic (loss of energy in the form of gravitational waves would lead to a shrinkage of the orbit). For PSR J1023+0038, for instance, [30], the positive value is due to mass loss in the system, while in other cases it is given by kinematic effects (acceleration in the Galactic potential and/or proper motion of the pulsar; e.g. [66]).

2.4 Pulsar Timing Basics

At the time of the discovery of a new pulsar, only the spin period P, dispersion measure DM and position are approximately measured. The latter, for instance, is initially determined with an uncertainty of the order of the size of the beam of the radio telescope used for the discovery. Combining the approximate value of DM and celestial position with a model for the electron distribution in the Galaxy (e.g. [4, 5, 67]), one can also get a rough estimate of the pulsar's distance.

To get a precise measurement of all the rotational, astrometric and, in case of a binary system, orbital parameters of any newly-discovered pulsar, it is necessary to start a follow-up procedure called *timing*, which will be briefly described in this section. The best results of pulsar timing are obtained for the most stable rotators, in general the fastest spinning pulsars. A high flux density, allowing a more precise determination of the "Times of Arrival" (ToAs) of the radio pulses, is also an advantage for accurate timing. As we will see, long-term timing studies of rapidly-rotating recycled pulsars are hence a powerful observational tool for performing a variety of experiments in fundamental physics, in particular for the investigation of relativistic gravity and for gravitational wave detection.

Chapters 7 and 8 of [68], and Chapters 4 and 5 of [69] can be used as references for a full description of pulsar timing procedures. In this chapter, we focus on the basic concepts and main operational steps.

2.4.1 Timing Procedure: Measurement of the ToAs

Experimentally, timing a pulsar means observing it semi-regularly to measure, for each observing epoch, one or more Times of Arrival (ToAs) of a specific recognisable feature—usually the peak—in the pulse phase-averaged radio light curve (the pulse profile). For different pulsars, depending on the specific aim of the experiment, a timing campaign can last from a year—the minimum time-span necessary to get a precise determination of the pulsar position by exploiting the motion of the Earth around the Sun (see Sect. 2.4.2)—to tens of years, if, for instance, pulsars are used as detectors for the nano-Hertz gravitational waves emitted by coalescing supermassive black holes (see Sect. 2.5.3). The spacing between observations also varies: initially, having a very rough determination of the pulsar ephemeris, a denser coverage (up to few observations a day) is needed; as the errors on timing parameters gets smaller (see Sect. 2.4.2), the cadence can be relaxed and one monthly observation is usually enough to derive a coherent timing solution over the entire dataset (see Sect. 2.4.2).

Single pulses from pulsars, with the exception of a few very bright sources, are usually below the level of the noise in the data. To get a precise measurement of a ToA, it is hence necessary to average several pulses. This, besides increasing the signal-to-noise ratio of the pulse, allows us to get a stable pulse shape, the peak of which always falls at the same rotational phase, which allows for a correct determination of the ToA. While single pulses do vary in shape, flux density and exact location of the peak within each given rotation, the pulse shape obtained by summing few hundreds of rotations is extremely stable,[3] so that the determination of the ToA of its peak can be used to precisely keep track of the rotation of the star.

A timing observation is hence obtained by summing a high number of pulse rotations modulo the expected apparent spin period, as predicted from the start time of the observation, the location of the telescope and the best pulsar ephemeris available. A correction for the dispersive delay within the instrument's bandwidth is also necessary and is performed on the basis of the pulsar's DM and the observing frequency. Each pulse profile is time-tagged using an accurate clock (e.g. an hydrogen-maser clock) present at the radio telescope, which is regularly corrected by comparing it with the time distributed by the Global Positioning System (GPS).

In order to remove radio frequency interference (RFI) from the data, the timing observation is usually divided into a number of time sub-integrations of short duration (a few seconds to a few minutes) and into frequency sub-bands (usually up to 1024): in this way, short-duration or narrow-band RFI can easily be mitigated by deleting only the frequency channel or the small section of time affected. Sub-integratios and sub-bands also allow us to verify if the folding period and DM are correct by checking that the pulse falls at the same rotational phase at all times and in all frequency channels. Once all sub-bands and sub-integrations contain only clean pulse profiles all in phase with each other, the data are further summed using the most recent ephemeris available and one (or more, depending on the length of the timing observation and on the brightness of the pulsar) high signal-to-noise profile is obtained.

To obtain the ToAs, the main ingredients of the timing process, these pulse profiles are finally compared—usually through a convolution method—with a

[3]There are rare examples in which even the shape of the integrated profile is variable in time, as for instance in the case of PSRJ0737−3039B in the double pulsar system [41, 42]; these cases require the application of more sophisticated timing methodologies

very high S/N standard profile, typically obtained from summing in phase many observations of the given pulsar. This comparison produces a set of so-called *topocentric* ToAs, which are calculated by adding at the reference time of any pulse profile the fraction of spin period by which the pulse profile is shifted with respect to the standard profile (e.g. when applying a convolution method, this is the fraction of rotational period at which the χ^2 of the convolution is minimised).

As an approximate rule-of-thumb, the characteristic rms uncertainty in the determination of a topocentric ToA scales as the ratio between the width of the pulse and the S/N of the pulse profile.

2.4.2 Timing Procedure: Modelling the ToAs

In order to fully exploit the precisely repetitive nature of the radio signals from a pulsar, it is necessary to be able to account for all the arrived pulses (i.e. all of the NS rotation) from a reference time $t_{\rm ref}$ to a generic time t and, consequently, to predict the times of arrival of all of the following pulses.

Assuming $t_{\rm ref}$ as the time of arrival of one pulse (this can always be done, since the exact reference phase in the pulse itself is not important), we can model the rotational evolution of a pulsar with a power series:

$$N(t) = \nu_{\rm ref} \times (t - t_{\rm ref}) + \frac{1}{2}\dot{\nu}_{\rm ref} \times (t - t_{\rm ref})^2 + \frac{1}{6}\ddot{\nu}_{\rm ref} \times (t - t_{\rm ref})^3 + \ldots . \quad (2.5)$$

Here $N(t)$ is the number of rotations occurred from $t_{\rm ref}$ to t. and $\nu_{\rm ref}, \dot{\nu}_{\rm ref}, \ddot{\nu}_{\rm ref}, \ldots$ are the star's spin frequency and its derivatives at the reference epoch $t_{\rm ref}$.

The objective of pulsar timing is to derive $\nu_{\rm ep}, \dot{\nu}_{\rm ref}, \ddot{\nu}_{\rm ref}, \ldots$ with high enough precision such that $N(t_{\rm next})$ will be very close to an integer (i.e. a complete rotation) for any future time $t_{\rm next}$ of appearance of a radio pulse. Through the timing procedure, in summary, we are able to predict ToAs, through the rotational model of Eq. (2.5), that will match the future observations within the observational uncertainties. To quantify the accuracy of timing, we can make use of the so-called *timing residuals* $R(t_{\rm i}) = N(t_{\rm i}) - n(t_{\rm i})$, where $n(t_{\rm i})$ is the nearest integer to the $N(t_{\rm i})$ derived by the model. If $R(t_{\rm i}) \ll 1$ for all the observed ToAs in a given time span $\Delta t_{\rm span}$, we have reached a satisfactory coherent timing solution over $\Delta t_{\rm span}$.

Note that $R(t_{\rm i}) = R(t_{\rm i}; \alpha_1, \alpha_2, \ldots, \alpha_m)$, where $\alpha_1, \alpha_2, \ldots, \alpha_m$ are the m parameters of the timing model ($\nu_{\rm ref}, \dot{\nu}_{\rm ep}, \ddot{\nu}_{\rm ref}$ in a simple 3-parameter case). Operationally, hence, the timing solution is determined and improved through a multi-parametric least-square fit, by minimising the residual's χ^2:

$$\chi^2 = \Sigma_i \left(\frac{R(t_{\rm i}; \alpha_1, \alpha_2, \ldots, \alpha_m)}{\epsilon_{\rm i}} \right)^2 \quad (2.6)$$

were $\epsilon_{\rm i}$ is the uncertainty on the i-th ToA.

A good timing model should therefore provide residuals randomly scattered around a zero mean value, when plotted against time.

If our telescope were in an inertial reference frame, the procedure described above would be sufficient to completely time an isolated pulsar. Since this is not the case, a preliminary step is needed to convert the topocentric ToAs to the Solar System Barycenter (SSB) reference frame and get *barycentric* ToAs $t_{i,bary}$. Because of the frequency-dependent dispersion delay introduced by the interstellar medium (see Sect. 2.2.1), all ToAs also need to be referred to infinite frequency. The adopted equation for these transformations is:

$$t_{i,bary} = t_i + t_{\text{clock}} - \frac{D}{f^2} + \Delta R_{\odot} + \Delta E_{\odot} + \Delta S_{\odot} \tag{2.7}$$

Here t_{clock} is a term used to convert the time derived from the reference clock at the radio telescope into the uniform atomic time provided by an ideal clock on the geoid. Its value is the sum of various terms (the *clock correction chain*) and it is added retroactively, using tabulated values published by the *Bureau International des Poids et Mesures* (BIPM). The next term in Eq. (2.7) is the correction for the dispersion effects. In particular:

$$D(t_i) = \frac{e^2}{2\pi m_e c} \int_0^d n_e \mathrm{d}l = \mathscr{D} \times \mathrm{DM}(t_i) \tag{2.8}$$

where $\mathscr{D}$=(4.148808$\pm$0.000003) $\times$ 10^3 MHz^2 pc^{-1}cm^3 s is the dispersion constant, d the distance to the pulsar, f the Doppler-corrected observing frequency, and we can note that the observed DM is time dependent.

The fourth term in Eq. (2.7) is the Roemer delay describing the extra path that an electromagnetic wave has to travel to reach the Earth at any given time. It can be written as

$$\Delta R_{\odot} = \frac{\mathbf{r} \cdot \mathbf{n}}{c} + \frac{(\mathbf{r} \cdot \mathbf{n})^2 - |r|^2}{2cd} \tag{2.9}$$

where **n** is the versor pointing from the SSB to the pulsar and **r** is the one pointing from SSB to Earth.

The fifth term is Einstein's delay, accounting for the relativistic time delay (due to the motion of the Earth) and for the gravitational red-shift (due to the masses of the other bodies in the Solar System). Its time derivative is given by:

$$\frac{d\Delta E_{\odot}}{dt} = \sum_k \frac{G m_k}{c^2 d_{k,\oplus}} + \frac{v_{\oplus}^2}{2c^2} - \text{constant}, \tag{2.10}$$

where G is the gravitational constant, m_k are the masses of the other k Solar System bodies, $d_{k,\oplus}$ their distances to the Earth and $v_{\oplus}$ the velocity of the Earth with respect to the SSB.

Finally, the last term is the Shapiro delay [70], which measures the extra time required for an electromagnetic wave to travel through the curved gravitational field of a celestial body. The amplitude of this effect is a function of the angle θ formed by the radius from the pulsar to the telescope with the radius from the telescope to the third body.

For the Sun, the formula is:

$$\Delta S_{\odot} = -\frac{2GM_{\odot}}{c^3} \ln(1 + \cos\theta), \tag{2.11}$$

where θ is the angle at the telescope between the pulsar and the Sun.

If the pulsar is in a binary system, the *pulsarcentric* ToAs (i.e. the ToAs in pulsar proper time) must also be corrected and transformed at the Binary System Barycenter (BSB). Equation 2.7 hence has to include four further terms:

$$t_{i,bary}^{Bin} = t_{i,bary} + \Delta R_{Bin} + \Delta E_{Bin} + \Delta S_{Bin} + \Delta A_{Bin} \tag{2.12}$$

In a Newtonian framework, fitting for the orbital modulations of the ToAs induced, mainly, by the Roemer delay, allows us to derive five Keplerian parameters of the orbit: the binary period P_b, the projected semi-major axis $x = a \sin i$, the eccentricity e, the longitude of the periastron ω and the epoch of the passage at periastron T_0. From these parameters we can derive the *mass function*:

$$f(M) = \frac{(M_c \sin i)^3}{(M_p + M_c)^2} = \frac{4\pi^2 (a \sin i)^3}{G P_b^2} \tag{2.13}$$

where M_p is the pulsar mass and M_c is the companion mass. Assuming a standard value for M_p and for an edge-on orbit ($i = 90°$), we can calculate a lower limit for M_c.

The other terms[4] in Eq. (2.12) describe the deviations from classical physics that occur in a pulsar which experiences strong gravitational fields and/or high orbital velocities in a relativistic binary system. A full description will follow in Sect. 2.5.

From the equations above, we can understand how the timing procedure allows us to also derive, besides the rotational parameters of Eq. (2.5), the astrometric parameters (position and proper motion, through the transformation to the SSB) and orbital parameters, in the case of a binary system (through the transformation to the BSB) of a pulsar.

As mentioned above, the timing procedure is made through a multi-parametric fit of all the spin, positional and orbital parameters, with the aim of minimising the χ^2 of the timing residuals. From a practical point of view, we first create a

[4] ΔA_{Bin}, a parameter describing the changing aberration along the orbit, is actually a classical effect. The parameters describing it, however, are degenerate with some relativistic parameter, hence it is usually described in the context of Post-Keplerian effects.

simple timing model including a minimal set of pulsar parameters (approximate spin period, DM and position). We then create timing residuals by comparing the observed ToAs with ToAs predicted on the basis of our model and we plot the residual as a function of time. We then search for trends in the timing residuals and fit for the model parameter(s) that can account for the trend(s) observed: a linear trend in the timing residuals (panel 2 of Fig. 2.11) means that the model spin period is over- or under-estimated, while a parabolic trend is related to a wrong estimate of the spin period derivative in the model (panel 3). If the position included in our timing model is wrong, the transformation to the SSB will be wrong and this will appear as a sinusoidal trend with a period of a year in the timing residuals. Note that the effect of an error in position and in $\dot{P}$ can be highly covariant until the data span covered by our ToAs is shorter than about 1 year.

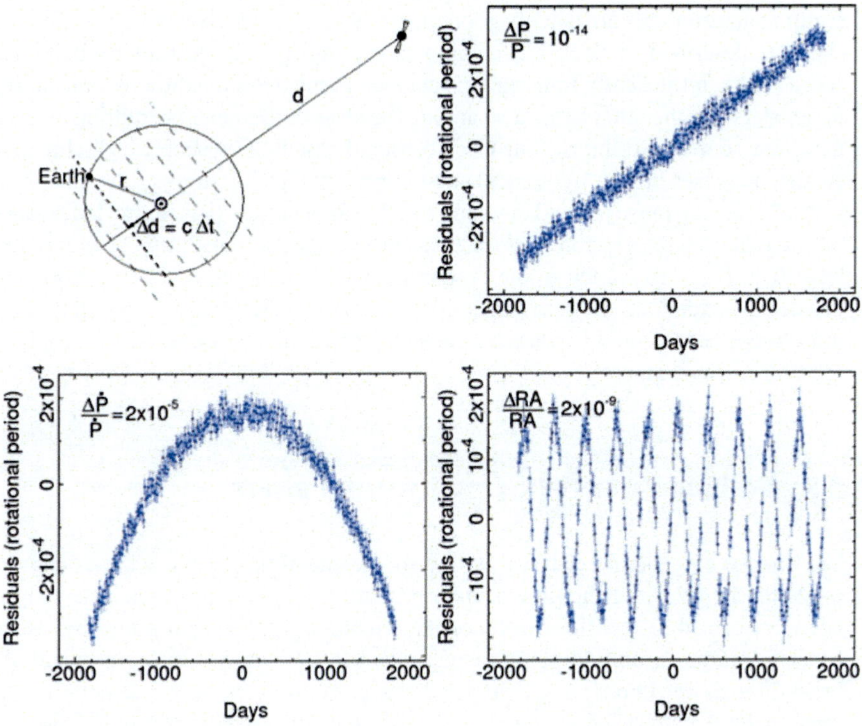

Fig. 2.11 *Upper left panel*: Graphical representation of the origin of the most important term—that due to the Roemer delay—in the formula for calculating the barycentric ToAs. The vector **r** points from the Solar System Barycenter to the radio telescope. In this case, the extra time Δt must be added to the topocentric ToAs to account for the wavefront from the pulsar to cover the distance Δd. *Other three panels*: Typical timing residuals signature for the occurrence of an error in the rotational period (*upper right panel*), in the spin period derivative (*lower left panel*) and in the Right Ascension (*lower right panel*) for the recycled pulsar PSR J0437−4715 (spin period of 5.8 ms)

Looking for these different trends (typically appearing over different time scales in our timing residual plot), one can fit different data-spans for the relevant parameters, recursively improving the timing model and, consequently, the residuals themselves. At the end of a successful timing procedure, we will have flat residuals that are randomly distributed around the zero with a small rms, and a set of precise timing parameters that correctly describe our pulsar. This allows us, for instance, to perform studies of the binary in which the neutron star is included, of its companion star and its gravitational effects, of the multiwavelength counterpart of our radio pulsar (by means of the precise—sub-milliarcsecond—localisation allowed by the timing), of the characteristics of the medium through which its signal has travelled etc. Pulsar timing is, in summary, a very powerful tool.

As a closing remark to this section on timing, we want to point out that, although the very high spin angular momentum of the compact object provides the basis for using the rotation of all the pulsars-like emitters as a clock, not all the radio emitting neutron stars are equally good time keepers. On one hand, there are neutron stars (as described in Sect. 2.2) shining in the radio band only intermittently or sporadically, intrinsically limiting the accuracy in the determination of their timing parameters. On the other hand, also among the class of the steadily emitting pulsars, there are sources exhibiting intrinsic rotational instabilities such as glitches (i.e. sudden increases in spin frequency) and *timing noise*. Observations show that these irregularities are mostly found in young and ordinary pulsars [71]. Recycled pulsars, on the other hand, very rarely show these effects and, at most, only at a very low level (e.g. [72]). Accuracy in the determination of the times of arrival, moreover, besides depending on the flux density of the pulsar, scales with the duration of the pulse itself, hence, roughly, with the star spin period; this, again, favours the rapidly-rotating recycled pulsars.

2.5 Probing Relativistic Gravity with Pulsars

The theory of general relativity is one of the most significant achievements of modern science. Since the time of its formulation (1915), many experiments have been performed to test the revolutionary scientific paradigm proposed by Albert Einstein, starting with Eddington's experiments in 1919, and most recently with the breakthrough detections of gravitational waves from black hole and neutron star binaries by the advanced LIGO and VIRGO detectors. To first approximation, the effects of general relativity can be parametrized as *deviations* from Newtonian gravity. The amplitude of these deviations is related to the strength of the gravitational field; this in turn can be defined by the gravitational potential $\epsilon \equiv GM/(Rc^2)$, for a body of mass M and radius R (c is the speed of light in vacuum and G is the gravitational constant). In the Solar System, ϵ is very small therefore Solar System experiments have only tested the *weak-field* limit of gravity [73]. General relativity has thus far passed all observation tests with flying colors, whether in the *weak-field limit* (Solar System tests) or the in the *strong-field limit* (i.e. in stronger gravitational fields) [74, 75]. As we probe astrophysical systems with yet stronger gravitational fields, we could continue to confirm general relativity (and achieve a higher precision in that confirmation) or find a *breakdown* of the theory.

Why consider a breakdown of general relativity? Since general relativity is, at a fundamental level, not compatible with quantum mechanics, one could expect for it to break down at small scales, or alternatively in strong gravitational fields. The idea is that an overarching theory (a *theory of everything*) could explain all forces of nature, including the electromagnetic, the weak and strong nuclear forces, and gravity. Since this overarching theory is still largely unknown, we cannot directly know what the 'correct' theory of gravity looks like. Instead, a number of *alternative theories* of gravity, some of them inspired by efforts to unify the forces of nature, have been proposed through the years and been constrained by observations (alternative theories contain additional parameters that are not present in general relativity, therefore this additional parameter space can be constrained). Some of these theories have been excluded entirely since they didn't pass observational tests. Other theories have not been excluded but their parameter space has been constrained by observations.

As mentioned earlier, general relativity has been extensively tested and confirmed in the weak-field limit; thus alternative theories of gravity need to be *behaving* like general relativity in that limit. General relativity could instead break down in the strong-field limit. In order to further test general relativity and alternative theories of gravity, we thus need to probe gravity in the strong-field limit (where ϵ is closer to one), that is in *extreme gravity* scenarios. Indeed, there exist alternative theories that pass all tests in the weak-field limit but could be excluded or constrained in the strong-field limit.

Relativistic binary pulsars, in particular where the pulsar is a recycled millisecond pulsar, can be great tools for probing gravity theories. Since the separation between the neutron star and its companion is large (as compared to the neutron star radius) in known binary pulsar systems, the orbital motion of any of the two components takes place in the weak gravitational field of the other binary component. While in general relativity, the orbital motion in these systems does not depend on the gravitational binding energy (or self-energy) of the components, that is no longer true in most alternative theories of gravity. Therefore the effects of these alternate theories can be detected in systems where the self-energy is large, which is the case at the neutron star surface (of the order of $\epsilon \sim 0.2$), therefore orbital motions involving neutron stars are ideal for testing strong gravity. In addition, pulsars, especially millisecond pulsars, are neutron stars that rotate with extreme precision, and their radio wave emission constitutes a stable clock with which to measure the orbital motion of the binary, using the timing methods described in Sect. 2.4. The combination of a large self-energy at the neutron star surface and the possibility of very precisely measuring orbital motions using pulsar timing techniques, makes binary pulsars perfect laboratories for testing strong gravity. The best binary pulsars for testing strong gravity involve those with short orbital periods (meaning high orbital velocities), as well as those whose companion is also a compact object (e.g. DNS or NS–WD binaries; ideally, one would also like to find a pulsar orbiting a black hole!), for which the spacetime curvature is high and which can be approximated to be a point-like mass (therefore simplifying the description). We are also especially interested in pulsars with high timing stability, usually rapidly-spinning millisecond pulsars with a narrow pulse, since they allow the precise testing of general relativity's effects.

General relativity and most alternative theories of gravity are both in the class of *metric theories of gravity* [75], however in alternative theories of gravity, additional fields are present. In these theories, while matter only responds to the curvature described by the spacetime metric (like in general relativity), the spacetime metric itself is influenced by these additional fields (whether of the scalar, vector or tensor form), which are associated with tunable parameters. These parameters can in turn be constrained by observations in the strong-field regime [73–75].

2.5.1 Tests Using PPN Parameters

Metric theories of gravity, whether general relativity or alternative theories of gravity, can be tested using the parametrized post-Newtonian (PPN) formalism. The post-Newtonian terms correspond to deviations from Newtonian physics; there are ten so-called PPN parameters [74] that can be constrained and that correspond to different physical effects such as the existence of preferred frames, preferred locations, the non-conservation of momentum, the non-linear superposition of gravitational effects or the spacetime curvature created by a unit mass. The formalism was initially used to constrain the weak-field limit of gravity in the Solar System [73–76]. It was extended to the strong-field limit (e.g. in the environment around compact objects where ($\epsilon \sim 0.2$)), for a class of tensor-scalar gravity theories [77, 78] with a partial redefinition of the original ten parameters.

Tests of the *Strong Equivalence Principle* (SEP) are particularly interesting: the SEP is satisfied by general relativity, meaning any violation of the SEP implies a violation of general relativity [73]. The SEP consists of both the *Weak Equivalence Principle* (WEP) and the *Einstein Equivalence Principle* (EEP). WEP corresponds to the universality of free fall, that is the trajectory of a free-falling body in a gravitational field should not depend on its internal structure. On the other hand, EEP corresponds to the Lorentz and positional invariance of non-gravitational experiments (i.e. the outcomes of these experiments should not depend on the time and place where they take place, or on the velocity of the reference frame). The SEP can be tested by monitoring the trajectories of two masses in a gravitational field, and check for any differences. The SEP would be violated if differences, or "polarizations" were detected due to different self-energies of the two bodies (as predicted in alternative theories of gravity), as for example the orbital trajectories of the Earth and the Moon in the gravitational field of the Sun: this is called the *Nordtvedt effect* or *gravitational Stark effect* [79]. Strong constraints on PPN parameters have been achieved in the Solar System using Lunar Laser Ranging (LLR) experiments. However this can be tested in the strong-field limit of gravity as well, for example looking at the possible "polarization" of NS–WD binaries in the gravitational field of the Galaxy. While the orbital parameters of single NS–WD binaries are usually not fully known, this test can instead be performed using a statistical approach with a number of binaries. The best constraint so far of the Nordtvedt effect was done using 21 NS–WD binaries [80], leading to a 95% confidence upper limit of 5.6×10^{-3} on $\Delta = (M_{gr}/M_{in})_{\rm NS} - (M_{gr}/M_{in})_{\rm WD}$, where $(M_{gr}/M_{in})_i$ is the ratio between gravitational and inertial mass of the i-th body. A more recent constraint of $\Delta < 4.6 \times 10^{-3}$ was placed using a sample of 27 binaries [81].

This is not the only way to test the SEP; other tests can be done to constrain various PPN parameters (a good summary of the PPN formalism can be found in [82]). For example, a non-zero α_3 implies the existence of a preferred frame and non-conservation of momentum. An upper limit on $\hat{\alpha_3}$ (the strong-field generalization of the original PPN parameter α_3) is 4×10^{-20} at the 95% confidence limit, and was achieved using an ensemble of NS–WD binaries [80]. This constraint is especially interesting since it is much more constraining (by 13 orders of magnitude!) than solar system tests using the Earth and Mercury. The timing analysis of the NS–WD binary PSR J1738+0333, coupled with the analysis of pulse profile stability from the isolated pulsars B1937+21 and J1744-1134, has led to constraints on the parameters $\hat{\alpha_1}$ and $\hat{\alpha_2}$, which involve a violation of the Lorentz invariance: $\hat{\alpha_1} < (-0.4^{+3.7}_{-3.1}) \times 10^{-5}$ and $\hat{\alpha_2} < 1.6 \times 10^{-9}$, both at the 95% confidence limit [60, 83–85]. Additionally, pulse profile stability from B1937+21 and J1744-1134 has also enabled the constraints on the parameter $\hat{\xi} < 3.9 \times 10^{-9}$ (95 % CL). The constraints on $\hat{\alpha_1}$, $\hat{\alpha_2}$, $\hat{\alpha_3}$ and $\hat{\xi}$ are all better than those achieved with solar system tests.

Other tests however are less constraining than in the solar system. For example, the parameter $\hat{\zeta_2}$, which represents non-conservation of momentum, has an upper limit of 4×10^{-5} using PSR B1913+16 [85]. The presence of preferred positions and times in the universe can also lead to variations in fundamental constants such as the gravitational constant G. The best constraint on variations of G is from a recent timing study of the fast and extremely precise millisecond pulsar J1713+0747: $|\dot{G}/G| < (-0.6 \pm 1.1) \times 10^{-12}\mathrm{yr}^{-1}$ at the 95% confidence limit [86].

The recent discovery of the triple system PSR J0337+1715 consisting of a white dwarf orbiting an inner NS–WD pair [87], will lead to significant improvements in our constraints on the SEP. The masses of all three bodies have been precisely determined, as well as the inclination of the orbits. The gravitational field of the outer white dwarf is larger than the Galaxy's gravitational field by approximately six orders of magnitude; therefore by studying the accelerations of the two inner compact objects in the gravitational field of the outer white dwarf, we can improve the constraint on the aforementioned parameter Δ by several orders of magnitude. This is therefore the best system known to date for constraining the SEP.

2.5.2 Tests Using PK Parameters

Relativistic effects are strongest when the pulsar's companion is another compact object, whether a neutron star or a white dwarf. In those binaries, the curvature of spacetime is large around each of the two compact objects, while in close orbits, the orbital velocity is very high. Both of these conditions lead to strong relativistic effects, which can be detected by the precise monitoring of times of arrival of radio pulses from the pulsar, and which can be compared to the predictions of general relativity and alternative theories. Damour and Deruelle [88, 89] have developed a successful framework for constraining gravity theories in relativistic binary systems. They introduced *post-Keplerian* (PK) parameters to describe the relativistic effects in the orbital motion of relativistic binaries, such as those composed of a pulsar and an additional compact object. PK parameters are determined by the chosen gravity theory, therefore the PK parameters of each theory can be constrained using precise measurements. They are a function of the masses of the system as well as the Keplerian parameters of the binary system. However since the Keplerian parameters are determined to very high precision, the PK parameters are essentially just a function of the two masses. Therefore, measuring two PK parameters leads to the complete determination of the two masses in the system (for any given gravity theory). In some binary systems, more than two PK parameters can be measured. In that case, the additional $N_{\rm PK} - 2$ parameters lead to independent tests of the chosen gravity theory. In a mass-mass plot for the two binary compact objects, lines are plotted for each measured PK parameter. If no overlap is found between allowed regions of each parameter, the gravity theory has to be rejected.

In general relativity, the five most commonly-used PK parameters are described by the following equations [89–91]:

$$\dot{\omega} = 3\left(\frac{P_b}{2\pi}\right)^{-5/3} (T_\odot M)^{2/3}\,(1-e^2)^{-1} \tag{2.14}$$

$$\gamma = e\left(\frac{P_b}{2\pi}\right)^{1/3} T_\odot^{2/3}\, M^{-4/3}\, m_2\,(m_1+2m_2) \tag{2.15}$$

$$\dot{P}_b = -\frac{192\pi}{5}\left(\frac{P_b}{2\pi T_\odot}\right)^{-5/3}\left(1+\frac{73}{24}e^2+\frac{37}{96}e^4\right)(1-e^2)^{-7/2}\,\frac{m_1 m_2}{M^{1/3}} \tag{2.16}$$

$$r = T_\odot m_2 \tag{2.17}$$

$$s = x\left(\frac{P_b}{2\pi}\right)^{-2/3} T_\odot^{-1/3}\, M^{2/3}\, m_2^{-1} \equiv \sin i \tag{2.18}$$

where m_1 and m_2 are the two star masses, $M = m_1 + m_2$, $x = a \sin i$ and $T_\odot \equiv GM_\odot/c^3 = 4.925490947\,\mu$s. The PK parameter $\dot{\omega}$ is associated with the advance of the periastron, γ accounts for gravitational red-shift and time dilation, $\dot{P}_b$ is the orbital damping which measures the rate at which the orbital period decreases due to emission of gravitational radiation. r and $s \equiv \sin i$ correspond to the *rate* and the *shape* of the Shapiro delay [70], which is caused by the spacetime deformations around the companion.

There are no extra parameters in these equations: the PK parameters only depend on the two masses in general relativity. However in alternative theories, extra theory-dependent parameters will appear in the definitions of the PK parameters (see [73, 75] for a few examples).

2.5.2.1 Double Neutron Star Binaries

Tests using PK parameters were applied for the first time to the so-called *Hulse-Taylor pulsar*, i.e. the binary system B1913+16, which was the first discovered double neutron star (DNS) binary [92]. One of the two neutron stars is a pulsar with a 59-ms period; the orbit is highly eccentric ($e = 0.61$) with an orbital period of 7.8 h. Observing this system over a number of years has allowed the precise determination of three PK parameters: $\dot{\omega}$, γ and $\dot{P}_b^{\rm int}$ (the observed, intrinsic value for the orbital decay). While the first two PK parameters ($\dot{\omega}$ and γ) allowed the determination of two neutron star masses [93], having a third PK parameter allows us to check the validity of general relativity. In particular, inserting the two determined masses into Eq. (2.16), we obtain a precise determination for $\dot{P}_b^{\rm GR}$, which is the expected value for the orbital decay according to general relativity. One can thus compare $\dot{P}_b^{\rm int}$ obtained from observations with $\dot{P}_b^{\rm GR}$. After more than 30 years of observing this binary pulsar, the agreement with general relativity is now at the 0.2% level [93]. This spectacular result provided the first indirect evidence for the emission of gravitational waves from any astrophysical system (see Fig. 2.12). It also showed that the internal structure of the neutron stars did not affect the dynamics and energy loss of the DNS binary system, lending more credence to the effacement of the neutron star interiors and therefore the Strong Equivalence Principle of general relativity [94].

We note that the intrinsic rate of orbital decay, while primarily determined from the fitting of observed TOAs to the pulsar model, has to be corrected for a number of factors, including vertical acceleration and differential rotation in the Galactic potential. As far as B1913+16 is concerned, our current limited knowledge of the shape of the Galactic potential is the largest limiting factor in improving the accuracy of general relativity tests [95–97].

As mentioned earlier, a minimum of 3 PK parameters is necessary for providing a test of relativistic gravity (the first 2 PK parameters help determine the 2 masses). Besides PSR B1913+16 for which 3 PK parameters were determined, PSR B1534+12 is a 38-ms pulsar in a relativistic, DNS binary [98] for which 5 PK parameters have been determined. Its longer orbital period and smaller eccentricity make it less relativistic than the Hulse-Taylor pulsar, but it has a higher orbital inclination, a stronger flux density and a narrower pulse width. An even more interesting, relativistic system for which 5 PK parameters have been determined is the so-called *Double Pulsar* PSR J0737-3039. This is the subject of the next section.

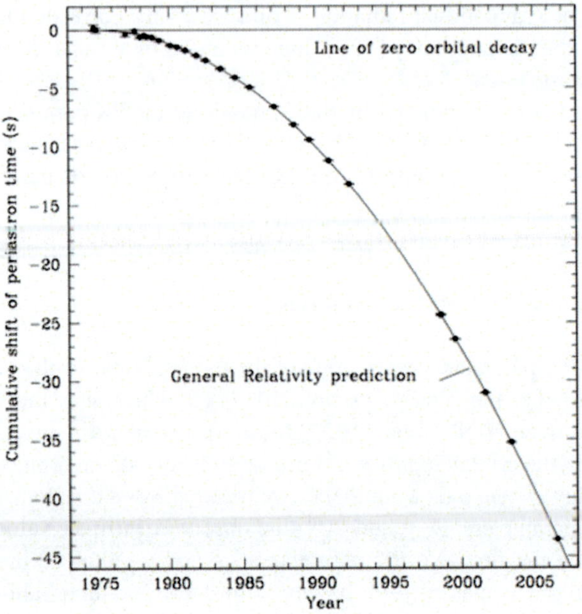

Fig. 2.12 Cumulative shift in the time of transit at periastron for the double neutron star system PSR B1913+16, as a function of the time since the discovery of the binary system. The solid line represents the prediction of general relativity based on the masses of the two neutron stars [$M_p = (1.4398 \pm 0.0002) M_\odot$; $M_c = (1.3886 \pm 0.0002) M_\odot$], as determined from the observation of the PK parameters $\dot{\omega}$ and γ. (Courtesy of J. Weisberg)

2.5.2.2 The Unique Case of the Double Pulsar

The 2003 breakthrough discovery of the first and only *Double Pulsar* [41, 42], a DNS binary where both neutron stars have been visible as pulsars, has greatly enhanced the study of compact objects and relativistic gravity. Its orbital period is only 2.4 h and its eccentricity is $e \sim 0.09$. The first pulsar PSR J0737-3039A (psrA) is a millisecond pulsar with a rotation period of 22 ms, while PSR J0737-3039B (psrB) is a young pulsar with a rotation period of 2.7 s. The Double Pulsar is thus far the best laboratory for testing the limits of general relativity in the strong-field regime, due to four main factors:

- The two neutron stars travel a high velocity along their orbit (at about one thousandth of the speed of light), therefore the system is highly relativistic.
- The orbit is really tight, that is the two neutron stars are close to each other (roughly double the Earth-Moon distance), therefore the system is highly relativistic (this is related to the first point)
- The orbital inclination is high (more than 88°). This makes it easy to determine certain PK parameters.

- Both neutron stars have been observed as pulsars: radio pulses have been monitored. This provides two clocks with which to determine PK parameters.

This particular set of circumstances enabled a quick determination (within 1 week) of the relativistic advance of periastron $\dot{\omega}$ of psrA [41]. Within 6 months of monitoring, an additional 3 PK parameters could be determined: γ, the range r and the shape s of the Shapiro delay (see Fig. 2.13). Monitoring both pulsars enabled the determination of the size of the orbit, which, using Kepler's third law, led to the determination of the mass ratio R [42]. This was the first time the mass ratio was determined in a double neutron star system. The mass ratio is usually not affected by self-field effects and is therefore independent of the chosen gravity theory, while the PK parameters do depend on the chosen theory [89]. After a year of observations, a fifth PK parameter (the orbital decay $\dot{P}_b$) was determined. This confirmed that this system loses energy by emitting gravitational waves in such a way that it shrinks at a pace of 7 mm/day, and should merge within 85 million years. This system is therefore useful for estimating the rate of DNS mergers [41].

This system thus provides 5 PK parameters and the mass ratio R. In addition, the Double Pulsar allows the measurement of an additional relativistic effect: the relativistic or *geodetic* precession of the spin axis of psrB. This has been measured using short eclipses in the radio emission from psrA (when psrB passes by). The dip in the signal from psrA was shown to vary with the rotational phase of psrB [99, 100]. Four years of observations led to a measured rate of: $\Omega_B = (4.77^{+0.66}_{-0.65})^\circ\ \mathrm{yr}^{-1}$ [101], which is compatible, at the 13% level, with the prediction of general relativity [102], $\Omega_B^{\mathrm{GR}} = (5.0734 \pm 0.0007)^\circ\ \mathrm{yr}^{-1}$. While this effect has been observed in other binary pulsars (like PSR B1534+12 [64], PSR B1913+16 [103–105], PSR J1141-6565 [106]), the Double Pulsar constraints allow a test of both general relativity and

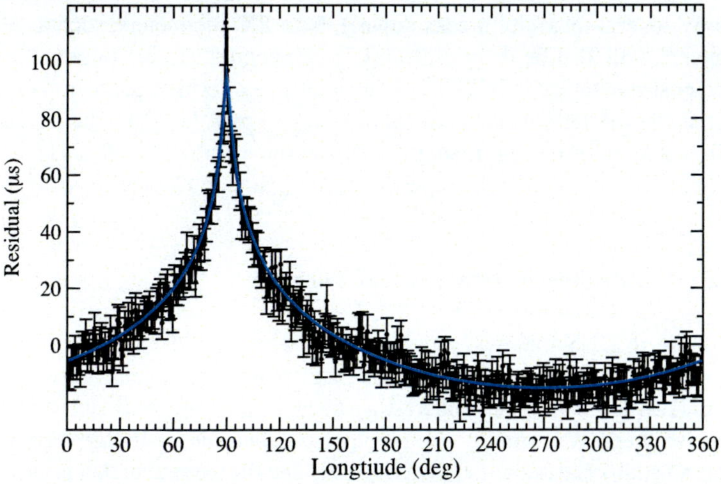

Fig. 2.13 The signature of Shapiro delay in a set of TOAs from the pulsar PSR J0737−3039A in the Double Pulsar system (figure courtesy of M. Kramer 2020)

alternative theories of gravity. We have:

$$\Omega_B = \left[\frac{x_A x_B}{s^2}\right]\left[\frac{8\pi^3}{(1-e^2)P_b^3}\right]\left[\frac{c^2\sigma_B}{G_{AB}}\right], \tag{2.19}$$

where x_A and x_B are the projected semi-major axes of the orbits of the two neutron stars, σ_B is a (theory dependent) *strong-field spin-orbit* coupling constant and G_{AB} is the (theory dependent) gravitational constant, while s and P_b are the usual PK parameters.

With precise measurements of s, P_b, x_A, x_B and Ω_B, one can solve for σ_B/G_{AB}. This is only possible if one of the two neutron stars has a high spin precession rate, a highly-inclined orbital plane and both neutron stars are observed as pulsars. So far, only the Double Pulsar allows this; we determine that: $\left[\frac{c^2\sigma_B}{G_{AB}}\right] = 3.38^{+0.49}_{-0.46}$. This therefore provides a new strong-field test, testing both general relativity and alternative theories of gravity.

The measurement of Ω_B can be added in the mass-mass diagram of the Double Pulsar, providing further constraints on theories of gravity. With 5 PK parameters, the mass ratio R and psrB's spin precession Ω_B, we obtain seven relativistic constraints for the Double Pulsar. Since two parameters are needed to determine the two neutron masses, we are left with 5 independent tests of general relativity! The mass functions for each of the two neutron stars provide additional, classical constraints on the mass-mass diagram (see Fig. 2.14). Taking into account all of these constraints, the Double Pulsar currently confirms general relativity with an uncertainty of 0.05%.

As the number of observations increases with time, the timing solution for the Double Pulsar keeps getting better, PK parameters are determined with higher precision and tests of general relativity are validated at yet a higher level (until, possibly, general relativity breaks down!). New PK parameters, such as aberration parameters, will also be determined [107]. We should be able to see effects at the second post-newtonian level (2PN) in the advance of orbital periastron $\dot{\omega}$, which in turn could help determine the moment of inertia of psrA [108] and help constrain the equation of state of nuclear matter in neutron star interiors [109]; this is especially attainable with new radio telescopes such as the Square Kilometre Array.

2.5.2.3 Constraints on Tensor-Scalar Theories

In addition to DNS binaries, pulsars with a white dwarf companion can also be interesting for testing theories of gravity, and provide complementary results. NS–WD binaries are particularly interesting because they are well-suited to constrain alternative theories of gravity such as the so-called *tensor-scalar* theories, which include an additional scalar field ψ [78, 110]. The PK parameter that is most affected by the presence of a scalar field is the orbital decay $\dot{P}_b$. The presence of scalar fields in the theory leads to the emission of *dipolar* gravitational waves, which

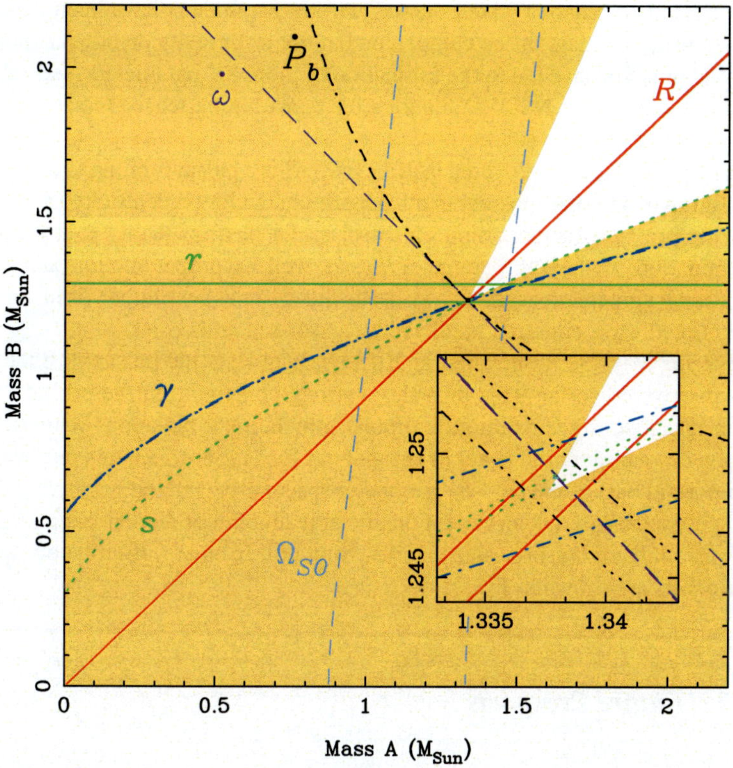

Fig. 2.14 Mass—Mass diagram for the Double Pulsar system J0737−3039. The shaded regions are those that are excluded by the Newtonian mass functions of the two neutron stars. Relativistic constraints are shown as pairs of lines enclosing permitted regions from the observation of the mass ratio R, the PK parameters and the precession rate Ω_B of the spin axis of psrB. The Double Pulsar currently confirms general relativity with an uncertainty of 0.05% (courtesy of Michael Kramer, 2017)

would affect $\dot{P}_b$. For two equal masses in the binary (such as in DNS binaries), the dipolar term almost vanishes; therefore double neutron star binaries are not the best systems for constraining tensor-scalar theories. Instead, the mass asymmetry in pulsar—white dwarf binaries makes them particularly well-suited.

Tensor-scalar theories include the *tensor-mono-scalar* class of theories, which are parametrized by the constants α_0 and β_0, which describe the couplings between matter and scalar field in the following coupling function:

$$a(\psi) = \alpha_0 \psi + 0.5 \beta_0 \psi^2 \tag{2.20}$$

We recover general relativity for $\alpha_0 = \beta_0 = 0$ and Jordan–Fierz–Brans–Dicke theory [111] for $\beta_0 = 0$ and $\alpha_0^2 = 1/(2\omega_{BD} + 3)$, where ω_{BD} is the Brans–Dicke parameter [112].

In the pulsar binaries where one or two PK parameters have been determined, one can infer the mass of the companion but not test gravity theories. A number of NS–WD binaries have led to the estimation of more PK parameters (see Table 2.1). The most interesting NS–WD binaries for constraining tensor-scalar theories are J0348+0432 and J1738+0333.

PSR J1738+0333, discovered in 2011 [60, 61] is a pulsar orbiting a low-mass WD companion (0.18 solar masses) in a 8.5 h orbit and a low eccentricity (3×10^{-7}). It currently provides the best limit on tensor-scalar theories thanks to a very precise determination of the PK parameter $\dot{P}_b$, as well as proper motion and parallax. The strong agreement with GR leads to the current best upper limit on dipolar gravitational wave emission predicted by tensor-scalar theories.

PSR J0348+0432, discovered in 2013 [113, 114], is the pulsar with the highest mass discovered so far. This provides a strong constraint on the orbital decay $\dot{P}_b$ and in turn a strong constraint on dipolar gravitational radiation. While the upper limit on dipolar radiation is not as stringent as for J1738+0333, this pulsar rules out an important part of the α_0—β_0 parameter space, especially at negative β_0's, since massive systems predict an exceptionally large amount of dipolar radiation.

Constraints on tensor-scalar theories from pulsar binary observations and solar system tests are shown in Fig. 2.15.

2.5.3 Future Prospects

Our constraints on general relativity and alternative theories of gravity will improve thanks to longer datasets for interesting pulsars (such as the Double Pulsar and NS–WD binaries) at current telescopes and at new telescopes such as the Large European Array for Pulsars (LEAP) [116], the Five-hundred meter A-spherical Single-aperture (FAST) radio telescope (China) [117] and the Square Kilometre Array (SKA) and precursosr, in particular MeerKAT [118]. These new telescopes will also allow the discovery of new binary systems (DNS binaries or NS–WD binaries), in particular at the SKA [119]. The most interesting type of binaries—that are yet to be discovered—are pulsar—black hole (PSR–BH) binaries. They are expected to be rare: we could find a pulsar in orbit around the supermassive black hole at the Galactic Center [120], or around a stellar-mass black hole [121]. We could also possibly find pulsars in globular clusters and which orbit an intermediate mass black hole [122].

As we have seen in the previous section (Fig. 2.15), PSR–BH binaries are expected to provide the best constraints on tensor-scalar theories. But that's not it. Using a pulsar (as a clock) around a spinning black hole will give us unprecedented knowledge about black holes, such as their mass $M_{\rm BH}$, angular momentum $S_{\rm BH}$ and quadrupole moment $Q_{\rm BH}$, allowing us to test the *Cosmic Censorship Conjecture* (which excludes the existence of naked singularities) and the *No Hair theorem* (which states that a black hole is only described by its mass, spin and charge) [123, 124].

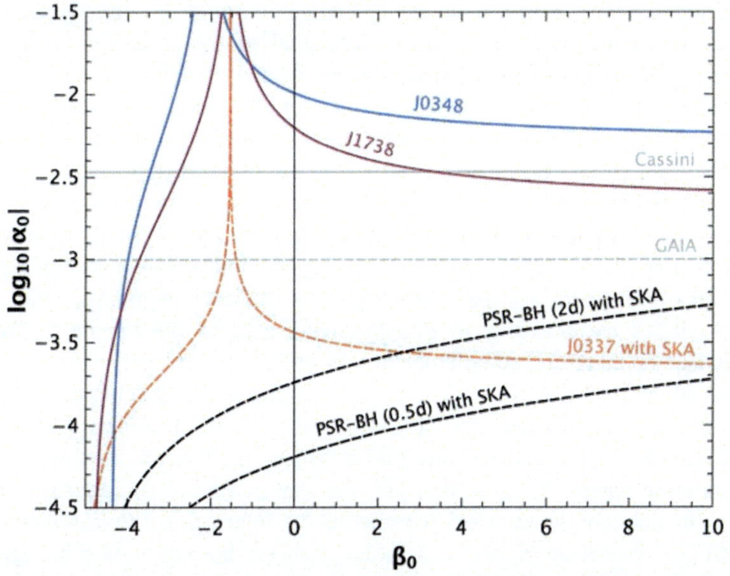

Fig. 2.15 Constraints on the $|\alpha_0|$ and β_0 parameters of the coupling function $a(\psi) = \alpha_0\psi + 0.5\beta_0\psi^2$ in tensor-scalar theories including a scalar field ψ. At ($|\alpha_0| \to 0$; $\beta_0 = 0$), we have general relativity. The vertical axis corresponds to Brans–Dicke theory. The allowed regions are shown below the solid lines. The two most stringent constraints from pulsars are from the NS–WD binaries J0348+0432 and J1738+0333. They are comparable to Solar System tests such as the Cassini spacecraft and the future GAIA astrometric satellite. Dashed lines are expected constraints from observations of the triple system PSR J0337+1715 with the SKA. We also plotted the expected limits from future discoveries and observations of pulsar—black hole binaries with the SKA (with orbital periods of $P_b = 2d$, and $P_b = 0.5d$) (figure courtesy of Norbert Wex, 2017) [115]

Since the event horizon of a black hole shrinks as the black hole spins up, there should be maximum spin for a black hole of any given mass; this is what is implied by the Cosmic Censorship Conjecture. On the other hand, the No Hair Theorem implies that the quadrupole moment of a black hole $Q_{\rm BH}$ must be expressed as a function of its mass $M_{\rm BH}$ and spin $S_{\rm BH}$. We define the dimensionless spin and quadrupole parameters χ and q as:

$$\chi = \frac{c}{G}\frac{S_{\rm BH}}{M_{\rm BH}^2} \; ; \; q = \frac{c^4}{G^2}\frac{Q_{\rm BH}}{M_{\rm BH}^3} \tag{2.21}$$

In general relativity, a black hole should satisfy:

$$\chi \leq 1 \; ; \; q = -\chi^2 \tag{2.22}$$

This can be tested by timing PSR–BH binaries.

Pulsars can also be used as *cosmic clocks* in order to directly detect low-frequency gravitational waves: this is the goal of the so-called *Pulsar Timing Arrays* (PTAs) [125]. PTAs consider the Earth and particular pulsars as pairs of test masses; the spacetime metric is perturbed by the passage of low-frequency (or long-wavelength) gravitational waves coming from distant galaxies. This metric perturbation causes delays in the arrival times of radio pulses from pulsars at radio telescopes on Earth, leaving a signature in the timing residuals of pulsars. We expect to detect a background of gravitational waves from an ensemble of supermassive black hole binaries by studying the correlations between the timing residuals of different Earth-pulsar pairs [126] of different angular size and forming a PTA. PTAs are sensitive to gravitational waves in the nanohertz frequency range, and is therefore complementary to other gravitational wave experiments such as advanced LIGO/VIRGO and LISA which are aimed at higher gravitational wave frequencies. The current PTA efforts include the European Pulsar Timing Array (EPTA), which manages the LEAP project, the NANOhertz Observatory for Gravitational waves (NANOGrav) and the Parkes Pulsar Timing Array (PPTA). These three collaborations work together as the International Pulsar Timing Array (IPTA) [8]. No detection has been achieved so far, but limits on a background of gravitational waves have been refined and a detection is predicted within the next few years [127–129]. The effort in detecting gravitational wave from supermassive black hole binaries is especially encouraged by the recent, breakthrough detection by advanced LIGO and VIRGO of gravitational waves from coalescing stellar mass black hole binaries (e.g. [130]) as well as a DNS binary [131]. If not detected earlier, we expect with high confidence that the SKA or FAST will lead to a detection of the background and help to understand these systems [132–134].

A century after Einstein's formulation of general relativity, pulsars are crucial to testing general relativity and alternative theories of gravity, testing fundamental physics thanks to its constraints on the equation of state of nuclear matter, and could soon lead to the first detection of low-frequency gravitational waves from supermassive black hole binaries.

Acknowledgments MB, DP and AP acknowledge the collaborators at INAF-Osservatorio di Cagliari and the international collaborators who are contributing to make the study of pulsars an always stimulating and often surprising activity.

References

1. R.N. Manchester, G.B. Hobbs, A. Teoh, M. Hobbs, The Australia telescope national facility pulsar catalogue. Astron. J. **129** 1993 (2005)
2. A. Pellizzoni et al., Discovery of new gamma-ray pulsars with AGILE. Astrophys. J. **695**, L115–L119 (2009)
3. A.A. Abdo et al., The second Fermi Large Area Telescope catalog of gamma-ray pulsars. Astrophys. J. Suppl. Ser. **208**, 17 (2013)
4. J.M. Cordes, T.J.W. Lazio, NE2001. I. A New Model for the Galactic Distribution of Free Electrons and its Fluctuations (2002). arXiv:astro-ph/0207156

5. J.M. Yao, R.N. Manchester, N. Wang, A new electron-density model for estimation of pulsar and FRB distances. Astrophys. J. **835**, 29 (2017)
6. J.L. Han, R.N. Manchester, A.G. Lyne, G.J. Qiao, W. van Straten, Pulsar rotation measures and the large-scale structure of the galactic magnetic field. Astrophys. J. **642**, 868–881 (2006)
7. M. Colpi, A. Possenti, A. Gualandris, The case of PSR J1911-5958A in the outskirts of NGC 6752: signature of a black hole binary in the cluster core? Astrophys. J. **570**, L85–L88 (2002)
8. G. Hobbs et al., The international pulsar timing array project: using pulsars as a gravitational wave detector. Classical and Quantum Gravity **27**, 084013 (2010)
9. G. Pavlov Central compact objects in supernova remnants, in *Neutron Stars at the Crossroads of Fundamental Physics* (2005). http://www.physics.ubc.ca/~heyl/ns2005
10. R. Turolla, Surface emission from isolated neutron stars, in *COSPAR, Plenary Meeting*, vol. 37 (2009), 3243
11. M.A. McLaughlin et al., Transient radio bursts from rotating neutron stars. Nature **439**, 817–820 (2006)
12. M. Kramer, A.G. Lyne, J.T. O'Brien, C.A. Jordan, D.R. Lorimer, A periodically active pulsar giving insight into magnetospheric physics. Science **312**, 549–551 (2006)
13. S. Mereghetti, The strongest cosmic magnets: soft gamma-ray repeaters and anomalous X-ray pulsars. Astron. Astrophys. Rev. **15**, 225 (2008)
14. A. Possenti, M. Burgay, The role of binary pulsars in testing gravity theories, in *Gravity: Where do we stand*, ed. by R. Peron, M. Colpi, V. Gorini, U. Moschella (Springer, Berlin, 2016), 279
15. K. Chen, M. Ruderman, Pulsar death lines and death valley. Astrophys. J. **402**, 264–270 (1993)
16. P. Goldreich, W.H. Julian, Pulsar electrodynamics. Astrophys. J. **157**, 869–880 (1969)
17. D.C. Backer, Pulsar nulling phenomena. Nature **228**, 42–43 (1970)
18. D.R. Lorimer, F. Camilo, K.M. Xilouris, Arecibo timing observations of 17 pulsars along the Galactic Plane. Astrophys. J. **123**, 1750–1756 (2002)
19. R.N. Manchester et al., The Parkes multi-beam pulsar survey—I. observing and data f analysis systems, discovery and timing of 100 pulsars. Mon. Not. R. Astron. Soc. **328**, 17–35 (2001)
20. J.M. Cordes, M.A. McLaughlin, Searches for fast radio transients. Astrophys. J. **596**, 1142–1154 (2003)
21. E.F. Keane and M.A. McLaughlin Rotating radio transients. BASI **39**, 333 (2011)
22. M.A. McLaughlin et al., Timing observations of rotating radio transients. Mon. Not. R. Astron. Soc. **400**, 1431–1438 (2009)
23. S. Reynolds et al., Discovery of the X-ray counterpart to the rotating radio transient J1819–1458. Astrophys. J. **639**, L71–L74 (2006)
24. M.A. McLaughlin, Rotating radio transients, in *Astrophysics and Space Science Library*, ed. by W. Becker, vol. 357 (2009), p. 41
25. E.F. Keane , M. Kramer, On the birthrates of Galactic neutron stars. Mon. Not. R. Astron. Soc. **391**, 2009–2016 (2008)
26. J.G. Hartnett, A. Luiten, Comparison of astrophysical and terrestrial frequency standards, in *International Frequency Control Symposium, Newport Beach, California* (2010). arXiv:1004.0115
27. M.A. Alpar, A.F. Cheng, M.A. Ruderman, J. Shaham, A new class of radio pulsars. Nature **300**, 728–730 (1982)
28. V. Urpin, U. Geppert, D. Konenkov, Magnetic and spin evolution of neutron stars in close binaries. Mon. Not. R. Astron. Soc. **295**, 907–920 (1998)
29. M.J. Miri, D. Bhattacharya, Evolution of the magnetic fields on neutron stars in low-mass binary systems. J. Astrophys. Astron. Suppl. **16**, 231–233 (1995)
30. A.M. Archibald et al., A Radio Pulsar/X-ray binary link. Science **324**, 1411 (2009)
31. A. Papitto et al., Swings between rotation and accretion power in a binary millisecond pulsar. Nature **501**, 517 (2013)
32. J.R. Roy et al., Discovery of PSR J1227-4853: a transition from a Low-mass X-ray binary to a redback millisecond pulsar. Astrophys. J. **800**, L12 (2015)

33. S.B. Popov, J.A. Pons, J.A. Miralles, P.A. Boldin, B. Posselt, Population synthesis studies of isolated neutron stars with magnetic field decay. Mon. Not. R. Astron. Soc. **401**, 2675 (2010)
34. C.-A. Faucher-Giguère, V.M. Kaspi, Birth and evolution of isolated radio pulsars. Astrophys. J. **643**, 332 (2006)
35. K. Belczynski, T. Bulik, V. Kalogera, Merger sites of double neutron stars and their host galaxies. Astrophys. J. **571**, L147–L150 (2002)
36. D.R. Lorimer et al., PSR J0609+2130: a disrupted binary pulsar? Mon. Not. R. Astron. Soc. **347**, L21–L25 (2004)
37. F. Verbunt, Origin and evolution of X-ray binaries and binary radio pulsars. Ann. Rev. Astr. Ap. **31**, 93 (1993)
38. P. Podsiadlowski, Electron-capture supernovae and accretion-induced collapse of ONeMg white dwarfs, in *Paths to Exploding Stars: Accretion and Eruption* (2007)
39. S. Rappaport, E. Pfahl, F.A. Rasio, P. Podsiadlowski, Formation of compact binaries in globular clusters, in *Evolution of Binary and Multiple Star Systems*, ed. by P. Podsiadlowski, S. Rappaport, A.R. King, F. D'Antona, L. Burderi. ASP Conference Series, vol. 229 (2001), p. 409
40. J.G. Martinez et al., A double neutron star system with a large mass asymmetry. Astrophys. J. **812**, 143 (2015)
41. M. Burgay et al., An increased estimate of the merger rate of double neutron stars from observations of a highly relativistic system. Nature **426**, 531–533 (2003)
42. A.G. Lyne et al., A double-pulsar system: a rare laboratory for relativistic gravity and plasma physics. Science **303**, 1153–1157 (2004)
43. D.J. Nice, R.W. Sayer, J.H. Taylor, PSR J1518+4904: a mildly relativistic binary pulsar system. Astrophys. J. **466**, L87–L90 (1996)
44. A. Wolszczan, PSR 1257+12 and PSR 1534+12, in *IAU Circ.*, ed. by D.W.E. Green, vol, 5073 (1990), p. 1
45. I.H. Stairs et al., Measurement of relativistic orbital decay in the PSR B1534+12 binary system. Astrophys. J. **505**, 352–357 (1998)
46. A.J. Faulkner et al., PSR J1756−2251: a new relativistic double neutron star system. Astrophys. J. **618**, L119–L122 (2005)
47. A.G. Lyne et al., The Parkes multibeam pulsar survey: PSR J1811−1736—a pulsar in a highly eccentric binary system. Mon. Not. R. Astron. Soc. **312**, 698–702 (2000)
48. A. Corongiu et al., The binary pulsar PSR J1811−1736: evidence of a low amplitude supernova kick. Astron. Astrophys. **462**, 703–709 (2007)
49. D.J. Champion et al., PSR J1829+2456: a relativistic binary pulsar. Mon. Not. R. Astron. Soc. 350, L61–L65 (2004)
50. D.R. Lorimer et al., Arecibo pulsar survey using ALFA. II. The young, highly relativistic binary pulsar J1906+0746. Astrophys. J. **640**, 428–434 (2006)
51. R.A. Hulse, J.H. Taylor, Discovery of a pulsar in a binary system. Astrophys. J. **195**, L51–L53 (1975)
52. J.M. Weisberg, J.H. Taylor, The Relativistic binary pulsar B1913+16, in *Radio Pulsars* eds. by M. Bailes, D.J. Nice, S. Thorsett (Astronomical Society of the Pacific, San Francisco, 2003), pp. 93–98
53. J.K. Swiggum et al., PSR J1930-1852: a pulsar in the widest known orbit around another neutron star. Astrophys. J. **805**, 156 (2015)
54. S.B. Anderson, P.W. Gorham, S.R. Kulkarni, T.A. Prince, Discovery of two radio pulsars in the globular cluster M15. Nature **346**, 42–44 (1990)
55. B.A. Jacoby et al., Measurement of orbital decay in the double neutron star binary PSR B2127+11C. Astrophys. J. **644**, L113–L116 (2006)
56. R.S. Lynch et al., The green bank telescope 350 MHz drift-scan survey II: data analysis and the timing of 10 new pulsars, including a relativistic binary. Astrophys. J. **763**, 81 (2013)
57. J.M. Antoniadis et al., Science **340**, 6131 (2013)
58. V.M. Kaspi et al., Discovery of a young radio pulsar in a relativistic binary orbit. Astrophys. J. **543** 321–327 (2000)

59. R.J. Dewey, J.H. Taylor, J.M. Weisberg, G.H. Stokes, A search for low-luminosity pulsars. Astrophys. J. **294**, L25–L29 (1985)
60. L. Shao et al., A new limit on local Lorentz invariance violation of gravity from solitary pulsars. Class. Quant. Grav. **30**, 165019 (2013). arXiv: 1307.2552
61. J. Boyles et al., The green bank telescope 350 MHz drift-scan survey I: survey observations and the discovery of 13 pulsars. Astrophys. J. **763**, 80 (2013). arXiv: 1209.4293
62. D. McConnell et al., Radio pulsars in the Magellanic Clouds. Mon. Not. R. Astron. Soc. **249**, 654–657 (1991)
63. I.H. Stairs et al., PSR J1740−3052—a pulsar with a massive companion. Mon. Not. R. Astron. Soc. **325**, 979–988 (2001)
64. I.H. Stairs, S.E. Thorsett, Z. Arzoumanian, Measurement of gravitational spin-orbit coupling in a binary-pulsar system. Phys. Rev. Lett. **93**, 141101 (2004)
65. P. Kumar, E.J. Quataert, On the Orbital Decay of the PSR J0045-7319 Binary. Astrophys. J. **493**, 412 (1998)
66. I.H. Stairs, S.E. Thorsett, J.H. Taylor, A. Wolszczan, Studies of the relativistic binary pulsar PSR B1534+12: I. timing analysis. Astrophys. J. **581**, 501–508 (2002)
67. J.H. Taylor, J.M. Cordes, Pulsar distances and the galactic distribution of free electrons. Astrophys. J. **411**, 674 (1993)
68. D.R. Lorimer, M. Kramer, in *Handbook of Pulsar Astronomy* (Cambridge University, Cambridge, 2005)
69. A.G. Lyne, G. Smith, in *Pulsar Astronomy*, 3rd edn. (Cambridge University, Cambridge, 2005)
70. I.I. Shapiro, Fourth test of general relativity. Phys. Rev. Lett. **13**, 789 (1964)
71. G. Hobbs, A.G. Lyne, M. Kramer, An analysis of the timing irregularities for 366 pulsars. Mon. Not. R. Astron. Soc. **402**, 1027 (2010)
72. I. Cognard, D.C. Backer, A Microglitch in the Millisecond Pulsar PSR B1821-24 in M28. Astrophys. J. **612**, 125 (2004)
73. C. Will, in *Theory and Experiments in Gravitational Physics* (Cambridge University, Cambridge, 1993)
74. C. Will, The confrontation between general relativity and experiment in *General Relativity and John Archibald Wheeler*, ed. by I. Ciufolini, R.A. Matzner. Advanced Solid State Lasers Conference (ASSL), vol. 367 (2010), p. 73
75. C. Will, *The Confrontation between General Relativity and Experiment*. LLR, vol. 17 (2014), p. 4
76. C. Will, K.J. Nordvedt, Conservation laws and preferred frames in relativistic gravity. I. Preferred-frame theories and an extended PPN formalism. Astrophys. J. **177**, 757 (1972)
77. T. Damour, G. Esposito-Farèse, Testing gravity to second post-Newtonian order: a field-theory approach. Phys. Rev. D. **53**, 5541 (1996)
78. T. Damour, G. Esposito-Farèse, Tensor-multi-scalar theories of gravitation. Class. Quantum Grav. **9**, 2093 (1992)
79. K. Nordtvedt, Testing relativity with laser ranging of the moon. Phys. Rev. **170**, 1186 (1968)
80. I.H. Stairs et al., Discovery of three wide-orbit binary pulsars: implications for binary evolution and equivalence principles. Astrophys. J. **632**, 1060 (2005)
81. P.C.C. Freire, M. Kramer, N. Wex, Tests of the universality of free fall for strongly self-gravitating bodies with radio pulsars. Class. Quant. Grav. **29**, 184007 (2012). arXiv: 1205.3751
82. I. Stairs, Testing general relativity with pulsar timing. LLR **6**, 5 (2003)
83. J. Antoniadis, et al., The relativistic pulsar-white dwarf binary PSR J1738+0333 I. Mass determination and evolutionary history. Mon. Not. R. Astron. Soc. **423**, 3316–3327 (2012). arXiv: 1204.3948
84. P.C.C. Freire et al., The relativistic pulsar-white dwarf binary PSR J1738+0333 II. The most stringent test of scalar-tensor gravity. Mon. Not. R. Astron. Soc. **423**, 3328–3343 (2012). arXiv: 1205.1450

85. C.M. Will, Is momentum conserved? A test in the binary system PSR B1913+16. Astrophys. J. **393**, L59 (1992)
86. L. Shao, N. Wex, New tests of local Lorentz invariance of gravity with small-eccentricity binary pulsars. Class. Quant. Grav. **29**, 215018 (2012). arXiv:1209.4503
87. L. Shao, Testing the strong equivalence principle with the triple pulsar PSR J0337+1715. Phys. Rev. D **93**, 0804023 (2016). arXiv: 1602.05725
88. T. Damour, N. Deruelle, General relativistic celestial mechanics of binary systems. I. The post-Newtonian motion. Ann. Inst. H. Poincaré (Phys. Théor.) **43**, 107 (1985)
89. T. Damour, N. Deruelle, General relativistic celestial mechanics of binary systems. II. The post-Newtonian timing formula. Ann. Inst. H. Poincaré (Phys. Théor.) **44**, 263 (1986)
90. J.H. Taylor, J. M. Weisberg, Further experimental tests of relativistic gravity using the binary pulsar PSR 1913+16. Astrophys. J. **345**, 434 (1989)
91. T. Damour, J.H. Taylor, Strong-field tests of relativistic gravity and binary pulsars. Phys. Rev. D **45**, 1840 (1992)
92. R.A. Hulse, J.H. Taylor, Discovery of a pulsar in a binary system. Astrophys. J. **195**, L51 (1975)
93. J.M. Weisberg, D.J. Nice, J.H. Taylor, Timing measurements of the relativistic binary pulsar PSR B1913+16. Astrophys. J. **722**, 1030 (2010)
94. T. Damour, The problem of motion in Newtonian and Einsteinian gravity, in *Three Hundred Years of Gravitation*, ed. by S.W. Hawking, W. Israel (Cambridge University, Cambridge, 1987), p. 128
95. T. Damour, J.H. Taylor, On the orbital period change of the Binary Pulsar PSR 1913+16. Astrophys. J. **366**, 501 (1991)
96. J.M. Weisberg, Y. Huang, Relativistic measurements from timing the Binary Pulsar PSR B1913+16. Astrophys. J. **829**, 65 (2016)
97. A.T. Deller, J.M. Weisberg, D.J. Nice, S. Chatterjee, A VLBI distance and transverse velocity for PSR B1913+16. Astrphys. J. **862**, 139 (2018)
98. A. Wolszczan, A nearby 37.9 ms radio pulsar in a relativistic binary system. Nature **350**, 688 (1991)
99. M. McLaughlin, et al., The double pulsar system J0737-3039: modulation of A by B at Eclipse. Astrophys. J. **616**, 131 (2004)
100. M. Lyutikov, C. Thompson, Magnetospheric Eclipses in the double pulsar system PSR J0737-3039. Astrophys. J. **634**, 1223 (2005)
101. R.P. Breton, et al., Relativistic spin precession in the double pulsar. Science **321**, 104 (2008)
102. B.M. Barker, R.F. O'Connell, Gravitational two-body problem with arbitrary masses, spins, and quadrupole moments . Phys. Rev. D **12**, 329 (1975)
103. M. Kramer Determination of the geometry of the PSR B1913+16 system by geodetic precession. Astrophys. J. **509**, 856 (1998)
104. J.M. Weisberg, J.H. Taylor, General relativistic geodetic spin precession in binary pulsar B1913+16: mapping the emission beam in two dimensions. Astrophys. J. **576** 942 (2002)
105. T. Clifton, J.M. Weisberg, A simple model for pulse profiles from precessing pulsars, with special application to relativistic binary PSR B1913+16. Astrophys. J. **679**, 687 (2008)
106. R.N. Manchester, et al., Observations and modeling of relativistic spin precession in PSR J1141-6545. Astrophys. J. **710**, 1694 (2010)
107. N. Wex, M. Kramer, A characteristic observable signature of preferred-frame effects in relativistic binary pulsars. Mon. Not. R. Astron. Soc. **380**, 455 (2007)
108. T. Damour, G. Schaefer, Higher-order relativistic periastron advances and binary pulsars. Nuovo Cim. **101**, 127 (1988)
109. J.M. Lattimer, B.F. Schutz, Constraining the equation of state with moment of Inertia measurements. Astrophys. J. **629**, 979 (2005)
110. T. Damour, G. Esposito-Farèse, Tensor-scalar gravity and binary pulsar experiments. Phys. Rev. D. **54**, 1474 (1996)
111. C. Brans, R.H. Dicke, Mach's principle and a relativistic theory of gravitation. Phys. Rev. **124**, 925 (1961)

112. C.M. Will, H.W. Zaglauer, Gravitational radiation, close binary systems, and the Brans–Dicke theory of gravity. Astrophys. J. **346**, 366 (1989)
113. R.S. Lynch et al., The Green Bank telescope 350 MHz drift-scan survey II: data analysis and the timing of 10 new pulsars, including a relativistic binary. Astrophys. J. **763**, 81 (2013). arXiv: 1209.4296
114. J. Alsing et al., Gravitational radiation from compact binary systems in the massive Brans–Dicke theory of gravity. Phys. Rev. D **85**, 064041 (2012). arXiv: 1112.4903
115. G. Esposito-Farèse, in *Motion in Alternative Theories of Gravity*. School on Mass (2009). arXiv:0905.2575
116. R.D. Ferdman et al., The European pulsar timing array: current efforts and a LEAP toward the future. Class. Quant. Gr. **27**, 084014 (2010)
117. R. Smits, Pulsar science with the five hundred metre aperture spherical telescope. Astron. Astrophys. **506**, 919 (2009)
118. D.B Davidson, MeerKAT ans SKA phase 1, in *Proceedings of the 10th International Symposium on Antennas, Propagation and EM Theory* (2012), p. 1279
119. R. Smits, Pulsar searches and timing with the square kilometre array. Astron. Astrophys. **493**, 1161 (2009)
120. I.S. Deneva, in *Elusive Neutron Star Populations: Galactic Center and Intermittent Pulsars*, Ph.D. dissertation (Cornell University, USA, 2010)
121. E. Pfahl, P. Podsiadlowski, S. Rappaport, Relativistic binary pulsars with black hole companions. Astrophys. J. **628**, 343 (2005)
122. B. Devecchi, M. Colpi, M. Mapelli, A. Possenti, *Millisecond pulsars around intermediate-mass black holes in globular clusters*. Mon. Not. R. Astron. Soc. **380**, 691 (2007)
123. M. Kramer et al., Strong-field tests of gravity using pulsars and black holes. New Astr. Rev. **48**, 993 (2004)
124. S.W. Hawking,, R. Penrose, The singularities of gravitational collapse and cosmology, in *Royal Society of Proceeding Series A*, vol. 314 (1970), p. 529
125. R. Foster, D. Backer, Constructing a pulsar timing array. Astrophys. J. **361**, 300 (1990)
126. R. Hellings, G. Downs, Upper limits on the isotropic gravitational radiation background from pulsar timing analysis. Astrophys. J. **265**, L39 (1983)
127. J. Verbiest et al., Timing stability of millisecond pulsars and prospects for gravitational-wave detection. Mon. Not. R. Astron. Soc. **400**, 951 (2009)
128. D.R.B. Yardley et al., The Sensitivity of the Parkes Pulsar timing array to individual sources of gravitational waves. Mon. Not. R. Astron. Soc. **407**, 669 (2010)
129. R. van Haasteren et al., Placing limits on the stochastic gravitational-wave background using European timing array data. Mon. Not. R. Astron. Soc. **414**(4), 3117–3128 (2011)
130. B.P. Abbott, R. Abbott, T.D. Abbott, M.R. Abernathy, F. Acernese, et al., Observation of gravitational waves from a binary black hole merger. Phys. Rev. Lett. **116**, 061102 (2016)
131. B.P. Abbott, R. Abbott, T.D. Abbott, F. Acernese, K. Ackley, et al., GW170817: observation of gravitational waves from a binary neutron star inspiral. Phys. Rev. Lett. **119**, 161101 (2017)
132. A. Sesana, A. Vecchio, Measuring the parameters of massive black hole binary systems with pulsar timing array observations of gravitational waves. Phys. Rev. D **81**, 104008 (2010)
133. X. Deng, L.S. Finn, Pulsar timing array observations of gravitational wave source timing parallax. Mon. Not. R. Astron. Soc. **414**(1), 50–58 (2011)
134. K.J. Lee, N. Wex, M. Kramer, B.W. Stappers, C.G. Bassa, et al., Gravitational wave astronomy of single sources with a pulsar timing array. Mon. Not. R. Astron. Soc. **414**, 3251 (2011)

Chapter 3
Magnetars: A Short Review and Some Sparse Considerations

Paolo Esposito, Nanda Rea, and Gian Luca Israel

Abstract We currently know about 30 magnetars: seemingly isolated neutron stars whose properties can be (in part) comprehended only acknowledging that they are endowed with magnetic fields of complex morphology and exceptional intensity—at least in some components of the field structure. Although magnetars represent only a small percentage of the known isolated neutron stars, there are almost certainly many more of them, since most magnetars were discovered in transitory phases called *outbursts*, during which they are particularly noticeable. In outburst, in fact, a magnetar can be brighter in X-rays by orders of magnitude and usually emit powerful bursts of hard-X/soft-gamma-ray photons that can be detected almost everywhere in the Galaxy with all-sky monitors such as those on board the Fermi satellite or the Neil Gehrels Swift Observatory. Magnetars command great attention because the large progress that has been made in their understanding is proving fundamental to fathom the whole population of isolated neutron stars, and because, due to their extreme properties, they are relevant for a vast range of different astrophysical topics, from the study of gamma-ray bursts and superluminous supernovae, to ultraluminous X-ray sources, fast radio bursts, and even to sources of gravitational waves. Several excellent reviews with different

P. Esposito (✉)
Scuola Universitaria Superiore IUSS Pavia, Pavia, Italy

INAF - Istituto di Astrofisica Spaziale e Fisica Cosmica di Milano, Milano, Italy
e-mail: paolo.esposito@iusspavia.it

N. Rea
Institute of Space Sciences (ICE, CSIC), Barcelona, Spain

Institut d'Estudis Espacials de Catalunya (IEEC), Barcelona, Spain

Astronomical Institute "Anton Pannekoek", University of Amsterdam, Amsterdam, The Netherlands
e-mail: rea@ice.csic.es

G. L. Israel
Osservatorio Astronomico di Roma, INAF, Monteporzio Catone, Rome, Italy
e-mail: gianluca@oa-roma.inaf.it

© Springer-Verlag GmbH Germany, part of Springer Nature 2021

T. M. Belloni et al. (eds.), *Timing Neutron Stars: Pulsations, Oscillations and Explosions*, Astrophysics and Space Science Library 461,
https://doi.org/10.1007/978-3-662-62110-3_3

focuses were published on magnetars in the last few years: among others, Israel and Dall'Osso, Bursts and flares from highly magnetic pulsars, in Proceedings of the First Session of the Sant Cugat Forum on Astrophysics High-Energy Emission from Pulsars and Their Systems, ed. by D.F. Torres, N. Rea. Astrophysics and Space Science Proceedings (Springer, Heidelberg, 2011), pp. 279–298, Rea and Esposito, Magnetar outbursts: an observational review, in High-Energy Emission from Pulsars and Their Systems. Proceedings of the First Session of the Sant Cugat Forum on Astrophysics, ed. by D.F. Torres, N. Rea. Astrophysics and Space Science Proceedings (Springer, Heidelberg, 2011), pp. 247–273, Turolla and Esposito Int J Mod Phys D 22:1330024-163, 2013, Mereghetti, et al. Space Sci Rev 191:315–338, 2015, Turolla et al., 78:116901, 2015; Kaspi and Beloborodov Annu Rev Astron Astrophys 55:261–301, 2017. Here, we quickly recall the history of these sources and travel through the main observational facts, trying to touch some recent and sometimes little-discussed ramifications of magnetars.

3.1 Historical Overview

The event at the birth of magnetar studies was the landmark giant flare observed in 1979 from SGR 0526–66 [34, 149]. Nothing alike had been observed before from a pulsar, and the flare was also different from and much brighter than any other gamma-ray burst. The extreme properties of the event forced the astronomer to conceive unusual scenarios. The most appealing one was the existence of a class of neutron stars with super-strong magnetic fields of 10^{14}–10^{15} G [56, 165]. In particular, the word *magnetar* was introduced to designate these objects by Duncan and Thompson [56]. The giant flare and the more common emission of short fainter bursts of these sources were explained by impulsive releases of magnetic energy stored in the neutron star, which could be triggered by fractures in the magnetically stressed crust, perhaps associated to sudden magnetic reconnections in the star's magnetosphere [56, 208, 209]. The handfuls of sources associated to these recurrent hard X-/gamma-ray transients were dubbed soft gamma repeaters (SGRs), to distinguish them from the gamma-ray bursts.

In the same years, another class of X-ray pulsars defying any easy pigeonholing was emerging: the anomalous X-ray pulsars (AXPs; Mereghetti and Stella [150], van Paradijs et al. [221]). They were characterised as persistent X-ray pulsars with periods of a few seconds and owed the adjective 'anomalous' to the fact that their X-ray luminosity exceeded that available from spin-down energy loss (while the accretion was ruled out by the lack of any trace of a stellar companion). Thompson and Duncan [209] noted that except for the emission of short bursts, which had never been observed in AXPs, these sources were very similar to the recently-discovered persistent soft X-rays counterparts of SGRs [159, 190, 223]. It was suggested that AXPs could be evolved SGRs, which ended or drastically reduced their explosive activity after having largely depleted their reservoir of magnetic energy, and the opposite possibility, too, was considered: that AXPs might be SGR progenitors in which the magnetic field decay has just begun (e.g. Gavriil et al. [73]).

The magnetar hypothesis was arguably the most successful in explaining the salient features of SGRs and AXPs, but some skepticism about the scenario persisted for long, in particular for the AXPs, for which several alternative models could not be excluded [31, 151]. The magnetar model really started to become the mainstream when the period derivative of an SGR was measured for the first time [130]. Using RossiXTE, it was established that SGR 1806–20 was pulsating at $P = 7.47$ s and the period was increasing at the rate $\dot{P} = 2.6 \times 10^{-3}\,\mathrm{s\,yr^{-1}}$: with the standard assumption of a magnetic dipole rotating in vacuum routinely used for standard pulsar, the values correspond to a surface magnetic field of $B \simeq 3.2 \times 10^{19}\sqrt{P\dot{P}} = 8 \times 10^{14}$ G at the neutron star's equator. Moreover, the release of magnetic energy was necessary to power the X-ray emission of SGR 1806–20, since the spin-down energy loss of the neutron star was two orders of magnitude lower than the observed X-ray luminosity.

Few years later, the detection of SGR-like bursts from the AXP 1E 1048-1–5937 [73] confirmed the suspect that also AXPs could harbour magnetars. Since then, with the discovery of many new magnetars showing a variegated phenomenology and observations of strong bursting activity and powerful flares from AXPs, as well as prolonged periods of quiescence in once-active SGRs, the AXP–SGR dichotomy seems completely obsolete [125, 153].

3.2 Observational Characteristics

3.2.1 Persistent Emission

3.2.1.1 X-Ray Emission

The X-ray emission from magnetars is gently modulated at the pulsar spin period, with generally one or two broad sine components and substantial pulsed fractions (the fraction of the flux that changes along the rotation cycle) of 10–30%. The pulse profiles may be energy dependent and may display dramatic changes with time, especially in connection with strong bursting/outbursting activity (see Sect. 3.2.2). As an example, Fig. 3.1 shows a multi-epoch and multi-instrument pulse profile of 1RXS J170849.0–400910 (see [89] for details).

Part of the X-ray luminosity of magnetars in quiescence has a thermal origin and can be fit by a blackbody with temperature $kT \approx 0.3$–1 keV, much higher than the typical values for rotation-powered pulsars; magnetar also tend to be more luminous than rotation-powered pulsars of similar characteristic age. Indeed, in magnetars the neutron star surface is believed to be particularly hot because of the extra-heating by the magnetic field decay (e.g. Aguilera et al. [1]).

The size of the region of the thermal emission inferred from a blackbody fit is generally much smaller than the surface of the star, possibly suggesting that a

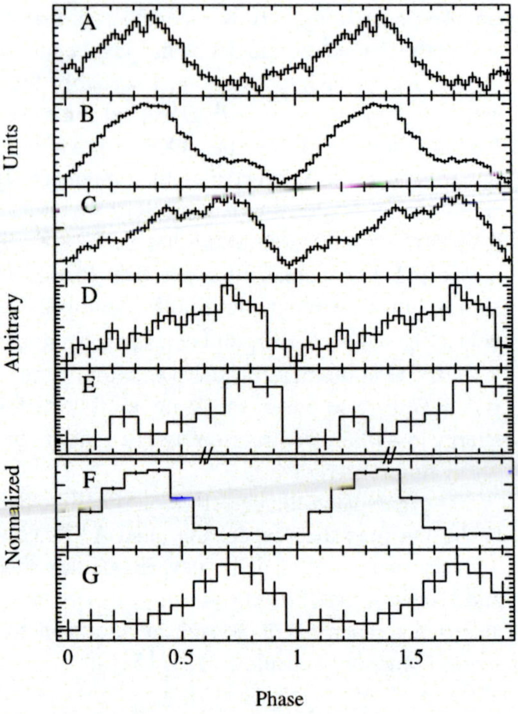

Fig. 3.1 Pulse profiles of 1RXS J170849.0–400910, obtained with RossiXTE/PCA (2004 data; panel A: 2.5–4 keV, B: 4–8 keV, C: 8–16 keV, D: 16–32 keV) and INTEGRAL/IBIS (2004 data; panel E, 20–200 keV). Panels F and G show the BeppoSAX MECS (1–10 keV) and PDS (20–200 keV) pulse profiles obtained during a single pointing in 2001. Note that data in panels A to E were folded using an RossiXTE timing solution, while panels F and G using the period measured in the MECS data: the profiles in the two groups are phase-aligned between themselves but not each other (From Götz et al. [89])

single-temperature blackbody is an oversimplification. Substantial anisotropies are indeed expected in the presence of a strong magnetic field in the crust and, since most magnetars are located in the Galactic plane, their spectra are generally heavily absorbed: it is possible that only 'hot spots' are detectable in the available X-ray spectra. Small and hot regions are also envisaged to result, rather than from internal heat transfer, from particle bombardment and heat deposition from magnetospheric currents induced by the globally twisted external magnetic field and/or by localised twists. In any case, the thermal emission in magnetars is expected to be significantly distorted by a magnetised atmosphere and also by magnetospheric effects, most likely resonant cyclotron scattering onto magnetospheric charges. Since the charged particles populate vast regions of the magnetosphere, with different magnetic field intensities, the scattering produces a hard tail instead than a narrow line or a set of distinct lines and harmonics.

Phenomenologically, in the 0.5–10 keV band magnetar spectra are always well described by the already mentioned blackbody and often one or more additionally harder blackbody or power-law components, the latter with photon index generally in the range $\Gamma \sim 2$–4 [126, 164]. Since in the spectral modelling the (usually large) interstellar absorption, the power-law slope and the blackbody temperature(s) are covariant, it is often difficult to disentangle the components or to tell whether a blackbody or a power-law component is to be preferred. For the magnetar with the lowest absorbing column, CXOU J0100–7211 in the Small Magellanic Cloud, a double-blackbody model provides a much better fit to the data than a blackbody-plus-power-law model [210]. On the other hand, a non-thermal component is certainly present at least in the magnetars detected in the hard-X-ray range.

Given the many complications and uncertainties, it is evident that prudence is necessary when drawing physical inferences from the spectral parameters. It is however worth noticing that physical models based on resonant cyclotron scattering (likely, repeated scatterings) of seed thermal photons on mildly relativistic electrons are quite successful in reproducing the general thermal-plus-power-law shape of the continuum and fit the spectra of most magnetars. We refer to Turolla et al. [219] for an overview of the state of the art of these models, their application, and the main open problems.

3.2.1.2 Hard-X-Ray Emission

In several magnetars, hard X-ray tails with power-law spectra with photon index $\Gamma \approx 0.5$–2 (flatter than the soft non thermal components) have been detected with BeppoSAX, RossiXTE, INTEGRAL, Suzaku and NuSTAR extending beyond ≈150 keV [3, 47, 48, 60, 88, 89, 133, 134, 204, 228]. Upper limits in the hundreds of keV and MeV regions obtained with CGRO and Fermi indicate that the tails do not extend above ≈500 keV (e.g. [48, 144]). Figure 3.2 shows the soft-to-hard-X-

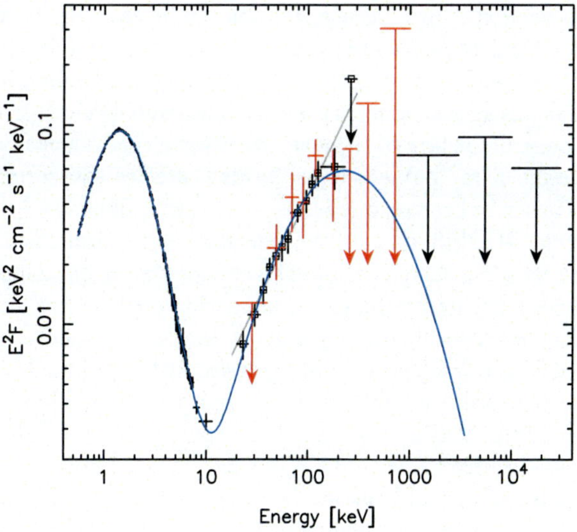

Fig. 3.2 Unabsorbed soft-to-hard-X-ray spectral energy distribution of 4U 0142+61 as observed with XMM–Newton (in black), INTEGRAL/ISGRI (black open squares), INTEGRAL/SPI (red), and CGRO/COMPTEL (black). Down arrows indicate upper limits. See den Hartog et al. ([48] from which the figure was taken) for details on the spectral modelling

ray spectrum of 4U 0142+61. Also searches in the GeV and Tev energy bands gave negative results [2, 144].

The hard components can be variable, but in general their luminosity is comparable with or larger than that measured below 10 keV; also in the sources in which the hard X-ray emission has not been detected, the upper limits do not exclude a substantial contribution to the total luminosity. A peculiar feature of the hard tails is that the pulsed emission has a harder spectrum than the averaged hard X-ray one and also shows phase dependent variations (also morphological changes in the pulse profiles and peak shifts with energy are observed; e.g. Hartog et al. [47], Götz et al. [89], see also Fig. 3.1).

The origin of the hard tails of magnetars is still poorly understood but in general the mechanism suggested, similarly to what proposed for the soft spectra, is the up-scattering on ultra-relativistic electrons with Lorentz factor $\gamma \gg 1$ [13, 67, 229], possibly associated to relativistic outflows near the neutron star [15, 16].

3.2.1.3 Optical or Infrared Emission

Optical or infrared counterparts have been found for about one-third of the known magnetars (e.g. Israel et al. [116], Mignani [155]). The search is complicated by the intrinsic faintness of magnetars at that wavelengths and by their location in crowded and heavily absorbed regions in the Galactic plane, but in most cases the associations are strengthened by the detection of long-term variability. In three cases, the association is firm because the spin modulation has been detected also in the optical band (4U 0142+61, Kern and Martin [129], Dhillon et al. [49]; 1E 1048.1–5937, Dhillon et al. [50]; SGR J0501+4516, Dhillon et al. [51]). As anticipated, magnetars are variable also in the infrared/optical range, but it is unclear (possibly because of the lack of adequate multi-wavelength campaigns) whether the changes trace the X-ray flux evolution, as also cases of anti-correlated or simply erratic variations have been reported [25, 51, 57, 201, 206].

The magnetar 4U 0142+61 has been detected in both infrared and optical bands and is the one for which the greatest wealth of data is available at these wavelengths [105, 106]. In near infrared, it shows an excess with respect to the blackbody that fits the optical data; indeed, a multi-temperature (700–1200 K) thermal model provides a better fit to the data. Wang et al. [230] suggested that the infrared component arises from an extended disk (possibly from supernova fallback) illuminated from the star's X-rays and passively heated. This interpretation is supported by the correlation observed in this source between the X-ray and infrared emissions [201]. On the other hand, infrared/optical emission is expected from the inner magnetosphere, a pair-dominated region where the curvature radiation should be able to produce the observed infrared/optical luminosity [239]. Moreover, a magnetospheric origin would account more easily for the observed optical pulsations, with profiles nearly aligned with those observed at X-rays and displaying similarly broad modulation and large pulsed fraction (20–50%). It is also possible that the infrared and optical emissions have different origins or that the infrared excess in 4U 0142+61 is not well understood. A handful of magnetar has been detected also as pulsating sources at longer wavelength, in the radio band. We give an overview of the properties of magnetars at radio frequencies in Sect. 3.2.2.3.

3.2.2 *Transient Activity*

Magnetars are certainly characterised by an extremely rich observational phenomenology, but this is particularly true when speaking of their transient activity: They display unpredictable and dramatic variations in their emission and timing properties in all the wavelengths at which they are detected, on time scales from milliseconds to months or years, and often with a dynamic range unparalleled by any other embodiment of isolated neutron stars. Their transient radiative events are usually outlined in two main categories: short-duration (ms–minutes) explosive events (giant flares and bursts) and outbursts, in which the X-ray luminosity rises to up to $\sim 10^3$ times the quiescent level and then decays in weeks to months/years. Perhaps, an outburst more than an event could be considered a 'syndrome', in the sense that the flux enhancement is generally accompanied by bursts, spectral changes and timing anomalies, including glitches.

3.2.2.1 Giant Flares

Giant flares are the rarest and most energetic events associated with magnetars. They are also the most important, at least historically, as it was the first giant flare, from SGR 0526–66 in the Large Magellanic Cloud on 1979 March 5 [34, 149], that brought magnetars on the astrophysical scene, provided clear-cut evidence of their neutron-star nature and propensity to produce multiple events (at variance with the gamma-ray bursts discovered by the Vela satellites), and prompted—among a multitude of different models (see e.g. Norris et al. [163], Woods and Thompson [232])—the concept of super-magnetic neutron stars [56, 165, 208]. Moreover, they still provide some of the most compelling clues for the presence of magnetic fields of 10^{14} G close to the neutron star surface in magnetars.

A total of three giant flares have been observed. In 1979 from SGR 0526–66, on 1998 August 27 from SGR 1900+14 [111] and on 2004 December 27 from SGR 1806–20 [112, 167]. It is worth noticing that the three giant flares were emitted from the three first SGRs discovered (but only SGR 0526–66 was discovered because of the event). Since a mere coincidence seems unlikely and a giant flare makes a significant dent in the total magnetic energy pool of a magnetar, a natural explanation would be that those three sources are at the magnetic-activity pinnacle of their life and their frequent bursting activity got them noticed earlier than other magnetars (see also Perna and Pons [169], Viganò et al. [226]).

All three giant flares started with a short ($\sim$0.1–0.2 s) flash of hard X-rays with peak luminosity $\gtrsim 10^{44}$–10^{45} erg s^{-1} ($\gtrsim 10^{47}$ erg s^{-1} in the case of SGR 1806–20;

Hurley et al. [112]). The spectrum of this impulsive blaze extends at least to the MeV range and can be described by a blackbody, with initial kT from ≈30 to 200 keV (for SGR 0526–66 and SGR 1806–20, respectively). These sudden releases of an immense amount of energy affected in a measurable way (at least for the two most recent events) the Earth's magnetic field [147] and ionosphere [113, 114]. Hurley et al. [112] argued that an extragalactic giant flare as bright as that of SGR 1806–20 could appear at Earth as a short gamma-ray burst up to a distance of several tens of Mpc an therefore magnetar flares might represent a non-negligible fraction of the population of these transients.

After the initial spikes, followed afterglows that were clearly modulated at the rotational period of the neutron stars. The afterglows were much softer than the flash and over a few minutes gradually further softened and faded (Fig. 3.3). It is extremely interesting that while the luminosity of the three peaks spans 2–3 orders of magnitude, the total energy of the oscillating tail was similar ($\approx 10^{44}$ erg) in all the events. Since the afterglow is believed to arise from a cloud of photon–pair plasma confined by the star's magnetic field that cools as the radiation gradually leaks out, this clearly indicates a similar value of the magnetic field in the three magnetars. To trap the fireball, the magnetic pressure must exceed that of the radiation and pairs at the external boundary of the cloud: $B_{\rm dip} > 2 \times 10^{14} (E_{\rm fireball}/(10^{44}\,{\rm erg}))^{1/2} (\Delta R/(10\,{\rm km}))^{-3/2} ((1 + \Delta R/R)/2)^3$ G, where R is the stellar radius and ΔR the characteristic size of the fireball [208].

In the SGR 1900+14 and SGR 1806–1920s giant flares, observations of transient nebular radio emissions provided evidence for outflows [71, 72, 76]. For the best-studied case of SGR 1806–20, the minimum energy in the extended radio emission was estimated at $\approx 10^{43}$ erg, which seems too much to be consistent with pair plasma leaked form the fireball. Indeed, the structure, which was observable for more than a year, is better explained in terms of an baryon-rich mildly-relativistic ejection interacting with matter surrounding the star [76, 96].

The detection of quasi-periodic oscillations (QPOs) in the tails of the giant flares from SGR 0526–66 (with detectors aboard the Prognoz 7 satellite and the Venera 11 and 12 space probes; Barat et al. [12]), SGR 1900+14 [199] and SGR 1806–20 (with RossiXTE and RHESSI; Israel et al. [117], Strohmayer and Watts [200], Watts and Strohmayer [231]), likely associated to seismic vibrations excited by the powerful explosion, started the field of asteroseismology for neutron stars (see gray box QPOs) and offered a new clue of the presence of a magnetic field $\gtrsim 10^{14}$ G in magnetars. Vietri et al. [224] observed that for any source there is a maximum rate of variation of the luminosity (ΔL) on a certain timescale (Δt): $\Delta L/\Delta t < \eta (2.8 \times 10^{18})/\sigma_{\rm T}\,{\rm erg\,s^{-2}}$, where $\sigma_{\rm T}$ is the Thomson cross section and η the energy extraction efficiency [29, 66]. The 1840-Hz QPO detected in SGR 1806–20 implies a $\Delta L/\Delta t$ exceeding this limit by a factor larger than $10/\eta$. However, a strong magnetic field suppresses the electron-scattering cross section (for one photon polarization mode) below the Thomson's value by a factor $\propto B^2$ ([101] see also van Putten et al. [222]). The presence of a magnetic field $B \gtrsim 2 \times 10^{15}$ G at the surface of SGR 1806–20 would reconcile the QPOs with the luminosity variability limit [224].

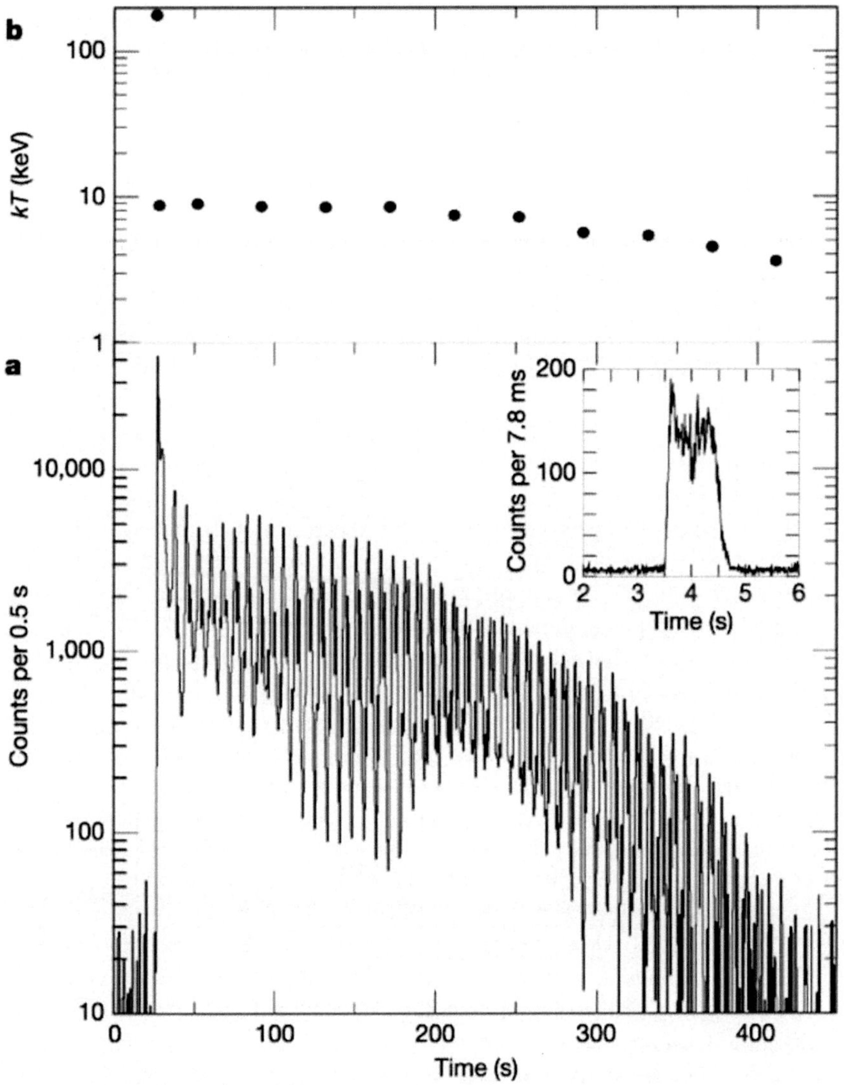

Fig. 3.3 The 2004 December 27 giant flare of SGR 1806–20 as observed by RHESSI (from Hurley et al. [112]). Panel a shows the 20–100-keV light curve. The (saturated) spike is at ~30 s and the inset shows the profile of a bright burst the preceded the flash by ~2 min. The modulation at the spin period of SGR 1806–20 (7.5 s) is apparent. Panel b shows the temporal evolution of the spectral temperature

Quasi Periodic Oscillations and Seismology of Magnetars

The tail of the 2004 giant flare from SGR 1806–20 displayed clear QPO signals at about 18, 30, 93, 150, 625 and 1840 Hz [117, 231]. QPOs around frequencies of 28, 54, 84 and 155 Hz were detected in the tail of the 1998 giant flare of SGR 1900+14 [199], while hints for a signal at ∼43 Hz were found in the 1979 event from SGR 0526–66 [12]. Some QPOs were excited simultaneously, others were detected only once in a very narrow time interval, some faded and were re-excited several times. All the detected QPOs are dependent on the phase of the spin period and show large variations of the amplitude with time. Their similarities suggest that the production mechanism is the same and the most obvious responsible are seismic vibrations induced by the giant flares. This is very exciting, as the QPOs provide a window on the neutron-star and magnetic field structures, and even on the dense matter equation of state.

The QPOs, in accordance with early theoretical suggestions [55], were initially interpreted in terms of torsional shear modes of the neutron star crust. However, it did not take long to realize that neutron stars can sustain many types of oscillation and that the identification of oscillatory modes of magnetars is conceptually (and computationally) extremely challenging because of the magnetic coupling between the crust and the core [82, 138]. Indeed, at the moment, the potential of magnetar asteroseismology is dampened by the high degeneracy because of the many uncertainty associated with the magnetic field and the superfluid state of matter (see for example [81, 139, 140] for detailed discussions and [219] for an overview of the current understanding of the field), but also by the paucity of new data. This motivated searches for QPOs in the more bountiful short bursts. Only one signal, an unusually broad but strong peak at 260 Hz, was identified in a burst, a 0.5-s-long event from 1E 1547.0–5408, while several candidates from ∼60 to 130 Hz were found in data sets combining many short busts [107–109]. Unfortunately, after the end of the RossiXTE mission and the non selection of LOFT [68] by ESA, the possibility of collecting a large number of photons with good time resolution and without saturation problems in case of exceptional events in the near future seems rather remote.

3.2.2.2 Short Bursts

The SGR/magnetar short bursts are the hallmark of magnetars and thousands of them have been recorded and studied, both individually and as samples [5, 87, 90, 110, 119, 220]. In the last few years, and in particular since the launch of Swift (which pairs a sensitive hard X-ray instrument with large field of view to an X-ray telescope that provides ~arcsec localisation), the short bursts have become the primary channel for the discovery of new magnetars and of the onset of magnetic activity from known sources. The bursting activity is unpredictable: Magnetars usually go through long stretches of quiescence (even decades) that can be interrupted by sparse bursts or by paroxysmal activity, during which hundreds or thousands of events are clustered in days. Or by everything in between.

The impulsive emission of X/gamma ray photons lasts from milliseconds to a few seconds and the peak luminosity is typically in the range $\approx 10^{36}$–10^{43} erg s^{-1}. Their profile is usually single-peaked and asymmetric, with a faster rise than decay. Two- or multi-peak bursts are not rare, but since series of bursts can be very rapid, the distinction between single events and multi-peak bursts may be tenuous. Figure 3.4 shows the light curves of 21 bursts collected from SGR 1806–20 with INTEGRAL in 2003 (from Götz et al. [86]).

Some bursts, in general the brightest ('intermediate flares'), can be followed by X-ray tails surviving from minutes to hours. These events resemble the overall shape of the giant flares and their afterglows, except that in some instances the energy in the tail exceed that emitted in the spike (see e.g. Pintore et al. [173]). Sometimes these tails show flux modulation at the neutron star spin period, linking directly the afterglow to the neutron star (e.g. to a region of the star surface that was heated by the burst), but a significant fraction of them may be due to (or receive a large contribution from) dust-scattering of the burst emission, in which a fraction of the photons of the burst is re-emitted after reprocessing by clouds/layers of interstellar dust between us and the source [61, 95, 137, 173]. In general, it is possible to disentangle the contribution of the delayed dust scattering from that intrinsic to the neutron star only when data sets with a large number of photons and good spatial resolution are available, but sometimes the phenomenon is truly spectacular: see Tiengo et al. [211] for the study of a scattering halo surrounding the magnetar 1E 1547.0–5408 that took the shape of at least three expanding symmetric rings.

A number of models have been used to model the spectral emission of short magnetar bursts, but a single-blackbody model, or a double-blackbody model when broad band data are available (e.g. 1–200 keV), are usually a safe bet [115, 119]. The blackbody temperature kT is typically in the range from ~2 to 12 keV and when the double-blackbody decomposition is viable, a bimodal distribution of radii end temperatures takes shape, with a soft blackbody and a hotter (~12 keV) blackbody with smaller surface.

The fluence (S) distribution for number (N) of bursts usually follows a power-law function $N(> S) \propto S^{-\alpha}$ over several orders of magnitude, with $\alpha \sim 0.6$–0.9, depending on the sources and the instruments [5, 87, 93, 94]. It has been pointed out many times that this behaviour is similar to what observed for earthquakes. The truth is that such distribution is rather ubiquitous in nature and also in artificial and human-influenced systems: classic examples are the sizes of landslides and avalanches, solar flares, forest fires, lunar craters and cities, or the frequency of use of words in human languages (e.g. Newman [162]). As observed by Paizis and Sidoli [166] for the distribution of hard-X luminosity of supergiant fast X-ray transient, the power-law distribution is characteristic (but not exclusive) of the self-organized

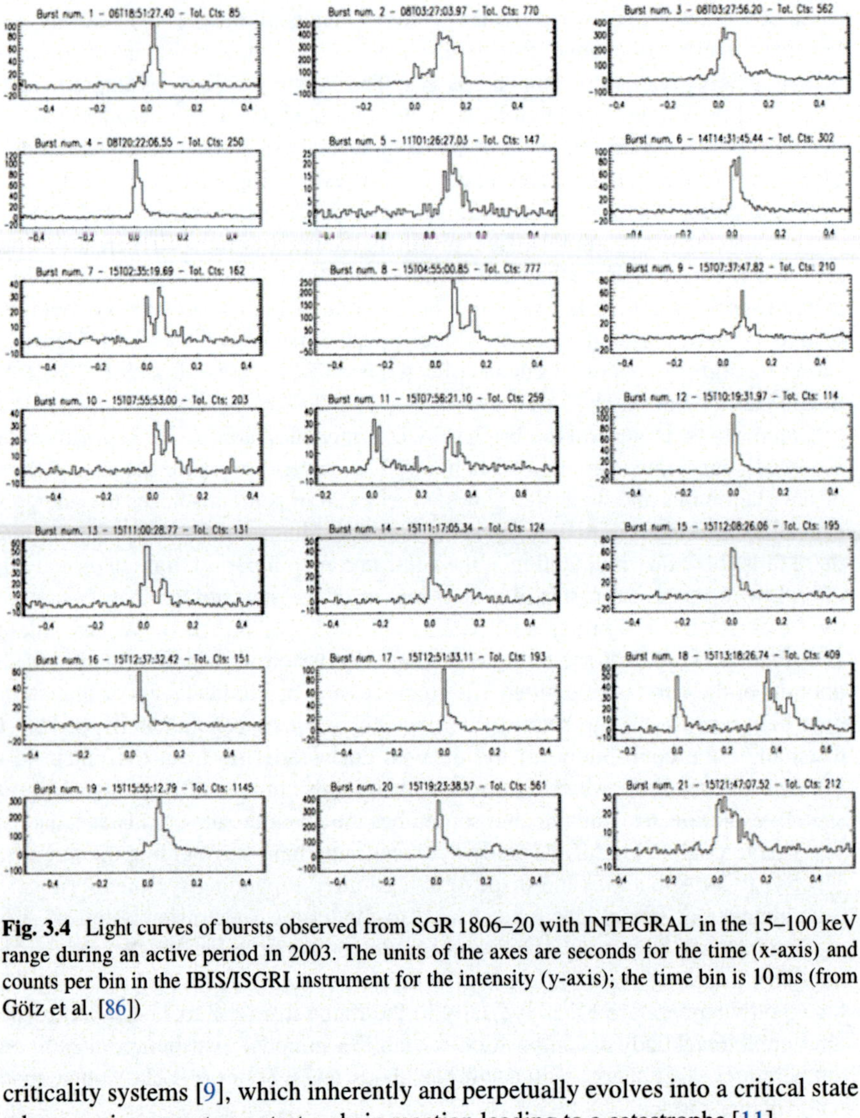

Fig. 3.4 Light curves of bursts observed from SGR 1806–20 with INTEGRAL in the 15–100 keV range during an active period in 2003. The units of the axes are seconds for the time (x-axis) and counts per bin in the IBIS/ISGRI instrument for the intensity (y-axis); the time bin is 10 ms (from Götz et al. [86])

criticality systems [9], which inherently and perpetually evolves into a critical state where a minor event can start a chain reaction leading to a catastrophe [11].

3.2.2.3 Outbursts

In the magnetar world, the term 'outburst' is usually used to denote a large enhancement (~10–1000) of the flux that eventually fades away over the course of months or even years (e.g. Rea and Esposito [181], Coti Zelati et al. [37]). A few magnetars have never been observed in outburst, others displayed a single outburst in decades, and some have gone through multiple events. Outbursts, and

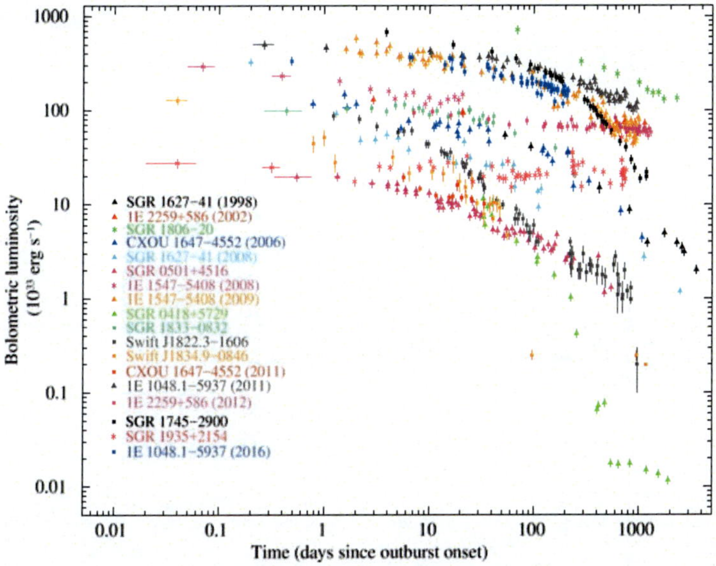

Fig. 3.5 Light curves of all the outbursts with good observed coverage (from Coti Zelati et al. [37]). The luminosities are bolometric (obtained from an extrapolation of the best-fit spectral models in the 0.01–100 keV range)

their onsets in particular, are generally associated to one or more short bursts. It is not clear, however, whether the bursts actually start the outbursts. In fact, the soft X-ray observations that catch a source in an enhanced-flux phase are generally carried out in response of the detection of a burst; on the other hand, for the outbursts discovered serendipitously it is not possible to pinpoint their exact start or exclude that bursts were missed. At any rate, the few cases in which observations were performed fortuitously shortly prior to a burst–outburst combination, indicate that the flux changes are rapid and happen close to the explosive activity, within $\sim$1–2 days [62, 118, 128, 238].

The decay pattern is usually complicated, but it often includes an initial rapid decay ($\lesssim$1 day; e.g. Woods et al. [235], Esposito et al. [62]) and a more extended phase that can be described by power-law or exponential functions. Sudden flux drops and periods of flux stability have also been observed. Figure 3.5 shows the long-term light curves of all the outbursts discovered up to the end of 2017 and followed with intense and long coverage, mainly using imaging instruments.

Despite the great variety of behaviours, all outbursts have some common features: At the beginning of the outburst, the X-ray spectrum is harder than in quiescence, and gradually softens as the flux decreases (Fig. 3.6); The larger the luminosity at the outburst peak, the larger the total energy released during the entire episode (generally in the range $\approx 10^{41}$–10^{43} erg); The larger the total energy of the outburst, the longer the time scale for the relaxation Coti Zelati et al. [37]. There is also an anticorrelation between the quiescent X-ray luminosity of a magnetar

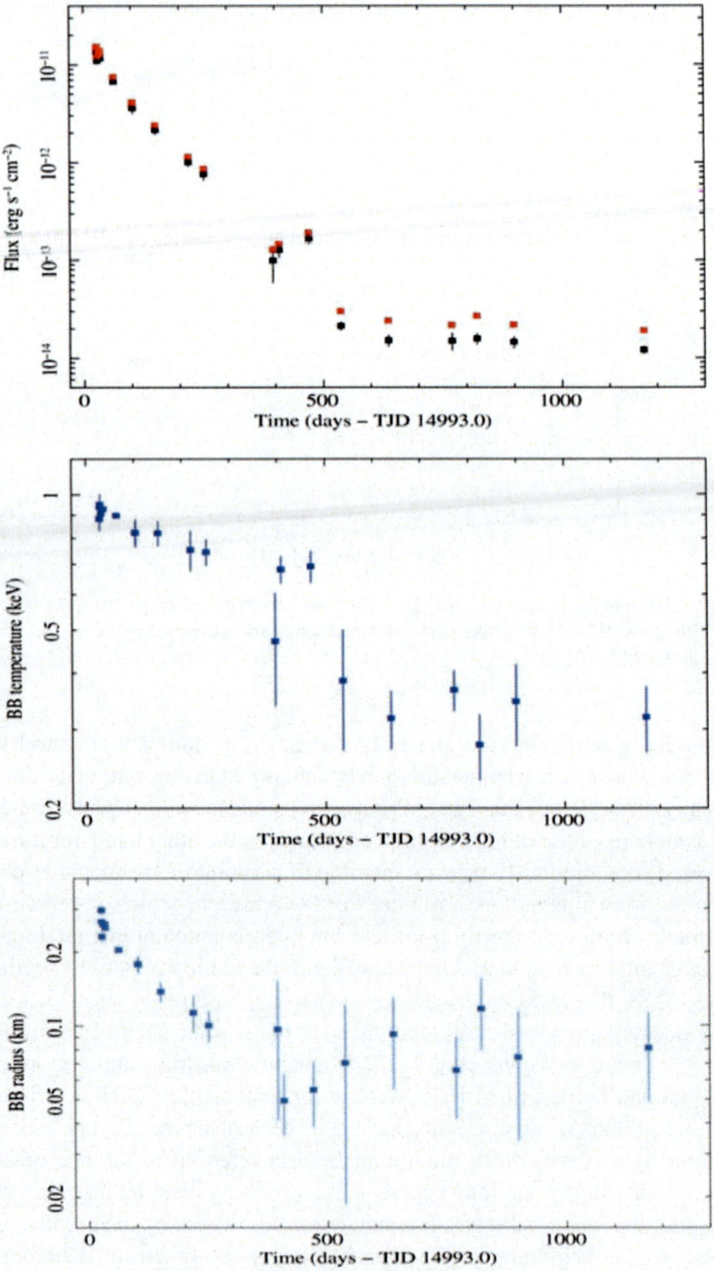

Fig. 3.6 Spectral evolution of SGR 0418+5729 with time during the outburst started in 2009 June (the time of the detection of the first burst is MJD 54987.9 ≡ TJD 14987.9). Top panel: flux evolution for the absorbed 0.5–10 keV flux (black), and for the bolometric unabsorbed flux (red). Middle and bottom panels: evolution of the blackbody temperature and radius, calculated at infinity and assuming a distance of 2 kpc (from Rea et al. [185])

($L_{X,q}$) and the dynamical range of its outbursts; Coti Zelati et al. [37] found that $L_{X,peak}/L_{X,q} \propto L_{X,q}^{-0.7}$, where $L_{X,peak}$ is the maximum X-ray luminosity achieved during the event (see also Pons and Rea [176]). Indeed, several magnetars that could be studied in detail when they were undergoing outbursts, are completely unnoticeable and hardly recognisable as magnetars while in quiescence (or even nondetectable without deep and targeted observations). For this reason, the strategy of the Swift mission [75] of slewing and pointing its X-ray telescope as soon as possible towards the transient events detected and localized by its wide-field coded-mask detector sensitive to hard X-rays, has proven tremendously successful in discovering new magnetars.

Outbursts are usually accompanied also by changes in the timing properties of the magnetar. The morphology of the pulse profiles, which generally are broad and with one or two main peaks per cycle, can vary dramatically during an outburst, both in shape (as a rule, becoming more complicate when the activity is high) and in pulsed fraction (e.g. Woods et al. [234], Israel et al. [120], Rodíguez Castillo et al. [188], Esposito et al. [63], Dib and Kaspi [52]). During outbursts and in general in period of enhanced activity, magnetars show a more efficient spin-down torque, with variations of a factor up to $\approx$10 (e.g. Mereghetti et al. [152], Olausen and Kaspi [164]); the spin-down evolution, however, does not trace or correlate well with the radiative or bursting behaviours [236, 237]. Among isolated pulsars, magnetars are particularly noisy rotators, and also this aspect is amplified during outbursts [52, 64].

In young radio pulsars ($\lesssim 10^5$ yr), timing noise has been often suggested to be linked to recovery from glitch events [104]. Magnetars are rather prolific glitchers, comparable to the most frequently glitching radio pulsars, and glitches often happen during outbursts (although in coincidence of some glitches, no X-ray flux enhancements were detected); also, while the amplitude distribution of their glitches peaks on larger $\Delta\nu/\nu$ with respect to the 'normal' pulsars, the values observed are in the same range ($\Delta\nu/\nu \sim 10^{-9}$–$10^{-5}$; Dib et al. [53], Dib and Kaspi [52]). Magnetars glitching behaviour appears to be different in the recovery, which is typically very strong, often resulting in an over-recovery, and in the fact that also *anti-glitches* (that is, episodes of sudden spin down) have been reported. The most eminent anti-glitch candidate was reported for 1E 2259+586 [6], where in less than 4 days, a spin-down of $\Delta\nu/\nu \sim -10^{-7}$ was achieved; the sudden variation was accompanied by a simultaneous short burst and by a small (factor $\sim$2) but long-lived (months) flux increase. A large 'braking glitch' could also have occurred in SGR 1900+14 in an 80-days interval including the epoch of its giant flare [233]; the observations were however too sparse to tell whether the abnormal increase of the period resulted from a sudden event or from a prolonged period of enhanced spin down.

Another magnetar activity associated to X-ray outbursts is the transient pulsed radio emission observed in a few of them. Until the first detection of radio pulses during the outburst of XTE J1810–197 [24], magnetars were (rather staunchly) believed to be radio quiet. Ironically, at the time of its radio activation, XTE J1810–197 was the brightest pulsar of the radio sky, with individual pulses reaching flux density of 10 Jy or more. A few other magnetars were subsequently detected in radio as pulsars: 1E 1547.0–5408 [26], PSR J1622–4950 (the only magnetar discovered at radio wavelengths so far; Levin et al. [141]), and SGR J1745–2900 [58, 184]. All the detections happened during an X-ray outburst but, interestingly, at least in some cases the radio emission outlived the X-ray flux enhancement [4, 28, 193]. Also, even in periods in which they are overall active in radio, magnetars seem to switch suddenly on and off [21].[1]

Apart from its temporary nature, the pulsed radio emission from magnetars shows in all four sources some clear differences with respect from that of ordinary rotation-powered pulsars [27, 132, 142, 196]. The radio emission has a very hard spectrum: $S \propto \nu^{-0.5}$ or flatter (where S is the flux density and ν is the frequency), while the typical spectral index of radio pulsars is approximately -1.8 (e.g Seiradakis and Wielebinski [194]). Another peculiar characteristic of magnetar's pulsed radio emission is the instability of the pulse shape. Most radio pulsars show some pulse-by-pulse variability, but the addition of a few hundred pulses is generally sufficient to attain a stable pulse profile; more rarely, radio pulsars switches on time scales from minutes to hours between a small number (usually two) of different pulse profiles [131, 146]. The individual pulses of magnetars have a spiky appearance and stable pulse profiles were never observed (e.g. Kramer et al. [132], Camilo et al. [28]); it is therefore unclear whether the profile stabilization would require for magnetars an unprecedented number of pulses or they simply do not have a stable characteristic pulse profile. Finally, high degrees of linear polarization have been measured in magnetars, up to very high frequencies [27, 215].

1E 161348–5055 in RCW 103: The CCO That Was Not

Central compact objects in supernova remnants (CCOs) are a small set of isolated neutron stars observed close to centres of young non-plerionic supernova remnants (see De Luca [43] for a review). CCOs are fairly steady X-ray sources with thermal-like spectra and no counterparts detected at other wavelengths. They owe their redundant designation to the fact that in the years the class emerged, there was not absolute certainty of their neutron-star nature, since no periodic modulations were find in their emission and also searches at radio frequencies failed to detect a pulsar. After numerous deep observations, there is now little doubt that CCOs are indeed neutron stars, and

(continued)

[1] In this respect, it is interesting to notice that when the radio pulsar PSR J1119–6127 emitted a number of SGR-like bursts and entered an outburst, initially it disappeared as a radio pulsar [22]. Furthermore, after the radio reactivation but still during the outburst, Archibald et al. [8] using simultaneous radio and X-ray observations observed that the radio emission shut down in coincidence with the X-ray bursts, with a recovery time of ~70 s.

spin periods between 0.1 and 0.5 s and their derivative have been measured in three of them. Interesting, for these sources the inferred dipolar magnetic fields are rather low: $\sim 10^{10}$–10^{11} G. This prompted for the CCOs an unifying scenario in which they are either born with weak magnetic fields or with a normal field that has been 'buried' beneath the neutron-star surface by a post-supernova stage of hypercritical accretion of fallback matter [85, 102, 225]. In the latter case, CCOs could in principle have magnetic field in the magnetar range [225].

1E 161348–5055 in the 2-kyr-old supernova remnant RCW 103 was one of the CCO prototypes. However, observations of large flux variations (about 2 orders of magnitude) and an unusual spin period of 6.7 h set it apart from CCOs or any other class of isolated pulsars [44]. No information or meaningful limit on its magnetic field are available from its rotational parameters [65], but it is interesting to notice that De Luca et al. [44] discussed as a possible mechanism to slow-down in ~2-kyr a pulsar born with a normal spin period to the rotation rate of 1E 161348–5055 the propeller interaction between an ultra-magnetised neutron star ($B \sim 10^{14}$–10^{15} G) and a surrounding supernova fallback debris disk.

A major breakthrough was when, on 2016 June 22, the Swift's Burst Alert Telescope detected an X-ray burst (see Fig. 3.7) resembling in all respects those of magnetars from the direction of 1E 161348–5055 [38, 187]. Its duration was ~10 ms, its luminosity $\sim 2 \times 10^{39}$ erg s^{-1} (15–150 keV), and the spectrum was well described by a blackbody with $kT \sim 9$ keV. Subsequent follow-up observations with Swift, Chandra and NuSTAR showed that 1E 161348–5055 was undergoing a magnetar-like outburst: Its luminosity was ~100 times higher than the level the source had maintained for several years and up at least to the last observation carried out before the burst (about 1 month earlier); The pulse profile from single- became double-peaked; A hard power-law component was observed up to ~30 keV for the first time in its energy spectrum, superimposed to its usual thermal emission [187]. In the first year from the onset of the outburst, the overall energy emitted was $\sim 3 \times 10^{42}$ erg [37]. Moreover, Hubble observations carried out in the summer of 2016 unveiled at the position of 1E 161348–5055 a faint infrared source that was not detected in older observations (implying a minimum brightening of 1.3 mag) and therefore can be assumed with a high degree of confidence to be the counterpart of the CCO [205].

While the features exhibited by 1E 161348–5055 during its outburst match precisely the distinguishing features of magnetars, its 6.7-h period remains puzzling. The infrared observations definitely ruled out any doubt about a binary system but could not confirm of exclude the presence of a fallback disk. Recent modelling of neutron star–debris disk interaction by Ho and Andersson ([103], see also Tong et al. [213]) has shown that a disk with mass

(continued)

of $\approx 10^{-9}$ $M_{\odot}$ could slow the neutron star period from milliseconds down to hours in ~1–3 kyr, if the dipole magnetic field is of 5×10^{15} G. Since the formation and survival of a supernova fallback disk are unclear [170], theoretical work is ongoing also on the possibility that the source was slowed down in a relatively short-lived phase of propeller during fallback accretion.

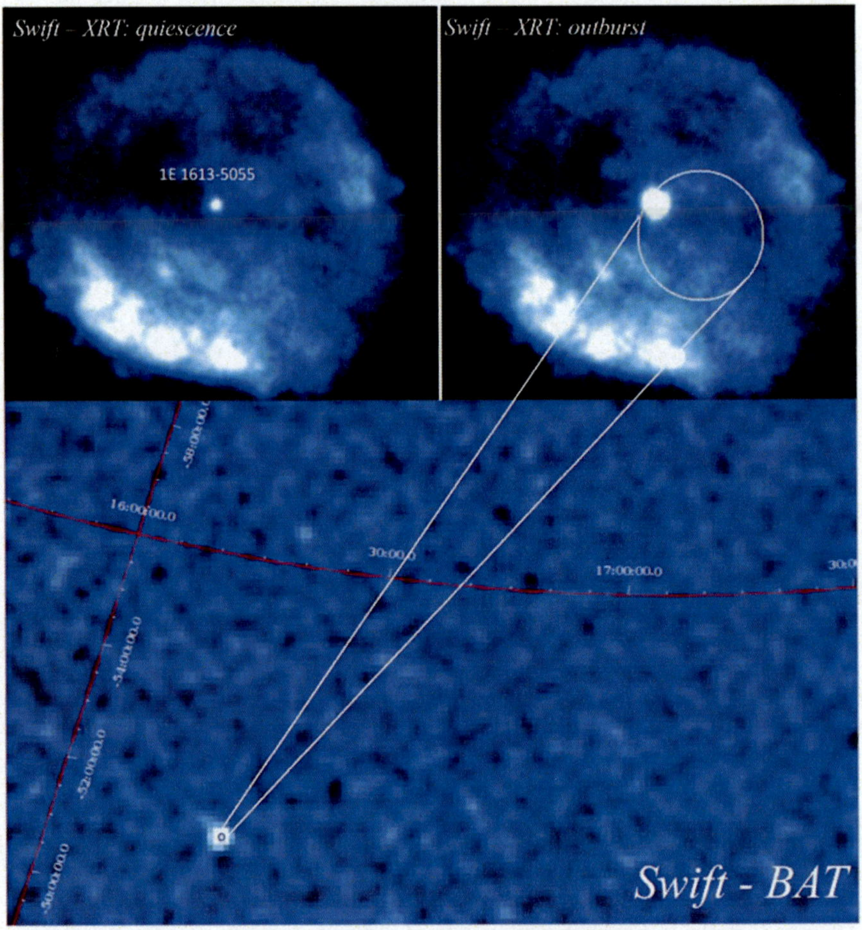

Fig. 3.7 Two Swift-XRT co-added 1–10 keV images of the supernova remnant RCW 103 during the quiescence state of 1E 161348–5055 (from 2011 April 18 to 2016 May 16; exposure time ~33 ks; top left) and in outburst (from 2016 June 22 to 2016 July 20; exposure time ~67 ks; top right). The white circle is the positional accuracy of the SGR-like burst detected by BAT (bottom), which has a radius of 1.5 arcmin (from Rea et al. [187])

SGR J1745–2900: The Galactic Centre Magnetar

When on 24 April 2013 Swift detected with its X-ray telescope a large flare from the region of Sgr A* [45], many minds went to the much anticipated pericenter passage of the object G2, which at the time was expected around mid-2013, and its possible tidal disruption by the Milky Way's 4×10^6-$M_\odot$ black hole [79, 80].

Two days later, however, a magnetar-like short burst was detected by the Swift's coded mask instrument and also a typical magnetar period of 3.8 s was detected, using NuSTAR [128, 157]. The situation was definitely settled when an observation carried out with Chandra (the only X-ray telescope with sufficient angular resolution) showed that a 3.8-s magnetar in outburst, SGR J1745–2900, very close to the position of Sgr A* (angular separation of (2.4 ± 0.3) arcsec) was responsible for all the X-ray luminosity increase measured by Swift ($\approx 2 \times 10^{35}$ erg s^{-1} at 8.3 kpc, while Sgr A* was not detected in the same exposure; Rea et al. [184]). (The closest approach of G2 to Sgr A* actually took place in early 2014; G2 survived and no flaring activity clearly associated to the event was observed; Phifer et al. [172], Pfuhl et al. [171], Ponti et al. [179], Plewa et al. [174].) On 28 April 2013, the new magnetar was also detected as a radio pulsar, the one with the highest dispersion measure and rotation measure, suggesting that the source is embedded in the dense and magnetized plasma of the Galactic center [19, 58, 145, 168, 184, 214].

The angular separation between the magnetar and Sgr A* corresponds to a projected distance of only 0.1 pc, and Rea et al. [184] estimated that if SGR J1745–2900 was born within 1 pc of Sgr A*, its probability of being in a bound orbit around the black hole is of ~90%. Bower et al. [19] measured a transverse velocity of the source relative to Sgr A* of (236 ± 11) km s^{-1} and provided further support to the possibility that the magnetar is bound to Sgr A*. Rea et al. [184] also noted that the high-energy emission produced by the past activity of the magnetar, passing through the molecular clouds surrounding the Galactic center region, might be responsible for a substantial fraction of the light echoes observed in the Fe fluorescence features.

SGR J1745–2900 is proving to be an important probe for the compact object population and the interstellar medium in the Galactic center, but is also exhibiting an interesting behaviour as a magnetar. About 3.5 years after the outset of the outburst, it has not reached the quiescent/pre-outburst luminosity level yet [36]. Its spectral evolution is difficult to reconcile with crustal cooling models, while a continuous particle bombardment from returning currents of the neutron star surface better explain the data. In this hypothesis, both temperature and size of the region at the footprint point of the bundle

(continued)

of current-carrying field lines decrease, as the magnetospheric twist gradually dissipates and the rate of particles impacting the surface consequently declines [35, 36].

3.2.3 *Magnetar Formation*

The generation of magnetar-like magnetic fields from the progenitor star is still a debated and relatively open problem. All along, preliminary calculations have shown that the effects of a turbulent dynamo amplification occurring in a newly born neutron stars can indeed result in a magnetic field of up to a few 10^{17} G. This dynamo effect is expected to operate only in the first $\sim$10 s after the supernova explosion of the massive progenitor, and if the proto-neutron star is born with sufficiently small rotational periods (of the order of 1–2 ms). The resulting amplified magnetic fields are expected to have a strong multipolar structure and toroidal component (Duncan and Thompson [56], Thompson and Duncan [207]). This formation scenario predicts two main observational consequences: (a) magnetars should have large kick velocities, of the order of 10^3 km s^{-1} and (b) their associated supernovae should be more energetic than ordinary core collapse-supernovae, because of the additional rotational energy loss of such fast spinning proto-neutron star.

However, this additional energy loss is not observed in the supernova remnants surrounding magnetars [148, 227], nor a large kick velocity is observed in the few cases where this could be measured. In particular, measured magnetar proper motions are $v = 212 \pm 35$ km s^{-1} for XTE J1810–197 [99], $v = 280 \pm 130$ km s^{-1} for 1E 1547–5408 [46], $v = 157 \pm 17$ km s^{-1} for 1E 2259+586 and $v = 102 \pm 26$ km s^{-1} for 4U 0142+61 [203], while candidate proper motion velocities are $v = 350 \pm 100$ km s^{-1} for SGR 1806–20 and $v = 130 \pm 30$ km s^{-1} for SGR 1900+14 [202]. All these values are well within the typical radio pulsar distribution (see also the gray box on the Galactic Centre magnetar SGR 1745–2900).

The fact that the former predictions did not seem to be fulfilled is however not sufficient to dismiss the dynamo formation mechanism (for example, about the lack of evidence for a particularly energetic supernova, Dall'Osso et al. [39] noted that most of the rotational energy of a proto-neutron star with internal toroidal field $\approx 10^{16}$ G should be released through gravitational waves, without supplying substantial additional energy to the ejecta), but it has lent some support to other formation scenarios. One alternative theory is based on magnetic flux conservation arguments and postulates that the distribution of field strengths in neutron stars simply reflects that of their progenitors. In this fossil field scenario, magnetars would be the descendant of the massive stars with the highest magnetic fields [69].

Counter-arguments have been put forward also in this scenario by Spruit [198], who argued that the number of highly magnetic massive stars with $B \gtrsim 1\,\text{kG}$ is not sufficient to explain the magnetar population. Furthermore, even assuming that most of the magnetic flux is indeed conserved, magnetic fields higher than 10^{14} G seem unattainable.

Recent surveys of very massive stars have shown how in our Galaxy massive stars tend to be in binaries [191]. Furthermore, a detailed radial velocity survey of Westerlund 1, an open cluster of very massive stars which contains a magnetar, CXOU J164710.2–455216, have discovered the possible companion massive star that might have resulted by the disruption of a massive binary progenitor (Clark et al. [33]; see Sect. 3.2.6 for more details). All these results are pointing to a further element in magnetar formation: the evolution in a binary system of massive stars. The binary scenario might overcome the problem of the spin down by the core–envelope coupling. In particular, both mass transfer and stellar merger in compact binaries may lead to substantial spin-up of the mass-gainer (or of the remnant of the merger), favoring the amplification of the magnetic field via dynamo effects [136]. Recent simulations have shown that gamma-ray bursts and hyper-luminous supernovae can indeed be powered by recently formed millisecond magnetar [154], although no direct or sound observational evidence of the existence of such fast spinning and strongly magnetic neutron stars has been collected thus far.

3.2.4 Magnetic Field Evolution and the Neutron Star Bestiary

The evolution of magnetic fields in neutron stars has been extensively studied by a number of authors in the past. In the neutron star solid crust, the field evolves under the influence of the Lorentz force (causing the Hall drift) and the Joule effect (responsible for Ohmic dissipation). The evolution in the liquid core is very uncertain. In the core, soon after the neutron star birth (from hours to days) protons undergo a transition to a type-II superconducting phase [14], in which the magnetic field is confined to tiny flux tubes surrounded by nonmagnetized matter. The dynamics of those flux tubes, likely coupled to the motion of superfluid neutron vortices, is a complex problem that makes the magnetic field evolution in the core formally difficult to tackle (see Elfritz et al. [59]). Most works (e.g. [77, 78, 83, 84, 177, 178, 226]) considered mainly the magnetic evolution in the solid crust, a ~1-km-thick lattice of ions, where the electrical conduction is governed by electrons.

The magnetic field evolution in a neutron star is strictly coupled to its thermal evolution. In fact, the magnetic field influences the heating rate and, secondarily, affects the rate of a few neutrino processes; on the other side, the conduction of heat becomes anisotropic in the presence of a strong magnetic field. The simultaneous study of the magnetic and temperature evolutions (*magneto-thermal evolution*) was started by Pons and Geppert [175] and Aguilera et al. [1] with simplifying assumptions, and later implemented in two-dimensional simulations of the fully coupled magneto-thermal evolution in Pons et al. [178], but including only the Ohmic dissipation. More recently, Viganò and Pons [225] presented the first two-dimensional magneto-thermal code able to manage arbitrarily large magnetic field intensities while self-consistently including the Hall term throughout the entire evolution.

The magneto-thermal evolution in the lifetime of the neutron star is governed by the Hall induction equation, for the magnetic evolution:

$$\frac{\partial \mathbf{B}}{\partial t} = -\nabla \times \left\{ \frac{c^2}{4\pi\sigma} \left[\nabla \times (e^{\nu}\mathbf{B}) - \omega_B \tau_e \left(\nabla \times (e^{\nu}\mathbf{B}) \right) \times \mathbf{B}/B \right] \right\} ,$$

where σ is the electrical conductivity, e^{ν} is the lapse function that accounts for redshift corrections, and $\omega_B \tau_e = B\sigma/en_e c$ is the magnetization parameter ($\omega_B = eB/m_e^* c$ is the gyration frequency of electrons, n_e is the electron number density, and τ_e and m_e^* are the relaxation time and effective mass of electrons), and the cooling, or energy-balance, equation for the thermal evolution of the crust:

$$c_v \frac{\partial T}{\partial t} + \nabla \cdot (-\hat{\kappa} \cdot \nabla T) = -Q_\nu + Q_j ,$$

where $c_v(T, \rho)$ is the specific heat (mainly driven by the neutrons), $\hat{\kappa}(T, \rho)$ is the thermal and electrical conductivity, and $Q_\nu(T, \rho)$ are the neutrino emissivities, and all these quantities depend on the temperature (T) and density (ρ).

For low temperatures ($T \lesssim 10^8$ K) or strong magnetic fields, $\omega_B \tau_e \gg 1$, the evolution is Hall-dominated. For high temperatures (large resistivity) or weak fields, $\omega_B \tau_e \ll 1$, we have instead what is called the *dissipative regime*.

During the past decades, the study of the neutron star field decay, thanks to the continual comparison between theoretical modelling and observations, proved to be fundamental for the understanding of the secular evolutions of pulsars. In particular, recent advances in the magneto-thermal evolutionary models and the availability of deep X-ray observations of many thermally emitting isolated neutron stars, allowed a significant improvement towards the unification of the "bestiary" of different classes of isolated neutron stars (see e.g. [124, 127] for an overview of the different observational manifestations of neutron stars). The sample of detected neutron stars with thermal emission consist of about 40 sources, ranging from magnetars, X-ray dim isolated neutron stars, and central compact objects, to rotational-powered pulsars and 'high-B' pulsars [226]. Figure 3.8 (upper panel) shows the thermal luminosity of all these neutron stars as a function of the age together with several cooling curves for magnetic fields in the range of 3×10^{14}–3×10^{15} G and for two envelope compositions, hydrogen and iron, as computed by Viganò et al. [226]. Note that for young neutron stars ($t < 100$ kyr, still in the neutrino cooling era), light-elements envelopes are able to maintain a higher luminosity (up to an order of magnitude) than iron envelopes.

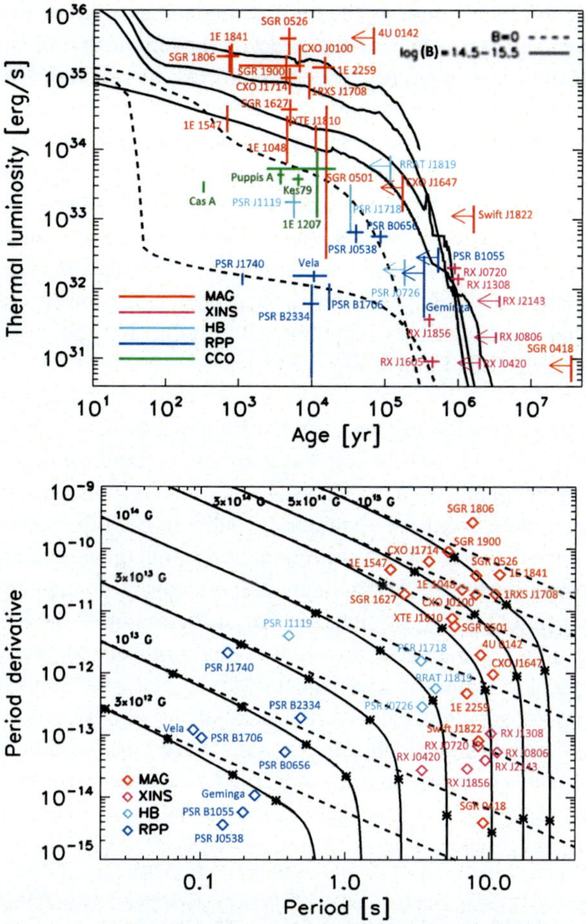

Fig. 3.8 Top: Comparison between observational data and theoretical cooling curves. The thermal luminosity was evaluated for each source using the best estimated distance available and ages were estimated from kinematic measurements, when available, or as the characteristic age derived from the timing parameters (in this case, an arrow was used for objects older than 10 kyr, for which the characteristic age is considered an upper limit on the real age and not a reasonable approximation). Dashed lines are non-magnetic cooling curves, the upper with $M = 1.10\,M_\odot$ and a light-element envelope, and the lower with $M = 1.76\,M_\odot$ and an iron envelope. The magneto-thermal evolutionary tracks (solid lines) were computed for magnetic fields in the range $B = 3\times10^{14}$–3×10^{15} G and both iron and light-element (upper) envelopes. Bottom: Evolutionary tracks in the P–$\dot{P}$ diagram for a 1.4-$M_\odot$ neutron star with $B_p^0 = 3\times10^{12},\ 10^{13},\ 3\times10^{13},\ 10^{14}, 3\times10^{14},$ and 10^{15} G. Asterisks mark the real ages $t = 10^3,\ 10^4,\ 10^5,\ 5\times10^5$ yr, while dashed lines show the tracks followed in absence of magnetic field decay (Adapted from Viganò et al. [226])

The inspection of a range of theoretical models, as well as observations, has shown that the magnetic field has little effect on the luminosity for 'weakly' magnetized neutron stars with $B < 10^{13}$ G. These objects, of which the radio pulsars are the most notable representatives, have thermal luminosities that are compatible with those predicted by standard non-magnetic cooling models. Overall, the magneto-thermal simulations can broadly reproduce the observed X-ray luminosities for a range of initial magnetic field strengths, envelope compositions, and neutron star masses. As the neutron stars age and become colder, they also spin down, primarily due to dipolar radiation losses. In the absence of field decay, pulsars should follow linear tracks in the P–$\dot{P}$ diagram (see dashed lines in the lower panel of Fig. 3.8). However, when magnetic field dissipation is taken into account, evolutionary tracks in the P–$\dot{P}$ diagram bend down (Fig. 3.8).

Comparing observations and theoretical modelling of the neutron star magneto-thermal evolution, considering both the luminosity and the rotational period properties, we can gather that objects like the traditional rotation-powered radio pulsars were born with magnetic fields in the range of a few 10^{12}–10^{13} G. When they cool and slow down, they eventually become invisible in both the radio and the X-ray bands, and hence they lack observable counterparts. On the other hand, pulsars born with fields exceeding the 10^{14} G, will be observed now as young magnetars or high-B pulsars (depending on the strength and the configuration of the field at birth), and have as descendants the objects known as X-ray dim isolated neutron stars. These simulations point to evolutionary connections (some of which have been suspected for long) between apparently different groups of pulsars: Most likely, they are all essentially the same kind of objects, but they were born with different magnetic field strength and geometry, and are observed at different evolutionary stages of their life.

3.2.5 *Low-B Magnetars and High-B Pulsars*

Recently, the long standing belief that magnetars must posses supercritical magnetic fields[2] has been challenged by the discovery of full-fledged magnetars with a dipole magnetic field well in the range of ordinary radio pulsars: SGR 0418+5729, Swift J1822.3−1606, and 3XMM J1852+0033 (Rea et al. [182, 183, 186]; see Turolla and Esposito [217] for a review). Those three magnetars are in fact not dissimilar from the other members of the class, except for the strength of the dipole magnetic field B_p estimated from the spin parameters, in the range $(0.6–4) \times 10^{13}$ G.

[2]The electron quantum critical magnetic field $B_Q = m_e^2 c^2/(\hbar e) \simeq 4.4 \times 10^{13}$ G was traditionally considered the threshold above which magnetars could be found.

Furthermore, this results also in a large characteristic age of $>10^6$ yr, two or three orders of magnitudes larger than the typical magnetar ages, suggesting that these low-field magnetars might be old objects. The small number of detected bursts (with comparatively low energetics) and the low persistent luminosity in quiescence have been taken as further hints that these might be worn-out magnetars, approaching the end of their active life [218]. The 'old magnetar' scenario sounds appealing since it offers an interpretation of the low-magnetic-field magnetars within an already well established framework, validating the magnetar model also for (surface) field strengths quite far away from those of canonical SGR/AXPs.

The crucial issue is whether a relatively low dipolar field is consistent with the starquake models, in which the primary cause of the outbursts is an internal deposition of energy following a crust failure once the magnetically induced shear stress exceeds a critical value. The magnetic stress needed to break the crust is strongly dependent on the density (it is much easier to break the outer crust than the inner crust); moreover, the crust thickness grows as the temperature drops with age. Detailed calculations show that a local magnetic field of $\approx 2 \times 10^{15}$ G should be necessary to break the crust, but closer to the surface of the crust, due to the smaller density, magnetic fields as low as $\sim 10^{14}$ G may lead to crust fractures [92, 135]. At any rate, the minimum requirement seems to be around 10^{14} G. So, can (and how) aged, cold and low-magnetic-field magnetars still produce bursts and outbursts? This depends on the internal toroidal component of its magnetic field. For this reason, objects with similar dipolar magnetic field strength as inferred from their period and period derivative can display very different behaviours. In general the toroidal component of the magnetic field is unmeasurable in a pulsar (but see the gray box for SGR 0418+5729), but this reasoning help us understanding and explaining the populations of active magnetars, low-magnetic-field magnetars, and high-B pulsars. A rough prediction of the expected outburst rate for different initial magnetic field configurations and life stages is given for standard assumptions in Fig. 3.9. For an object similar to SGR 0418+5729, a rate of $\approx 10^{-3}$ starquakes yr^{-1} is expected [169, 226]. Assuming that there are about 10^4 neutron stars in the Galaxy with similar age, and that a (very approximatively) 10% of them were born as magnetars, a naive extrapolation of this event rate to the whole neutron star population leads to the occurrence of ~ 1 low-magnetic-field-magnetar outburst per year. Therefore, we expect that more and more objects of this class will be discovered in the upcoming years. Similarly, we can anticipate some magnetar-like events from only-sporadically-active sources labelled as belonging to other classes of isolated neutron stars, as perhaps shown by the outbursts of PSR J1846–0258 and PSR J1119–6127 [7, 74, 91].

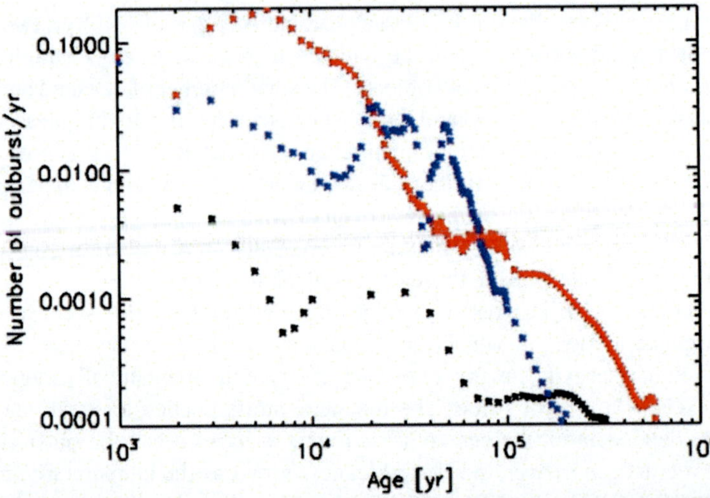

Fig. 3.9 Outburst rate as a function of time for neutron stars with different magnetic field configurations and strength: $B_p^0 = 3 \times 10^{14}$ G in black, $B_p^0 = 10^{15}$ G in red, and in blue $B_p^0 = 10^{14}$ G but with an initial strong toroidal field of $B_t^0 = 5 \times 10^{15}$ G (from [226])

An Energy-Dependant Absorption Line in the Cornerstone Low-Magnetic-Field Magnetar: SGR 0418+5729

The most searched-for indicator of the magnetic field strength in neutron stars are cyclotron features in their spectra. The cyclotron energy for a particle of charge and mass e and m is $E_{\rm cycl} = 11.6\,(m_e/m)/(1+z)\ B_{12}$ keV, where $z \approx 0.8$ is the gravitational redshift, m_e is the mass of the electron, and B_{12} is the magnetic field in units of 10^{12} G. For magnetic fields of $\approx 10^{14}$ G, magnetospheric protons can produce cyclotron lines in the soft X-ray range .

The most convincing of such features in a magnetar was reported by Tiengo et al. [212], who observed a phase-dependent absorption feature in the spectrum of the low-magnetic-field magnetar SGR 0418+5729 during its 2009 outburst [63, 185]. The feature was more prominent in a deep XMM–Newton observation but was present also in data collected with RossiXTE and Swift ([212], see also Esposito et al. [63]). The line energy was varying between ~1 and 5 keV in approximately one-fifth of the rotation cycle (Fig. 3.10), corresponding to magnetic field strength values of 10^{14} to 10^{15} G if due to protons ([212] devised a toy model in which the cyclotron line is due to thermal photons crossing protons localised in a magnetic loop with magnetic field of 10^{14}–10^{15} G close to the surface of the star). An electron cyclotron feature is indeed implausible, as the electrons, considered the dipole magnetic

(continued)

field of $B_p \simeq 6 \times 10^{12}$ G, should be confined in a small volume at a few stellar radii from the neutron star surface (while the alternative explanation in terms of atomic transition lines is ruled out by the variability of the line energy with the spin phase). If the feature is really a proton cyclotron line, it demonstrates in SGR 0418+5729 the presence of nondipolar magnetic field components strong enough to break the neutron star crust and give rise to magnetar outbursts.

Rodríguez Castillo et al. [189] reported the presence of a similar feature in Swift J1822.3–1606, the magnetar with the second lower magnetic field ($B_p \sim$ (1–3) $\times 10^{13}$ G Olausen and Kaspi [164], Scholz et al. [192]), again indicating the presence of localised magnetic fields of 10^{14}–10^{15} G. Interestingly, spectral features with analogous characteristics have been detected also in two X-ray dim isolated neutron stars, RX J0720.4–3125. and RX J1308.6+2127 [17, 18]; in these cases, the magnetic fields deduced in the hypothesis of a proton cyclotron line are of ~2×10^{14} G, in both cases around 5 times the values inferred from the spin parameters.

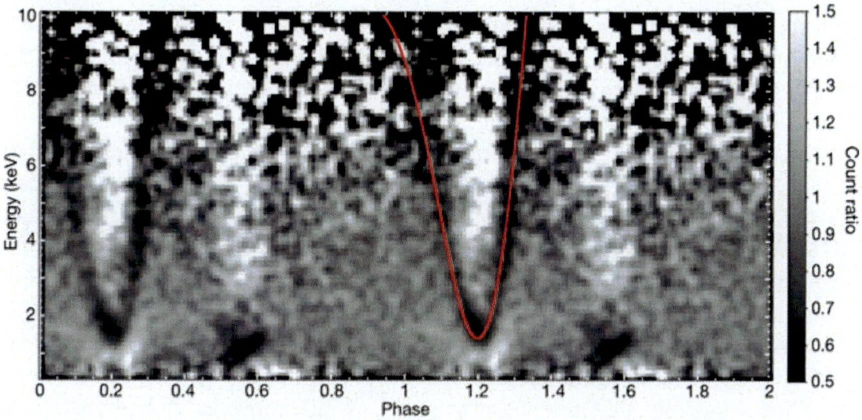

Fig. 3.10 Normalised energy-versus-phase 'spectral image' of SGR 0418+5729. It was obtained from the XMM–Newton data binning the source photons into 100 phase bins and 100-eV-width energy channels and normalising the counts first by the phase-averaged energy spectrum and then by the pulse profile (normalised to the average count rate). The red line shown for one of the cycles represents the proton cyclotron model of Tiengo et al. ([212] from which the image was taken) and highlights the V-shaped absorption feature

3.2.6 *Magnetars in Binary Systems*

It is not clear whether magnetar exist in binary systems but, although it is possible that magnetars to form have to sacrifice the companion, binarity may be an important ingredient for the production of magnetars for at least two reasons. Firstly, if magnetars descend from particularly massive stars ($\gtrsim 20\,M_\odot$), the binary interaction may prevent the formation of a black hole instead of a neutron star. Secondly, in the case the magnetic fields of magnetars are formed through dynamo amplification, even if magnetars form from stars with lower masses, a binary system may help the stellar core to maintain the angular momentum necessary for the dynamo mechanisms. On the other hand, in the fossil field scenario, in which the neutron star magnetic field reflects that of the stellar precursor, progenitors belonging to binary systems may be disfavoured, since magnetic hot stars in compact binary systems are very rare [161].

Both scenarios for the origin of magnetar magnetic field require a very massive progenitor, heavier than $\sim 20\,M_\odot$. For the fossil field, because there seems to be in main sequence stars a trend of stronger magnetic fields with larger masses (Ferrario and Wickramasinghe [70] and references therein). In the dynamo hypothesis, because to produce neutron stars rotating fast enough to generate a magnetar field via α-ω dynamo, a star more massive than $\sim$20–35 $M_\odot$ star is necessary [98]. There are also observational facts that favour very massive stars as the magnetar precursors. Considering the spatial distribution of the known magnetars, it has been noted that their height on the Galactic plane is smaller than that of OB stars. This suggests that they were produced by the most massive O stars [164]. Furthermore, there are some tentative associations of magnetars with massive star clusters.

In the most convincing case, CXOU J164710.2–455216 in Westerlund 1, the young age of the cluster ($\sim$4 Myr) implies a progenitor with minimum mass of $\sim 40\,M_\odot$ [158].[3] To allow such a massive star to produce a neutron star, Clark et al. [33] suggested a binary with $(41 + 35)\,M_\odot$ stars and an orbital period shorter than 8 days.[4] The idea is that the binary interaction drives the primary in a Wolf–Rayet star that through its powerful stellar winds loses mass to the point that the formation of a neutron star is possible. The star also avoids the supergiant stage, during which the core would lose angular momentum because of the core–envelope coupling.

Observationally, the presence of magnetars is often invoked in high-mass X-ray binaries (e.g. [20]) and in particular for some persistent Be systems with long-spin-period ($\gtrsim$1000 s) neutron stars. The issue is that, according to the standard picture, after a short propeller phase, the neutron star enters the accretor stage and its spin period quickly settles at an equilibrium value. To have an equilibrium

[3]For the progenitor of SGR 1900+14, a lower mass of $\sim 17\,M_\odot$ has been inferred, suggesting that magnetars could form from stars with a wide spectrum of initial masses [32, 41].

[4]They also found a candidate for the other member of the pre-supernova system: Wd1–5, a $\sim$9-$M_\odot$ runaway star that is escaping the cluster at high velocity and has a peculiar carbon excess that may be due to the binary evolution.

period longer than 1000 s, the neutron star dipole field must be $\gtrsim 10^{14}$ G [42]. This is unless the accretion rate is very low, but it would be orders of magnitude below the rate necessary to account for the luminosity of these persistent Be X-ray binary systems, hence the puzzle. However, in the recent model of quasi-spherical settling accretion in wind-fed high-mass X-ray binaries by Shakura et al. [195], the equilibrium period can be of $\sim$1000 s even for 'ordinary' magnetic fields of $\sim 10^{12}$–10^{13} G and acceptable accretion rates. The model has been applied successfully in populations studies and to model samples and singles sources (e.g. [30, 143]), including the slowest known accreting pulsar, in the high-mass X-ray binary system AX J1910.7+0917 (spin period of 36.2 ks; [197]).

There is however an exception, the Be X-ray binary SXP 1062 in the Small Magellanic Cloud, which has a spin period of 1062 s and is robustly associated to a supernova remnant with kinematic age of (2–4) $\times$ 10^4 yr (Hénault-Brunet et al. [100]; Haberl et al. [97] propose an even younger age using a temperature–size relationship). In fact, for typical values of magnetic field and accretion rate, it would have been impossible for the neutron star to enter the propeller stage and become an accretor within the time constraint of a few $\times 10^4$ yr dictated by the age of the supernova remnant. After considering several possible scenarios, [180] proposed a neutron star born with an initial magnetic field of $\gtrsim 10^{14}$ G that then decayed to a present value of $\sim 10^{13}$ G (derived in the assumption that the star is rotating close to the equilibrium period).

Magnetars are increasingly popular also in the field of ultraluminous X-ray sources (ULXs). These sources—a few hundreds of them are known—are observed in off-nucleus regions of nearby galaxies at X-ray luminosities exceeding a few 10^{39} erg s^{-1} [123]. Since this threshold for isotropic luminosity is larger than the Eddington limit for spherical accretion of fully ionized hydrogen onto a $\sim$10-$M_\odot$ compact object (a scale value of the black holes of stellar origin observed in our Galaxy), ULXs were considered the observational manifestation of massive black holes of stellar origin ($\lesssim$80–100 $M_\odot$) and, the brightest ones in particular, promising candidates of intermediate-mass black holes of 10^3–10^5 $M_\odot$ [156]. For these reason, the recent discovery in three ULXs (M82 X–2, NGC 5907 ULX, and NGC 7793 P13) of pulsars with spin periods from 0.4 to 1.4 s has been a blow, showing both that some ULXs (even in the high side of their luminosity distribution) may host neutron stars and that neutron stars can achieve extreme supper-Eddington luminosities [10, 121, 122]. The most luminous of the bunch, NGC 5907 ULX, which has a period of $\sim$1.1 s, was observed at a maximum X-ray luminosity of $\sim 10^{41}$ erg s^{-1}, more than 500 times the Eddington limit for a 1.4-$M_\odot$ neutron star [121]. In principle, a neutron star with a magnetic field of $\gtrsim 10^{15}$ G could attain such a super-Eddington luminosity [40, 160], since the magnetic field reduces the electron scattering cross section (see also Sect. 3.2.2.1). However, this explanation is not viable in the cases of NGC 5907 ULX and NGC 7793 P13, because such a huge magnetic field coupled with the rapid spinning of the star would inhibit the accretion via the propeller mechanisms. A possible solution proposed by Israel et al. [121] is that the magnetic field is indeed of a few 10^{14} G at the neutron star surface, but is dominated (∼90%) by multipolar components, which vanish rapidly with the distance, so that at the magnetospheric radius (≈100 neutron star radii) the magnetic field is virtually the dipolar one, low enough for the accretion to proceed. It is interesting to note that this hypothesis does not require for the neutron star only a superstrong magnetic field (a purely dipolar magnetic field would not work), but a complex magnetic field configuration similar to that envisaged for magnetars.

Finally, there is a peculiar binary source in which the presence of an ultra-magnetized neutron star has been suggested not as a wildcard to explain its properties, but because it actually showed one of the specific characteristics of magnetars: LS I +61°303. It is one of the few gamma-ray (TeV) binaries [54] and is in a orbit of 27 days with a 10–15-$M_{\odot}$ Be star. LS I +61°303 is also a periodic (27 d) and variable radio source (it is often referred to as a *microquasar*) and, since no direct evidence for the presence of a neutron star has been obtained so far (for example, pulsations, mass limits or thermonuclear bursts), the nature of its compact object is still debated. In 2008, Swift triggered on a short, SGR-like burst from LS I +61°303. With a duration of ∼0.2–0.3 s, a blackbody spectrum with temperature of 7.5 keV, and luminosity of $\sim 2 \times 10^{37}$ erg s^{-1}, the event had the characteristics of a magnetar burst [216]. A second magnetar-like burst was detected again by Swift in 2012 [23]. Torres et al. [216] discussed the implication of the presence of a magnetar in LS I +61°303 and showed that it would be compatible with the properties of the source.

3.3 Final Remarks

Until about a decade or little more ago, magnetars were regarded as a sort of astrophysical oddities and only comparatively few small groups of astronomers were interested in them. In recent times however, a number of surprising observational discoveries connected more strictly magnetars to the other classes of neutron stars and made them hard to be ignored by the large community studying pulsars at different wavelengths. In fact, it has been discovered that magnetars can be radio pulsars and also that 'ordinary' X-ray and radio pulsars, as well as other sources, such as the peculiar neutron star in the supernova remnant RCW 103, can behave like magnetars, showing the whole array of magnetar activity: bursts, outbursts, dramatic and abrupt pulse profile changes and other timing anomalies. We have also learnt that magnetars can populate unexpected regions of the P–$\dot{P}$ diagram, with dipole magnetic fields measured from the rotation parameters that are comparable to or lower than those of the radio pulsars, disguising at the same time much stronger nondipolar magnetic field components. On the other hand, there are mounting evidences that magnetar (nondipolar) magnetic fields may be present also in other classes of isolated and binary neutron stars; in the future, they might manifest magnetar behaviour. Summarising, episodes of magnetar activity and their frequency seem to be related to the total magnetic energy stored in the internal field of a neutron star but, while a huge tank of this energy is typically associated to the pulsars in the up–right corner of the P–$\dot{P}$ diagram, the external dipolar magnetic field inferred from the spin period and the slow-down rate is not necessary a good proxy for it. For this reason, maybe it would be more appropriate to speak of magnetar-like activity (or 'magnetic restlessness') in neutron stars rather than of magnetars.

Perhaps even more importantly, magnetars are proving to be key objects in understanding the puzzling observational diversity among the different classes of isolated neutron stars. After all, neutron stars are relatively simply, collapsed, objects, presumably all governed by the same equation of state: Why do they come in so many flavours? Theoretical progresses, chiefly about the complexity and the evolution of their magnetic field, are paving the way to a unifying view, in which the age, the different magnetic field strength and geometry at birth, and few other pivotal parameters, such as the mass and the chemical composition of the envelope, can explain the neutron star diversity.

Magnetars are being increasingly invoked in a variety of astrophysical sources and phenomena, from high-mass X-ray binaries and ULXs, to gamma-ray bursts, superluminous supernovae, fast radio bursts, sources of gravitational waves, and many others. While in some cases they are used in a lighthearted way as jacks-of-all-trades, because a huge magnetic field offers an easy way to solve an observational or theoretical problem, magnetar are finally receiving the attention they deserve. This will trigger more and more observational and theoretical efforts which, together with new forthcoming powerful and innovative instruments, such as CTA, SKA, Athena+, X-ray polarimeters, space interferometers, and giant optical telescopes, are bound to deliver many important and surprising discoveries. Magnetars enthusiasts are well positioned to enjoy the next few decades.

Acknowledgments PE and NR acknowledge funding in the framework of the NWO Vidi award A.2320.0076.

References

1. D.N. Aguilera, J.A. Pons, J.A. Miralles, 2D Cooling of magnetized neutron stars. Astron. Astrophys. **486**, 255–271 (2008). https://doi.org/10.1051/0004-6361:20078786
2. J. Aleksić, L.A. Antonelli, P. Antoranz, M. Asensio, U. Barres de Almeida, J.A. Barrio, J. Becerra González, W. Bednarek, K. Berger, E. Bernardini, A. Biland, O. Blanch, R.K. Bock, A. Boller, G. Bonnoli, D. Borla Tridon, T. Bretz, E. Carmona, A. Carosi, P. Colin, E. Colombo, J.L. Contreras, J. Cortina, L. Cossio, S. Covino, P. Da Vela, F. Dazzi, A. De Angelis, G. De Caneva, E. De Cea del Pozo, B. De Lotto, C. Delgado Mendez, A. Diago Ortega, M. Doert, D. Dominis Prester, D. Dorner, M. Doro, D. Eisenacher, D. Elsaesser, D. Ferenc, M.V. Fonseca, L. Font, C. Fruck, R.J. García López, M. Garczarczyk, D. Garrido Terrats, M. Gaug, G. Giavitto, N. Godinović, A. González Muñoz, S.R. Gozzini, A. Hadamek, D. Hadasch, D. Häfner, A. Herrero, J. Hose, D. Hrupec, B. Huber, F. Jankowski, T. Jogler, V. Kadenius, S. Klepser, M.L. Knoetig, T. Krähenbühl, J. Krause, J. Kushida, A. La Barbera, D. Lelas, E. Leonardo, N. Lewandowska, E. Lindfors, S. Lombardi, M. López, R. López-Coto, A. López-Oramas, E. Lorenz, M. Makariev, G. Maneva, N. Mankuzhiyil,

K. Mannheim, L. Maraschi, B. Marcote, M. Mariotti, M. Martínez, D. Mazin, M. Meucci, J.M. Miranda, R. Mirzoyan, J. Moldón, A. Moralejo, P. Munar-Adrover, A. Niedzwiecki, D. Nieto, K. Nilsson, N. Nowak, R. Orito, S. Paiano, M. Palatiello, D. Paneque, R. Paoletti, J.M. Paredes, S. Partini, M. Persic, M. Pilia, J. Pochon, F. Prada, P.G. Prada Moroni, E. Prandini, I. Puljak, I. Reichardt, R. Reinthal, W. Rhode, M. Ribó, J. Rico, S. Rügamer, A. Saggion, K. Saito, T.Y. Saito, M. Salvati, K. Satalecka, V. Scalzotto, V. Scapin, C. Schultz, T. Schweizer, S.N. Shore, A. Sillanpää, J. Sitarek, I. Snidaric, D. Sobczynska, F. Spanier, S. Spiro, V. Stamatescu, A. Stamerra, B. Steinke, J. Storz, S. Sun, T. Surić, L. Takalo, H. Takami, F. Tavecchio, P. Temnikov, T. Terzić, D. Tescaro, M. Teshima, O. Tibolla, D.F. Torres, T. Toyama, A. Treves, M. Uellenbeck, P. Vogler, R.M. Wagner, Q. Weitzel, V. Zabalza, F. Zandanel, R. Zanin, N. Rea, M. Backes, Observations of the magnetars 4U 0142+61 and 1E 2259+586 with the MAGIC telescopes. Astron. Astrophys. **549**, A23 (2013). https://doi.org/10.1051/0004-6361/201220275

3. H. An, V.M. Kaspi, R. Archibald, M. Bachetti, V. Bhalerao, E.C. Bellm, A.M. Beloborodov, S.E. Boggs, D. Chakrabarty, F.E. Christensen, W.W. Craig, F. Dufour, K. Forster, E.V. Gotthelf, B.W. Grefenstette, C.J. Hailey, F.A. Harrison, R. Hascoët, T. Kitaguchi, C. Kouveliotou, K.K. Madsen, K. Mori, M.J. Pivovaroff, V.R. Rana, D. Stern, S. Tendulkar, J.A. Tomsick, J.K. Vogel, W.W. Zhang, NuSTAR Team, NuSTAR results and future plans for magnetar and rotation-powered pulsar observations. Astron. Nachr. **335**, 280–284 (2014). https://doi.org/10.1002/asna.201312032
4. G.E. Anderson, B.M. Gaensler, P.O. Slane, N. Rea, D.L. Kaplan, B. Posselt, L. Levin, S. Johnston, S.S. Murray, C.L. Brogan, M. Bailes, S. Bates, R.A. Benjamin, N.D.R. Bhat, M. Burgay, S. Burke-Spolaor, D. Chakrabarty, N. D'Amico, J.J. Drake, P. Esposito, J.E. Grindlay, J. Hong, G.L. Israel, M.J. Keith, M. Kramer, T.J.W. Lazio, J.C. Lee, J.C. Mauerhan, S. Milia, A. Possenti, B. Stappers, D.T.H. Steeghs, Multi-wavelength Observations of the Radio Magnetar PSR J1622-4950 and discovery of its possibly associated supernova remnant. Astrophys. J. **751**, 53 (2012). https://doi.org/10.1088/0004-637X/751/1/53
5. R.L. Aptekar, D.D. Frederiks, S.V. Golenetskii, V.N. Il'inskii, E.P. Mazets, V.D. Pal'shin, P.S. Butterworth, T.L. Cline, Konus catalog of soft gamma repeater activity: 1978 to 2000. Astrophys. J. Suppl. **137**, 227–277 (2001)
6. R.F. Archibald, V.M. Kaspi, C.-Y. Ng, K.N. Gourgouliatos, D. Tsang, P. Scholz, A.P. Beardmore, N. Gehrels, J.A. Kennea, An anti-glitch in a magnetar. Nature **497**, 591–593 (2013). https://doi.org/10.1038/nature12159
7. R.F. Archibald, V.M. Kaspi, S.P. Tendulkar, P. Scholz, A magnetar-like outburst from a high-b radio pulsar. Astrophys. J. Lett. **829**, L21 (2016). https://doi.org/10.3847/2041-8205/829/1/L21
8. R.F. Archibald, M. Burgay, M. Lyutikov, V.M. Kaspi, P. Esposito, G. Israel, M. Kerr, A. Possenti, N. Rea, J. Sarkissian, P. Scholz, S.P. Tendulkar, Magnetar-like X-ray bursts suppress pulsar radio emission. Astrophys. J. Lett. **849**, L20 (2017). https://doi.org/10.3847/2041-8213/aa9371
9. M.J. Aschwanden, (ed.). *Self-Organized Criticality Systems* (Open Academic Press, Berlin, 2013), 483pp.
10. M. Bachetti, F.A. Harrison, D.J. Walton, B.W. Grefenstette, D. Chakrabarty, F.Fürst, D. Barret, A. Beloborodov, S.E. Boggs, F.E. Christensen, W.W. Craig, A.C. Fabian, C.J. Hailey, A. Hornschemeier, V. Kaspi, S.R. Kulkarni, T. Maccarone, J.M. Miller, V. Rana, D. Stern, S.P. Tendulkar, J. Tomsick, N.A. Webb, W.W. Zhang, An ultraluminous X-ray source powered by an accreting neutron star. Nature **514**, 202–204 (2014). https://doi.org/10.1038/nature13791
11. P. Bak, K. Chen, Self-organized criticality. Sci. Am. **264**, 46–53 (1991). https://doi.org/10.1038/scientificamerican0191-46
12. C. Barat, R.I. Hayles, K. Hurley, M. Niel, G. Vedrenne, U. Desai, V.G. Kurt, V.M. Zenchenko, I.V. Estulin. Fine time structure in the 1979 March 5 gamma ray burst. Astron. Astrophys. **126**, 400–402 (1983)
13. M.G. Baring, A.K. Harding, Resonant compton upscattering in anomalous X-ray pulsars. Astrophys. Space Sci. **308**, 109–118 (2007). https://doi.org/10.1007/s10509-007-9326-x.

14. G. Baym, C. Pethick, D. Pines, Superfluidity in neutron stars. Nature **224**, 673–674 (1969). https://doi.org/10.1038/224673a0
15. A.M. Beloborodov, Electron-positron flows around magnetars. Astrophys. J. **777**, 114 (2013). https://doi.org/10.1088/0004-637X/777/2/114
16. A.M. Beloborodov, On the mechanism of hard X-Ray emission from magnetars. Astrophys. J. **762**, 13 (2013). https://doi.org/10.1088/0004-637X/762/1/13
17. A. Borghese, N. Rea, F. Coti Zelati, A. Tiengo, R. Turolla, Discovery of a strongly phase-variable spectral feature in the isolated neutron star RX J0720.4-3125. Astrophys. J. Lett. **807**, L20 (2015). https://doi.org/10.1088/2041-8205/807/1/L20
18. A. Borghese, N. Rea, F. Coti Zelati, A. Tiengo, R. Turolla, S. Zane, Narrow phase-dependent features in X-ray dim isolated neutron stars: a new detection and upper limits. Mon. Not. Roy. Astron. Soc. **468**, 2975–2983 (2017). https://doi.org/10.1093/mnras/stx632
19. G.C. Bower, A. Deller, P. Demorest, A. Brunthaler, H. Falcke, M. Moscibrodzka, R.M. O'Leary, R.P. Eatough, M. Kramer, K.J. Lee, L. Spitler, G. Desvignes, A.P. Rushton, S. Doeleman, M.J. Reid, The proper motion of the galactic center pulsar relative to sagittarius a*. Astrophys. J. **798**, 120 (2015). https://doi.org/10.1088/0004-637X/798/2/120
20. E. Bozzo, M. Falanga, L. Stella, Are there magnetars in high-mass X-ray binaries? The case of supergiant fast X-ray transients. Astrophys. J. **683**, 1031–1044 (2008). https://doi.org/10.1086/589990
21. M. Burgay, G.L. Israel, A. Possenti, N. Rea, P. Esposito, S. Mereghetti, A. Tiengo, D. Götz, Back to radio: parkes detection of radio pulses from the transient AXP 1E1547.0-5408 Astron. Tel. 1913 (2009)
22. M. Burgay, A. Possenti, M. Kerr, P. Esposito, N. Rea, F.C. Zelati, G.L. Israel, S. Johnston, Pulsed radio emission from PSR J1119-6127 disappeared. Astron. Tel. 9286 (2016)
23. D.N. Burrows, M.M. Chester, V. D'Elia, D.M. Palmer, P. Romano, C.J. Saxton, E. Sonbas, M. Stamatikos, G. Stratta, Swift detection of a burst from LS I +61° 303. GCN Circ. 12914 (2012)
24. F. Camilo, S.M. Ransom, J.P. Halpern, J. Reynolds, D.J. Helfand, N. Zimmerman, J. Sarkissian, Transient pulsed radio emission from a magnetar. Nature **442**, 892–895 (2006). https://doi.org/10.1038/nature04986
25. F. Camilo, I. Cognard, S.M. Ransom, J.P. Halpern, J. Reynolds, N. Zimmerman, E.V. Gotthelf, D.J. Helfand, P. Demorest, G. Theureau, D.C. Backer, The Magnetar XTE J1810-197: variations in torque, radio flux density, and pulse profile morphology. Astrophys. J. **663**, 497–504 (2007). https://doi.org/10.1086/518226
26. F. Camilo, S.M. Ransom, J.P. Halpern, J. Reynolds, 1E 1547.0-5408: a radio-emitting magnetar with a rotation period of 2 seconds. Astrophys. J. Lett. **666**, L93–L96 (2007). https://doi.org/10.1086/521826
27. F. Camilo, J. Reynolds, S. Johnston, J.P. Halpern, S.M. Ransom, The magnetar 1E 1547.0-5408: radio spectrum, polarimetry, and timing. Astrophys. J. **679**, 681–686 (2008). https://doi.org/10.1086/587054
28. F. Camilo, S.M. Ransom, J.P. Halpern, J.A.J. Alford, I. Cognard, J.E. Reynolds, S. Johnston, J. Sarkissian, W. van Straten, Radio disappearance of the magnetar XTE J1810-197 and continued X-ray timing. Astrophys. J. **820**, 110 (2016). https://doi.org/10.3847/0004-637X/820/2/110
29. G. Cavallo, M.J. Rees, A qualitative study of cosmic fireballs and gamma-ray bursts. Mon. Not. Roy. Astron. Soc. **183**, 359–365 (1978). https://doi.org/10.1093/mnras/183.3.359
30. A. Chashkina, S.B. Popov, Magnetic field estimates for accreting neutron stars in massive binary systems and models of magnetic field decay. New Astron. **17**, 594–602 (2012). https://doi.org/10.1016/j.newast.2012.01.004
31. P. Chatterjee, L. Hernquist, R. Narayan, An accretion model for anomalous X-ray pulsars. Astrophys. J. **534**, 373–379 (2000). https://doi.org/10.1086/308748
32. J.S. Clark, M.P. Muno, I. Negueruela, S.M. Dougherty, P.A. Crowther, S.P. Goodwin, R. de Grijs, Unveiling the X-ray point source population of the young massive cluster westerlund 1. Astron. Astrophys. **477**, 147–163 (2008). https://doi.org/10.1051/0004-6361:20077186

33. J.S. Clark, B.W. Ritchie, F. Najarro, N. Langer, I. Negueruela, A VLT/FLAMES survey for massive binaries in Westerlund 1. IV. Wd1-5 - binary product and a pre-supernova companion for the magnetar CXOU J1647-45? Astron. Astrophys. **565**, A90 (2014). https://doi.org/10.1051/0004-6361/201321771
34. T.L. Cline, U.D. Desai, G. Pizzichini, B.J. Teegarden, W.D. Evans, R.W. Klebesadel, J.G. Laros, K. Hurley, M. Niel, G. Vedrenne, Detection of a fast, intense and unusual gamma-ray transient. Astrophys. J. Lett. **237**, L1–L5 (1980). https://doi.org/10.1086/183221
35. F. Coti Zelati, N. Rea, A. Papitto, D. Viganò, J.A. Pons, R. Turolla, P. Esposito, D. Haggard, F.K. Baganoff, G. Ponti, G.L. Israel, S. Campana, D.F. Torres, A. Tiengo, S. Mereghetti, R. Perna, S. Zane, R.P. Mignani, A. Possenti, L. Stella, The X-ray outburst of the galactic centre magnetar SGR J1745-2900 during the first 1.5 year. Mon. Not. Roy. Astron. Soc. **449**, 2685–2699 (2015). https://doi.org/10.1093/mnras/stv480
36. F. Coti Zelati, N. Rea, R. Turolla, J.A. Pons, A. Papitto, P. Esposito, G.L. Israel, S. Campana, S. Zane, A. Tiengo, R.P. Mignani, S. Mereghetti, F.K. Baganoff, D. Haggard, G. Ponti, D.F. Torres, A. Borghese, J. Elfritz, Chandra monitoring of the galactic centre magnetar SGR J1745-2900 during the initial 3.5 years of outburst decay. Mon. Not. Roy. Astron. Soc. **471**, 1819–1829 (2017). https://doi.org/10.1093/mnras/stx1700
37. F. Coti Zelati, N. Rea, J.A. Pons, S. Campana, P. Esposito, Systematic study of magnetar outbursts. Mon. Not. Roy. Astron. Soc. **474**, 961–1017 (2018). https://doi.org/10.1093/mnras/stx2679
38. A. D'Aì, P.A. Evans, D.N. Burrows, N.P.M. Kuin, D.A. Kann, S. Campana, A. Maselli, P. Romano, G. Cusumano, V. La Parola, S.D. Barthelmy, A.P. Beardmore, S.B. Cenko, M. De Pasquale, N. Gehrels, J. Greiner, J.A. Kennea, S. Klose, A. Melandri, J.A. Nousek, J.P. Osborne, D.M. Palmer, B. Sbarufatti, P. Schady, M.H. Siegel, G. Tagliaferri, R. Yates, S. Zane, Evidence for the magnetar nature of 1E 161348-5055 in RCW 103. Mon. Not. Roy. Astron. Soc. **463**, 2394–2404 (2016). https://doi.org/10.1093/mnras/stw2023
39. S. Dall'Osso, S.N. Shore, L. Stella, Early evolution of newly born magnetars with a strong toroidal field. Mon. Not. Roy. Astron. Soc. **398**, 1869–1885 (2009). https://doi.org/10.1111/j.1365-2966.2008.14054.x
40. S. Dall'Osso, R. Perna, L. Stella, NuSTAR J095551+6940.8: a highly magnetized neutron star with super-Eddington mass accretion. Mon. Not. Roy. Astron. Soc. **449**, 2144–2150 (2015). https://doi.org/10.1093/mnras/stv170
41. B. Davies, D.F. Figer, R.-P. Kudritzki, C. Trombley, C. Kouveliotou, S. Wachter, The progenitor mass of the magnetar SGR1900+14. Astrophys. J. **707**, 844–851 (2009). https://doi.org/10.1088/0004-637X/707/1/844
42. R.E. Davies, J.E. Pringle, Spindown of neutron stars in close binary systems. II. Mon. Not. Roy. Astron. Soc. **196**, 209–224 (1981)
43. A. De Luca, Central compact objects in supernova remnants, in *International Conference Physics of Neutron Stars - 2017. 50 Years After (St. Petersburg, Russian Federation)*, ed. by G.G. Pavlov, J.A. Pons, P.S. Shternin, D.G. Yakovlev. Journal of Physics Conference Series, vol. 932 (2018), p. 012006
44. A. De Luca, P.A. Caraveo, S. Mereghetti, A. Tiengo, G.F. Bignami, A long-period, violently variable X-ray source in a young supernova remnant. Science **313**, 814–817 (2006). https://doi.org/10.1126/science.1129185
45. N. Degenaar, M.T. Reynolds, J.M. Miller, J.A. Kennea, R. Wijnands, Large Flare from Sgr A* Detected by Swift. The Astronomer's Telegram 5006 (2013)
46. A.T. Deller, F. Camilo, J.E. Reynolds, J.P. Halpern, The proper motion of PSR J1550-5418 measured with VLBI: a second magnetar velocity measurement. Astrophys. J. Lett. **748**, L1 (2012). https://doi.org/10.1088/2041-8205/748/1/L1
47. P.R. den Hartog, L. Kuiper, W. Hermsen, Detailed high-energy characteristics of AXP 1RXS J170849-400910. Probing the magnetosphere using INTEGRAL, RXTE, and XMM-Newton. Astron. Astrophys. **489**, 263–279 (2008). https://doi.org/10.1051/0004-6361:200809772

48. P.R. den Hartog, L. Kuiper, W. Hermsen, V.M. Kaspi, R. Dib, J. Knödlseder, F.P. Gavriil, Detailed high-energy characteristics of AXP 4U 0142+61. Multi-year observations with INTEGRAL, RXTE, XMM-Newton, and ASCA. Astron. Astrophys. **489**, 245–261 (2008). https://doi.org/10.1051/0004-6361:200809390
49. V.S. Dhillon, T.R. Marsh, F. Hulleman, M.H. van Kerkwijk, A. Shearer, S.P. Littlefair, F.P. Gavriil, V.M. Kaspi, High-speed, multicolour optical photometry of the anomalous X-ray pulsar 4U 0142+61 with ULTRACAM. Mon. Not. Roy. Astron. Soc. **363**, 609–614 (2005). https://doi.org/10.1111/j.1365-2966.2005.09465.x
50. V.S. Dhillon, T.R. Marsh, S.P. Littlefair, C.M. Copperwheat, P. Kerry, R. Dib, M. Durant, V.M. Kaspi, R.P. Mignani, A. Shearer, Optical pulsations from the anomalous X-ray pulsar 1E1048.1-5937. Mon. Not. Roy. Astron. Soc. **394**, L112–L116 (2009). https://doi.org/10.1111/j.1745-3933.2009.00623.x
51. V.S. Dhillon, T.R. Marsh, S.P. Littlefair, C.M. Copperwheat, R.D.G. Hickman, P. Kerry, A.J. Levan, N. Rea, C.D.J. Savoury, N.R. Tanvir, R. Turolla, K. Wiersema, The first observation of optical pulsations from a soft gamma repeater: SGR 0501+4516. Mon. Not. Roy. Astron. Soc. **416**, L16–L20 (2011). https://doi.org/10.1111/j.1745-3933.2011.01088.x
52. R. Dib, V.M. Kaspi, 16 yr of RXTE monitoring of five anomalous X-ray pulsars. Astrophys. J. **784**, 37 (2014). https://doi.org/10.1088/0004-637X/784/1/37
53. R. Dib, V.M. Kaspi, F.P. Gavriil, Glitches in anomalous X-ray pulsars. Astrophys. J. **673**, 1044–1061 (2008). https://doi.org/10.1086/524653
54. G. Dubus, Gamma-ray binaries and related systems. Astron. Astrophys. Rev. **21**, 64 (2013). https://doi.org/10.1007/s00159-013-0064-5
55. R.C. Duncan, Global seismic oscillations in soft gamma repeaters. Astrophys. J. Lett. **498**, L45–L49 (1998). https://doi.org/10.1086/311303
56. R.C. Duncan, C. Thompson, Formation of very strongly magnetized neutron stars - implications for gamma-ray bursts. Astrophys. J. Lett. **392**, L9–L13 (1992)
57. M. Durant, M.H. van Kerkwijk, The broadband spectrum and infrared variability of the magnetar AXP 1E 1048.1-5937. Astrophys. J. 627, 376–382 (2005). https://doi.org/10.1086/429714
58. R.P. Eatough, H. Falcke, R. Karuppusamy, K.J. Lee, D.J. Champion, E.F. Keane, G. Desvignes, D.H.F.M. Schnitzeler, L.G. Spitler, M. Kramer, B. Klein, C. Bassa, G.C. Bower, A. Brunthaler, I. Cognard, A.T. Deller, P.B. Demorest, P.C.C. Freire, A. Kraus, A.G. Lyne, A. Noutsos, B. Stappers, N. Wex, A strong magnetic field around the supermassive black hole at the centre of the Galaxy. Nature **501**, 391–394 (2013). https://doi.org/10.1038/nature12499
59. J.G. Elfritz, J.A. Pons, N. Rea, K. Glampedakis, D. Viganò, Simulated magnetic field expulsion in neutron star cores. Mon. Not. Roy. Astron. Soc. **456**, 4461–4474 (2016). https://doi.org/10.1093/mnras/stv2963
60. T. Enoto, S. Shibata, T. Kitaguchi, Y. Suwa, T. Uchide, H. Nishioka, S. Kisaka, T. Nakano, H. Murakami, K. Makishima, Magnetar broadband X-ray spectra correlated with magnetic fields: suzaku archive of SGRs and AXPs combined with NuSTAR, swift, and RXTE. Astrophys. J. Suppl. **231**, 8 (2017). https://doi.org/10.3847/1538-4365/aa6f0a
61. P. Esposito, S. Mereghetti, A. Tiengo, L. Sidoli, M. Feroci, P. Woods, Five years of SGR 1900+14 observations with BeppoSAX. Astron. Astrophys. **461**, 605–612 (2007). https://doi.org/10.1051/0004-6361:20065529
62. P. Esposito, G.L. Israel, S. Zane, F. Senziani, R.L.C. Starling, N. Rea, D.M. Palmer, N. Gehrels, A. Tiengo, A. De Luca, D. Götz, S. Mereghetti, P. Romano, T. Sakamoto, S.D. Barthelmy, L. Stella, R. Turolla, M. Feroci, V. Mangano, The 2008 May burst activation of SGR1627-41. Mon. Not. Roy. Astron. Soc. **390**, L34–L38 (2008). https://doi.org/10.1111/j.1745-3933.2008.00530.x
63. P. Esposito, G.L. Israel, R. Turolla, A. Tiengo, D. Götz, A. De Luca, R.P. Mignani, S. Zane, N. Rea, V. Testa, P.A. Caraveo, S. Chaty, F. Mattana, S. Mereghetti, A. Pellizzoni, P. Romano, Early X-ray and optical observations of the soft gamma-ray repeater SGR 0418+5729. Mon. Not. Roy. Astron. Soc. **405**, 1787–1795 (2010). https://doi.org/10.1111/j.1365-2966.2010.16551.x

64. P. Esposito, G.L. Israel, R. Turolla, F. Mattana, A. Tiengo, A. Possenti, S. Zane, N. Rea, M. Burgay, D. Götz, S. Mereghetti, L. Stella, M.H. Wieringa, J.M. Sarkissian, T. Enoto, P. Romano, T. Sakamoto, Y.E. Nakagawa, K. Makishima, K. Nakazawa, H. Nishioka, C. François-Martin, Long-term spectral and timing properties of the soft gamma-ray repeater SGR 1833-0832 and detection of extended X-ray emission around the radio pulsar PSR B1830-08. Mon. Not. Roy. Astron. Soc. **416**, 205–215 (2011). https://doi.org/10.1111/j.1365-2966.2011.19022.x
65. P. Esposito, R. Turolla, A. De Luca, G.L. Israel, A. Possenti, D.N. Burrows, Swift monitoring of the central X-ray source in RCW 103. Mon. Not. Roy. Astron. Soc. **418**, 170–175 (2011). https://doi.org/10.1111/j.1365-2966.2011.19473.x.
66. A.C. Fabian, Theories of the nuclei of active galaxies. Proc. Roy. Soc. Lond. Ser. A **366**, 449–459 (1979). https://doi.org/10.1098/rspa.1979.0064
67. R. Fernández, C. Thompson, Resonant cyclotron scattering in three dimensions and the quiescent nonthermal X-ray emission of magnetars. Astrophys. J. **660**, 615–640 (2007). https://doi.org/10.1086/511810
68. M. Feroci, E. Bozzo, S. Brandt, M. Hernanz, M. van der Klis, L.-P. Liu, P. Orleanski, M. Pohl, A. Santangelo, S. Schanne, et al. The LOFT mission concept: a status update, in *Space Telescopes and Instrumentation 2016: Ultraviolet to Gamma Ray*. Proceedings of the SPIE, vol. 9905 (2016), p. 99051R. https://doi.org/10.1117/12.2233161
69. L. Ferrario, D. Wickramasinghe, Modelling of isolated radio pulsars and magnetars on the fossil field hypothesis. Mon. Not. Roy. Astron. Soc. **367**, 1323–1328 (2006). https://doi.org/10.1111/j.1365-2966.2006.10058.x
70. L. Ferrario, D. Wickramasinghe, Origin and evolution of magnetars. Mon. Not. Roy. Astron. Soc. **389**, L66–L70 (2008). https://doi.org/10.1111/j.1745-3933.2008.00527.x
71. D.A. Frail, S.R. Kulkarni, J.S. Bloom, An outburst of relativistic particles from the soft gamma-ray repeater SGR 1900+14. Nature **398**, 127–129 (1999)
72. B.M. Gaensler, C. Kouveliotou, J.D. Gelfand, G.B. Taylor, D. Eichler, R.A.M.J. Wijers, J. Granot, E. Ramirez-Ruiz, Y.E. Lyubarsky, R.W. Hunstead, D. Campbell-Wilson, A.J. van der Horst, M.A. McLaughlin, R.P. Fender, M.A. Garrett, K.J. Newton-McGee, D.M. Palmer, N. Gehrels, P.M. Woods, An expanding radio nebula produced by a giant flare from the magnetar SGR 1806-20. Nature **434**, 1104–1106 (2005). https://doi.org/10.1038/nature03498
73. F.P. Gavriil, V.M. Kaspi, P.M. Woods, Magnetar-like X-ray bursts from an anomalous X-ray pulsar. Nature **419**, 142–144 (2002)
74. F.P. Gavriil, M.E. Gonzalez, E.V. Gotthelf, V.M. Kaspi, M.A. Livingstone, P.M. Woods, Magnetar-like emission from the young pulsar in Kes 75. Science **319**, 1802–1805 (2008). https://doi.org/10.1126/science.1153465
75. N. Gehrels, G. Chincarini, P. Giommi, K.O. Mason, J.A. Nousek, A.A. Wells, N.E. White, S.D. Barthelmy, D.N. Burrows, L.R. Cominsky, K.C. Hurley, F.E. Marshall, P. Mészáros, P.W.A. Roming, L. Angelini, L.M. Barbier, T. Belloni, S. Campana, P.A. Caraveo, M.M. Chester, O. Citterio, T.L. Cline, M.S. Cropper, J.R. Cummings, A.J. Dean, E.D. Feigelson, E.E. Fenimore, D.A. Frail, A.S. Fruchter, G.P. Garmire, K. Gendreau, G. Ghisellini, J. Greiner, J.E. Hill, S.D. Hunsberger, H.A. Krimm, S.R. Kulkarni, P. Kumar, F. Lebrun, N.M. Lloyd-Ronning, C.B. Markwardt, B.J. Mattson, R.F. Mushotzky, J.P. Norris, J. Osborne, B. Paczynski, D.M. Palmer, H.-S. Park, A.M. Parsons, J. Paul, M.J. Rees, C.S. Reynolds, J.E. Rhoads, T.P. Sasseen, B.E. Schaefer, A.T. Short, A.P. Smale, I.A. Smith, L. Stella, G. Tagliaferri, T. Takahashi, M. Tashiro, L.K. Townsley, J. Tueller, M.J.L. Turner, M. Vietri, W. Voges, M.J. Ward, R. Willingale, F. M. Zerbi, W.W. Zhang, The swift gamma-ray burst mission. Astrophys. J. **611**, 1005–1020 (2004). https://doi.org/10.1086/422091
76. J.D. Gelfand, Y.E. Lyubarsky, D. Eichler, B.M. Gaensler, G.B. Taylor, J. Granot, K.J. Newton-McGee, E. Ramirez-Ruiz, C. Kouveliotou, R.A.M.J. Wijers, A rebrightening of the radio nebula associated with the 2004 December 27 giant flare from SGR 1806-20. Astrophys. J. Lett. **634**, L89–L92 (2005). https://doi.org/10.1086/498643

77. U. Geppert, M. Rheinhardt, Non-linear magnetic field decay in neutron stars. Theory and observations. Astron. Astrophys. **392**, 1015–1024 (2002). https://doi.org/10.1051/0004-6361:20020978
78. U. Geppert, M. Rheinhardt, Magnetars versus radio pulsars. MHD stability in newborn highly magnetized neutron stars. Astron. Astrophys. **456**, 639–649 (2006). https://doi.org/10.1051/0004-6361:20065144
79. S. Gillessen, R. Genzel, T.K. Fritz, E. Quataert, C. Alig, A. Burkert, J. Cuadra, F. Eisenhauer, O. Pfuhl, K. Dodds-Eden, C.F. Gammie, T. Ott, A gas cloud on its way towards the supermassive black hole at the Galactic Centre. Nature **481**, 51–54 (2012). https://doi.org/10.1038/nature10652
80. S. Gillessen, R. Genzel, T.K. Fritz, F. Eisenhauer, O. Pfuhl, T. Ott, J. Cuadra, M. Schartmann, A. Burkert, New observations of the gas cloud G2 in the galactic center. Astrophys. J. **763**, 78 (2013). https://doi.org/10.1088/0004-637X/763/2/78
81. K. Glampedakis, D.I. Jones, Three evolutionary paths for magnetar oscillations. Mon. Not. Roy. Astron. Soc. **439**, 1522–1535 (2014). https://doi.org/10.1093/mnras/stu017
82. K. Glampedakis, L. Samuelsson, N. Andersson, Elastic or magnetic? A toy model for global magnetar oscillations with implications for quasi-periodic oscillations during flares. Mon. Not. Roy. Astron. Soc. **371**, L74–L77 (2006). https://doi.org/10.1111/j.1745-3933.2006.00211.x
83. P. Goldreich, A. Reisenegger, Magnetic field decay in isolated neutron stars. Astrophys. J. **395**, 250–258 (1992). https://doi.org/10.1086/171646
84. D. Gonzalez, A. Reisenegger, Internal heating of old neutron stars: contrasting different mechanisms. Astron. Astrophys. **522**, A16 (2010). https://doi.org/10.1051/0004-6361/201015084
85. E.V. Gotthelf, J.P. Halpern, J. Alford, The spin-down of PSR J0821-4300 and PSR J1210-5226: confirmation of central compact objects as anti-magnetars. Astrophys. J. **765**, 58 (2013). https://doi.org/10.1088/0004-637X/765/1/58
86. D. Götz, S. Mereghetti, I.F. Mirabel, K. Hurley, Spectral evolution of weak bursts from SGR 1806-20 observed with INTEGRAL. Astron. Astrophys. **417**, L45–L48 (2004). https://doi.org/10.1051/0004-6361:20040080
87. D. Götz, S. Mereghetti, S. Molkov, K. Hurley, I.F. Mirabel, R. Sunyaev, G. Weidenspointner, S. Brandt, M. del Santo, M. Feroci, E. Göğüş, A. von Kienlin, M. van der Klis, C. Kouveliotou, N. Lund, G. Pizzichini, P. Ubertini, C. Winkler, P.M. Woods, Two years of INTEGRAL monitoring of the soft gamma-ray repeater SGR 1806-20: from quiescence to frenzy. Astron. Astrophys. **445**, 313–321 (2006). https://doi.org/10.1051/0004-6361:20053648.
88. D. Götz, S. Mereghetti, A. Tiengo, P. Esposito, Magnetars as persistent hard X-ray sources: INTEGRAL discovery of a hard tail in SGR 1900+14. Astron. Astrophys. **449**, L31–L34 (2006). https://doi.org/10.1051/0004-6361:20064870
89. D. Götz, N. Rea, G.L. Israel, S. Zane, P. Esposito, E.V. Gotthelf, S. Mereghetti, A. Tiengo, R. Turolla, Long term hard X-ray variability of the anomalous X-ray pulsar 1RXS J170849.0-400910 discovered with INTEGRAL. Astron. Astrophys. **475**, 317–321 (2007). https://doi.org/10.1051/0004-6361:20078291
90. E. Göğüş, C. Kouveliotou, P.M. Woods, C. Thompson, R.C. Duncan, M.S. Briggs, Temporal and spectral characteristics of short bursts from the soft gamma repeaters 1806-20 and 1900+14. Astrophys. J. **558**, 228–236 (2001)
91. E. Göğüş, L. Lin, Y. Kaneko, C. Kouveliotou, A.L. Watts, M. Chakraborty, M.A. Alpar, D. Huppenkothen, O.J. Roberts, G. Younes, A.J. van der Horst, Magnetar-like X-ray bursts from a rotation-powered pulsar, PSR J1119-6127. Astrophys. J. Lett. 829, L25 (2016). https://doi.org/10.3847/2041-8205/829/2/L25
92. K.N. Gourgouliatos, A. Cumming, Hall drift and the braking indices of young pulsars. Mon. Not. Roy. Astron. Soc. **446**, 1121–1128 (2015). https://doi.org/10.1093/mnras/stu2140
93. E. Göğüş , P.M. Woods, C. Kouveliotou, J. van Paradijs, M.S. Briggs, R.C. Duncan, C. Thompson, Statistical properties of SGR 1900+14 bursts. Astrophys. J. Lett. **526**, L93–L96 (1999). https://doi.org/10.1086/312380

94. E. Göğüş, P.M. Woods, C. Kouveliotou, J. van Paradijs, M.S. Briggs, R.C. Duncan, C. Thompson, Statistical properties of SGR 1806-20 bursts. Astrophys. J. Lett. **532**, L121–L124 (2000). https://doi.org/10.1086/312583
95. E. Göğüş, P.M. Woods, C. Kouveliotou, M.H. Finger, V. Pal'shin, Y. Kaneko, S. Golenetskii, D. Frederiks, C. Airhart, Extended tails from SGR 1806-20 bursts. Astrophys. J. **740**, 55 (2011). https://doi.org/10.1088/0004-637X/740/2/55
96. J. Granot, E. Ramirez-Ruiz, G.B. Taylor, D. Eichler, Y.E. Lyubarsky, R.A.M.J. Wijers, B.M. Gaensler, J.D. Gelfand, C. Kouveliotou, Diagnosing the outflow from the SGR 1806-20 giant flare with radio observations. Astrophys. J. **638**, 391–396 (2006). https://doi.org/10.1086/497680
97. F. Haberl, R. Sturm, M.D. Filipović, W. Pietsch, E.J. Crawford, SXP 1062, a young be X-ray binary pulsar with long spin period. Implications for the neutron star birth spin. Astron. Astrophys. **537**, L1 (2012). https://doi.org/10.1051/0004-6361/201118369
98. A. Heger, S.E. Woosley, H.C. Spruit, Presupernova evolution of differentially rotating massive stars including magnetic fields. Astrophys. J. **626**, 350–363 (2005). https://doi.org/10.1086/429868
99. D.J. Helfand, S. Chatterjee, W.F. Brisken, F. Camilo, J. Reynolds, M.H. van Kerkwijk, J.P. Halpern, S.M. Ransom, VLBA measurement of the transverse velocity of the magnetar XTE J1810-197. Astrophys. J. **662**, 1198–1203 (2007). https://doi.org/10.1086/518028
100. V. Hénault-Brunet, L.M. Oskinova, M.A. Guerrero, W. Sun, Y.-H. Chu, C.J. Evans, J.S. Gallagher, III, R.A. Gruendl, J. Reyes-Iturbide, Discovery of a Be/X-ray pulsar binary and associated supernova remnant in the wing of the small magellanic cloud. Mon. Not. Roy. Astron. Soc. **420**, L13–L17 (2012). https://doi.org/10.1111/j.1745-3933.2011.01183.x
101. H. Herold, Compton and Thomson scattering in strong magnetic fields. Phys. Rev. D **19**, 2868–2875 (1979). https://doi.org/10.1103/PhysRevD.19.2868
102. W.C.G. Ho, Evolution of a buried magnetic field in the central compact object neutron stars. Mon. Not. Roy. Astron. Soc. **414**, 2567–2575 (2011). https://doi.org/10.1111/j.1365-2966.2011.18576.x
103. W.C.G. Ho, N. Andersson, Ejector and propeller spin-down: how might a superluminous supernova millisecond magnetar become the 6.67 h pulsar in RCW 103. Mon. Not. Roy. Astron. Soc. **464**, L65–L69 (2017). https://doi.org/10.1093/mnrasl/slw186
104. G. Hobbs, A.G. Lyne, M. Kramer, An analysis of the timing irregularities for 366 pulsars. Mon. Not. Roy. Astron. Soc. **402**, 1027–1048 (2010). https://doi.org/10.1111/j.1365-2966.2009.15938.x
105. F. Hulleman, M.H. van Kerkwijk, S.R. Kulkarni, An optical counterpart to the anomalous X-ray pulsar 4U0142+61. Nature **408**, 689–692 (2000)
106. F. Hulleman, M.H. van Kerkwijk, S.R. Kulkarni, The anomalous X-ray pulsar 4U 0142+61: variability in the infrared and a spectral break in the optical. Astron. Astrophys. **416**, 1037–1045 (2004). https://doi.org/10.1051/0004-6361:20031756
107. D. Huppenkothen, A.L. Watts, P. Uttley, A.J. van der Horst, M. van der Klis, C. Kouveliotou, E. Göğüş, J. Granot, S. Vaughan, M.H. Finger, Quasi-periodic oscillations and broadband variability in short magnetar bursts. Astrophys. J. **768**, 87 (2013). https://doi.org/10.1088/0004-637X/768/1/87
108. D. Huppenkothen, C. D'Angelo, A.L. Watts, L. Heil, M. van der Klis, A.J. van der Horst, C. Kouveliotou, M.G. Baring, E. Göğüş, J. Granot, Y. Kaneko, L. Lin, A. von Kienlin, G. Younes, Quasi-periodic oscillations in short recurring bursts of the soft gamma repeater J1550-5418. Astrophys. J. **787**, 128 (2014). https://doi.org/10.1088/0004-637X/787/2/128
109. D. Huppenkothen, L.M. Heil, A.L. Watts, E. Göğüş, Quasi-periodic oscillations in short recurring bursts of magnetars SGR 1806-20 and SGR 1900+14 observed with RXTE. Astrophys. J. **795**, 114 (2014). https://doi.org/10.1088/0004-637X/795/2/114
110. D. Huppenkothen, B.J. Brewer, D.W. Hogg, I. Murray, M. Frean, C. Elenbaas, A.L. Watts, Y. Levin, A.J. van der Horst, C. Kouveliotou, Dissecting magnetar variability with Bayesian hierarchical models. Astrophys. J. **810**, 66 (2015). https://doi.org/10.1088/0004-637X/810/1/66

111. K. Hurley, T. Cline, E. Mazets, S. Barthelmy, P. Butterworth, F. Marshall, D. Palmer, R. Aptekar, S. Golenetskii, V. Il'Inskii, D. Frederiks, J. McTiernan, R. Gold, J. Trombka, A giant periodic flare from the soft gamma-ray repeater SGR 1900+14. Nature **397**, 41–43 (1999)
112. K. Hurley, S.E. Boggs, D.M. Smith, R.C. Duncan, R. Lin, A. Zoglauer, S. Krucker, G. Hurford, H. Hudson, C. Wigger, W. Hajdas, C. Thompson, I. Mitrofanov, A. Sanin, W. Boynton, C. Fellows, A. von Kienlin, G. Lichti, A. Rau, T. Cline, An exceptionally bright flare from SGR 1806-20 and the origins of short-duration γ-ray bursts. Nature **434**, 1098–1103 (2005). https://doi.org/10.1038/nature03519
113. U.S. Inan, N.G. Lehtinen, S.J. Lev-Tov, M.P. Johnson, T.F. Bell, K. Hurley, Ionization of the lower ionosphere by γ-rays from a magnetar: detection of a low energy (3–10 keV) component. Geophys. Res. Lett. **26**, 3357–3360 (1999)
114. U.S. Inan, N.G. Lehtinen, R.C. Moore, K. Hurley, S. Boggs, D.M. Smith, G.J. Fishman, Massive disturbance of the daytime lower ionosphere by the giant γ-ray flare from magnetar SGR 1806-20. Geophys. Res. Lett. **34**, L08103 (2007). https://doi.org/10.1029/2006GL029145
115. G. Israel, S. Dall'Osso, Bursts and flares from highly magnetic pulsars, in *Proceedings of the First Session of the Sant Cugat Forum on Astrophysics High-Energy Emission from Pulsars and Their Systems*, ed. by D.F. Torres, N. Rea. Astrophysics and Space Science Proceedings (Springer, Heidelberg, 2011), pp. 279–298. ISBN 978-3-642-17250-2. https://doi.org/10.1007/978-3-642-17251-9_23.
116. G. Israel, L. Stella, S. Covino, S. Campana, L. Angelini, R. Mignani, S. Mereghetti, G. Marconi, R. Perna, Unveiling the multi-wavelength phenomenology of anomalous X-ray pulsars, in *Young Neutron Stars and Their Environments, IAU Symposium no. 218*, ed. by F. Camilo, B.M. Gaensler (Astronomical Society of the Pacific, San Francisco, 2004), p. 247
117. G.L. Israel, T. Belloni, L. Stella, Y. Rephaeli, D.E. Gruber, P. Casella, S. Dall'Osso, N. Rea, M. Persic, R.E. Rothschild, The discovery of rapid X-ray oscillations in the tail of the SGR 1806-20 hyperflare. Astrophys. J. Lett. 628, L53–L56 (2005). https://doi.org/10.1086/432615
118. G.L. Israel, D. Götz, S. Zane, S. Dall'Osso, N. Rea, L. Stella, Linking the X-ray timing and spectral properties of the glitching AXP 1RXS J170849-400910. Astron. Astrophys. **476**, L9–L12 (2007). https://doi.org/10.1051/0004-6361:20078215
119. G.L. Israel, P. Romano, V. Mangano, S. Dall'Osso, G. Chincarini, L. Stella, S. Campana, T. Belloni, G. Tagliaferri, A.J. Blustin, T. Sakamoto, K. Hurley, S. Zane, A. Moretti, D. Palmer, C. Guidorzi, D.N. Burrows, N. Gehrels, H.A. Krimm, A swift gaze into the 2006 March 29 burst forest of SGR 1900+14. Astrophys. J. **685**, 1114–1128 (2008). https://doi.org/10.1086/590486
120. G.L. Israel, P. Esposito, N. Rea, S. Dall'Osso, F. Senziani, P. Romano, V. Mangano, D. Götz, S. Zane, A. Tiengo, D.M. Palmer, H. Krimm, N. Gehrels, S. Mereghetti, L. Stella, R. Turolla, S. Campana, R. Perna, L. Angelini, A. De Luca, The 2008 October swift detection of X-ray bursts/outburst from the transient SGR-like AXP 1E1547.0-5408. Mon. Not. Roy. Astron. Soc. **408**, 1387–1395 (2010). https://doi.org/10.1111/j.1365-2966.2010.17001.x
121. G.L. Israel, A. Belfiore, L. Stella, P. Esposito, P. Casella, A. De Luca, M. Marelli, A. Papitto, M. Perri, S. Puccetti, G.A.R. Castillo, D. Salvetti, A. Tiengo, L. Zampieri, D. D'Agostino, J. Greiner, F. Haberl, G. Novara, R. Salvaterra, R. Turolla, M. Watson, J. Wilms, A. Wolter, An accreting pulsar with extreme properties drives an ultraluminous x-ray source in NGC 5907. Science **355**, 817–819 (2017). https://doi.org/10.1126/science.aai8635
122. G.L. Israel, A. Papitto, P. Esposito, L. Stella, L. Zampieri, A. Belfiore, G.A. Rodríguez Castillo, A. De Luca, A. Tiengo, F. Haberl, J. Greiner, R. Salvaterra, S. Sandrelli, G. Lisini, Discovery of a 0.42-s pulsar in the ultraluminous X-ray source NGC 7793 P13. Mon. Not. Roy. Astron. Soc. **466**, L48–L52 (2017). https://doi.org/10.1093/mnrasl/slw218.
123. P. Kaaret, H. Feng, T.P. Roberts, Ultraluminous X-ray sources. Annu. Rev. Astron. Astrophys. **55**, 303–341 (2017). https://doi.org/10.1146/annurev-astro-091916-055259
124. V.M. Kaspi, Grand unification of neutron stars. Proc. Natl. Acad. Sci. **107**, 7147–7152 (2010). https://doi.org/10.1073/pnas.1000812107

125. V.M. Kaspi, A.M. Beloborodov, Magnetars. Annu. Rev. Astron. Astrophys. **55**, 261–301 (2017). https://doi.org/10.1146/annurev-astro-081915-023329
126. V.M. Kaspi, K. Boydstun, On the X-ray spectra of anomalous X-ray pulsars and soft gamma repeaters. Astrophys. J. Lett. **710**, L115–L120 (2010). https://doi.org/10.1088/2041-8205/710/2/L115
127. V.M. Kaspi, M. Kramer, Radio pulsars: the neutron star population & fundamental physics, in *Astrophysics and Cosmology*, ed. by R. Blandford, D. Gross, A. Sevrin. Proceedings of the 26th Solvay Conference on Physics (World Scientific, Singapore, 2016), pp. 22–61
128. J.A. Kennea, D.N. Burrows, C. Kouveliotou, D.M. Palmer, E. Göğüş, Y. Kaneko, P.A. Evans, N. Degenaar, M.T. Reynolds, J.M. Miller, R. Wijnands, K. Mori, N. Gehrels, Swift discovery of a new soft gamma repeater, SGR J1745–29, near Sagittarius A*. Astrophys. J. Lett. **770**, L24 (2013). https://doi.org/10.1088/2041-8205/770/2/L24
129. B. Kern, C. Martin, Optical pulsations from the anomalous X-ray pulsar 4U0142+61. Nature **417**, 527–529 (2002). https://doi.org/10.1038/417527a
130. C. Kouveliotou, S. Dieters, T. Strohmayer, J. van Paradijs, G.J. Fishman, C.A. Meegan, K. Hurley, J. Kommers, I. Smith, D. Frail, T. Murakami, An X-ray pulsar with a superstrong magnetic field in the soft gamma-ray repeater SGR 1806-20. Nature **393**, 235–237 (1998)
131. M. Kramer, I.H. Stairs, R.N. Manchester, M.A. McLaughlin, A.G. Lyne, R.D. Ferdman, M. Burgay, D.R. Lorimer, A. Possenti, N. D'Amico, J.M. Sarkissian, G.B. Hobbs, J.E. Reynolds, P.C.C. Freire, F. Camilo, Tests of general relativity from timing the double pulsar. Science **314**, 97–102 (2006). https://doi.org/10.1126/science.1132305
132. M. Kramer, B.W. Stappers, A. Jessner, A.G. Lyne, C.A. Jordan, Polarized radio emission from a magnetar. Mon. Not. Roy. Astron. Soc. **377**, 107–119 (2007). https://doi.org/10.1111/j.1365-2966.2007.11622.x
133. L. Kuiper, W. Hermsen, M. Mendez, Discovery of hard nonthermal pulsed X-Ray emission from the anomalous X-ray pulsar 1E 1841-045. Astrophys. J. **613**, 1173–1178 (2004). https://doi.org/10.1086/423129
134. L. Kuiper, W. Hermsen, P.R. den Hartog, W. Collmar, Discovery of luminous pulsed hard X-ray emission from anomalous X-ray pulsars 1RXS J1708-4009, 4U 0142+61, and 1E 2259+586 by INTEGRAL and RXTE. Astrophys. J. **645**, 556–575 (2006). https://doi.org/10.1086/504317
135. S.K. Lander, N. Andersson, D. Antonopoulou, A.L. Watts, Magnetically driven crustquakes in neutron stars. Mon. Not. Roy. Astron. Soc. **449**, 2047–2058 (2015). https://doi.org/10.1093/mnras/stv432
136. N. Langer, Presupernova evolution of massive single and binary stars. Annu. Rev. Astron. Astrophys. **50**, 107–164 (2012). https://doi.org/10.1146/annurev-astro-081811-125534
137. G.T. Lenters, P.M. Woods, J.E. Goupell, C. Kouveliotou, E. Göğüş, K. Hurley, D. Frederiks, S. Golenetskii, J. Swank, An extended burst tail from SGR 1900+14 with a thermal X-ray spectrum. Astrophys. J. **587**, 761–770 (2003)
138. Y. Levin, QPOs during magnetar flares are not driven by mechanical normal modes of the crust. Mon. Not. Roy. Astron. Soc. **368**, L35–L38 (2006). https://doi.org/10.1111/j.1745-3933.2006.00155.x
139. Y. Levin, On the theory of magnetar QPOs. Mon. Not. Roy. Astron. Soc. **377**, 159–167 (2007). https://doi.org/10.1111/j.1365-2966.2007.11582.x
140. Y. Levin, M. van Hoven, On the excitation of f modes and torsional modes by magnetar giant flares. Mon. Not. Roy. Astron. Soc. **418**, 659–663 (2011). https://doi.org/10.1111/j.1365-2966.2011.19515.x
141. L. Levin, M. Bailes, S. Bates, N.D.R. Bhat, M. Burgay, S. Burke-Spolaor, N. D'Amico, S. Johnston, M. Keith, M. Kramer, S. Milia, A. Possenti, N. Rea, B. Stappers, W. van Straten, A radio-loud magnetar in X-ray quiescence. Astrophys. J. Lett. **721**, L33–L37 (2010). https://doi.org/10.1088/2041-8205/721/1/L33
142. L. Levin, M. Bailes, S.D. Bates, N.D.R. Bhat, M. Burgay, S. Burke-Spolaor, N. D'Amico, S. Johnston, M.J. Keith, M. Kramer, S. Milia, A. Possenti, B. Stappers, W. van Straten, Radio emission evolution, polarimetry and multifrequency single pulse analysis of the radio

magnetar PSR J1622-4950. Mon. Not. Roy. Astron. Soc. **422**, 2489–2500 (2012). https://doi.org/10.1111/j.1365-2966.2012.20807.x
143. T. Li, Y. Shao, X.-D. Li, Can the subsonic accretion model explain the spin period distribution of wind-fed X-ray pulsars? Astrophys. J. **824**, 143 (2016). https://doi.org/10.3847/0004-637X/824/2/143
144. J. Li, N. Rea, D.F. Torres, E. de Oña-Wilhelmi, Gamma-ray upper limits on magnetars with six years of fermi-LAT observations. Astrophys. J. **835**, 30 (2017). https://doi.org/10.3847/1538-4357/835/1/30
145. R.S. Lynch, R.F. Archibald, V.M. Kaspi, P. Scholz, Green bank telescope and swift X-ray telescope observations of the galactic center radio magnetar SGR J1745-2900. Astrophys. J. **806**, 266 (2015). https://doi.org/10.1088/0004-637X/806/2/266
146. A. Lyne, G. Hobbs, M. Kramer, I. Stairs, B. Stappers, Switched magnetospheric regulation of pulsar spin-down. Science **329**, 408 (2010). https://doi.org/10.1126/science.1186683
147. M. Mandea, G. Balasis, FAST TRACK PAPER: the SGR 1806-20 magnetar signature on the earth's magnetic field. Geophys. J. Int. **167**, 586–591 (2006). https://doi.org/10.1111/j.1365-246X.2006.03125.x
148. J. Martin, N. Rea, D.F. Torres, A. Papitto, Comparing supernova remnants around strongly magnetized and canonical pulsars. Mon. Not. Roy. Astron. Soc. **444**, 2910–2924 (2014). https://doi.org/10.1093/mnras/stu1594
149. E.P. Mazets, S.V. Golentskii, V.N. Ilinskii, R.L. Aptekar, I.A. Guryan, Observations of a flaring X-ray pulsar in Dorado. Nature **282**, 587–589 (1979)
150. S. Mereghetti, L. Stella, The very low mass X-ray binary pulsars: a new class of sources? Astrophys. J. Lett. **442**, L17–L20 (1995)
151. S. Mereghetti, L. Chiarlone, G.L. Israel, L. Stella, The anomalous X-ray pulsars, in *Neutron Stars, Pulsars, and Supernova Remnants*, ed. by W. Becker, H. Lesch, J. Trümper. MPE Report 278 (2002), pp. 29–43
152. S. Mereghetti, A. Tiengo, P. Esposito, D. Götz, L. Stella, G.L. Israel, N. Rea, M. Feroci, R. Turolla, S. Zane, An XMM-newton view of the soft gamma repeater SGR 1806-20: long-term variability in the pre-giant flare epoch. Astrophys. J. **628**, 938–945 (2005). https://doi.org/10.1086/430943
153. S. Mereghetti, J.A. Pons, A. Melatos, Magnetars: properties, origin and evolution. Space Sci. Rev. **191**, 315–338 (2015). https://doi.org/10.1007/s11214-015-0146-y
154. B.D. Metzger, D. Giannios, T.A. Thompson, N. Bucciantini, E. Quataert, The protomagnetar model for gamma-ray bursts. Mon. Not. Roy. Astron. Soc. **413**, 2031–2056 (2011). https://doi.org/10.1111/j.1365-2966.2011.18280.x
155. R.P. Mignani, Optical, ultraviolet, and infrared observations of isolated neutron stars. Adv. Space Res. **47**, 1281–1293 (2011). https://doi.org/10.1016/j.asr.2009.12.011
156. M.C. Miller, E.J.M. Colbert, Intermediate-mass black holes. Int. J. Mod. Phys. D **13**, 1–64 (2004). https://doi.org/10.1142/S0218271804004426
157. K. Mori, E.V. Gotthelf, S. Zhang, H. An, F.K. Baganoff, N.M. Barrière, A.M. Beloborodov, S.E. Boggs, F.E. Christensen, W.W. Craig, F. Dufour, B.W. Grefenstette, C.J. Hailey, F.A. Harrison, J. Hong, V.M. Kaspi, J.A. Kennea, K.K. Madsen, C.B. Markwardt, M. Nynka, D. Stern, J.A. Tomsick, W.W. Zhang, NuSTAR discovery of a 3.76 s transient magnetar near sagittarius A*. Astrophys. J. Lett. **770**, L23 (2013). https://doi.org/10.1088/2041-8205/770/2/L23
158. M.P. Muno, J.S. Clark, P.A. Crowther, S.M. Dougherty, R. de Grijs, C. Law, S.L.W. McMillan, M.R. Morris, I. Negueruela, D. Pooley, S. Portegies Zwart, F. Yusef-Zadeh, A neutron star with a massive progenitor in Westerlund 1. Astrophys. J. Lett. **636**, L41–L44 (2006). https://doi.org/10.1086/499776
159. T. Murakami, Y. Tanaka, S.R. Kulkarni, Y. Ogasaka, T. Sonobe, Y. Ogawara, T. Aoki, A. Yoshida, X-ray identification of the soft gamma-ray repeater 1806-20. Nature **368**, 127 (1994)

160. A.A. Mushtukov, V.F. Suleimanov, S.S. Tsygankov, J. Poutanen, On the maximum accretion luminosity of magnetized neutron stars: connecting X-ray pulsars and ultraluminous X-ray sources. Mon. Not. Roy. Astron. Soc. **454**, 2539–2548 (2015). https://doi.org/10.1093/mnras/stv2087
161. C. Neiner, J. Morin, E. Alecian, The "Binarity and Magnetic Interactions in various classes of stars" (BinaMIcS) project, in *SF2A-2015: Proceedings of the Annual Meeting of the French Society of Astronomy and Astrophysics* (2015), pp. 213–216
162. M.E.J. Newman, Power laws, pareto distributions and Zipf's law. Contemp. Phys. **46**, 323–351 (2005). https://doi.org/10.1080/00107510500052444
163. J.P. Norris, P. Hertz, K.S. Wood, C. Kouveliotou, On the nature of soft gamma repeaters. Astrophys. J. **366**, 240–252 (1991). https://doi.org/10.1086/169556
164. S.A. Olausen, V.M. Kaspi, The McGill magnetar catalog. Astrophys. J. Suppl. **212**, 6 (2014). https://doi.org/10.1088/0067-0049/212/1/6
165. B. Paczynski, GB 790305 as a very strongly magnetized neutron star. Acta Astron. **42**, 145–153 (1992)
166. A. Paizis, L. Sidoli, Cumulative luminosity distributions of supergiant fast X-ray transients in hard X-rays. Mon. Not. Roy. Astron. Soc. **439**, 3439–3452 (2014). https://doi.org/10.1093/mnras/stu191
167. D.M. Palmer, S. Barthelmy, N. Gehrels, R.M. Kippen, T. Cayton, C. Kouveliotou, D. Eichler, R.A.M.J. Wijers, P.M. Woods, J. Granot, Y.E. Lyubarsky, E. Ramirez-Ruiz, L. Barbier, M. Chester, J. Cummings, E.E. Fenimore, M.H. Finger, B.M. Gaensler, D. Hullinger, H. Krimm, C.B. Markwardt, J.A. Nousek, A. Parsons, S. Patel, T. Sakamoto, G. Sato, M. Suzuki, J. Tueller, A giant γ-ray flare from the magnetar SGR 1806 - 20. Nature **434**, 1107–1109 (2005). https://doi.org/10.1038/nature03525
168. T.T. Pennucci, A. Possenti, P. Esposito, N. Rea, D. Haggard, F.K. Baganoff, M. Burgay, F. Coti Zelati, G.L. Israel, A. Minter, Simultaneous multi-band radio and X-ray observations of the galactic center magnetar SGR 1745-2900. Astrophys. J. **808**, 81 (2015). https://doi.org/10.1088/0004-637X/808/1/81
169. R. Perna, J.A. Pons, A unified model of the magnetar and radio pulsar bursting phenomenology. Astrophys. J. Lett. **727**, L51 (2011). https://doi.org/10.1088/2041-8205/727/2/L51
170. R. Perna, P. Duffell, M. Cantiello, A.I. MacFadyen, The fate of fallback matter around newly born compact objects. Astrophys. J. **781**, 119 (2014). https://doi.org/10.1088/0004-637X/781/2/119
171. O. Pfuhl, S. Gillessen, F. Eisenhauer, R. Genzel, P.M. Plewa, T. Ott, A. Ballone, M. Schartmann, A. Burkert, T.K. Fritz, R. Sari, E. Steinberg, A.-M. Madigan, The galactic center cloud G2 and its gas streamer. Astrophys. J. **798**, 111 (2015). https://doi.org/10.1088/0004-637X/798/2/111
172. K. Phifer, T. Do, L. Meyer, A.M. Ghez, G. Witzel, S. Yelda, A. Boehle, J.R. Lu, M.R. Morris, E.E. Becklin, K. Matthews, Keck observations of the galactic center source G2: gas cloud or star? Astrophys. J. Lett. **773**, L13 (2013). https://doi.org/10.1088/2041-8205/773/1/L13
173. F. Pintore, S. Mereghetti, A. Tiengo, G. Vianello, E. Costantini, P. Esposito, The effect of X-ray dust scattering on a bright burst from the magnetar 1E 1547.0-5408. Mon. Not. Roy. Astron. Soc. **467**, 3467–3474 (2017). https://doi.org/10.1093/mnras/stx368
174. P.M. Plewa, S. Gillessen, O. Pfuhl, F. Eisenhauer, R. Genzel, A. Burkert, J. Dexter, M. Habibi, E. George, T. Ott, I. Waisberg, S. von Fellenberg, The post-pericenter evolution of the galactic center source G2. Astrophys. J. **840**, 50 (2017). https://doi.org/10.3847/1538-4357/aa6e00
175. J.A. Pons, U. Geppert, Magnetic field dissipation in neutron star crusts: from magnetars to isolated neutron stars. Astron. Astrophys. **470**, 303–315 (2007). https://doi.org/10.1051/0004-6361:20077456
176. J.A. Pons, N. Rea, Modeling magnetar outbursts: flux enhancements and the connection with short bursts and glitches. Astrophys. J. Lett. **750**, L6 (2012). https://doi.org/10.1088/2041-8205/750/1/L6
177. J.A. Pons, B. Link, J.A. Miralles, U. Geppert, Evidence for heating of neutron stars by magnetic field decay. Phys. Rev. Lett. **98**, 071101 (2007)

178. J.A. Pons, J.A. Miralles, U. Geppert, Magneto-thermal evolution of neutron stars. Astron. Astrophys. **496**, 207–216 (2009). https://doi.org/10.1051/0004-6361:200811229
179. G. Ponti, B. De Marco, M.R. Morris, A. Merloni, T. Muñoz-Darias, M. Clavel, D. Haggard, S. Zhang, K. Nandra, S. Gillessen, K. Mori, J. Neilsen, N. Rea, N. Degenaar, R. Terrier, A. Goldwurm, Fifteen years of XMM-Newton and Chandra monitoring of Sgr A*: evidence for a recent increase in the bright flaring rate. Mon. Not. Roy. Astron. Soc. **454**, 1525–1544 (2015). https://doi.org/10.1093/mnras/stv1537
180. S.B. Popov, R. Turolla, Probing the neutron star spin evolution in the young small magellanic cloud Be/X-ray binary SXP 1062. Mon. Not. Roy. Astron. Soc. **421**, L127–L131 (2012). https://doi.org/10.1111/j.1745-3933.2012.01220.x
181. N. Rea, P. Esposito, Magnetar outbursts: an observational review, in *High-Energy Emission from Pulsars and Their Systems. Proceedings of the First Session of the Sant Cugat Forum on Astrophysics*, ed. by D.F. Torres, N. Rea. Astrophysics and Space Science Proceedings (Springer, Heidelberg, 2011), pp. 247–273. ISBN 978-3-642-17250-2. https://doi.org/10.1007/978-3-642-17251-9_21
182. N. Rea, P. Esposito, R. Turolla, G.L. Israel, S. Zane, L. Stella, S. Mereghetti, A. Tiengo, D. Götz, E. Göğüş, C. Kouveliotou, A low-magnetic-field soft gamma repeater. Science **330**, 944 (2010). https://doi.org/10.1126/science.1196088
183. N. Rea, G.L. Israel, P. Esposito, J.A. Pons, A. Camero-Arranz, R.P. Mignani, R. Turolla, S. Zane, M. Burgay, A. Possenti, S. Campana, T. Enoto, N. Gehrels, E. Göğüş, D. Götz, C. Kouveliotou, K. Makishima, S. Mereghetti, S.R. Oates, D.M. Palmer, R. Perna, L. Stella, A. Tiengo, A new low magnetic field magnetar: the 2011 outburst of swift J1822.3-1606. Astrophys. J. **754**, 27 (2012). https://doi.org/10.1088/0004-637X/754/1/27
184. N. Rea, P. Esposito, J.A. Pons, R. Turolla, D.F. Torres, G.L. Israel, A. Possenti, M. Burgay, D. Viganò, A. Papitto, R. Perna, L. Stella, G. Ponti, F.K. Baganoff, D. Haggard, A. Camero-Arranz, S. Zane, A. Minter, S. Mereghetti, A. Tiengo, R. Schödel, M. Feroci, R. Mignani, D. Götz, A strongly magnetized pulsar within the grasp of the milky way's supermassive black hole. Astrophys. J. Lett. **775**, L34 (2013). https://doi.org/10.1088/2041-8205/775/2/L34
185. N. Rea, G.L. Israel, J.A. Pons, R. Turolla, D. Viganò, S. Zane, P. Esposito, R. Perna, A. Papitto, G. Terreran, A. Tiengo, D. Salvetti, J.M. Girart, A. Palau, A. Possenti, M. Burgay, E. Göğüş, G.A. Caliandro, C. Kouveliotou, D. Götz, R.P. Mignani, E. Ratti, L. Stella, The outburst decay of the low magnetic field magnetar SGR 0418+5729. Astrophys. J. 770, 65 (2013). https://doi.org/10.1088/0004-637X/770/1/65
186. N. Rea, D. Viganò, G.L. Israel, J.A. Pons, D.F. Torres, 3XMM J185246.6+003317: another low magnetic field magnetar. Astrophys. J. Lett. **781**, L17 (2014). https://doi.org/10.1088/2041-8205/781/1/L17
187. N. Rea, A. Borghese, P. Esposito, F. Coti Zelati, M. Bachetti, G.L. Israel, A. De Luca, Magnetar-like Activity from the central compact object in the SNR RCW103. Astrophys. J. Lett. **828**, L13 (2016). https://doi.org/10.3847/2041-8205/828/1/L13
188. G.A. Rodríguez Castillo, G.L. Israel, P. Esposito, J.A. Pons, N. Rea, R. Turolla, D. Viganò, S. Zane, Pulse phase-coherent timing and spectroscopy of CXOU J164710.2-45521 outbursts. Mon. Not. Roy. Astron. Soc. **441**, 1305–1316 (2014). https://doi.org/10.1093/mnras/stu603
189. G.A. Rodríguez Castillo, G.L. Israel, A. Tiengo, D. Salvetti, R. Turolla, S. Zane, N. Rea, P. Esposito, S. Mereghetti, R. Perna, L. Stella, J.A. Pons, S. Campana, D. Götz, S. Motta, The outburst decay of the low magnetic field magnetar SWIFT J1822.3-1606: phase-resolved analysis and evidence for a variable cyclotron feature. Mon. Not. Roy. Astron. Soc. **456**, 4145–4155 (2016). https://doi.org/10.1093/mnras/stv2490.
190. R.E. Rothschild, S.R. Kulkarni, R.E. Lingenfelter, Discovery of an X-ray source coincident with the soft gamma-ray repeater 0525-66. Nature **368**, 432 (1994)
191. H. Sana, G. Rauw, Y. Nazé, E. Gosset, J.-M. Vreux, An XMM-Newton view of the young open cluster NGC 6231 - II. The OB star population. Mon. Not. Roy. Astron. Soc. **372**, 661–678 (2006). https://doi.org/10.1111/j.1365-2966.2006.10847.x

192. P. Scholz, C.-Y. Ng, M.A. Livingstone, V.M. Kaspi, A. Cumming, R.F. Archibald, Post-outburst X-ray flux and timing evolution of swift J1822.3-1606. Astrophys. J. **761**, 66 (2012). https://doi.org/10.1088/0004-637X/761/1/66
193. P. Scholz, F. Camilo, J. Sarkissian, J.E. Reynolds, L. Levin, M. Bailes, M. Burgay, S. Johnston, M. Kramer, A. Possenti, Spin-down evolution and radio disappearance of the magnetar PSR J1622-4950. Astrophys. J. **841**, 126 (2017). https://doi.org/10.3847/1538-4357/aa73de
194. J.H. Seiradakis, R. Wielebinski, Morphology and characteristics of radio pulsars. Astron. Astrophys. Rev. **12**, 239–271 (2004). https://doi.org/10.1007/s00159-004-0022-3
195. N. Shakura, K. Postnov, A. Kochetkova, L. Hjalmarsdotter, Theory of quasi-spherical accretion in X-ray pulsars. Mon. Not. Roy. Astron. Soc. **420**, 216–236 (2012). https://doi.org/10.1111/j.1365-2966.2011.20026.x
196. R.M. Shannon, S. Johnston, Radio properties of the magnetar near Sagittarius A* from observations with the Australia telescope compact array. Mon. Not. Roy. Astron. Soc. **435**, L29–L32 (2013). https://doi.org/10.1093/mnrasl/slt088
197. L. Sidoli, G.L. Israel, P. Esposito, G.A. Rodríguez Castillo, K. Postnov, AX J1910.7+0917: the slowest X-ray pulsar. Mon. Not. Roy. Astron. Soc. **469**, 3056–3061 (2017). https://doi.org/10.1093/mnras/stx1105
198. H.C. Spruit, Origin of neutron star magnetic fields, in *40 Years of Pulsars: Millisecond Pulsars, Magnetars and More*, ed. by C. Bassa, Z. Wang, A. Cumming, V.M. Kaspi. American Institute of Physics Conference Series, vol. 983, pp. 391–398 (2008). https://doi.org/10.1063/1.2900262
199. T.E. Strohmayer, A.L. Watts, Discovery of fast X-ray oscillations during the 1998 giant flare from SGR 1900+14. Astrophys. J. Lett. **632**, L111–L114 (2005). https://doi.org/10.1086/497911
200. T.E. Strohmayer, A.L. Watts, The 2004 hyperflare from SGR 1806-20: further evidence for global torsional vibrations. Astrophys. J. **653**, 593–601 (2006). https://doi.org/10.1086/508703
201. C.R. Tam, V.M. Kaspi, M.H. van Kerkwijk, M. Durant, Correlated infrared and X-ray flux changes following the 2002 June outburst of the anomalous X-ray pulsar 1E 2259+586. Astrophys. J. Lett. **617**, L53–L56 (2004). https://doi.org/10.1086/426963
202. S.P. Tendulkar, P.B. Cameron, S.R. Kulkarni, Proper motions and origins of SGR 1806-20 and SGR 1900+14. Astrophys. J. **761**, 76 (2012). https://doi.org/10.1088/0004-637X/761/1/76
203. S.P. Tendulkar, P.B. Cameron, S.R. Kulkarni, Proper motions and origins of AXP 1E 2259+586 and AXP 4U 0142+61. Astrophys. J. **772**, 31 (2013). https://doi.org/10.1088/0004-637X/772/1/31
204. S.P. Tendulkar, R. Hascöet, C. Yang, V.M. Kaspi, A.M. Beloborodov, H. An, M. Bachetti, S.E. Boggs, F.E. Christensen, W.W. Craig, S. Guillot, C.A. Hailey, F.A. Harrison, D. Stern, W. Zhang, Phase-resolved NuSTAR and Swift-XRT observations of magnetar 4U 0142+61. Astrophys. J. **808**, 32 (2015). https://doi.org/10.1088/0004-637X/808/1/32
205. S.P. Tendulkar, V.M. Kaspi, R.F. Archibald, P. Scholz, A near-infrared counterpart of 2E1613.5–5053: the central source in supernova remnant RCW103. Astrophys. J. **841**, 11 (2017). https://doi.org/10.3847/1538-4357/aa6d0c
206. V. Testa, N. Rea, R.P. Mignani, G.L. Israel, R. Perna, S. Chaty, L. Stella, S. Covino, R. Turolla, S. Zane, G. Lo Curto, S. Campana, G. Marconi, S. Mereghetti, Adaptive optics, near-infrared observations of magnetars. Astron. Astrophys. **482**, 607–615 (2008). https://doi.org/10.1051/0004-6361:20078692
207. C. Thompson, R.C. Duncan, Neutron star dynamos and the origins of pulsar magnetism. Astrophys. J. **408**, 194 (1993). https://doi.org/10.1086/172580
208. C. Thompson, R.C. Duncan, The soft gamma repeaters as very strongly magnetized neutron stars - I. Radiative mechanism for outbursts. Mon. Not. Roy. Astron. Soc. **275**, 255–300 (1995).
209. C. Thompson, R.C. Duncan, The soft gamma repeaters as very strongly magnetized neutron stars. II. Quiescent neutrino, X-ray, and Alfven wave emission. Astrophys. J. **473**, 322–342 (1996)

210. A. Tiengo, P. Esposito, S. Mereghetti, XMM-Newton observations of CXOU J010043.1-721134: the first deep look at the soft X-ray emission of a magnetar. Astrophys. J. Lett. **680**, L133–L136 (2008). https://doi.org/10.1086/590078
211. A. Tiengo, G. Vianello, P. Esposito, S. Mereghetti, A. Giuliani, E. Costantini, G.L. Israel, L. Stella, R. Turolla, S. Zane, N. Rea, D. Götz, F. Bernardini, A. Moretti, P. Romano, M. Ehle, N. Gehrels, The dust-scattering X-ray rings of the anomalous X-ray pulsar 1E 1547.0-5408. Astrophys. J. **710**, 227–235 (2010). https://doi.org/10.1088/0004-637X/710/1/227
212. A. Tiengo, P. Esposito, S. Mereghetti, R. Turolla, L. Nobili, F. Gastaldello, D. Götz, G.L. Israel, N. Rea, L. Stella, S. Zane, G.F. Bignami, A variable absorption feature in the X-ray spectrum of a magnetar. Nature **500**, 312–314 (2013). https://doi.org/10.1038/nature12386
213. H. Tong, W. Wang, X.W. Liu, R.X. Xu, Rotational evolution of magnetars in the presence of a fallback disk. Astrophys. J. **833**, 265 (2016). https://doi.org/10.3847/1538-4357/833/2/265
214. P. Torne, R.P. Eatough, R. Karuppusamy, M. Kramer, G. Paubert, B. Klein, G. Desvignes, D.J. Champion, H. Wiesemeyer, C. Kramer, L.G. Spitler, C. Thum, R. Güsten, K.F. Schuster, I. Cognard, Simultaneous multifrequency radio observations of the galactic centre magnetar SGR J1745-2900. Mon. Not. Roy. Astron. Soc. **451**, L50–L54 (2015). https://doi.org/10.1093/mnrasl/slv063
215. P. Torne, G. Desvignes, R.P. Eatough, R. Karuppusamy, G. Paubert, M. Kramer, I. Cognard, D.J. Champion, L.G. Spitler, Detection of the magnetar SGR J1745-2900 up to 291 GHz with evidence of polarized millimetre emission. Mon. Not. Roy. Astron. Soc. **465**, 242–247 (2017). https://doi.org/10.1093/mnras/stw2757
216. D.F. Torres, N. Rea, P. Esposito, J. Li, Y. Chen, S. Zhang, A magnetar-like event from LS I +61°303 and its nature as a gamma-ray binary. Astrophys. J. **744**, 106 (2012). https://doi.org/10.1088/0004-637X/744/2/106
217. R. Turolla, P. Esposito, Low-magnetic magnetars. Int. J. Mod. Phys. D **22**, 1330024-163 (2013). https://doi.org/10.1142/S0218271813300243
218. R. Turolla, S. Zane, J.A. Pons, P. Esposito, N. Rea, Is SGR 0418+5729 indeed a waning magnetar? Astrophys. J. **740**, 105 (2011). https://doi.org/10.1088/0004-637X/740/2/105
219. R. Turolla, S. Zane, A.L. Watts, Magnetars: the physics behind observations. A review. Rep. Progress Phys. **78**, 116901 (2015). http://stacks.iop.org/0034-4885/78/i=11/a=116901
220. A.J. van der Horst, C. Kouveliotou, N.M. Gorgone, Y. Kaneko, M.G. Baring, S. Guiriec, E. Göğüş, J. Granot, A.L. Watts, L. Lin, P.N. Bhat, E. Bissaldi, V.L. Chaplin, M.H. Finger, N. Gehrels, M.H. Gibby, M.M. Giles, A. Goldstein, D. Gruber, A.K. Harding, L. Kaper, A. von Kienlin, M. van der Klis, S. McBreen, J. Mcenery, C.A. Meegan, W.S. Paciesas, A. Pe'er, R.D. Preece, E. Ramirez-Ruiz, A. Rau, S. Wachter, C. Wilson-Hodge, P.M. Woods, R.A.M.J. Wijers, SGR J1550-5418 bursts detected with the fermi gamma-ray burst monitor during its most prolific activity. Astrophys. J. **749**, 122 (2012). https://doi.org/10.1088/0004-637X/749/2/122
221. J. van Paradijs, R.E. Taam, E.P.J. van den Heuvel, On the nature of the 'anomalous' 6-s X-ray pulsars. Astron. Astrophys. **299**, L41 (1995)
222. T. van Putten, A.L. Watts, C.R. D'Angelo, M.G. Baring, C. Kouveliotou, Models of hydrostatic magnetar atmospheres at high luminosities. Mon. Not. Roy. Astron. Soc. **434**, 1398–1410 (2013). https://doi.org/10.1093/mnras/stt1093
223. G. Vasisht, S.R. Kulkarni, D.A. Frail, J. Greiner, Supernova remnant candidates for the soft gamma-ray repeater 1900+14. Astrophys. J. Lett. **431**, L35–L38 (1994)
224. M. Vietri, L. Stella, G.L. Israel, SGR 1806-20: evidence for a superstrong magnetic field from quasi-periodic oscillations. Astrophys. J. **661**, 1089–1093 (2007). https://doi.org/10.1086/517506
225. D. Viganò, J.A. Pons, Central compact objects and the hidden magnetic field scenario. Mon. Not. Roy. Astron. Soc. **425**, 2487–2492 (2012). https://doi.org/10.1111/j.1365-2966.2012.21679.x
226. D. Viganò, N. Rea, J.A. Pons, R. Perna, D.N. Aguilera, J.A. Miralles, Unifying the observational diversity of isolated neutron stars via magneto-thermal evolution models. Mon. Not. Roy. Astron. Soc. **434**, 123–141 (2013). https://doi.org/10.1093/mnras/stt1008

227. J. Vink, L. Kuiper, Supernova remnant energetics and magnetars: no evidence in favour of millisecond proto-neutron stars. Mon. Not. Roy. Astron. Soc. **370**, L14–L18 (2006). https://doi.org/10.1111/j.1745-3933.2006.00178.x
228. J.K. Vogel, R. Hascoët, V.M. Kaspi, H. An, R. Archibald, A.M. Beloborodov, S.E. Boggs, F.E. Christensen, W.W. Craig, E.V. Gotthelf, B.W. Grefenstette, C.J. Hailey, F.A. Harrison, J.A. Kennea, K.K. Madsen, M.J. Pivovaroff, D. Stern, W.W. Zhang, NuSTAR observations of the magnetar 1E 2259+586. Astrophys. J. **789**, 75 (2014). https://doi.org/10.1088/0004-637X/789/1/75
229. Z. Wadiasingh, M.G. Baring, P.L. Gonthier, A.K. Harding, Resonant inverse compton scattering spectra from highly-magnetized neutron stars. Astrophys. J. **854**, 98 (2018)
230. Z. Wang, D. Chakrabarty, D.L. Kaplan, A debris disk around an isolated young neutron star. Nature **440**, 772–775 (2006). https://doi.org/10.1038/nature04669
231. A.L. Watts, T.E. Strohmayer, Detection with RHESSI of high-frequency X-ray oscillations in the tailof the 2004 hyperflare from SGR 1806-20. Astrophys. J. Lett. **637**, L117–L120 (2006). https://doi.org/10.1086/500735
232. P.M. Woods, C. Thompson, in *Compact Stellar X-ray Sources*, ed. by W.H.G. Levin, M. van der Klis (Cambridge University Press, Cambridge, 2006), p. 547
233. P.M. Woods, C. Kouveliotou, J. van Paradijs, M.H. Finger, C. Thompson, R.C. Duncan, K. Hurley, T. Strohmayer, J. Swank, T. Murakami, Variable Spin-down in the soft gamma repeater SGR 1900+14 and correlations with burst activity. Astrophys. J. Lett. **524**, L55–L58 (1999)
234. P.M. Woods, C. Kouveliotou, E. Göğüş, M.H. Finger, J. Swank, D.A. Smith, K. Hurley, C. Thompson, Evidence for a sudden magnetic field reconfiguration in soft gamma repeater 1900+14. Astrophys. J. **552**, 748–755 (2001)
235. P.M. Woods, V.M. Kaspi, C. Thompson, F.P. Gavriil, H.L. Marshall, D. Chakrabarty, K. Flanagan, J. Heyl, L. Hernquist, Changes in the X-ray emission from the magnetar candidate 1E 2259+586 during its 2002 outburst. Astrophys. J. **605**, 378–399 (2004)
236. P.M. Woods, C. Kouveliotou, M.H. Finger, E. Göğüş, C.A. Wilson, S.K. Patel, K. Hurley, J.H. Swank, The prelude to and aftermath of the giant flare of 2004 December 27: persistent and pulsed X-ray properties of SGR 1806–20 from 1993 to 2005. Astrophys. J. **654**, 470–486 (2007). https://doi.org/10.1086/507459
237. G. Younes, M.G. Baring, C. Kouveliotou, A. Harding, S. Donovan, E. Göğüş, V. Kaspi, J. Granot, The sleeping monster: NuSTAR observations of SGR 1806-20, 11 years after the giant flare. Astrophys. J. **851**, 17 (2017). https://doi.org/10.3847/1538-4357/aa96fd
238. G. Younes, C. Kouveliotou, A. Jaodand, M.G. Baring, A.J. van der Horst, A.K. Harding, J.W.T. Hessels, N. Gehrels, R. Gill, D. Huppenkothen, J. Granot, E. Göğüş, L. Lin, X-ray and radio observations of the magnetar SGR J1935+2154 during its 2014, 2015, and 2016 outbursts. Astrophys. J. **847**, 85 (2017). https://doi.org/10.3847/1538-4357/aa899a
239. S. Zane, L. Nobili, R. Turolla, The magnetar emission in the IR band: the role of magnetospheric currents, in *High-Energy Emission from Pulsars and Their Systems. Proceedings of the First Session of the Sant Cugat Forum on Astrophysics*, ed. by D.F. Torres, N. Rea. Astrophysics and Space Science Proceedings (Springer, Heidelberg, 2011), pp. 329–335

Chapter 4
Accreting Millisecond X-ray Pulsars

Alessandro Patruno and Anna L. Watts

Contents

A. Patruno (✉)
Institute of Space Sciences, CSIC (Consejo Superior de Investigaciones Cientificas), Barcelona, Spain
e-mail: patruno@ice.csic.esl

A. L. Watts
Astronomical Institute "Anton Pannekoek", University of Amsterdam, Amsterdam, The Netherlands
e-mail: a.l.watts@uva.nl

© Springer-Verlag GmbH Germany, part of Springer Nature 2021
T. M. Belloni et al. (eds.), *Timing Neutron Stars: Pulsations, Oscillations and Explosions*, Astrophysics and Space Science Library 461,
https://doi.org/10.1007/978-3-662-62110-3_4

Abstract Accreting Millisecond X-ray Pulsars (AMXPs) are astrophysical laboratories without parallel in the study of extreme physics. In this chapter we review the past 15 years of discoveries in the field. We summarize the observations of the 15 known AMXPs, with a particular emphasis on the multi-wavelength observations that have been carried out since the discovery of the first AMXP in 1998. We review accretion torque theory, the pulse formation process, and how AMXP observations have changed our view on the interaction of plasma and magnetic fields in strong gravity. We also explain how the AMXPs have deepened our understanding of the thermonuclear burst process, in particular the phenomenon of burst oscillations. We conclude with a discussion of the open problems that remain to be addressed in the future.

4.1 Introduction

Neutron stars (NSs), amongst the most extreme astrophysical objects in the Universe, allow us to study physics in regimes that cannot be accessed by terrestrial laboratories. They play a key role in the study of fundamental problems including the equation of state (EoS) of ultra-dense matter, the production of gravitational waves, dense matter superfluidity and superconductivity, and the generation and evolution of ultra-strong magnetic fields. Since the discovery of NSs as radio pulsars in 1967 [136], many different classes have been discovered including more than $\sim$130 NSs in low mass X-ray binaries (LMXBs). In LMXBs the NS accretes matter from a non-collapsed stellar companion (with mass $M \lesssim 1\,M_\odot$) via an accretion disk. This chapter focuses on a subgroup of the LMXBs, the accreting millisecond X-ray pulsars (AMXPs). In the AMXPs, the gas stripped from the companion is channeled out of the accretion disk and onto the magnetic poles of the rotating neutron star, giving rise to X-ray pulsations at the spin frequency. We will explore the details of this process, why it is so rare (the AMXPs are a small class), the physics of the disk-magnetosphere interaction, and how AMXPs can be used to probe extreme physics.

Immediately after the discovery of the first millisecond radio pulsar in 1982 [12], LMXBs were identified as possible incubators for millisecond pulsars. It was suggested that LMXBs might be responsible for the conversion of slow NSs with high magnetic field ($B{\sim}10^{12}$ G), into a rapidly spinning objects with a relatively weak magnetic field ($B \sim 10^8$ G). Two independent papers published in 1982 [1, 281] proposed transfer of angular momentum through accretion as the mechanism responsible for the spin-up of pulsars. This is known as the *recycling scenario* (see [312] for an excellent review). This name originates from the fact that radio pulsars switch off their pulsed radio emission after entering the so-called "pulsar graveyard". If this happens while the NS is in a binary with a non-collapsed low or intermediate mass stellar companion, binary evolution of the system [23, 328] can bring the companion into Roche lobe contact and trigger a prolonged epoch of mass transfer from the companion (donor) towards the NS (accretor). The mass is transferred with large specific angular momentum and the NS is spun-up by the resulting accretion torques. Once the mass transfer episode terminates, the NS might eventually switch on again as a "recycled" millisecond radio pulsar.

The first AMXP (SAX J1808.4-3658), found in 1998 with the *Rossi X-ray Timing Explorer* (*RXTE*) [360] provided a beautiful confirmation of the recycling scenario. Fourteen more AMXPs have since been found, with spin frequencies from 182 to 599 Hz. Another important milestone came with the discovery (in 2007) of a binary radio millisecond pulsar (PSR J1023+0038) for which archival optical observations, taken ~7 years before the radio pulsar discovery, showed evidence for an accretion disk[1] [9]. This is the first NS observed to have switched on as a radio pulsar after being an X-ray binary and another system, XSS J12270-4859 [78, 79], has been recently discovered to behave in a very similar way [16, 29, 297]. A final confirmation that indeed AMXPs and radio pulsars are related has recently come with the discovery of the system IGR J18245-2452 which has shown both an AMXP and a radio millisecond pulsar phase (see [247] and Sect. 4.3.15 for a detailed discussion).

The *RXTE* observatory has played an extraordinary role by discovering many systems of this kind and by collecting extensive data records of each outburst detected during its 15 year lifetime. The excellent timing capabilities of *RXTE* have brought new means to study NSs with coherent X-ray timing, and helped to constrain the long term properties of many AMXPs over a baseline of more than a decade. Observation of the orbital Doppler shift of the AMXP pulse frequency contains information on the orbital parameters of the binary and their evolution in time. Binary evolution has benefited from the study of AMXPs [25, 76, 230] which are now known to include ultra-compact systems (orbital period $P_b \lesssim 80$ min) with white dwarf companions, compact systems ($P_b \simeq 1.5$–3 h) with brown dwarf donors and wider systems ($P_b \simeq 3.5$–20 h) with main sequence stars. Other X-ray and gamma ray space missions like *XMM-Newton*, *INTEGRAL*, *Chandra*, *Swift* and *HETE* have also played an important role in discovering and understanding the spectral and timing properties of these objects. Multiwavelength observations covering radio, infrared, optical and UV wavelengths have also illuminated different aspects of these fascinating systems. Several optical and infrared counterparts have

[1] A new transition from a radio pulsar to a LMXB has happened and is currently ongoing at the moment of writing this review [265, 313].

been identified with ground based observations and in some cases have led to the discovery of the spectral type of the donor, while radio and infrared observations have revealed the possible presence of jets.

This chapter is structured as follows:

Section 4.2. An overview of the AMXP family.
Section 4.3. Observations of individual AMXPs
Section 4.4. Coherent timing analysis and accretion torques.
Section 4.5. X-ray pulse properties, their formation and use.
Section 4.6. Long-term evolution of spins and orbital parameters.
Section 4.7. Thermonuclear bursts and burst oscillations.
Section 4.8. Aperiodic phenomena including kilohertz QPOs.
Section 4.9. Future developments and open questions.

4.2 The Accreting Millisecond X-ray Pulsar Family

An AMXP is an accretion powered X-ray pulsar spinning at frequencies $\nu \gtrsim 50$ Hz, with weak surface magnetic fields ($B \sim 10^{8-9}$ G) and bound in a binary system with a donor companion of mass $M \lesssim 1\,M_{\odot}$. All known AMXPs have donor stars that transfer mass via Roche lobe overflow. This definition excludes binary millisecond pulsars, which are detached systems with a pulsar powered by its rotational energy, X-ray pulsars in high mass X-ray binaries, symbiotic X-ray binaries and the slow X-ray pulsars in LMXBs with $\nu < 50$ Hz. Several LMXBs show millisecond oscillations during thermonuclear bursts that ignite on their surface. These nuclear powered X-ray pulsars (NXPs), or burst oscillation sources, are not AMXPs, since they are powered by nuclear burning rather than channeled accretion (although note that some AMXPs also show burst oscillations).

Table 4.1 reports the main characteristics of the 23 known pulsating NSs in LMXBs: 15 AMXPs, 3 slow accreting X-ray pulsars in LMXBs, 1 slow accreting X-ray pulsar in an intermediate mass X-ray binary (IMXB), 3 symbiotic X-ray binaries accreting wind from a K/M giant companion (compatible with being of low mass type) and 1 mildly recycled accreting pulsar recently discovered in the globular cluster Terzan 5. This latter source is an 11 Hz X-ray pulsar with a magnetic field of 10^{9}–10^{10} G and is of particular interest because it might be the only known accreting pulsar in the process of becoming an AMXP on a short timescale [264]. The other four slow pulsars are very different systems, in the sense that they have most likely followed a completely different evolutionary history, have strong magnetic fields ($B \sim 10^{12}$ G) and may never reach millisecond periods. Symbiotic X-ray binaries are also different from the AMXPs since they are wide binaries ($P_b \gtrsim 50$ days) and their NSs have extremely long spin periods caused by a prolonged phase of spin-down during a wind accretion process.

All of the AMXPs are transient systems, with accretion disks that run through cycles of outburst and quiescence. X-ray pulsations have never been observed in

Table 4.1 Accreting X-ray pulsars in low mass X-ray binaries

Source	ν_s (Hz)	P_b (h)	f_x ($M_\odot$)	$M_{c,min}$ ($M_\odot$)	Companion type	Ref.
Accreting millisecond pulsars						
SAX J1808.4-3658	401	2.01	3.8×10^{-5}	0.043	BD	[41, 123, 125, 242, 262]
XTE J1751-305	435	0.71	1.3×10^{-6}	0.014	He WD	[204, 241]
XTE J0929-314	185	0.73	2.9×10^{-7}	0.0083	C/O WD	[99, 229, 285]
XTE J807-294	190	0.67	1.5×10^{-7}	0.0066	C/O WD	[60, 163, 205, 259, 287]
XTE J1814-338	314	4.27	2.0×10^{-3}	0.17	MS	[202, 239]
IGR J00291+5934	599	2.46	2.8×10^{-5}	0.039	BD	[37, 88, 100, 126, 245, 249]
HETE J1900.1-2455	377	1.39	2.0×10^{-6}	0.016	BD	[158]
Swift J1756.9-2508	182	0.91	1.6×10^{-7}	0.007	He WD	[170, 260]
Aql X-1	550	18.95	N/A	0.6[a]	MS	[44, 355]
SAX J1748.9-2021	442	8.77	4.8×10^{-4}	0.1	MS/SubG ?	[2, 253]
NGC6440 X-2	206	0.95	1.6×10^{-7}	0.0067	He WD	[3]
IGR J17511-3057	245	3.47	1.1×10^{-3}	0.13	MS	[211, 289]
Swift J1749.4-2807	518	8.82	5.5×10^{-2}	0.59	MS	[6]
IGR J17498-2921	401	3.84	2.0×10^{-3}	0.17	MS	[244]
IGR J18245-2452	254	11.03	2.3×10^{-3}	0.17	MS	[247]
Mildly recycled X-ray pulsar						
IGR J17480-2466	11	21.27	2.1×10^{-2}	0.4	SubG	[46, 239, 321]
Slow pulsars in LMXBs & IMXBs						
2A 1822-371	1.7	5.57	2×10^{-2}	0.33	?	[152]
4U 1626-67	0.13	0.69	1.3×10^{-6}	0.06	WD or He star	[26, 62, 187]
GRO 1744-28	2.14	282.24	1.3×10^{-4}	< 0.3	Giant	[26, 66, 367]
Her X-1	0.81	40.80	0.85	~ 2[a]	MS	[26, 62]

(continued)

Table 4.1 (continued)

Source	ν_s (Hz)	P_b (h)	f_x ($M_\odot$)	$M_{c,min}$ ($M_\odot$)	Companion type	Ref.
Symbiotic X-ray binaries						
GX 1+4	6.3×10^{-3}	1161 days			M5 Giant	[66, 138, 274]
4U 1954+31	5.5×10^{-5}				M4 Giant	[64, 274]
IGR J16358-4724[b]	1.7×10^{-4}				K/M Giant	[64, 274]

ν_s is the spin frequency, P_b the orbital period, f_x is the X-ray mass function, $M_{c,min}$ is the minimum companion mass for an assumed NS mass of 1.4 $M_\odot$.
The companion types are: *WD* White Dwarf, *BD* Brown Dwarf, *MS* Main Sequence, *SubG* Sub-Giant, *He Core* Helium Star
[a]The donor mass is inferred from photometric data and does represent the most likely mass
[b]Binary with parameters that are still compatible with an intermediate/high mass donor

quiescence, although the upper limits are unconstraining due to the poor photon flux. The shortest outburst recurrence time is one month for the globular cluster source NGC 6440 X-2, with an outburst duration of less than 4–5 days, whereas the longest outburst, from HETE J1900.1-2455, has so far lasted for almost 7 years (see Sect. 4.3). The small group of AMXPs therefore form a rather heterogeneous class of objects. There are, however, some common features shared by all AMXPs:

- Outburst luminosities are usually faint, suggesting low time averaged mass accretion rates (Sect. 4.3).
- Spin frequencies appear to be uniformly distributed with an abrupt cutoff at about ~700 Hz (Sect. 4.6).
- Ultra-compact binaries are rather common, comprising about 40% of the total AMXP population.
- Very small donors are preferred, with masses almost always below 0.2 $M_{\odot}$.
- The orbital periods are always relatively short, with $P_b < 1$ day. Therefore the known AMXPs are probably **not** the progenitors of wide orbit binary millisecond radio pulsars.

4.2.1 Intermittency

Until 2007 it was believed that AMXPs showed X-ray pulsations throughout outbursts. The seventh AMXP (HETE J1900.1-2455), however, showed an unexpected new behaviour [101]. For the first ≈20 days of its outburst (which started in 2005) it showed typical AMXP pulsations. The pulsations then became intermittent, appearing and disappearing on different timescales for the next ≈2.5 years. Pulsation disappeared and never reappeared again after MJD 54,499 [250], with the most stringent upper limits of 0.05% rms on the fractional pulse amplitude (i.e., a sinusoidal fractional amplitude of 0.07%, see Eq. (4.15)).

This discovery was exciting, since it may help to bridge the gap between non-pulsating LMXBs and AMXPs. It also became immediately clear that other LMXBs might show sporadic episodes of pulsations during their outbursts. This has now been found to be the case for two other sources: Aql X-1 and SAX J1748.9-2021. Aql X-1 is perhaps the most striking case recorded so far: coherent pulsations were discovered in 1998 *RXTE* archival data and appeared in only one 120s data segment out of a total exposure time of 1.5 Ms from more than 10 years of observations. The extremely short pulse episode has raised discussions about whether it really originated from magnetic channeled accretion, and whether Aql X-1 can truly be considered an AMXP. However, the high coherence of the signal leaves little doubt about the presence of pulsations, and the accretion-powered origin appears, so far, the most promising explanation [44]. The case of SAX J1748.9-2021 is slightly different: pulsations were detected sporadically in several data segments and in three (2001, 2005 and 2009–2010 [2, 253, 261]) out of four outbursts observed (the first being in 1998). It is unclear why these three systems show pulsations intermittently. In HETE J1900.1-2455, an increase in pulse fractional amplitude was reported approximately in coincidence with the occurrence of Type I X-ray bursts [101], followed by a steady decrease. On other occasions the pulsations appeared a few hours before or after a burst, indicating that pulsations might be linked somehow with some yet to be identified property of the NS envelope. In SAX J1748.9-2021 the pulsed amplitudes showed some abrupt changes in amplitude and/or phase in coincidence with about 30% of the observed Type I X-ray bursts [2, 253]. The pulsations, however, displayed a more diverse behaviour than in HETE J1900.1-2455, without the typical steady decrease of fractional amplitudes. A period of global surface activity during which both Type I bursts and pulsations are produced might be at the origin of this link [253]. The single pulsating episode of Aql X-1 had instead no clear connection with Type I bursts, even though Aql X-1 is a bursting LMXB.

The spin frequency derivative of HETE J1900.1-2455 was measured over a baseline of 2.5 years, an unprecedented long baseline for an AMXP (whose outbursts last usually less than 100 days). This has indirectly provided hints on the physical origin of such a period of global surface activity in intermittent sources. The spin frequency derivative exhibited an exponential decay in time that was interpreted as evidence of the screening of the NS magnetic field [250]. It was proposed that intermittency originates because the magnetic field strength drops by almost three orders of magnitude on a timescale of few hundred days so that the disk cannot be truncated and only a (very) shallow layer of gas can be channeled to form (weak) pulsations (see also Sect. 4.6.3 for a discussion of the model).

One feature that intermittent pulsars share is that the long term average mass accretion rate $\langle \dot{M} \rangle$ is higher than for the persistent AMXPs and smaller than for the bright non-pulsating systems (like Sco X-1 and the other Z sources). Calculating the precise value of $\langle \dot{M} \rangle$ is rather difficult, since it depends on poorly constrained parameters in most LMXBs, like the distance d, the X-ray to bolometric flux conversion and the recurrence time of the outburst. However, even considering these caveats, it seems clear that at least the brightest systems have never showed pulsations. Whether it is the high mass accretion rate that determines the lack of pulsations or some other properties shared by the brightest systems is still unclear. See Sect. 4.6 for an extended discussion on the mechanism that might prevent the formation of pulsations in LMXBs.

4.3 Observations of the AMXPs

In this section we discuss the main characteristics of each AMXP, including number of outbursts observed, luminosity, distance, orbital parameters and (at the end of each subsection) radio/IR/optical counterparts. We devote Sects. 4.4 and 4.6 to the discussion of the results on the pulsar rotational parameters and their secular evolution, respectively. We briefly mention results relating to X-ray bursts, burst oscillations and aperiodic variability but refer to Sects. 4.7 and 4.8 for a general discussion of these topics. The astrometric position of each AMXP is given in Table 4.2.

4.3.1 *SAX J1808.4-3658*

The source SAX J1808.4-3658 was discovered with the X-ray satellite *BeppoSAX* during an outburst in September 1996 [147]. The source showed via three type I X-ray bursts that the compact object in the binary is a NS. The 2–28 keV peak luminosity of $3 \times 10^{36}\,\mathrm{erg\,s^{-1}}$ [149] is rather faint and below the average peak luminosity reached by other LMXBs. X-ray pulsations were not detected during this outburst, with poorly constraining upper limits of 20% on the pulsed fraction. Coherent pulsations at 401 Hz were discovered instead in 1998, during the second observed outburst [360], thanks to the better sensitivity of the Proportional Counter Array (PCA) on *RXTE*. An orbital modulation of 2.01 h was detected using Doppler delays in the coherent timing data [50] and it was suggested that the companion of the pulsar is a heated brown dwarf with mass of $\approx 0.05\,M_\odot$ [27]. No thermonuclear X-ray bursts were observed during the 1998 outburst, but re-analysis of the 1996 data provided marginal evidence for burst oscillations at the NS spin frequency [149].

SAX J1808.4-3658 went into outburst again in 2000, 2002, 2005, 2008 and 2011, with an approximate recurrence time of about 1.6–3.3 yr, and is the best sampled and studied of all AMXPs. All of the outbursts showed faint luminosities, with peaks always below $10^{37}\,\mathrm{erg\,s^{-1}}$, even when considering broad energy bands. The 2000 outburst was poorly sampled due to solar constraints, and was observed only during the "flaring tail", a peculiar outburst phase displayed by only a very small number of X-ray sources. The typical SAX J1808.4-3658 outburst can be split in five phases: a fast rise, with a steep increase in luminosity lasting only a few days, a peak, a slow decay stage, a fast decay phase and the flaring tail. The first four phases are typical of several X-ray binaries and dwarf novae and can in principle be partially explained with the disk instability model. The flaring tail instead shows bumps, called "reflares", with a quasi-oscillatory behaviour of a few days and a variation in luminosity of up to three orders of magnitude on timescales ~1–2 days [255, 358, 361]. The flaring tail has no clear explanation within the disk instability model [181], but has been observed in all outbursts [98, 255]. Pulsations

Table 4.2 Astrometric position of AMXPs

Source	Right ascension [HH:MM:SS] (J2000)	Declination [DD:MM:SS] (J2000)	Error (90% c.l.)	Observatory	Reference
SAX J1808.4-3658	18:08:27.62	−36:58:43.3	0″.15	6.5 m Baade (Magellan I)	[123]
XTE J1751-305	17:51:13.49(5)	−30:37:23.4(6)		Chandra	[204]
XTE J1807-294	18:06:59.80	−29:24:30	0″.6	Chandra	[205]
IGR J00291+5934	00:29:03.05(1)	+59:34:18.93(5)		Multiple Optical Obs.	[334]
XTE J1814-338	18:13:39.04	−33:46:22.3	0″.2	Magellan	[169]
XTE J0929-314	09:29:20.19	−31:23:03.2	0″.1	Mt. Canopus	[114]
Swift J1756.9-2508	17:56:57.35	−25:06:27.8	3″.5	Swift/XRT	[170]
Aql X-1	19:11:16.0245341	+00:35:05.879384	0″.0005	e-EVN	[337]
SAX J1748.9-2021	17:48:52.163	−20:21:32.40	0″.6	Chandra	[272]
NGC6440 X-2	17:48:52.76(2)	−20:21:24.0(1)		Chandra	[131]
Swift J1749.4-2807	17:49:31.83	−28:08:04.7	1″.6	*Swift*/XRT	[73]
IGR J17511-3057	17:51:08.66(1)	−30:57:41.0(1)	0″.6	Chandra	[232]
IGR J17498-2921	17:49:55.38	−29:19:19.7	0″.6	Chandra	[53]
HETE J1900.1-2455	19:00:08.65	−24:55:13.7	0″.2	Palomar	[94]
IGR J18245-2452	18:24:32.51	−24:52:07.9	0″.5	ATCA	[267]

The errors in parentheses refer to R.A. and DEC separately, whereas those that appear in the column "Error" refer to both coordinates

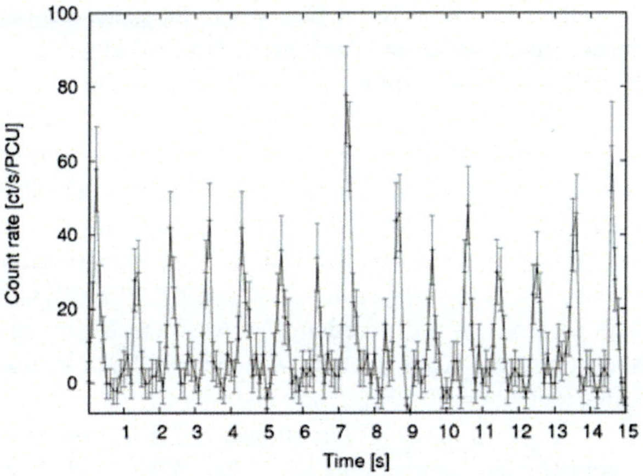

Fig. 4.1 Small portion (15 s) of the 2005 X-ray light-curve of SAX J1808.4-3658, showing the 1 Hz flaring phenomenon during a reflare. The amplitude of the 1 Hz oscillation reaches 125% rms (Figure from [255])

are still detected during this low luminosity stage, and a strong ~1 Hz oscillation is observed to modulate the reflares (see Fig. 4.1; [255, 341]). This has been interpreted as a possible signature of the onset of a disk instability with radial perturbation of the magnetospheric boundary (the so-called "Spruit and Taam instability" [70, 311]) close to the onset of the"propeller stage" [255]. The 1 Hz modulation, however, has not appeared in all the flaring tails of the different outbursts, but only in the years 2000, 2002 and 2005. The presence of the 1 Hz modulation during the 1998 flaring tail cannot be excluded, however, due to lack of observations.

The study of the high frequency aperiodic variability of SAX J1808.4-3658 led to a breakthrough when twin kHz Quasi Periodic Oscillations (QPOs) at a frequency of approximately 500 and 700 Hz were discovered for the first time in a source with a well established spin frequency [362]. The twin kHz QPOs (Sect. 4.8) were observed when SAX J1808.4-3658 reached the peak luminosity during the 2002 outburst, and their frequency separation of 196 ± 4 Hz is consistent with being at half the spin frequency. This suggested a possible link between spin frequency and kHz QPOs and hence refuted the beat-frequency model for the formation of kHz QPOs. In this model the upper kHz QPO at 700 Hz reflects an orbital frequency in the inner accretion disk, whereas the lower kHz QPO should appear as a beat frequency between the upper kHz QPO and the spin frequency. The twin kHz QPOs must be separated by a frequency equal to the spin frequency and not half its value, as observed. The upper kHz QPO was observed during most of the outburst, and another QPO at 410 Hz also appeared during a few observations [362]. The origin of this latter QPO is unclear and it has been suggested [362] that it might be related to the side-band phenomenon observed in other LMXBs [153]. In particular, it was suggested that the 410 Hz QPO might be a sideband of the pulsation, created by a resonance occurring at the radius in the disk where the general relativistic vertical epicyclic frequency matches the spin frequency of the pulsar.

The 3–150 keV X-ray spectrum of the 1998 outburst had a remarkably stable power law shape with photon index of ~2 and a high energy cutoff at ~100 keV [115]. Further spectral analysis provided evidence for a two-component model: a blackbody at soft energies and a hard Comptonization component at higher energies [112]. The blackbody is interpreted as the heated hot spot on the NS surface, whereas the Comptonization is produced in the accretion shock created at the bottom of the magnetic field lines as the plasma abruptly decelerates close to the NS surface. The presence of an accretion disk was detected much later at lower energies, with observations taken in 2008 with the EPIC-pn camera on the *XMM-Newton* telescope. Its large sensitivity at soft energies (down to about 0.5 keV) well below the nominal 2 keV limit of *RXTE*, allowed the detection of the typical cold accretion disk signature at a temperature of 0.2 keV [242, 257]. The signature of a fluorescent relativistic iron Kα emission line profile was also found. A similar result was obtained with combined *XMM-Newton* and *Suzaku* data [41]. Spectral modeling of the iron line constrained the magnetic field of the pulsar to be $\sim 3 \times 10^8$ G at the poles [41, 242]. Simultaneous spectral modeling of the inner disk radius and pulse profile shapes of the 2002 outburst [142] lead to a similar constraint of the magnetic field ($B \sim 10^8$ G).

Thermonuclear bursts were observed in 1996, 2002, 2005, 2008 and 2011 [51, 98, 149]. Most of the bursts exhibit photospheric radius expansion (PRE), where the luminosity reaches the Eddington limit, lifting the photosphere off the surface of the NS until the flux dies down. Such bursts can be used as standard candles [174]. The distance estimated using this method is 2.5–3.6 kpc [98, 149]. Note that a different lower limit of 3.4 kpc is reported in [98]. This lower limit is based on the assumption that the long-term mass transfer rate is driven purely by loss of angular momentum in the binary via emission of gravitational radiation, which may not be a good approximation (see Sect. 4.6). All bursts observed in the *RXTE* era have shown burst oscillations (with a possible marginal detection in one burst in 1996 observed with *BeppoSAX*) with an amplitude of a few percent rms. SAX J1808.4-3658 provided the first robust confirmation that burst oscillation frequency was, to within a few Hz, the spin frequency of the star ([51] and Fig. 4.2).

An optical/IR counterpart (V4584 Sagittarii) was discovered during the 1998 outburst, coincident with the position of SAX J1808.4-3658 [291]. The reported magnitudes of the candidate were V = 16.6, R = 16.1, I = 15.6, J = 15.0, H = 14.4, K = 13.8, with an uncertainty of 0.2 mag in VRI and <0.1 mag in JHK. The V band was further monitored [113] and a possible sinusoidal modulation at the 2 h period of the binary identified, together with a decay in the luminosity as the outburst progressed in its decay stage. A multiband optical/IR photometric study of the optical counterpart during the 1998 outburst revealed an optical flux consistent with an X-ray heated accretion disk and an inclination of $\cos i = 0.65^{+0.23}_{-0.33}$ (90% c.l.) [346]. The IR observations, however, showed an excess with respect to an accretion disk plus irradiated donor star model. This excess was transient in nature, as it was detected only during one observation, the rest being consistent with the irradiated

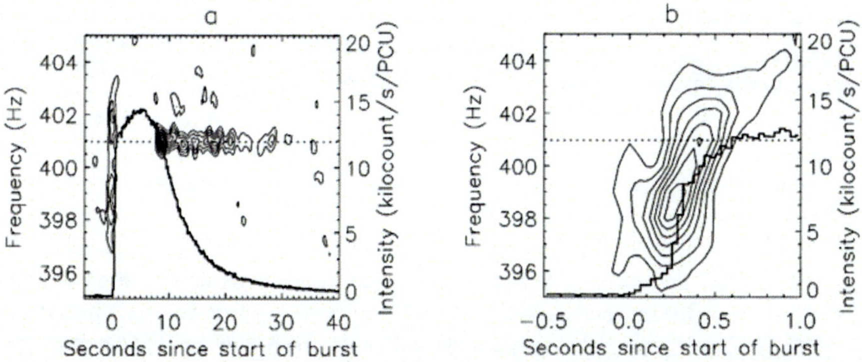

Fig. 4.2 Left panel: a typical X-ray burst observed in the 2–60 keV energy band on 18 October 2002 with a dynamical power spectrum of barycentered data superimposed. The contours refer to power levels of 2 s power spectra spaced by 0.25 s. The burst oscillations are clearly detected during the rise, overshooting the spin frequency of SAX J1808.4-3658 at 401 Hz. The oscillations then disappear during the photospheric radius expansion close to the burst peak and reappear in the burst tail at a frequency consistent with the 401 Hz of the pulsar. Right panel: a zoom in of the burst rise and peak, with finer power spectra resolution of 0.25 s spaced by 0.03125 s. The frequency overshoot is clearly visible, with the BO frequency reaching ∼403 Hz (Figure from [51])

accretion disk plus donor star model. Radio observations carried out one week later (April 27, 1998) with the Australia Telescope Compact Array (ATCA) revealed a transient radio counterpart with a flux of ∼0.8 mJy with no further detection at later epochs during the 1998 outburst or during quiescence. This observation was interpreted as synchrotron emission which might in principle constitute a viable explanation also for the IR transient excess. This suggests that material can be ejected from the binary via relativistic jets and/or outflows. A similar IR excess was detected during 2005 June 5 [119] and radio Very Large Array (VLA) observations carried at 4.86 and 8.46 GHz showed a transient ∼0.4 mJy flux, again interpreted as possible synchrotron emission [301].

Optical observations in *quiescence* have been performed extensively for over a decade [77, 140, 348]. In the first observation, a clear optical excess was detected, well above the level expected from residual X-ray irradiation of the donor star that is still present during quiescence [140]. A clear modulation at the orbital period of the binary in IR bands was also identified [77, 347, 348]. To explain the observations, a strong source of extra irradiation was required. In particular, the initial donor mass estimate of $0.05\,M_{\odot}$ [27] was revised to 0.07–$0.11\,M_{\odot}$, to account for the large entropy of the donor in quiescence [77]. The optical excess was first interpreted as evidence for the turning on of an active radio pulsar when accretion halts [35, 43]. Whether this is the correct interpretation is still to be verified, but it appears a plausible explanation even though no radio pulsations have been detected [141].

4.3.2 *XTE J1751-305*

The AMXP XTE J1751-305 was discovered by *RXTE* on April 3, 2002 and coherent pulsations at 435 Hz were observed immediately [204]. The orbital period of 42 min makes this AMXP an ultra-compact binary with a heated He or C/O white dwarf of minimum donor mass 0.013–0.017 $M_{\odot}$[2] (depending on the NS mass) [76]. Archival observations taken with the All-Sky-Monitor (ASM) aboard *RXTE* showed that a faint outburst had already occurred in 1998. Three other outbursts were observed in 2005, 2007 and 2009. The first two had very faint peak luminosities of about 10–20% the value reached in 2002. The 2005 outburst in particular was observed by *INTEGRAL* [117] and lasted only 2 days, with only one follow-up *RXTE* observation [326]. No high resolution timing data were taken, and the association of the outburst with XTE J1751-305 remains dubious given the large source position error of the two X-ray observatories. On April 5, 2007, XTE J1751-305 was observed with *RXTE* during routine galactic bulge monitoring observations [203]. No pulsations were seen because of the dim flux (10 mCrab in the 2–10 keV) and the short bulge scan exposure ($\sim$50 s). However, a follow-up *Swift* observation made the identification of the source certain thanks to the high angular resolution of the XRT telescope [208]. The outburst had a fluence of about 10^{-4} erg cm^{-2}, comparable to the already low 2002 outburst fluence of 2.5×10^{-3} erg cm^{-2} [204]. In 2009 *INTEGRAL* recorded a new outburst [55] and *RXTE* observed it extensively, with pulsations immediately detected[210]. This outburst has a duration and brightness similar to the 2002 event.

XTE J1751-305 shows no Type I X-ray bursts and its distance is therefore difficult to determine. The X-ray spectrum taken with *XMM-Newton* in 2002 showed a power law with spectral index 1.44 ± 0.01 contributing 83% of the total 0.5–10.0 keV flux and a blackbody at temperature $kT = 1.05 \pm 0.01$ keV [214]. A broadband spectrum analysis of the 2002 outburst was performed by combining *RXTE* (PCA and HEXTE) and *XMM-Newton* (EPIC-pn) data [111]. The analysis revealed the presence of three components, two blackbodies and a hard component, which were interpreted as a thermal emission from a cold accretion disk ($kT \sim 0.6$ keV), a hotter ($kT \sim 1$ keV) hot spot on the NS surface and a thermal Comptonization in plasma of temperature $kT \sim 40$ keV (see Fig. 4.3). These findings were very similar to the spectral modeling described above for SAX J1808.4-3658. Unsuccessful searches for an optical/near IR counterpart were carried out during the 2002 outburst and in quiescence [154]. However the upper limits are not particularly constraining if one assumes that the source is close to the Galactic center ($d = 8.5$ kpc).

[2] Hydrogen rich donor stars never reach orbital periods below $\sim$80 min [282].

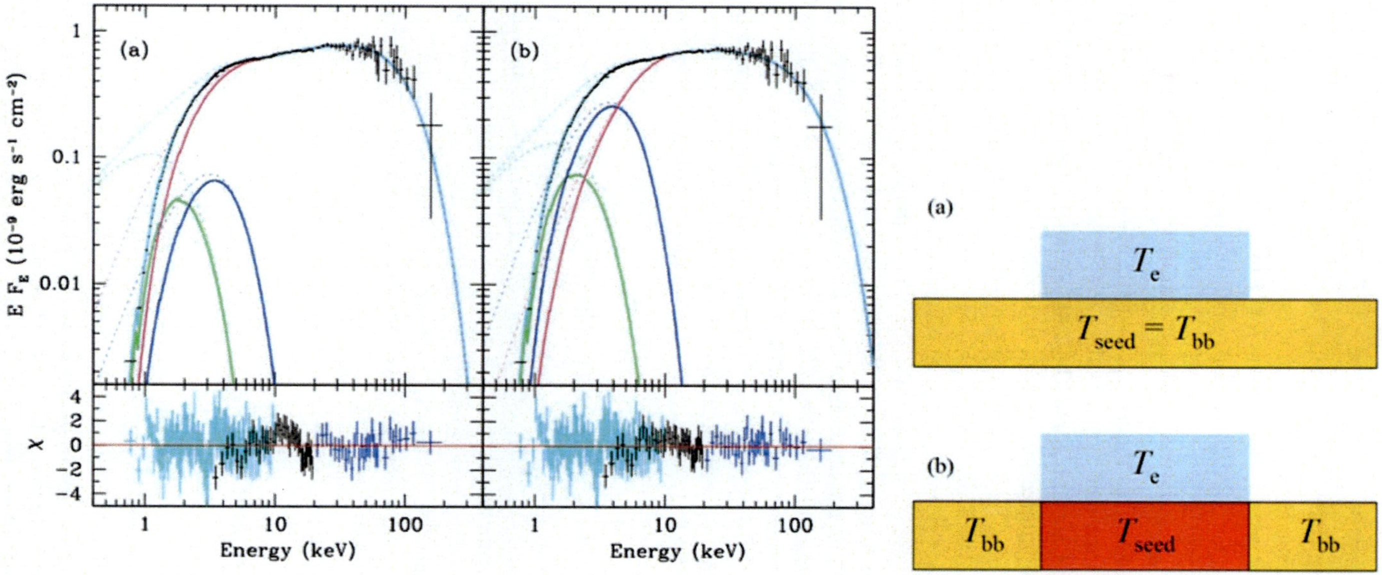

Fig. 4.3 Broadband X-ray spectrum of XTE J1751-305 obtained by combining *XMM-Newton* (light blue points) and *RXTE* data (black and dark blue points). Three components are used to model the spectrum: a cold blackbody identified with the thermal disk emission (green line), a hot blackbody (hot spot; blue line)) and a hard tail created via thermal Comptonization in a plasma (pink line). The left spectrum refers to the emission geometry "(a)" in which a small temperature gradient is present in the hot spot. The right spectrum uses the "(b)" model, with a large temperature gradient between the hot spot and the remaining illuminated NS surface around the accretion shock (image taken from [111])

4.3.3 XTE J0929-314

With a spin of 185 Hz and an orbital period of 43.6 min, XTE J0929-314 was the third AMXP to be discovered, and the second in an ultra-compact binary. The source was first detected on April 13, 2002 by the *RXTE*/ASM [285]. Coherent pulsations were first detected in an observation performed on May 2, 2002 [285]. Doppler modulation of the pulsations revealed the ultra-compact nature of the binary [99]. The very small mass function ($2.7 \times 10^{-7}\,M_{\odot}$) implied a tiny donor star of minimum mass of about $0.01\,M_{\odot}$. This requires a cold He white dwarf companion with a possible H-rich envelope that would explain the detection of an H_{α} line at $\lambda = 6563$ Å [45] during the 2002 outburst. The 2002 outburst lasted at least 52 days, but only 34 days were monitored by the high time resolution *RXTE*/PCA instrument. The X-ray spectrum of XTE J0929-314, taken with *Chandra*, showed a power-law plus blackbody similar to XTE J1751-305 [156].

XTE J0929-314 has been detected in several wavebands: radio, near IR and optical counterparts were identified during the 2002 outburst. An optical counterpart consistent with the position of XTE J0929-314 was detected in V, B, R and I bands [118]. Broad band BVRI observations at the Mt Canopus 1-m telescope carried in May 2002 revealed a modulation of ≈10% in the I band with periodicity consistent with the orbit of the binary [114]. An anomalous I-band excess was also observed in the first few days of the outburst and was interpreted as synchrotron radiation, in addition to the accretion disk/irradiated donor emission. This behavior is similar to that reported for SAX J1808.4-3658. The optical spectrum of the source (in outburst) shows a strong emission line at 4640 $\AA$ compatible with a C/O or a He/N disc [229]. Optical spectroscopy also lead to the conclusion that the most likely donor of XTE J0929-314 is a C/O white dwarf. A radio counterpart (during the outburst) was detected with the VLA at 4.86 GHz [299]. A source flux of 0.31 ± 0.07 mJy (May 3) and 0.36 ± 0.05 mJy (May 7) was reported from a position only 0.8 arcseconds from the previously detected optical position [118]. The source has also a very faint optical counterpart ($R = 27.2$) in quiescence [218]. PSF VRI photometry in quiescence [72] also suggested that the donor is irradiated by a source which emits in excess (by a factor of ≈8) of the quiescent X-ray luminosity of XTE J0929-314 ($L_X \sim 4.5 \times 10^{31}$ erg s^{-1}).

4.3.4 XTE J1807-294

The fourth AMXP was discovered on February 21, 2003 by *RXTE* [205]. The source is the third ultra-compact AMXP, with the shortest orbital period known to date, 40.1 min, and a minimum companion mass of $0.0066\,M_{\odot}$. Simultaneous *INTEGRAL*, *XMM-Newton* and *RXTE* spectral analysis [87] constrains the companion mass to be less than $0.022\,M_{\odot}$, implying that the donor is likely a C/O white dwarf rather than a He white dwarf, which would require unlikely a-priori low inclinations. With the exception of the quasi-persistent intermittent AMXP HETE J1900.1-2455, XTE J1807-294 had the longest AMXP outburst observed so far: ≈120 days, of which ≈109 days had sufficiently high flux to allow the detection of pulsations. Twin kHz QPOs were detected in several observations and the separation $\Delta\nu$ of the kHz QPOs is consistent with the spin frequency of the AMXP [191]. No counterparts have been reported at any wavelength during either outburst or quiescence. A candidate optical counterpart detection was reported in 2009 [72] with $V\sim22.1$ and $R\sim21.4$ along with a tentative detection in the J band. The counterpart falls within the $0''.6$ *Chandra* (90% confidence level) error circle. No variability has been observed, with upper limits of 0.1 mag.

4.3.5 XTE J1814-338

The fifth AMXP was found during routine Galactic center observations by the *RXTE*/PCA on June 5, 2003 [202]. The pulse frequency is 314 Hz and the orbital period 4.3 h, giving a minimum companion mass of 0.17 $M_{\odot}$. The outburst lasted for ≈50 days and pulsations were detectable for ≈45 days. The source showed 28 Type I X-ray bursts and *RXTE* detected burst oscillations at the spin frequency in all bursts. The 28th burst showed signs of PRE, allowing a determination of the source distance of ≈8 kpc. The X-ray position of XTE J1814-338 is consistent with the *EXOSAT* source EXMS B1810-337, detected in September 1984 [359]. If this identification is correct, then XTE J1814-338 has a recurrence time of ~20 yr.

The optical counterpart was first tentatively identified using BVRI data taken with the Magellan 6.5 m telescope [168]. An observation of H and He emission lines, with a double-peaked H_{α} line typical of interacting binaries, was reported soon after the discovery of the binary [314]. The counterpart was studied in more detail and finally unambiguously identified [169]. The B, V and R counterparts during the 2003 outburst had magnitudes between 18 and 19 mag, whereas the I band was observed with an excess flux reaching about 17.4 mag. This excess is again similar to that reported for SAX J1808.4-3658 and XTE J0929-314 and suggests the presence of jets and/or outflows. A faint object with $V\sim23.3$ and $R\sim22.5$ was identified in ESO Very Large Telescope (VLT) archival data taken in May 20–21, 2004 [72]. A multiband Very Large Telescope (VLT) campaign carried in 2009 (during quiescence) shows an irradiated companion star that requires an energy source compatible with the spin-down luminosity of a millisecond pulsar [13]. This proves further evidence that AMXPs might turn on as radio pulsars when in quiescence.

4.3.6 IGR J00291+5934

IGR J00291+5934 is the fastest AMXP known, with a spin of 599 Hz. The source was discovered by *INTEGRAL* on 2004 December 2 [82] and prolonged *RXTE* observations lead to the discovery of pulsations [206] and an orbital period of 2.46 h, with a minimum donor mass of about $0.04\,M_{\odot}$ [207]. The 2004 outburst had a duration of ≈14 days from the *INTEGRAL* discovery, with a smooth flux decay. On 2008 August 13 the source started a new outburst [52] and was observed for about a week before reaching an X-ray flux level below the sensitivity limit of *RXTE*. The peak flux observed was about half the value reached in 2004. Follow up *Swift*/XRT observations on August 21 revealed a flux level below 10^{-11} – $10^{-12}\,\mathrm{erg\,s^{-1}}$ and an *XMM-Newton* observation on August 25 provided a 2–10 keV flux of $(1.4 \pm 0.3) \times 10^{-14}\,\mathrm{erg\,s^{-1}}$ [189], compatible with the $10^{-13}\,\mathrm{erg\,s^{-1}}$ (0.5–10 keV) quiescent flux level observed two months after the end of the 2004 outburst [155]. On 2008 September 21, the source re-brightened again, showing another outburst lasting for about 14 days. The fluence of the second and third outbursts were very similar and it is unclear whether the two 2008 outbursts were distinct [126] or part of the same outburst. An analysis of archival *RXTE*/ASM data revealed possible outbursts in 1998 November and 2001 September, suggesting a recurrence time of 3–4 yr [284].

The pulsar has shown no thermonuclear bursts despite very similar orbital parameters to SAX J1808.4-3658. The donor star of IGR J00291+5934 is also probably a X-ray heated brown dwarf as in SAX J1808.4-3658 [100]. Broadband spectral observations performed during the 2004 outburst with *INTEGRAL* and *RXTE* were well fitted with a thermal Comptonization model with an electron temperature of 50 keV and Thomson optical depth $\tau_{\mathrm{T}} \sim 1$ in a slab geometry [88]. The *RXTE* data analysis of the first 12 days of the outburst revealed a power law plus blackbody (with $kT \sim 1$ keV) interpreted as emission from the hot spot on the NS surface. Further simultaneous *RXTE* and *Chandra*/HETGS (High Energy Transmission Grating Spectrometer) spectral observations collected towards the end of the 2004 outburst revealed a power law plus a cold black body at 0.42 keV, interpreted as the cold accretion disk, with no signature of the hot spot blackbody [236]. This is consistent with the lack of detected pulsations towards the end of the outburst [100].

A tentative optical counterpart was identified during the 2004 outburst [95], with an R magnitude of 17.4. Follow up spectral measurements with the 4.2-m William Herschel Telescope on La Palma strengthened the association with the identification of a weak HeII and H_{α} line [292]. An optical/NIR photometric study in quiescence was then performed in 2005 with the 3.6-m Telescopio Nazionale Galileo (TNG) [71]. The VRIJH counterparts of IGR J00291+5934 were detected, with a strong upper limit in the K-band. The optical light curve shows variability consistent with the 2.46 h orbital period. The radio to X-ray Spectral Energy Distributions (SEDs) revealed a blue component indicative of an irradiated disc [189]. The SED contained also a transient NIR excess similar to that found in SAX J1808.4-3658, XTE J0929-

314 and XTE J1814-338. An H_α line was observed in the optical spectrum of the 2004 and 2008 outbursts. Evidence was also found for an *I* band excess during quiescence in an observation performed with the 4.2-m William Herschel Telescope (WHT) on 2006 September 13 and 14 [155].

Radio observations taken on 2004 December 4 with the Ryle Telescope in Cambridge at 15 GHz resulted in a detection at 1.1 mJy [271]. Follow up observations taken one day later gave a non detection, with upper limits of 0.6 mJy at 15 GHz [90]. The Westerbork Synthesis Radio Telescope (WSRT) observed IGR J00291+5934 between 2004 December 6 and 7, and a radio counterpart was seen at 5 GHz with a flux of 0.250 ± 0.035 mJy [90]. Very Large Array (VLA) observations performed on 2004 December 9 at 4.86 GHz confirmed the reported radio detection, with a flux of 0.17 ± 0.05 mJy. The source was not detected in radio during the 2008 outburst, with upper limits of 0.16 mJy at 5 GHz in an observation taken between August 15 and 16 [192].

4.3.7 HETE J1900.1-2455

The *High Energy Transient Explorer* (HETE-2) discovered HETE J1900.1-2455 on 2005 June 14 via an X-ray burst [342]. Follow up observations taken on June 16 by *RXTE*/PCA showed pulsations at 377 Hz [219]. A source distance ∼4–5 kpc (assuming a typical NS mass of $1.4\,M_\odot$ and a He rich NS atmosphere) was determined using the first type I X-ray burst detected with HETE-2 [161, 325], which showed PRE. The orbital period was determined to be 1.4 h with a minimum companion mass of $0.016\,M_\odot$[158]. Unlike the other AMXPs with brown dwarf companions discussed so far, the donor star in HETE J1900.1-2455 is most likely a brown dwarf which does not require X-ray irradiation or a non-standard evolution to fill its Roche lobe.

A single 882.8 Hz QPO was detected in the *RXTE* lightcurve during a bright flare occurred on MJD 53,559.45 [158]. Simultaneous *RXTE* and *INTEGRAL* observations showed that the 2–300 keV spectrum is well fit by a blackbody component peaking at 0.8 keV plus thermal Comptonization with electron temperature of 30 keV and optical depth $\tau_T \sim 2$ for a slab geometry [86]. The source was detected up to energies of 200 keV and the 0.1–200 keV luminosity was inferred to be $5 \times 10^{36}\,\mathrm{erg\,s^{-1}}$. Although the peak luminosity is not particularly high compared to other AMXPs, the average mass accretion rate $\langle \dot{M} \rangle$ of HETE J1900.1-2455 is among the highest observed due to its prolonged X-ray outburst that has lasted for 8 years (and is still active as of November 2013) [97, 102]. This makes HETE J1900.1-2455 a so called "quasi-persistent source", i.e., an X-ray binary with a prolonged outburst duration of several years. Despite the long baseline provided by the outburst, pulsations were *continuously* detected only in the first 2 months of the observations although sporadic pulse detections occurred over a baseline of about 2.5 years (Sect. 4.2.1).

An optical counterpart was identified with the Robotic Palomar 60-inch Telescope on June 18 [94]. The source had a brightness $R{\sim}18.4$. Near IR observations taken with the 1.3 m robotic PAIRITEL telescope on Mt. Hopkins indicated a counterpart with $J = 17.6 \pm 0.2$, with no variability between observations taken on June 16 and 18 [315]. The counterpart was observed again with the FLWO 1.2-m telescope at Mt. Hopkins with $V = 18.09 \pm 0.03$, $R = 18.02 \pm 0.03$ and $I = 17.88 \pm 0.02$ [316]. A spectral analysis of the optical counterpart showed a HeII 4686 Å, suggesting correct identification of the binary's optical counterpart. Further optical observations taken with the 8.4-m Large Binocular Telescope (LBT) on Mt. Graham, Arizona, showed a brightness of $R = 18.64 \pm 0.02$ and 18.54 ± 0.02 on 2007 June 13, after an initial decline of 1.7 mag in optical, and a non-detection in X-rays which occurred two weeks prior [75, 333]. Phase resolved optical photometry and spectroscopy obtained in September 2006 at the 2.3-m ATT, 3.5-m WIYN and 8.4-m LBT [84], revealed a sinusoidal modulation in the R-band, close to the orbital period. However, further analysis showed that the modulation is probably due to a super-hump of a precessing accretion disk, and not irradiation of the donor star. A broad H_{α} emission line in a 10-m Keck I spectrum taken by the same authors confirmed this interpretation. No radio counterpart was identified in VLA observations on 2005 June 19 and 24 at 8.46 GHz [302].

4.3.8 Swift J1756.9-2508

Swift J1756.9-2508 was discovered by the Burst Alert Telescope (BAT) aboard *Swift* on June 7, 2007 [170]. High time resolution data from the *RXTE*/PCA, showed that the source was a 182 Hz AMXP [209]. This was the fourth ultra-compact AMXP discovered, with an orbital period of 54.7 min. The minimum donor star mass, given by the X-ray mass function, is 0.0067 $M_{\odot}$. It has been proposed that Swift J1756.9-2508 harbors a He rich white dwarf which is irradiated by X-ray flux from the accretor (see Fig. 4.4 and [170]). The spectrum of the source, taken with *Swift*/XRT, BAT and *RXTE*/PCA, is well fit by a single absorbed power law with spectral index $\Gamma \sim 2$ [170]. The 2007 outburst length was 13 days, with no prior outbursts seen in archival data. A second outburst was detected on 2009 July 8 by *Swift*/BAT and *RXTE*/PCA [256]. This second outburst showed a very similar length and shape as the 2007 outburst. No thermonuclear bursts were observed. There is evidence for a fluorescent Fe line in the X-ray spectrum, although the poor spectral resolution of *RXTE* prevented to constrain the physical parameters of the system [260]. A possible NIR K_s counterpart was identified [38] by inspecting the variability of 70 objects within the 3.5 arcsec error circle of the *Swift*/XRT position. The candidate counterpart exhibited variability of about 1.3 mag in two observations taken close to the end of the outburst (2007 June 19) and during quiescence (2007 July 1). No radio counterpart has been identified [134, 273] and Chandra observations taken on 2007 July 7 detected one single photon in a 11 ks long observations [240].

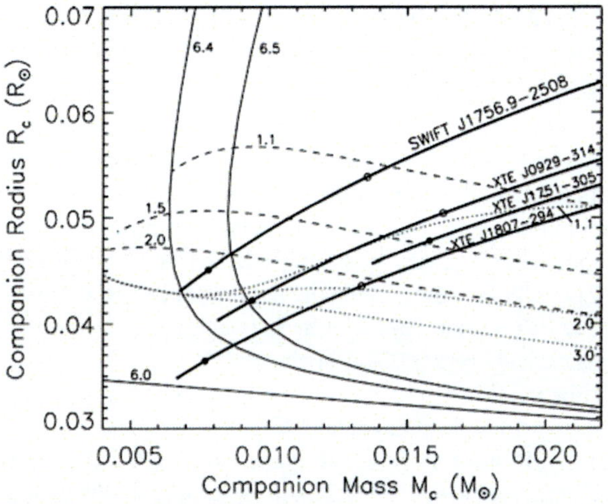

Fig. 4.4 Mass-Radius relation (solid line) for the companions of the four ultra-compact AMXPs, as determined from the mass function. Open and filled circles indicate an inclination of 60° and 30°, respectively. Dotted and dashed curves are non-irradiated/irradiated white dwarf models, respectively. The numbers next to the curves indicate the degeneracy parameter. The solid thin curves are pure C white dwarf models and the numbers refer to the log of the temperature (in Kelvin) (Figure from [170])

4.3.9 Aql X-1

Aql X-1 is perhaps the most peculiar AMXP since it has showed pulsations at 550.27 Hz in just 120 s of X-ray data [44]: the short baseline means that no coherent timing solution can be obtained. As a consequence, no orbital modulation or mass function has yet been measured with X-rays. The source shows thermonuclear bursts with burst oscillations whose asymptotic frequency is ≈0.5 Hz smaller than the coherent accretion powered pulsations observed in 1998 [368]. The recurrence time of the outbursts is about 1 yr and they have been extensively monitored at different wavelengths. Twin kHz QPOs were observed at a frequency of ∼800 and ∼1080 Hz [15], so that their frequency separation is consistent with being equal to half the NS spin frequency. The optical counterpart of Aql X-1, named V1333 Aql, was identified in 1978 [330] as a star with $B{\sim}17$ mag (in outburst) and $B > 20$ mag (in quiescence). The spectral type of the source was identified as a K6-M0 star [59]. An orbital modulation at a period of 18.95 h was reported from an optical modulation of the counterpart of Aql X-1 [57, 58, 355]. Despite several attempts, the radio counterpart has been observed in only a few outbursts at a level of ∼0.1–0.5 mJy [139, 300, 303]. In November 2009, radio observations provided evidence for the emission to arise from steady jets triggered at state transitions from the soft to hard state [216].

4.3.10 SAX J1748.9-2021

SAX J1748.9-2021 was discovered with the *BeppoSAX*/WFC on August 22, 1998 in the globular cluster NGC 6440 [148, 343]. The distance of the source is well constrained thanks to the knowledge of the globular cluster distance of 8.5 ± 0.4 kpc [233]. *RXTE*/PCA and *BeppoSAX*/WFC monitoring revealed no pulsations, but thermonuclear bursts were observed with no burst oscillations. The position of the source was coincident with the position of MX 1746-20 detected by *Uhuru* and *OSO*-7 in 1971-1972. However, given the large error box of these missions, it is not possible to establish a secure association between the two sources. SAX J1748.9-2021 showed a typical average X-ray spectrum with a power-law with photon index 1.5 and a soft blackbody component at $kT = 0.9$ keV [148]. However, other spectral models also gave a good fit to the data [148].

In August 2001, after 3 years of quiescence, SAX J1748.9-2021 started a new outburst observed with the *RXTE*/ASM. *Chandra* observations allowed the establishment of a precise position in the globular cluster and confirmed that the 1998 outburst was from the same source. Thermonuclear bursts without burst oscillations were again observed. The beginning of the third outburst was detected in 2005 March 7 during a routinely *RXTE*/PCA Galactic Bulge Scan Survey. The source was monitored until 2005 July 21 and several pointed *RXTE*/PCA observations were performed. The first pulsations reported for this source were seen in the 2005 outburst [104, 105]. A coherent signal at ≃442 Hz was detected during a flux decay reminiscent of the tail of a superburst. However, neither the temporal profile nor the energetics of the tail were consistent with a super-burst and this lead to the suggestion that this was a new intermittent AMXP [105]. An independent analysis of *RXTE*/PCA archival data from the 2001 and 2005 outburst revealed intermittent coherent X-ray pulsations at 442 Hz, appearing and disappearing on timescales of few hundred seconds [2, 253]. A further outburst was observed in December 2009 with *RXTE* and intermittent pulsations were observed again [261]. Observations in quiescence carried out in July 2000 and June 2003 revealed an X-ray spectrum with a thermal component detected in both observations. The 2000 observation, however, required also a power law component possibly because of residual accretion on the NS [40].

Thanks to *ROSAT*/HRI archival observations and a 3.5 m New Technology Telescope (NTT) observation on 1998 August 26–27, two candidate optical counterparts, dubbed V1 and V2, were identified [343]. V2, with $B \simeq 22.7$ mag, has now been recognized as the true counterpart thanks to the precise astrometric position obtained with the 2001 *Chandra* observations [150]. VLA radio observations provided stringent upper limits during the 2009 outburst and implied that the radio emission is quenched at high X-ray luminosities [217].

4.3.11 NGC 6440 X-2

On 2009 June 28, the X-ray transient NGC 6440 X-2 was discovered [129] in a *Chandra* observation of the globular cluster NGC 6440 (the same cluster that harbors the AMXP SAX J1748.9-2021). Coherent pulsations at 206 Hz were detected in a subsequent outburst in *RXTE*/PCA observations on 2009 August 30 [3]. The orbital period is 0.95 h, making this the fifth ultra-compact AMXP discovered. One of the most striking features of this pulsar is the very low long-term average mass accretion rate $\langle \dot{M} \rangle < 2 \times 10^{-12}\,\mathrm{M_\odot\,yr^{-1}}$, which could be determined thanks to the well-defined distance to the globular cluster [131]. Peak flux is one of the faintest observed from the AMXPs ($L_{2-10\,\mathrm{keV}} < 1.5 \times 10^{36}\,\mathrm{erg\,s^{-1}}$) and the outburst length is extremely short, with an duration of ≈3–5 days. It should be noted that the *RXTE*/PCA Galactic Bulge Scan flux limit corresponds to a luminosity $\sim 10^{35}\,\mathrm{erg\,s^{-1}}$ [3] so that the presence of a longer and very faint outburst cannot be ruled out, although archival *Chandra* observations did not reveal any counterpart in quiescence with an upper limit on the X-ray luminosity of $L_X \lesssim 10^{31}\,\mathrm{erg\,s^{-1}}$. This raises the question of how many faint sources of this kind might have been missed, particularly important given that AMXPs seem to be associated with faint LMXBs, and systems of this type may therefore contain undiscovered AMXPs.

The recurrence time of the outburst is also the shortest among AMXPs, ∼1 month. An *RXTE*/PCA and *Swift*/XRT monitoring campaign revealed the occurrence of 10 outbursts between 2009 and 2011 [131, 252]). In November 2009, visibility constraints and a new outburst of the other AMXP SAX J1748.9-2021 prevented further observations of NGC 6440 X-2. A strong (∼50% rms) 1 Hz modulation was observed in the light-curve of at least 6 outbursts [252], which strongly resembled the 1 Hz modulation seen in SAX J1808.4-3658 [255, 341].

Despite the precise astrometric X-ray position of the source, the optical counterpart of NGC 6440 X-2 has not yet been identified, with $B > 22$ and $V > 21$ from archival *Hubble Space Telescope* (HST) imaging of the globular cluster when NGC 6440 X-2 was in quiescence. During the 2009 August outburst, Gemini-South observations did not reveal any counterpart with $g' > 22$. Observations carried with the CTIO 4-m telescope during the 2009 July outburst also did not reveal any counterpart with $J > 18.5$ and $K > 17$ [131].

4.3.12 IGR J17511-3057

The X-ray binary IGR J17511-3057 was discovered during Galactic Bulge Monitoring by the *INTEGRAL* satellite in 2009 September 12 [14]. Coherent pulsations at 245 Hz and an orbital modulation of 3.5 h were measured with follow up observations by the *RXTE*/PCA [211]. Given the X-ray mass function, the minimum donor star mass is $0.13\,\mathrm{M_\odot}$. The X-ray light-curve was also extensively monitored by *Swift*/XRT, *XMM-Newton*, *RXTE*/PCA and *Chandra*. The light-curve showed a typical fast rise with a slow decay lasting for about 20 days, before entering a fast decay phase [4, 243]. Curiously, while the source was in the fast decay phase, a very short and faint outburst of the ultra-compact AMXP XTE J1751-305 appeared in the same field of view (FoV) of *RXTE*. The mini-outburst of XTE J1751-305 lasted less than 3 days, and pulsations were clearly visible at 435 Hz [210]. Thermonuclear bursts were observed in IGR J17511-3057 with *Swift* [30], and burst oscillations were detected in *RXTE* data [4, 353]. None of the bursts showed PRE, so there are several different distance determinations in the literature. The most constraining gives an upper limit on distance of $\lesssim$5 kpc [4].

Swift spectral analysis revealed a power law ($\Gamma \sim 2$) plus a blackbody at 0.9 keV [30]. Broadband spectral data from simultaneous *RXTE*/PCA and *Swift*/XRT observations could be well fitted with a three model component [143]: a disk with low temperature ($kT \sim 0.24$ keV), a hot black body (hot spot with $kT \sim 1$ keV) and a Comptonization component originating from the accretion shock (electron temperature $T_e \sim 30$ keV; $\tau_T \sim 2$). The spectral fitting also required a fluorescent iron line at 6.4 keV with Compton reflection and provided an interstellar absorption column of $N_H = 0.88^{+0.21}_{-0.24} \times 10^{22}$ cm^{-2}. Very similar results were obtained using simultaneous *XMM-Newton* and *RXTE*/PCA spectral data [243]. Combined *RXTE*/PCA, *Swift*/XRT and *INTEGRAL* data also gave consistent results using a thermal Comptonization model [89]. Twin kHz QPOs were observed at the beginning and at the end of the outburst, and had $\Delta\nu \simeq 120$ Hz, about half the value of the spin frequency [160]. A NIR counterpart with brightness $K_s = 18.0 \pm 0.1$ was identified with the 6.5 m Magellan Baade telescope on 2009 September 22 [335]. A second observation on October 7, while the X-ray flux was fading rapidly, showed no counterpart with 3σ upper limits of $K_s > 18.8$. No radio counterpart has been found, with upper limits of 0.10 mJy [215].

4.3.13 Swift J1749.4-2807

The first observation of Swift J1749.4-2807 was made on June 2, 2006 with the *Swift*/BAT telescope, which recorded a high level activity consistent with an unidentified source [305]. This was later identified as a "burst-only" accreting NS binary (i.e., a NS that is bright enough to be detectable only during the occurrence of thermonuclear explosions) thanks to a spectral analysis of the *Swift*/BAT data [363]. The analysis was consistent with the observation of a thermonuclear burst for a source distance of 6.7 $\pm$ 1.3 kpc. Further *Swift*/XRT data analysis revealed an X-ray counterpart of the burst and archival *XMM-Newton* data showed a faint point source coincident with the *Swift*/XRT position [121, 363]. Increased activity was reported in 2010 by *INTEGRAL*, and was linked to the accretion powered emission process [56, 266]. Pulsations were discovered immediately afterwards at 518 Hz in *RXTE*/PCA follow-up observations [5, 6]. A very strong second harmonic was detected at 1036 Hz [31] and orbital modulation was measured at 8.82 h, giving a minimum donor mass of about 0.6 $M_\odot$[18, 320]. Swift J1749.4-2807 was finally identified as the first eclipsing AMXP [201] and the inclination of the system was constrained to be $i \sim 74$–$78°$ [6, 201]. The binary showed three eclipses with a duration of 2172 $\pm$ 13 s, allowing the first attempted detection of Shapiro delay effects in X-ray timing data for an object outside the Solar System [201].

To fit the 0.5–40 keV spectrum, an absorption column density of $N_H = 3.0 \times 10^{22}\,\mathrm{cm}^{-2}$ and a power-law with spectral index $\Gamma \simeq 1.7$ are required [91]. The absorption column is about 3 times larger than the expected Galactic column density in the direction of the source. The X-ray light-curve showed an exponential decay with a *Swift*/XRT non detection after 11 days since its first 2010 observation. Due to the crowded field (the source is on the Galactic plane), no single counterpart has been identified in NIR counterpart searches [73]. More than forty counterparts were identified in the X-ray error circle with ESO's Very Large Telescope (VLT).

4.3.14 IGR J17498-2921

IGR J17498-2921 was discovered by *INTEGRAL* on 2011 August 11 [109]. Soon after, coherent X-ray pulsations were discovered in *RXTE*/PCA data at 401 Hz with an orbital period modulation of 3.8 h [244]. Type I X-ray bursts were observed with *INTEGRAL* [92] and *RXTE*. Burst oscillations were detected [49, 194] and evidence of PRE placed the source at a distance of 7.6 kpc [194]. *Swift*/XRT observed the source returning to quiescence on 2011 September 19 [195]. A candidate NIR counterpart of IGR J17498-2921 was first detected in archival data taken with the 4-m VISTA telescope at the Paranal observatory [120]. The optical counterpart was detected on 2011 August 25 and 26 with the 2-m Faulkes Telescope South [304]. The position detected was consistent with both the *Chandra* and the NIR counterparts. However, further observations suggested that the NIR/optical counterpart is a foreground star aligned by chance with the X-ray source [339]. This counterpart was confirmed to be a foreground source with observations taken at the 2.5-m Irenee du Pont telescope. The analysis suggested that, given the distance and extinction of IGR J17498-2921, the expected magnitudes are $R \sim 28$ and $J \sim 23.3$, both well above the limit reached by the optical/NIR observations [336].

4.3.15 IGR J18245-2452

The most recent addition to the AMXP family was discovered by INTEGRAL on 2013 March 28 [83] during an X-ray outburst (20–100 keV luminosity of $3 \times 10^{36}\,\mathrm{erg\,s}^{-1}$) and is perhaps the most spectacular of all the AMXPs. The source, at a distance of 5.5. kpc is located in the core of the globular cluster M28 [132]. *XMM-Newton* observations revealed an AMXP whose spin (254 Hz) and orbital parameters were identical to those of a previously known radio millisecond pulsar located in the same region (PSR J1824-2452I, a.k.a. M28I) [247]. This makes this source the first known radio millisecond pulsar that has switched to an AMXP state. The 0.5–10 keV luminosity shows a very peculiar X-ray flickering with flux variations of up to 2 orders of magnitude happening within a few seconds[93, 247]. The two flux states can be described by two very different spectral models with power-law index $\Gamma \sim 1.7$ (high flux) and $\Gamma \sim 0.7$ (low flux) [93]. This has been interpreted as evidence for the onset of a propeller phase *during* the outburst, rapidly alternating with a normal accretion phase [93]. A similar flickering was also observed in quiescence[197] when the source switches from a stable low luminosity state of $\sim 10^{32}$ erg s^{-1} to a flickering state with luminosity $\sim 10^{33} - 10^{34}$ erg s^{-1}.

An iron K_α line was also observed in the *XMM-Newton* spectrum [247]. Thermonuclear X-ray bursts were observed by *Swift*/XRT and MAXI [190, 246, 306] and burst oscillations were later identified [251]. Archival optical observations obtained with *HST* revealed a faint counterpart on April 2009 and 2010. However, on August 2009 the *HST* detected a blue counterpart, brighter by ~2 mag in several filters (F390W = 20.37±0.06, F606W = 19.51±0.04, and F656N = 17.26±0.04)[63, 237], with a strong H_α emission indicative of the presence of an accretion disk four years before the 2013 outburst. Radio observations performed with ATCA on 2013 April 5, show a bright radio continuum counterpart (0.62 and 0.75 mJy at 5.5 and 9 GHz, respectively) [207]. The source was last detected in X-rays on 2013 May 1 and soon after it turned back on in radio as a millisecond pulsar [248].

4.4 Accretion Torques

Once the donor star overflows its Roche lobe, gas flowing through the inner Lagrangian point L_1 carries large specific angular momentum and hence forms an accretion disk around the NS. The type of disk depends on the microphysical conditions governing the gas dynamics [106–108, 280]. If gas pressure dominates, the disk will be geometrically thin and optically thick [106, 231, 307] with material moving in Keplerian orbits with orbital frequency:

$$\nu_K = \frac{1}{2\pi}\sqrt{\frac{GM}{R^3}} \simeq 767\,\mathrm{Hz}\left(\frac{M}{1.4\,\mathrm{M}_\odot}\right)^{1/2}\left(\frac{R}{20\,\mathrm{km}}\right)^{-3/2} \tag{4.1}$$

At a distance of a few tens of kilometers from the NS, the gas orbits several hundred cycles per second and flows almost undisturbed until the magnetic field of the NS (that for AMXPs is of the order of 10^8 G) is strong enough to perturb its orbit. At the magnetospheric-radius r_m, the kinetic energy of the free-falling gas becomes comparable to the magnetic energy of the NS magnetosphere:

$$\begin{aligned} r_m &= \xi\, r_A = \xi\left(\frac{\mu^4}{2GM\dot{M}^2}\right)^{1/7} \\ &= 35\,\mathrm{km}\,\xi\left(\frac{\mu}{10^{26}\,\mathrm{G\,cm^3}}\right)^{4/7} \times \left(\frac{\dot{M}}{10^{-10}\,\mathrm{M}_\odot\,\mathrm{yr}^{-1}}\right)^{-2/7}\left(\frac{M}{1.4\,\mathrm{M}_\odot}\right)^{-1/7} \end{aligned} \tag{4.2}$$

where $\dot{M}$ is the mass accretion rate at the inner disk boundary, μ is the dipole magnetic moment of the NS and r_A is the Alfvén radius. This latter parameter is calculated assuming spherical accretion:

$$\frac{B^2\left(r_A\right)}{8\pi} \simeq \frac{1}{2}\rho\left(r_A\right)\upsilon^2\left(r_A\right). \tag{4.3}$$

The parameters ρ and υ are gas density and velocity, respectively. The term $\xi \approx$ 0.3–1.0 is a correction factor due to the non-spherical geometry of the problem and is required because the gas orbits in a disk rather than falling radially from every direction. In the disk geometry, magnetic and fluid stresses balance when:

$$B_p B_\phi r^2 = \dot{M}\frac{\partial(\upsilon_\phi r)}{\partial r} \tag{4.4}$$

where B_p and B_ϕ are the poloidal and toroidal components of the NS magnetic filed, r the radial coordinate measured from the NS center and υ_ϕ the azimuthal velocity of the plasma at r. If one assumes that the transition region Δr connecting the unperturbed plasma flow far from the NS and the magnetospheric flow is much smaller than r_m, then the above equation takes the form:

$$B_p B_\phi r_m^2 \Delta r = \dot{M}\upsilon_\phi r_m \tag{4.5}$$

Once the gas reaches the transition region Δr, it stops flowing in Keplerian orbits and starts to co-rotate with the magnetosphere. The gas exchanges angular momentum with the magnetosphere and changes the NS spin frequency. The NS is spun up if its specific angular momentum is smaller than that of the accreting gas, and otherwise spun down. The spin-up/spin-down condition can be thought of in terms of characteristic radii: if r_m is smaller than the radius where the Keplerian frequency equals the NS spin frequency (the co-rotation radius r_{co}), the NS is spun up, otherwise it is spun down. The co-rotation radius can be defined as:

$$r_{co} = 1683\left(\frac{M}{1.4\,\mathrm{M}_\odot}\right)^{1/3}\nu_s^{-2/3}\,\mathrm{km}. \tag{4.6}$$

It is important to stress that this is an over-simplified picture of the true physical conditions close in the inner disk. In this simple description, the accretion torque exerted on the NS for a Keplerian disk truncated at r_m, with $r_m < r_{co}$, is:

$$N = 2\pi I\dot{\nu}_s = \dot{M}\sqrt{GMr_m} \tag{4.7}$$

where I is the moment of inertia of the NS, $\dot{\nu}_s$ the NS spin frequency derivative and G the universal gravitational constant. At radii larger than the co-rotation radius, the magnetic field lines are threaded into the accretion disk and dragged by the high conductivity plasma so that an extra torque due to magnetic stresses has to be expected [107, 108, 283, 345] in addition to the torques due to the matter flow. Spin down due to dipole emission is also always present. A possible way to express the total torque acting on the pulsar is [327]:

$$N = \left(\dot{M}\sqrt{GMr_m} + \frac{\mu^2}{9r_m^3}\right) n(\omega) - \frac{\dot{E}_{dipole}}{2\pi \nu_s} \tag{4.8}$$

where $\dot{E}_{dipole}$ is the energy loss due to dipole radiation and $n(\omega) \approx \pm 1$ is a dimensionless function that depends on the fastness parameter $\omega = (r_m/r_{co})^{3/2}$:

$$n(\omega) = \tanh\left(\frac{1-\omega}{\Delta \mathrm{r}}\right) \tag{4.9}$$

This term takes account of the gradual transition from the spin-down to spin-up zone in the accretion disk [283, 327]. The two extra terms in Eq. (4.8) have a minor effect during most of the outburst, when the mass accretion has the largest weight in determining the net torque. The expression in Eq. (4.7) is therefore a good approximation most of the time.

As the AMXP is spun up, r_{co} moves towards and eventually reaches r_m. When this happens, the pulsar is said to have reached the "equilibrium spin period" $P_{\rm eq}$:

$$P_{\rm eq} \simeq 2.7 \left(\frac{B}{10^8\,\mathrm{G}}\right)^{6/7} \left(\frac{M}{1.4\,\mathrm{M}_\odot}\right)^{-5/7} \left(\frac{\dot{M}}{10^{-10}\,\mathrm{M}_\odot\,\mathrm{yr}^{-1}}\right)^{-3/7} \left(\frac{R}{10\,\mathrm{km}}\right)^{18/7} \mathrm{ms} \tag{4.10}$$

Substituting into Eq. (4.10) the surface magnetic field (at the poles) derived from dipole spin down [200]:

$$B = \sqrt{\frac{6c^3 I P\dot{P}}{4\pi^2 R^6}\frac{1}{\sin\alpha}} \simeq 6.4 \times 10^{19}\,\mathrm{G}\sqrt{P\dot{P}}\left(\frac{M}{1.4\,\mathrm{M}_\odot}\right)^{3/2}\frac{1}{\sin\alpha} \tag{4.11}$$

(where we have assumed $I = 10^{45}\,\mathrm{g\,cm}^2$, $R = 10$ km and α is the misalignment angle between spin and magnetic axes), one obtains a relation between $P_{\rm eq}$ and $\dot{P}$. When the accretion rate reaches the maximum Eddington rate, this relation defines a "spin-up line" in the $P - \dot{P}$ diagram of radio pulsars (see Fig. 4.5), above which millisecond radio pulsars should not be found. Indeed, if millisecond pulsars are created in LMXBs via accretion torques, then the maximum possible torque is set by the Eddington limit[3] It is important to stress that the numerical solution of the force-free relativistic MHD equations lead to a similar (but slightly different) result

[3]Note, however, that the spin-up line depends on several parameters which are difficult to constrain like the angle α. Its position in the $P - \dot{P}$ diagram is therefore subject to uncertainties (see [329] for a discussion).

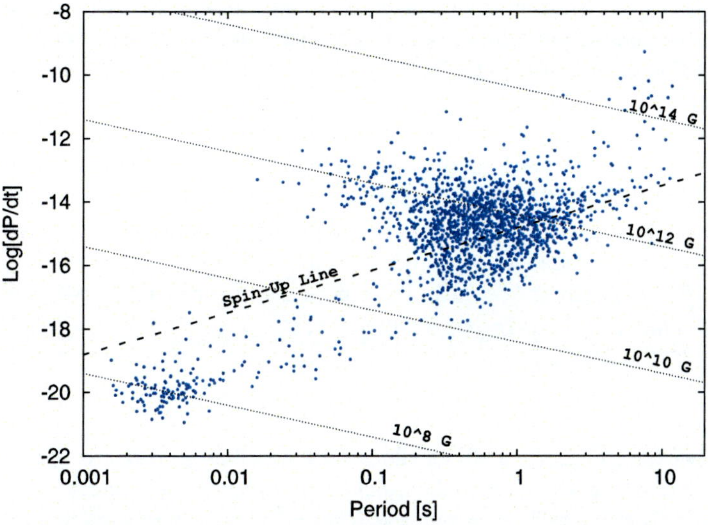

Fig. 4.5 $P-\dot{P}$ diagram of radio pulsars with a spin up line (dashed black line) above which radio pulsars should not be found if they are "recycled" via accretion. The black dotted lines identify magnetic fields of different strengths

for the dipolar magnetic field (again, at the poles) [309]:

$$B=\sqrt{\frac{c^3 I P\dot{P}}{\pi^2 R^6}}\frac{1}{\left(1+\sin^2\alpha\right)^{1/2}}\simeq 5.2\times10^{19}\,\mathrm{G}\sqrt{P\dot{P}}\left(\frac{M}{1.4\,\mathrm{M}_\odot}\right)^{3/2}\frac{1}{\left(1+\sin^2\alpha\right)^{1/2}} \tag{4.12}$$

To detect the effect of accretion torques in AMXPs one simply needs a measurement of $\dot{\nu}_s$, which in principle is straightforward when using coherent timing techniques. Unfortunately, however, determining the location of the magnetospheric radius is a non-trivial problem.

The main reason is that $\Delta r/r_m$ is assumed to be much smaller than unity, which is a good approximation only for slow accreting pulsars with high magnetic fields. This might not be the case for AMXPs with rapid rotation and weak magnetic fields (see for example Eq. (4.5) and the underlying assumptions).

Furthermore we have considered so far only gas-pressure dominated accretion disks, considered to be a good assumption at the low average accretion rates of AMXPs. However, when the outbursts are close to their peak luminosities, radiation pressure may play a role in the exchange of angular momentum between the gas and the NS. At these accretion rates accreting plasma and magnetic field lines may couple and modify the amount of enhanced angular momentum [7, 263], thus invalidating the use of Eqs. (4.7)–(4.8).

Noting these caveats, theoretical expectations for the average spin frequency derivative of an AMXP can be calculated with the simplest accretion model (Eq. (4.7)) once one has a reasonable estimate of the mean mass accretion rate $\dot{M}$ *during an outburst*. As discussed, such a measure is difficult to obtain, but several estimates of M are present in the literature which can be taken as a first step to compare the measured $\dot{\nu}_s$ with the expected values (see for example [130] and [263]). The explicit expression for the expected $\dot{\nu}_s^{exp}$ can be obtained by substituting Eq. (4.2) into Eq. (4.7). The substitution gives:

$$\dot{\nu}_s^{\mathrm{exp}} = 2.3 \times 10^{-14} \xi^{1/2} \dot{M}_{-10}^{6/7} M_{1.4}^{3/7} B_8^{2/7} R_{10}^{6/7} \,\mathrm{Hz\,s^{-1}} \tag{4.13}$$

where we have normalized all variables with their typical values as in Eq. (4.10).

As one can easily verify, the calculated values of $\dot{\nu}_s^{exp}$ are all of the order of $10^{-13}\,\mathrm{Hz\,s^{-1}}$ when neglecting the fact that at low mass accretion rates, when $r_m > r_{co}$ the "propeller regime" can set in, with the spin-down terms dominating in Eq. (4.8). In this phase, the specific angular momentum of the accreting plasma is insufficient for spin up to occur, and the centrifugal barrier removes angular momentum from the AMXP, slowing it down. It was originally suggested that centrifugal inhibition by the rotating magnetosphere would expel gas from the system and shut down the accretion process [144]. In fact for this to happen, r_m must exceed r_{co} by a margin of at least 1.3 so that matter can be accelerated to the escape velocity and can be flung out of the disk [70, 283, 311]. Accretion can in fact still take place, and two propeller regimes have been observed in 3D-MHD simulations of accreting pulsars [294, 338]: a "weak propeller" with no outflows and a "strong propeller" with expulsion of material. In both cases channeled accretion is still ongoing so that a spinning down AMXP could be in principle observed via coherent timing measurements in either propeller regime. The propeller is therefore expected to affect $\dot{\nu}_s$, but determining its onset is a very difficult task. Indeed, AMXP observations do not help much to constrain the theory in this case, since only weak evidence for possible propeller phases exists [93, 126, 252, 255]. Coherent timing does not provide further insights, since the propeller is expected to start when the mass accretion (and thus the source luminosity) drops substantially, with a consequent decrease of the S/N of the pulsations. Furthermore the propeller may not last sufficiently long to allow measurement of the spin frequency derivative.

4.4.1 Coherent Timing Technique

Coherent timing analysis of the phase evolution of AMXPs can be performed after converting photon times of arrival from the spacecraft non-inertial reference frame to the Solar System barycenter (an approximate inertial reference frame) and correcting for general and special relativistic perturbations due to planets and minor bodies in the Solar System that affect photon propagation. Before discussing observational tests of accretion theory it is useful to clarify the observables that play a role in coherent timing studies. AMXPs show pulsations that are too weak to detect as single pulses: to reconstruct the signal it is necessary to fold the data in segments of several hundred seconds to obtain a pulse profile that is the average of several hundred thousand NS cycles. It is then possible to measure the fractional amplitude of the pulsations, and the pulse phase. Pulse profiles of AMXPs are generally highly sinusoidal, with little or no harmonic content beyond the fundamental frequency. In this case the pulse phases can be measured by choosing a fiducial point (e.g. the pulse peak) and tracking the variation of the phase at this point over time. Sometimes, however, strong harmonic content is observed, with the pulse profile shape varying during an outburst. This means that unlike in radio pulsar timing, where average profiles are often very stable, there is no stable fiducial point on which to base timing analysis. To avoid this problem it has become standard to decompose the pulse profile into its harmonic components and measure pulse amplitude b_k and phase ϕ_k of the k-th harmonic ($k = 1$ for the fundamental, $k = 2$ for the second harmonic and so on) via the expression:

$$x_j = b_0 + \sum_k b_k \cos\left[2\pi\left(\frac{\mathrm{k(j-0.5)}}{N} - \phi_k\right)\right] \tag{4.14}$$

where b_0 is the unpulsed component, x_j is the number of counts detected in the j-th bin of the pulse profile, with $j = 1, 2, \ldots, N$. This way the fiducial point of each harmonic is well defined since all harmonic components are pure sinusoids. The fractional amplitude can be measured by adding in quadrature the fractional amplitudes of each harmonic:

$$R = \left(\sum_k R_k^2\right)^{1/2} = \sum_k \frac{N b_k}{N_{ph} - B} \tag{4.15}$$

where R_k is the fractional amplitude of the k-th harmonic, $N_{ph} = \sum_j x_j$ the total number of photons in a profile and B the total number of background photons. Note that this definition gives a fractional amplitude larger by a factor $\sqrt{2}$ than the often reported rms fractional amplitude. We use this definition of fractional amplitude since it has an immediate physical meaning as the pulsed flux fraction. To avoid confusion, we always refer to the fractional amplitude as "sinusoidal fractional amplitude" when we use the definition given in Eq. (4.15) or otherwise to "rms fractional amplitude". Note that in the AMXP literature "rms fractional amplitude", "sinusoidal fractional amplitude" and "peak-to-peak fractional amplitude" are also used. This latter is calculated by measuring the flux at the peak and at the minimum of the pulse and dividing by the average flux.

Pulse phase can be measured as a function of time and each harmonic analyzed separately. We drop the subscript k from the phase symbols, since the following equations are valid for each harmonic. The pulse phase series can be decomposed into several terms encoding different physical effects:

$$\phi(t) = \phi_L + \phi_Q + \phi_{orb} + \phi_A + \phi_M + \phi_N \tag{4.16}$$

- $\phi_L = \phi_0 + \nu t$ is the linear term due to a *constant* spin of the NS (ϕ_0 is an initial reference phase)
- $\phi_Q = \frac{1}{2}\dot{\nu}t^2$ is the quadratic variation due to *constant* spin up or spin down
- ϕ_{orb} contains the effect of the orbital motion of the NS around the companion and, as a first order approximation, can be calculated by measuring the delays in the time of arrival of photons:

$$t_{em} = t_{arr} - \frac{A_1}{c} \sin i \sin\left[\frac{2\pi}{P_b}(t_{arr} - T_{asc})\right] \tag{4.17}$$

  where A_1 is the semi-major axis of the NS orbit, c the speed of light, i the inclination of the orbit with respect to the observer, T_{asc} the time of passage through the ascending node, P_b the orbital period and t_{em} and t_{arr} the photon emission and arrival time [238]. In this expression we have assumed that the orbit is perfectly circular, a good first order approximation in all AMXPs. More sophisticated models for nearly circular orbits can be found for example in [180].
- ϕ_A gives phase variations related to uncertainty in astrometric source position, which introduces a spurious frequency and frequency derivative offset. These offsets can be expressed as [123]:

$$\Delta\nu = \nu_0\epsilon\,(a_\oplus \cos\beta/c)\,(2\pi/P_\oplus)\,\cos\tau \tag{4.18}$$

$$\Delta\dot{\nu} = -\nu_0\epsilon\,(a_\oplus \cos\beta/c)\,(2\pi/P_\oplus)^2\,\sin\tau \tag{4.19}$$

  Here ν_0 is the true pulse frequency, ϵ the position error parallel to the plane of the ecliptic, β the ecliptic latitude of the AMXP, $a_\oplus$ and $P_\oplus$ the Earth semi-major axis and orbital period and $\tau = 2\pi\, t/P_\oplus$ the orbital phase of the Earth. Phase zero is defined as the point where the Earth is closest to the AMXP and for order of magnitude estimates can be assumed such that $\cos\tau = \sin\tau = 1$.
- ϕ_M refers to unavoidable phase wandering due to measurement errors, and is normally distributed with an amplitude predictable by propagating the Poisson uncertainties due to counting statistics.
- ϕ_N is a subtle term covering residual phase variations that do not fall into any of the previous categories. These are usually called "X-ray timing noise", by analogy with the timing noise often observed in radio pulsars. Note that if the spin-up (or spin-down) process is not constant in time we **do expect** terms higher than the quadratic (ϕ_Q) and these are considered, in our definition, as part of ϕ_N

even if they are true variations of the NS rotation. This should be expected if, for example, the accretion torques exhibit stochastic variations.

When performing coherent timing after having removed the ϕ_{orb} term and assuming that ϕ_A is negligible, the observer measures a pulse frequency and its time derivatives. These can be determined using a Taylor expansion:

$$\phi(t) = \phi_0 + \frac{\partial \phi}{\partial t}(t - t_0) + \frac{\partial^2 \phi}{\partial t^2}\frac{(t - t_0)^2}{2!} + \ldots \tag{4.20}$$

The parameters $\nu = \frac{\partial \phi}{\partial t}$, $\dot{\nu} = \frac{\partial^2 \phi}{\partial t^2}$, etc., can be determined by fitting the pulse phases with standard χ^2 minimization techniques.

The *observable* quantities here are *not* the spin frequency ν_s and derivatives, but the pulse frequency ν and derivatives which are encoded as a combination of ϕ_L, ϕ_Q and ϕ_N. The pulse frequency is the frequency of the pulsations detected by the distant observer, whereas the spin frequency is the rotational rate of the NS as measured by the distant observer. The assumption that pulse and spin frequency (and derivatives) are identical may not always be true. For pulse and spin frequencies to be identical, ϕ_N must have no linear component. Similarly for the pulse and spin frequency derivative: only if ϕ_N has no quadratic component will the two be the same.

4.4.2 Observations: Accretion Torques in AMXPs

With the exception of the intermittent pulsar Aql X-1, all of the AMXPs have shown pulsations of sufficient quality and with a sufficiently long baseline to constrain the pulse frequency derivatives. As a rule of thumb, the condition that must be met to detect a pulse frequency derivative $\dot{\nu}$ in a data segment of length Δt is that $\sigma_{rms} < \dot{\nu}\,(\Delta t)^2$, where σ_{rms} is the root-mean-square error on pulse phases. Since all AMXPs (bar Aql X-1) have $\sigma_{rms} \sim 0.01$ cycles, and baselines of several days, we are sensitive to $\dot{\nu} \sim 10^{-15}$–10^{-13} Hz s^{-1}. These values overlap the range of expected $\dot{\nu}$ given by Eq. (4.7) for typical $\dot{M}$ and weak dipolar B fields $\sim 10^8$–10^9 G.

Several papers have reported pulse frequency derivatives in AMXPs. The first measure was made for the source IGR J00291+5934 [88] with a reported $\dot{\nu} = 8.4 \times 10^{-13}$ Hz s^{-1} during its 2004 outburst. Table 4.3 summarizes the measurements for all AMXPs. However none of these values, including the measurement made for IGR J00291+5934, take into account the presence of X-ray timing noise in the pulse phases. One must therefore bear in mind that what is reported is the combined effect of ϕ_Q and ϕ_N, as explained in Sect. 4.4.1. Nevertheless timing noise has different strength in different sources, so that pulse and spin frequency

Table 4.3 Pulse frequency and spin frequency derivatives

Source	Pulse freq. deriv. $\dot{\nu}$ (PFD) [$Hz\,s^{-1}$]	Timing noise strength	PFD reference	Spin freq. deriv $\dot{\nu}_s$ (SFD) [$Hz\,s^{-1}$]	SFD reference
SAX J1808.4-3658	4.4×10^{-13}; -7.6×10^{-14}	s	[36]	$<|2.5| \times 10^{-14}$	[123, 125]
XTE J1807-294	2.5×10^{-14}	s	[287]	$<|4| \times 10^{-14}$	[254]
IGR J00291+5934	$[5; 11] \times 10^{-13}$	w	[37, 88, 245]	$+5.5 \times 10^{-13}$	[37, 88, 126, 245, 249]
XTE J1814-338	-6.7×10^{-14}	s	[239]	$<|1.5| \times 10^{-14}$	[127]
XTE J0929-314	$-9.2(4) \times 10^{-14}$	s	[99]		
XTE J1751-305	5.6×10^{-13}	w	[241]		
IGR J17511-3057	1.6×10^{-13}	s	[289]		
IGR J17498-2921	-6.3×10^{-14}	w	[244]		
HETE J1900.1-2455		s		$[2.3; 0.4] \times 10^{-13}$	[250]
Swift J1756.9-2508		w		$<|3| \times 10^{-13}$	[260]
Swift J1749.4-2807		s		$<|1.2| \times 10^{-12}$	[201]
IGR J18245-2452		?		$<|1.3| \times 10^{-12}$	[247]

The AMXPs Aql X-1, SAX J1748.9-2021 and NGC 6440 X-2 do not appear in the table above as the observting baseline is too short to provide meaningful upper limits derivatives may in some cases not be so dissimilar. We have marked each AMXP in Table 4.3 with a "w" for weak timing noise and "s" for strong. This distinction is somehow arbitrary, but is useful to understand the probability of the pulse frequency derivatives differing from the true spin frequency derivatives.

To verify whether the pulse frequency derivative is a robust indicator of the spin frequency derivative $\dot{\nu}_s$, one has to investigate the strength of ϕ_N on the timescales over which $\dot{\nu}$ is measured. To do this, one can use Monte Carlo (MC) simulations to estimate the true uncertainty on the pulse frequency derivative [123]. The MC method works as follows: one fits a second-order polynomial to measure ν and $\dot{\nu}$ with standard χ^2 minimization techniques after having removed all other effects (such as orbital variations) and obtains an estimate of $\nu \pm \sigma_\nu$ and $\dot{\nu} \pm \sigma_{\dot{\nu}}$. If in the phase residuals there are still substantial variations of the pulse phase, in excess of that expected from measurement errors alone (i.e., ϕ_M)—which can be verified by checking whether χ^2 gives a statistically unacceptable fit—then the statistical errors σ_ν and $\sigma_{\dot{\nu}}$ are not good representations of the true uncertainties of the spin parameters. Instead one can take the phase residuals, calculate a power spectrum and simulate several thousand time series with nearly identical noise content as the original phase residual time series [123]. At this point ν and $\dot{\nu}$ can be measured for each simulated time series, and a distribution of parameters constructed. The standard deviation of the distribution of ν and $\dot{\nu}$ provides a good representation of the true uncertainties on the spin parameters. In this way one can immediately check whether the noise content ϕ_N affects the measured value of ν and $\dot{\nu}$. Applying this technique has revealed discrepancies between pulse and spin parameters in several AMXPs (Table 4.3). The most striking finding has been the non-detection of spin frequency derivatives in several AMXPs, with upper limits below the expected $\dot{\nu}_s$.

The use of MC simulations is not the final word on this problem, since the origin of timing noise remains unexplained. It also has its limitations: if the lowest Fourier frequency component of timing noise is comparable to the length of the data segment over which ν and $\dot{\nu}$ are measured, MC simulations cannot distinguish pulse and spin parameters. It has been noted that the phase residuals obtained after removing ν and $\dot{\nu}$ from the pulse phases of two AMXPs (XTE J1814-338 and XTE 1807-294) are anti-correlated with variations in X-ray flux [239, 287]. The anti-correlation improves substantially if one fits a simple $\nu = const$ model, suggesting that timing noise, which is related to the X-ray flux variations, is almost entirely responsible for the measured $\dot{\nu}$ [258]. Correlations (or anti-correlations) between pulse phase residuals (with respect to a $\nu = const$ model) and X-ray flux have now been found in at least six AMXPs where such studies have been carried out [258]. In some cases the correlation was striking (e.g., in XTE J1814-338 [127]) leaving little doubt that the pulse frequency derivative $\dot{\nu}$ is **not** the spin frequency derivative $\dot{\nu}_s$. This discovery also suggests that **it is the pulse phase ϕ and not its second derivative ($\dot{\nu}$) that correlates with the flux** (and thus the mass accretion rate), as would instead be expected from Eq. (4.7). One way to test these findings is to measure "instantaneous" short timescale spin frequencies and check whether these scale with the bolometric flux as $\dot{\nu}_s \propto F_{bol}^{\gamma}$, where γ is a scale factor that depends on accretion disk structure. The bolometric flux is, however, almost never available, since observations usually cover only a narrow (high) energy band. If we assume that the X-ray flux is $F_X \propto F_{bol}$ then we expect to see $\dot{\nu}_s \propto F_X^{\gamma}$. However this assumes that the mass accretion rate $\dot{M} \propto F_X$, which has been shown to be untrue in some LMXBs [340]. There is one additional complication: AMXPs have outbursts that are almost always too short to split data segments long enough to yield more than one short-term "instantaneous" spin frequency derivative.

If one keeps these caveats in mind, then it is possible to test the relation between $\dot{\nu}_s$ and F_X for some AMXPs, such as XTE J1807-294 and XTE J1814-338 which have shown long outbursts ($\sim$100 and $\sim$40 days, respectively) and high S/N ratio for the pulsations. In XTE J1807-294, the "instantaneous" $\dot{\nu}$ changes sign several times during the outburst decay and all "instantaneous" spin frequency derivatives $\dot{\nu}_s$ are insignificant and consistent with being part of the underlying timing noise process [254]. In XTE J1814-338, the instantaneous $\dot{\nu}$ are too large (up to 10^{-11} Hz s^{-1}) to be physically meaningful, requiring accretion rates well above the Eddington limit [352]. This is at odds with the fact that AMXPs are faint and rarely reach accretion rates above 10% Eddington, suggesting that $\dot{\nu}$ is different to $\dot{\nu}_s$ and is completely dominated by timing noise. This does not prove that $\dot{\nu}_s$ does not scale with flux, but only that measurements of "instantaneous" $\dot{\nu}_s$ are contaminated by timing noise, which prevents testing of the relation between $\dot{\nu}_s$ and F_X.

Theoretical expectations for $\dot{\nu}_s$ mostly exceed observed values measured by taking into account the contamination of timing noise (Table 4.3). This means that most AMXPs do not behave in accordance with the predictions of accretion torque theory (Eq. (4.8)). In addition, once gas attaches to the magnetic field lines of the AMXP, an exchange of angular momentum is inevitable unless $r_m = r_{co}$. But this cannot be the case throughout an outburst: $r_m \propto \dot{M}^{-2/7}$ and the mass accretion rate varies, so r_m must at some point differ from r_{co}. Larger torques, which are expected close to the outburst peak, are clearly ruled out in some AMXPs [123, 127, 259]. The angular momentum of the accreting plasma must however be transferred somewhere so that the mismatch between expected and observed $\dot{\nu}_s$ appears problematic.

4.4.2.1 The Origin of X-ray Timing Noise

The phase wandering of pulsations in coherent timing analysis requires careful consideration if we are to understand how a NS responds to accretion. Theoretically the phenomenon is easy to explain if one assumes that the accretion hot spot is not completely anchored to one location on the surface. One possibility is that the magnetic and rotational axes of the NS are almost aligned and the hot spot wanders around the magnetic axis by small displacements, generating large variations in phase and amplitude [178]. In this case pulsed fractions and times of arrivals should be anti-correlated. A moving hot spot has been observed in 3D-MHD simulations of magnetized accreting NSs [11, 293]. In particular, it was shown that the usual assumption of a fixed hot spot is valid only for large misalignment angles between the magnetic and spin axes (Fig. 4.6, [11]), in good agreement with the magnetic and rotational axis alignment model [178, 298]. This strengthens the idea

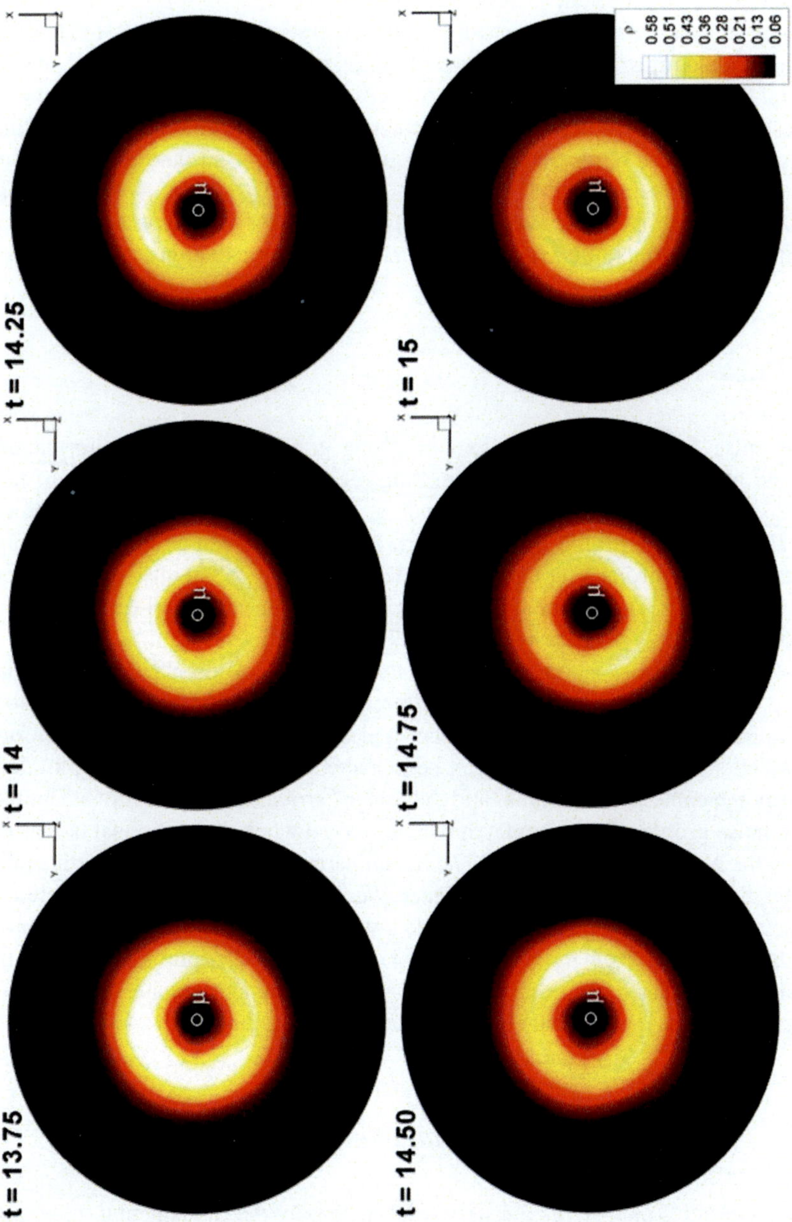

Fig. 4.6 Evolution of the position of the hot spot during accretion (ν_s = 4.1 ms). μ represents the magnetic dipole axis. Note how the densest portion of the circular hot spot (encoded by the ρ scale, bottom right) shifts during time even if the observer is in the rotating reference frame of the NS (Figure from [11])

that most X-ray timing noise may be related to a moving hot spot configuration. Alternatively, the phase wandering might be related to complexity in the structure of the pulsar's magnetic field, with different multipole components dominating the accretion process at different accretion rates [199].

Timing noise may also be due to fluctuating accretion torques [176, 177]. This has been studied for strongly magnetized accreting pulsars like Vela X-1 [74], and it is natural to expect something similar in AMXPs. However as discussed, this model requires unphysical large accretion rate fluctuations to explain all timing noise or the large jumps in phase observed for example in XTE J1814-338 (and probably also SAX J1808.4-3658 and XTE J1807-294, [123, 239, 258, 287, 352]). Nonetheless some X-ray timing noise, or even its entirety for AMXPs with weak timing noise content, could in principle be related to accretion torque fluctuations.

4.5 Pulse Profiles

AMXP pulse profiles form in regions of strong gravity and encode information about the physical properties of the emitting regions and the compactness of the NS [185, 186, 220, 276, 278]. Light bending, aberration and relativistic Doppler shifts all affect pulse profile shape [42, 225, 275, 277] and can be measured using coherent timing analysis. This can be used to constrain the EoS of ultra-dense matter (Fig. 4.7) although large uncertainties still exist: in part due to model-dependencies, but more importantly because of the low S/N of the relativistic features.

Pulse profile shapes can also constrain the geometry of the NS magnetic field [34, 167, 365] since multipoles generate different accretion columns and hot spots on the NS surface (see, however, [8], for a recent discussion of the problem). The information contained in pulse profiles can be extracted by studying pulse amplitudes, the harmonic content of the signal and the time of arrival of the pulsations. Double peaked pulse profiles, for example, may be observed when two antipodal hot-spots exist on the NS surface [178, 225, 277]; a high harmonic content is suggestive of a complex field geometry [198] and the energy dependence of the pulse amplitudes and the time lags of the pulsations provide a way to explore the angular pattern of the radiation emitted from the NS surface [111, 277, 278]. The temporal evolution of the pulse shapes also reveals details of the complex disk/magnetospheric interaction responsible for magnetic channeling of the accreting plasma [142, 171].

4.5.1 Pulse Fractional Amplitudes and Phase Lags

The pulse profiles of AMXPs are sinusoidal, with only the fundamental frequency detected in most cases (including IGR J00291+5934, XTE J0929-314, XTE J1751-305, SAX J1748.9-2021 and IGR J17498-2921). In some AMXPs there is a first overtone (e.g. XTE J1814-338, IGR J18245-2452 and NGC 6440 X-2), and

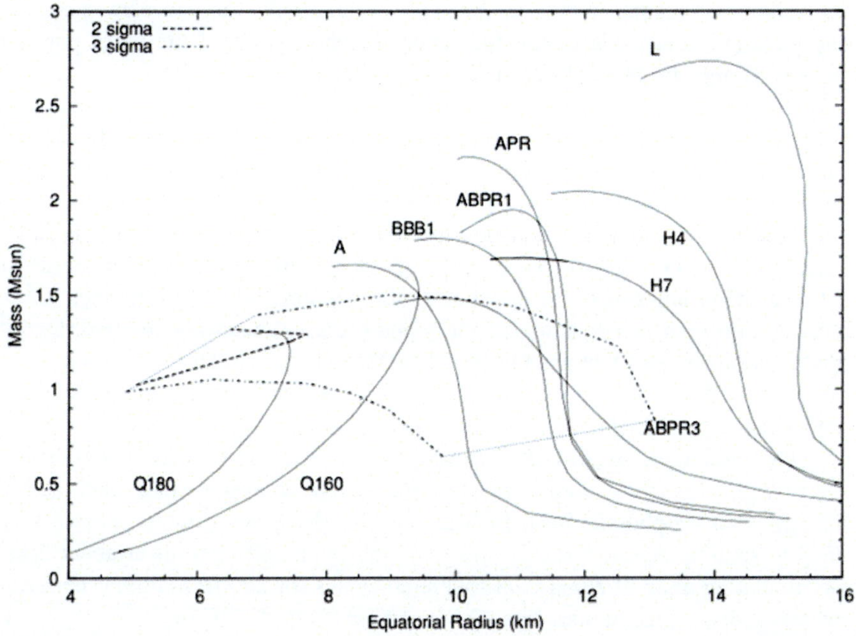

Fig. 4.7 Mass-Radius relation for several different EoS of ultra-dense matter, and constraints on the M-R relation of the AMXP SAX J1808.4-3658. These result from modeling pulse profiles observed in the 1998 and 2002 outbursts (Figure from [220])

higher overtones are seen only sporadically in just three AMXPs (SAX J1808.4-3648, Swift J1749.4-2807 and XTE J1807-294). Pulse fractional amplitudes reach values of a few percent, with the highest sinusoidal amplitude measured being 30–40% (Swift J1749.4-2807 and XTE J1807-294) and the lowest being 0.3% (HETE J1900.1-2455). With the exception of Swift J1749.4-2807, overtones do not generally contribute to the amplitude of the pulsations by more than about ~5%.

All AMXPs show an energy dependence of the pulse fractional amplitudes. In some sources the fractional amplitude increases steeply with energy, e.g. Aql X-1 [44] Swift J1756.9-2508 [259] and SAX J1748.9-2021 [253]. In this latter source, the fractional amplitude increases by a factor ~10 between 2 and ~20 keV. Some AMXPs have fractional amplitudes that drop with energy (SAX J1808.4-3658 [67], XTE J1751-305 [111], XTE J0929-314 [101], HETE J1900.1-2455 [101], IGR J17511-3057 [89, 142]). Other AMXPs (IGR J00291+5934 and XTE J1807-294) are more complex, with the amplitude rising and falling in different energy bands [88, 259]. The energy dependence of the pulsed fractions provides insights into the pulse formation process, since a simple hot spot emitting blackbody radiation with a temperature contrast with respect to the NS surface would produce pulsations which increase in amplitude at higher energies in the observer reference frame [225].

Another interesting finding has been the correlation or anti-correlation between the fractional amplitude of the fundamental (where most of the pulse power is observed) and the time of arrival of the pulsations of some AMXPs. The most prominent example is XTE J1807-294 where high/low fractional amplitude pulses arrive systematically earlier/later than predicted by the timing model. This finding, and a similar anti-correlation between the time of arrivals and the X-ray flux, may be evidence for a hot spot moving on the NS surface [259]. Similar conclusions were reached by other authors to explain the timing noise observed in the timing residuals of several AMXPs [123, 239, 253, 287]. Some of these observations can be understood in terms of the hot spot wandering model described in Sect. 4.4.2.1 although no final confirmation of the models has yet been put forward. Despite many open problems that need to be solved (such as why such correlations are not observed in AMXPs like SAX J1808.4-3658) the moving hot spot model help us to understand the AMXP pulse formation process.

Soon after the first AMXP discovery, it was realized that the lowest energy ("soft") photons that compose the pulsations arrive on average later than the high energy ("hard") photons [67]. In SAX J1808.4-3658, hard photon arrival times tend to saturate at some energy $E \gtrsim 10$ keV, with soft photons at about 2 keV accumulating a lag of about 0.08 rotational cycles (or 200 μs). Similar behaviour was soon discovered in other five AMXPs: XTE J0929-314 (0.14 cycles [99]), XTE J1751-305 (0.06 cycles, [111]), XTE J1814-338 (0.016 cycles, [350]), XTE J1807-294 (0.1 cycles, [60]) and in the intermittent pulsar HETE J1900.1-2455 (0.07 cycles, [101]). In IGR J00291+5934, soft lags were observed with hard photons leading by 0.06 cycles at 6–8 keV [88]. Above this energy, harder photons ($E \gtrsim 8$ keV) start to gradually reduce their lead until at energies of 30–100 keV there are no phase lags with respect to the soft photons (at ~2 keV). The origin of these lags is poorly understood, and has been discussed in terms of the different angular distribution ("fan" and "pencil" beams) of a two spectral component model [278] or in terms of Compton down-scattering of hard X-ray photons from the cold disk plasma or the NS surface [86]. Detailed study of the soft lags in SAX J1808.4-3658 [124] suggests that they have a flux dependence, with the lag being almost zero at high fluxes, increasing steadily at lower fluxes and then decreasing again below a flux threshold coincident with the onset of the rapid decay. This may be linked with the changing properties of the accretion disk as it transitions towards the propeller phase.

4.5.2 *Pulse Shape Evolution*

Most AMXPs with strong harmonic content show pulse shape variability that often correlates with the stage of the outburst. The pulse profiles observed in *different* outbursts of SAX J1808.4-3658 have characteristic shapes that can be associated with the specific stage of the outburst (e.g., rise, peak, decay, etc.; see [123, 142, 159]). There is a very strong linear anti-correlation between the fractional amplitude of the first overtone and the X-ray flux (something very similar was also observed in XTE J1807-294; [259]), suggesting that the secondary hot spot becomes visible or more prominent as the accretion process becomes less intense. This may be explained if the inner disk radius moves in or out with respect to the accretion rate, rendering the secondary hot-spot visible [123, 142]. If the inner disk radius r_{in} is assumed to vary at different outburst stages, then the pulse profile changes of SAX J1808.4-3658 can be explained in this way [279]. Variations in the disk-magnetosphere coupling may also explain explain some of the pulse shape and phase variations [159]. Calculation of the light-curve of accreting NSs in global 3D-MHD simulations showed that if an octupolar magnetic field dominates the field configuration, double peaked pulse profiles can be reproduced [199]. Similar pulse shape variations were seen to correlate with the outburst stage of Swift J1756.9-2508 [259] IGR J17511-3057 [142] and to some extent XTE J1807-294 [249]. However, to date it is still unclear why pulse profiles change abruptly and unpredictably in some sources during certain outbursts but at other times have limited variability.

4.6 Long Term Evolution and Pulse Formation Process

If AMXP pulsations are observed in different outbursts one can follow the long-term spin and orbital evolution. To date only five AMXPs have been monitored with high time resolution instruments in different outbursts: SAX J1808.4-3658, IGR J00291+5934, XTE J1751-305, Swift J1756.9-2508 and NGC6440 X-2 (although with relatively low S/N and short outburst duration, it is difficult to constrain the long-term evolution of the latter two).

4.6.1 Specific Sources

SAX J1808.4-3658 The best constrained AMXP is SAX J1808.4-3658, for which secular spin evolution and orbital period variation have now been measured over a 13 year baseline [123, 125, 262]. As discussed previously, accretion torques have a small effect on the spin of SAX J1808.4-3658 and have never been conclusively detected during any of the six outbursts monitored by *RXTE*. However, if one compares the constant spin frequency measured in each outburst, SAX J1808.4-3658 is clearly spinning down at a constant rate (bottom panel in Fig. 4.8; [262]). The stability of the spin-down rate suggests that its origin is unrelated to propeller onset, since spin-down would then depend on the amount of mass propelled and on the duration of this phase, which can vary from outburst to outburst. Loss of angular momentum via emission of gravitational waves was also considered [123], and is interesting since this might also balance accretion torques, explaining the lack of measurable spin-up during the outbursts. However, fine tuning is also required in

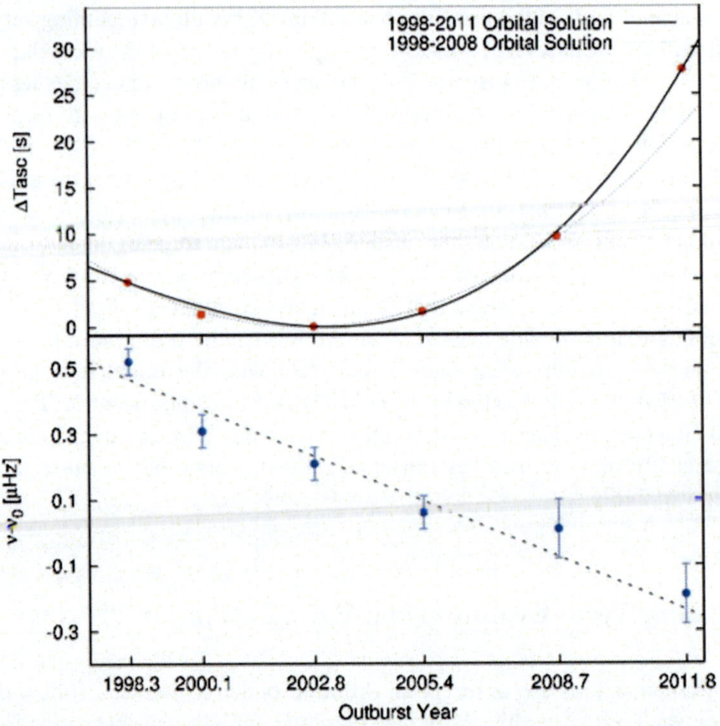

Fig. 4.8 Orbital (top panel) and spin frequency evolution (bottom panel) of SAX J1808.4-3658 over a baseline of 13 years. The top panel shows the delays accumulated by the pulsar as it passes through the ascending node. The data were consistent until 2008 with a parabolic increase which is equivalent to a steady widening of the orbit (dotted line). The 2011 data revealed instead an acceleration of the orbit, as evidenced by the solid line in the top panel. The spin frequency has shown a steady decrease in between outbursts (bottom panel; ν_0 is a frequency offset of 400.975210 Hz), compatible with a spin down of the order of $-10^{-15}\,\mathrm{Hz\,s^{-1}}$ (Figure from [262])

this case since the mechanism producing gravitational waves has to be triggered for the right amount of time so as to guarantee a constant spin-down between outbursts [125, 127, 262]. The most likely scenario appears to be a loss of angular momentum via magnetic-dipole radiation. This is expected for any rapidly rotating NS with a magnetic field, even if the NS does not turn on as a radio pulsar. The measured spin-down is of the order of $-10^{15}\,\mathrm{Hz\,s^{-1}}$ and is consistent with a polar magnetic field of 2×10^8 G [123, 125, 262].

The long-term orbital evolution of SAX J1808.4-3658 is also exciting. Theory states this should be driven by loss of angular momentum via emission of gravitational waves from the tight binary orbit. For SAX J1808.4-3658, the resulting orbital evolution should proceed on a timescale $\sim 10^9$ years [235]:

$$\tau_{\mathrm{GW}} = 0.01 \frac{(M_{NS} + M_2)^{1/3}}{M_{NS}\, M_2} P_b^{8/3} \mathrm{Gyr} \tag{4.21}$$

where P_b is in hours and M_{NS} and M_2 are the NS and donor masses. The observed value, however, implies an expansion of the orbit on a timescale of ~70 Myr [39, 81, 123, 125]. Further observations in 2011 provided a long enough baseline to detect an orbital period acceleration (top panel in Fig. 4.8; [262]). SAX J1808.4-3658 may therefore behave like the "black widow" binary millisecond pulsars observed at radio frequencies [184] with a short-term exchange of angular momentum between the donor star and the orbit. A similar suggestion was also previously proposed for this AMXP [39, 81], invoking a "hidden" black widow scenario in which the AMXP switches on as a radio pulsar during quiescence and ablates its donor.

IGR J00291+5934 Long-term spin down has also been measured between the 2004 and 2008 outbursts of IGR J00291+5934. The AMXP rotation behaves like a saw-tooth, spinning up during the 2004 outburst then spinning down until the 2008 outburst. The spin-down rate is $\approx -3 \times 10^{-15}\,\mathrm{Hz\,s^{-1}}$ [126, 245, 259], comparable to that observed in SAX J1808.4-3658. Although spin down between outbursts has been observed only once and it is not possible (until the next outburst) to verify whether it is constant, the most likely explanation also involves magnetic dipole radiation spin-down. If this saw-tooth behaviour is a good indicator of the *secular* spin evolution of this AMXP, then the net spin frequency is *increasing* at $\approx +2.5 \times 10^{-15}\,\mathrm{Hz\,s^{-1}}$. Indeed, the spin frequency at the beginning of the 2004 outburst was less than that observed in the 2008 outburst: the 2004 spin-up accelerated the pulsar more than it subsequently decelerated in quiescence. The relatively small baseline did not permit any useful constraints on orbital period evolution: this should be possible after the observation of the next outburst.

XTE J1751-305 Comparing the spin frequencies from 2002, 2005, 2007 and 2009 reveals a secular spin down of $-5.5 \times 10^{-15}\,\mathrm{Hz\,s^{-1}}$ [288]. This is again consistent with magnetic dipole spin-down. If the spin-up measured during the 2005 outburst reported in [241] is not strongly affected by timing noise, then its behaviour resembles IGR J00291+5934, with spin frequency increasing in outburst and decreasing between outbursts in a saw-tooth. The inferred magnetic field is $\sim 4 \times 10^8$ G, in line with the other two AMXPs constrained in this way. The low quality of the data taken in the 2002 outburst does not permit stringent constraints on orbital evolution.

Swift J1756.9-2508 Observations during the outbursts in 2007 and 2009 give upper limits of $|\dot{\nu}| \lesssim 2 \times 10^{-15}\,\mathrm{Hz\,s^{-1}}$ on the secular spin evolution of this AMXP [259]. This requires a magnetic field $<9 \times 10^8$ G. Magnetic field estimates for all four of these AMXPs are given in Table 4.4.

Table 4.4 Secular evolution and inferred magnetic fields in AMXPs

Source	Spin-down in quiescence [$\mathrm{Hz\,s^{-1}}$]	Magnetic field [10^8 G]	Secular evolution	Reference
SAX J1808.4-3658	-10^{-15}	1.5–2.5	Spin-down	[123, 125, 262]
IGR J00291+5934	-3×10^{-15}	1.5–2.0	Spin-up	[126, 245, 259]
XTE J1751-305	-5.5×10^{-15}	4	Spin-down	[288]
Swift J1756.9-2508	$<\|2\| \times 10^{-15}$	<9	?	[253]

The magnetic field is inferred from the spin down observed in quiescence, and refers to field at the poles of the NS for a pure dipolar configuration, $R = 10$ km and $M = 1.4\,\mathrm{M_\odot}$

4.6.2 The Maximum Spin Frequency of Neutron Stars

None of the AMXPs discussed above has a spin rate that *increases* on long timescales, except for IGR J00291+5934. Even this source has a *net* acceleration so small that its spin frequency will change significantly only on timescales of several billion years. This raises the question on whether this behavior is the norm for AMXPs.

For all realistic EoS of ultra-dense matter, NSs are stable at spin frequencies well in excess of 1000 Hz. The break-up frequency is well approximated by

$$\nu_{\max} = 1230 \left(\frac{M_{NS}}{1.4\,\mathrm{M_\odot}}\right)^{1/2} \left(\frac{R_{NS}}{10\,\mathrm{km}}\right)^{-3/2} \;\mathrm{Hz} \tag{4.22}$$

(where M_{NS} and R_{NS} refer to the non-rotating mass and radius of the NS under consideration, expression valid for arbitrary NS mass and EoS as long as the mass is not too close to the maximum permitted for that EoS [182, 183]). However the distribution of spin frequencies of the ensemble of AMXPs and NXPs (nuclear powered X-ray pulsars, see Sects. 4.1 and 4.7) has an abrupt cutoff at about 730 Hz. This was first noticed in 2003 [51] and has been confirmed in later works with larger sample size [47, 249]. Figure 4.9 shows the current distribution of spin frequencies for the 15 AMXPs and the 10 NXPs known (see Table 1 in [349] for a complete list of NXPs). So far no AMXP or radio millisecond pulsar has been found above this cutoff (the fastest radio millisecond pulsar has a spin of 716 Hz [135]). So not only there are no pulsars with $\nu_s \gtrsim 700$ Hz, but also at least four AMXPs that should be accelerating in response to accretion are instead either decelerating or spinning up very slowly, on timescales of billions of years.

Gravitational waves have been invoked to explain this cutoff, but as shown recently [126, 127, 245], this cannot be the explanation for all AMXPs as it would require substantial fine-tuning to explain the X-ray timing observations. The cutoff might instead be related to the magnetic field evolution of NSs [263]. However, further investigation is needed to assess this fascinating question in a robust way since it is still unclear how the magnetic field of NSs evolves in response to accretion [17, 61, 165, 296, 356, 357].

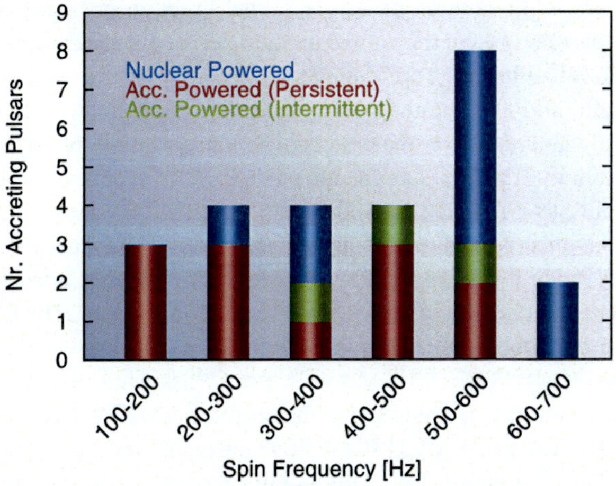

Fig. 4.9 Histogram of the spin frequency of AMXPs (both intermittent and persistent pulsators) and NXPs. The histogram is empty for frequencies larger than about 700 Hz. This is similar to what is found in radio pulsar data, where the sample size is considerably larger (with hundreds of millisecond radio pulsars)

4.6.3 Why Do Most Low Mass X-ray Binaries Not Pulsate?

Even after 15 years of high time resolution X-ray observations, only a few LMXBs have shown millisecond pulsations. Several LMXBs have pulsations at long periods (Table 4.1) but the number of pulsating systems is still small compared to the entire LMXB population. Several mechanisms have been proposed to solve this problem, including burial of the magnetic field by accretion [28, 68, 296], smearing of the pulsations by an optically thick corona [32, 331, 332], smearing of pulsations due to gravitational light bending [234, 364], alignment of the NS magnetic and rotational axes [178, 298] and onset of MHD instabilities at the disk/magnetospheric boundary [295]. None of these models have yet been confirmed, although several new findings have helped to test their predictions.

Magnetic Field Screening In this model, fresh unmagnetized accreted plasma diamagnetically screens the NS magnetic field as it is deposited on the surface [28, 68, 296]. The field then re-emerges via Ohmic diffusion and competition between the Ohmic diffusion and accretion timescale sets the behaviour of the magnetic field. If the accretion timescale is shorter than the Ohmic diffusion timescale, the magnetic field is buried deep in the ocean and outer crust. The two timescales can be defined as:

$$\tau_{\text{acc}} = \frac{y}{\dot{m}} \tag{4.23}$$

$$\tau_{\text{diff}} = \frac{H^2}{\eta} \tag{4.24}$$

where $y = \rho dz$ is the column density, z is the direction orthogonal to the surface in the plane parallel geometry approximation, $\dot{m}$ is the local mass accretion rate per unit area (in units of $\mathrm{g\,cm^{-2}\,s^{-1}}$), $H = \frac{dlnP}{dz}$ is the pressure (defined as P) scale-height, and η is the magnetic diffusivity coefficient. This model predicts that LMXBs with very low mass accretion rates should be AMXPs, and that the most luminous LMXBs should not pulsate. This seems to be the case, since persistently pulsating AMXPs are indeed all faint LMXBs. However, the discovery of intermittent pulsations as short as ~100 s in Aql X-1 poses severe problems for the model since it is unclear how a buried B field could emerge for this brief episode. Another interesting observational result comes from HETE J1900.1-2455, which, with a mean bolometric luminosity of $4.4 \times 10^{36}\,\mathrm{erg\,s^{-1}}$, has a mean mass accretion rate $\dot{M} \approx 8 \times 10^{-10}\,\mathrm{M_\odot\,yr^{-1}}$ for an assumed accretion efficiency of 20%. This accretion rate is about twice that of most AMXPs, but is smaller than the bolometric luminosity of SAX J1808.4-3658 during observations taken close to the peak of the 2008 outburst [159]. This might indicate that the screening timescale is long compared to the outburst duration of most AMXPs. In HETE J1900.1-2455 observational evidence for magnetic field screening has been found [250], with the magnetic field possibly reducing by orders of magnitude on a timescale of about 100 days during which channeled accretion becomes less and less efficient until it stops working. If confirmed this indicates that indeed most LMXBs might be not pulsating because of the lack of an extended magnetosphere. Recent 2-D and 3-D numerical simulations have started to address this problem and results seems to suggest that submergence of the magnetic field is possible at least when the accretion rate is high ([20] but see also [222, 268, 269] for different results). Despite still open problems, the screening model remains therefore one of the most promising, although further efforts are necessary to formulate new sophisticated predictions and tests.

Smearing from an Optically Thick Corona This model suggests that X-ray emission from AMXPs is formed primarily via Comptonization processes in a relatively optically thin medium. If the Compton cloud is thick, then pulsations can be washed out. This scenario has been criticized on the grounds that spectral analyses of most LMXBs show small optical depths, well below the limit required to smear out pulsations [116], which should therefore be visible (but see [332]). It is not clear why in this model the intermittent AMXP SAX J1748.9-2021 shows pulsations that appear and disappear on timescales of a few hundred seconds [2, 253], uncorrelated with spectral changes. However it provides a reasonable explanation for the time lags and pulse energy dependence, especially for the pulses of IGR J00291+5934 which have a turnover at 7 keV that is otherwise difficult to explain [86, 88].

Gravitational Light Bending For large stellar compactness (M/R), gravitational light bending can distort the path of X-ray photons up to the point where pulsations are strongly reduced [19, 234, 364]. Persistent sources may therefore be less likely to show pulsations because they accrete more mass than transient systems. However, the total amount of mass that a NS has accreted does not depend on the current mass accretion rate and on the persistence of the X-ray flux, but on the evolutionary stage of the donor star and on the average mass transfer rate from the donor towards the accretor. This model also requires that the NSs in most LMXBs are more massive that those in AMXPs. Although possible, this is difficult to reconcile with the fact that most AMXPs have probably accreted for a long time (see for example [230]). The model also requires some special configuration of the accretion column, which existed for only a very short time and has never recurred, to explain the short pulsed episode of Aql X-1. Finally, the light bending effect never suppresses completely oscillation amplitudes since the Doppler shift due to the rapid NS rotation introduces anisotropies in the emission pattern and thus pulsations [277, 344].

Magnetic and Rotational Axes Alignment The possibility that the magnetic poles of NS migrate was first suggested by Ruderman in 1991 [298]. Vortex/fluxoid coupling in the NS interior links the spin and magnetic axes. Once the NS is spun-up by accretion, crustal shear stress breaks the crust into "plates" that drift towards the rotational poles. The phenomenology of an AMXP with a nearly aligned magnetic and rotational axis was developed recently [178] and explains fairly well the observed correlations between pulse amplitude, pulse phase and X-ray flux. The model can explain intermittency [179], but implies that the persistent AMXPs should also show occasional "oscillation dropouts", a sudden and brief disappearance of pulsations due to sporadic alignment of the hot spot and rotational axis. Such dropouts have not been observed, although the timescale over which this might happen may be too short compared to timescale over which the pulsations are measured (i.e., hundreds of seconds). The model also predicts correlations between excess background noise produced by the hot spot motion which is anti-correlated with the pulse amplitude. This prediction could in principle be easily verified in the future with observational tests.

MHD Instabilities Accretion onto a magnetized NS can proceed via channeled accretion but also via the Schwarzschild-Kruskal instability (the magnetic version of the Rayleigh-Taylor instability) in the disk equatorial plane [10, 85]. This latter mode of accretion has now been observed in 3D-MHD simulations [172] and predicts a decoherence of the X-ray signal at high accretion rates, when several "tongues" of plasma penetrate the magnetosphere and impact the NS surface at random positions. This model explains why bright LMXBs do not pulsate, and might explain why intermittent sources are on average brighter than persistent AMXPs. However, the intermittent AMXP SAX J1748.9-2021 has shown pulsations that appear and disappear in a broad range of accretion rates [253] (rather than at a sharp threshold). It is also hard to explain why tongues never develop in any other AMXP when the outbursts reach peak luminosity, during which the accretion rates are comparable or higher than observed in intermittent AMXPs.

4.7 Thermonuclear Bursts

Type I X-ray bursts are thermonuclear explosions triggered by unstable burning of hydrogen or helium on the layers that build up on the surfaces of accreting NSs. The basic cause is an imbalance between nuclear heating and radiative

Table 4.5 Burst oscillation properties for the nine AMXPs that have shown thermonuclear bursts

Source	Burst osc.	Phase locking	References
Accreting millisecond X-ray pulsars			
SAX J1808.4-3658	Yes	No	[21, 22, 51, 98, 103, 147, 149]
XTE J1814-338	Yes	Yes	[24, 103, 324, 350–352]
Aql X-1	Yes	No	[69, 103, 166, 188, 223, 227, 368]
SAX J1748.9-2021	No		[2, 103, 148, 157]
HETE J1900.1-2455	Yes	No	[101, 103, 158, 325, 354]
IGR J17511-3057	Yes	No	[4, 30, 89, 289]
IGR J17498-2921	Yes	Unknown	[49, 194]
Swift J1749.4-2807	No		[91, 363]
IGR J18245-2452	Yes	Possible	[247, 251, 290]
Mildly recycled X-ray pulsar			
IGR J17480-2466	Yes	Yes	[46, 48, 54, 193, 196, 221]

The table indicates whether a source shows burst oscillations and whether their phases match and follow the evolution of the accretion-powered pulses observed before and after the bursts ("Phase Locking")

cooling, which leads to the runaway of temperature-dependent thermonuclear reactions [188]. There are however many aspects of burst physics that remain puzzling, such as discrepancies between predicted and observed recurrence times, and the phenomenon of burst oscillations [319, 349]. The latter are high frequency oscillations (11–600 Hz) often detected in the X-ray flux produced by thermonuclear bursts [103], and whose cause remains unknown. AMXPs can offer particular insight into burst physics and the burst oscillation mechanism, for two reasons. Firstly, they are the only sources where we have an independent measurement of the stellar spin, allowing us to measure the dependence on rotation. Secondly, they are the only sources where we can make a reasonable assessment of the effects of the magnetic field, and the role of uneven fuel distribution (due to magnetic channeling of material onto the poles of the star).

Table 4.5 summarizes the AMXPs that have exhibited bursts, and which of these have shown burst oscillations. Bursts are expected and observed to occur only within a range of accretion rates from ~2 to 30% of the Eddington rate [65, 96, 228], and in this regard the occurrence of bursts in the AMXP population accords with the behaviour seen in the non-pulsing bursters. The sources that do not show bursts are the five ultracompact sources (which are thought to have rather low accretion rates [319], likely below the minimum required for regular bursting [151]) and IGR J00291+5934, which is also thought to have a rather low accretion rate [100]. None of the AMXPs has yet shown a super-burst, longer duration bursts thought to be due to unstable carbon burning [173].

Burst Properties Bursting behaviour is known to be extremely variable, and the bursts of the AMXPs are no exception to this rule. To date there has not been a comprehensive study assessing whether the bursts from the AMXPs have properties that differ in any statistically significant way from the non-pulsing sources. However just as for the non-pulsing bursters, bursts from the AMXPs show variable recurrence times, peak fluxes and fluences, and durations, even when accretion rate seems to be relatively stable. There are nonetheless some features of AMXP bursts that are worthy of note, in particular when they are compared with bursting sources with higher magnetic fields like the mildly recycled pulsar IGR J17480-2466.

Magnetic Fields and Bursting Bursts from AMXPs confirm that magnetic fields $\sim 10^8$–10^{10} G (Table 4.1) are no impediment to thermonuclear bursting.

Burst Recurrence Times Many neutron star low mass X-ray binaries show burst recurrence times of less than an hour, something that is problematic for theoretical models [162]. The only AMXP in this short recurrence time group is the highly intermittent source Aql X-1. By contrast the mildly recycled pulsar IGR J17480-2466 has the shortest recurrence time between thermonuclear bursts yet recorded, at 3.3 min [221], a possible sign of confined burning.

Rotation and Bursting The AMXPs, which have independent estimates of spin rate, confirm that bursting can occur for spin rates of up to a few hundred Hz (although the fastest rotating burster, 4U 1608-522, is a burst oscillation source not an accretion-powered pulsar, so the spin measure is more indirect [122]). The mildly-recycled accreting pulsar IGR J17480-2466 is instead the bursting source with burst oscillations with the lowest measured rotation rate, at 11 Hz. For a star rotating at $\sim$10 Hz, the Coriolis force is not dynamically relevant [46, 349], indicating that Coriolis force induced confinement of the igniting patch is not essential to the development of X-ray bursts [310] or to very short burst recurrence times (see previous point).

Cooling as a Signature of Thermonuclear Bursts Cooling during the tails of bursts (inferred from blackbody fits to the burst spectra) has long been regarded as a signature trait of thermonuclear Type I (rather than accretion-powered Type II) bursts. Detailed studies of the bursts from the mildly recycled pulsar IGR J17480-2466 have now shown that cooling is a sufficient but not necessary condition for the identification of thermonuclear bursts: under certain circumstances, particularly at high accretion rates, the cooling signature may not be readily detectable [48, 54, 193, 196]. Bursts from the AMXPs do however show cooling.

Marginally Stable Burning The mildly recycled pulsar IGR J17480-2466 shows mHz QPOs that are interpreted as marginally stable burning of H and He on the NS surface [128, 196]. The mHz QPOs are observed when the accretion rate rises and the source transitions from unstable burning (X-ray bursts) to stable nuclear burning (no bursts). The only AMXP to have shown mHz QPOs is the highest accretion rate source, the intermittent pulsar Aql X-1 [286].

Pure Helium Bursts SAX J1808.4-3658 has very bright, short bursts that are thought to be the only secure example of bursting occurring in a layer of pure helium, for a source where the accreted fuel is a mix of hydrogen and helium [98]. The only other sources that are thought to have pure helium bursts are ultra-compact sources where the donor star is too small for there to be significant hydrogen in the accreted material. Pure helium bursting is thought to occur in mixed fuel sources only in a very narrow range of accretion rates where hydrogen can burn stably to form helium before the helium ignites unstably [96].

Bursting and Intermittent Accretion-Powered Pulsations Two of the intermittent AMXPs, HETE J1900.1-2455 and SAX J1748.9-2021, show intermittent pulsations with variable amplitudes at times when the source is also bursting [2, 101]. There does not appear to be any clear-cut causal relationship between the two phenomena (and the third intermittent AMXP, Aql X-1, does not show bursts at the time of its one intermittent pulsation episode, [44]). Nonetheless there has been speculation in the literature as to whether the two might be related. Radiation from the burst might perhaps lead to a temporary disruption of the inner edge of the disk, perturbing the flow of material onto the magnetic poles and moving the accretion footprint to a location more favorable for the formation of pulsations [179].

Burst Oscillations Where the AMXPs have had the biggest impact is in our understanding of burst oscillations (see [349] for a more comprehensive review of this topic). Burst oscillations were first observed in 1996, from the non-pulsar 4U 1728-34 [322]. The oscillations manifested as a coherent signal at about 363 Hz, with an upwards drift in frequency of $\sim$1 Hz as the bursts progressed, towards an asymptotic maximum in the tails. The fact that the same frequency was seen in multiple bursts from the same source, and the stability of the asymptotic maximum frequency, suggested a link to a stable clock such as the stellar spin rate [323]. Firm identification of burst oscillation frequency with the spin frequency however had to wait until the first robust detection of burst oscillations from an AMXP [51]. Although the two frequencies appear to be very close (within a few Hz), however, the separation does differ markedly for different AMXPs. The AMXP XTE J1814-338 and the mildly recycled pulsar IGR J17480-2466 have burst oscillation frequencies that agree with the spin frequency very closely (within the error bars, $\sim 10^{-8}$ Hz for XTE J1814-338 [324, 351, 352], $\sim 10^{-4}$ Hz for IGR J17480-2466 [46]). SAX J1808.4-3658 and IGR J17511-3057, by contrast, have burst oscillations that show a rapid increase in frequency in the burst rise, slightly overshooting the spin frequency, and then stabilizing to within $\sim$10 mHz of the spin frequency in the tails [4, 51]. The intermittent pulsars Aql X-1 and HETE J1900.1-2455 show burst oscillations with slow drifts throughout the bursts, more similar to those of the non-pulsars, and in these cases the asymptotic maximum frequency is $\sim$1 Hz below the spin frequency [44, 224, 354, 368]. Any model for the burst oscillation mechanism must be able to explain this diversity in frequency drift and the small variations in offset from the spin frequency.

There are interesting differences in the properties of the burst oscillations from the pulsars as compared to those of the non-pulsars. For the non-pulsing bursters, burst oscillations tend only to be seen when the source is in the soft (high accretion rate) state [103]. The two intermittent pulsars conform to this rule: indeed HETE

J1900.1-2455 has shown burst oscillations in only one burst, when the source entered an unusually soft state [354]. By contrast five of the six persistent pulsars with bursts have had burst oscillations in all of their bursts, even though these sources tend to be in the hard (low accretion rate) state. For the fifth, IGR J17498-2921, the data are of much poorer quality: oscillations are seen only in the brightest bursts but the upper limits on the presence of detections in the weaker bursts are comparable to the amplitudes detected in the brighter bursts [49]. The duration of burst oscillation trains also appears to differ. For SAX J1808.4-3658, XTE 1814-338 and the mildly recycled pulsar IGR J17480-2466, burst oscillations persist throughout the bursts (except during episodes of Photospheric Radius Expansion in the bright bursts from SAX 1808.4-3658). For the non-pulsars, although burst oscillations are sometimes detected throughout bursts they are more commonly detected in the tails [103], and this is also the case for the intermittent accretion-powered pulsars. In this regard the persistent pulsar IGR J17511-3057 seems to be a transitional object: at lower accretion rates oscillations are detectable throughout bursts, but as accretion rate rises the burst oscillation signal vanishes from the burst rise [4]. Burst oscillations from the pulsars have higher harmonic content than burst oscillations from the intermittent pulsars and non-pulsars [4, 46, 51, 225, 324, 351, 354]. They also have rather different amplitude-energy relations. Burst oscillations from the persistent pulsars show the same amplitude-energy relationship as their accretion-powered pulsations, irrespective of whether the latter rise or fall with energy but with the latter being more common ([4, 46, 49, 51, 239, 324, 354] and see also Sect. 4.5.1). The intermittent pulsars and non-pulsars, by contrast, have burst oscillation amplitudes that rise with energy [226, 353]. The cause of these various differences in burst oscillation properties between the pulsars and non-pulsars remains as yet unclear, but the most obvious hypothesis is that the differences are due to the effects of a dynamically important magnetic field.

One can also use the pulsars to compare the properties of the accretion-powered pulsations (where the peak of the emission is presumed to be centered on the magnetic poles) and their burst oscillations. Burst oscillation amplitudes vary substantially from source to source, but are in most cases within a few percent of (and most lower than) the amplitude of the accretion-powered pulsations at the time of the burst [4, 46, 51, 239, 324, 351, 354]. The transitional AMXP/radio pulsar IGR J18245-2452 is an exception: for the one burst detected, the burst oscillation amplitude was much higher than the amplitude of accretion-powered pulsations immediately before the burst [247].

Pulsar burst oscillations show no statistically significant phase lags as a function of energy [226, 350, 354], unlike accretion-powered pulsations which show soft lags ([67, 99, 110, 111, 163, 243]; see also Sect. 4.5.1). Of particular interest is the strong phase-locking between accretion-powered pulsations and burst oscillations seen in XTE J1814-338 and IGR J17480-2466 [46, 352], with very small phase offset ($<3^\circ$ and $\lesssim 10^\circ$ respectively), implying that the hot spots responsible for the two sets of pulsations are longitudinally coincident in these two sources.

So what do these results tell us about the burst oscillation mechanism ? Although the mechanism is not yet understood, substantial progress has been made and the accretion-powered pulsars have played a key role. Their primary contribution, of course, has been to highlight the very close relationship between the burst oscillation frequency and the spin frequency of the star. This fact underpins the two main classes of model: hot spot models and global mode models. Small temperature variations in the surface layers mean that ignition is expected to begin at a point, with a flame front then spreading out across the star [308]. This flame may then either stall, confining the burning to a small region (the hot spot model) or excite large-scale waves in the surface layers (global mode models). The resulting temperature asymmetry gives rise to the burst oscillations. In hot spot models, the relationship with the spin rate is straightforward since the hot spot should be near stationary in the rotating frame of the star [322]. The main open theoretical question for hot spot models is what might cause the flame to stall, with both magnetic and Coriolis forces under consideration [33, 46, 310]. For global mode models, the requirement that burst oscillation frequency be very close to spin frequency puts very stringent restrictions on the types of surface modes that might be responsible [133, 137, 270]. Efforts are now ongoing to develop both classes of model to determine whether they could match the rest of the observed properties [349]. The differences between the burst oscillations from the pulsars and the non-pulsars are informative, and suggest a role for the magnetic field. What is still unclear, however, is whether we are seeing one mechanism with a continuum of properties set by stellar parameters such as the magnetic field—or whether two different mechanisms may be required. The apparent gradual changes in burst oscillation properties from the non-pulsars and intermittent pulsars (where magnetic field is likely to be weak) to the persistent pulsars (with stronger fields) had until recently favoured the former. However new results from the pulsars point increasingly towards the possibility that we are indeed seeing two different mechanisms. The development of burst oscillations in the slowly rotating mildly recycled pulsar IGR J17480-2466 is particularly hard to understand in the context of global modes or Coriolis force induced hot spots [46]. Magnetic confinement of the flame front is at present the only plausible explanation for this source, and would also explain the extraordinary phase-locking between accretion-powered pulsations and burst oscillations. This mechanism might also operate in the other strong magnetic field source with phase-locked burst oscillations, the AMXP XTE J1814-338 [46, 352]. However for the other sources, with weaker magnetic fields, magnetic confinement is unlikely to be effective. For these sources global mode models remain a good possibility, with the differences in the pulsar burst oscillations being perhaps due to magnetic modifications to the mode structure.

4.8 Aperiodic Variability and kHz QPOs

The study of aperiodic signals in power spectra of AMXPs reveals a rich phenomenology which has helped us to understand the physics of accretion disks and compact objects. All AMXPs are "atoll sources", so-called because of their pattern

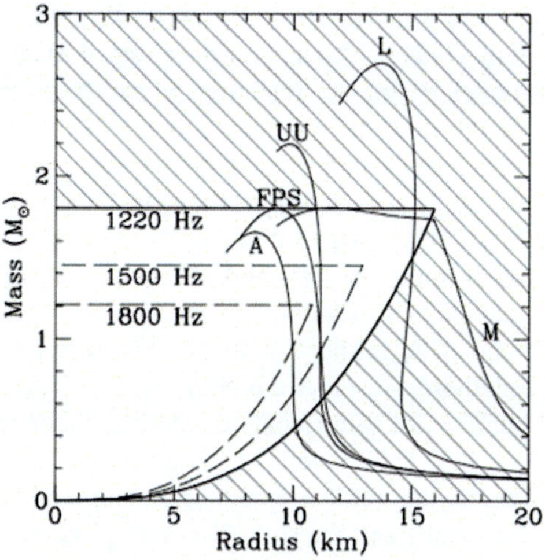

Fig. 4.10 Constraints on the Mass-Radius relation of NSs from kHz QPOs. The solid curves refer to a sample of different EoS models (for non-rotating NSs), and the thick solid and dashed lines enclose regions of the M-R diagram compatible with the observation of kHz QPOs at a certain Keplerian frequency (image from [213])

in the X-ray colour-colour diagram. Atoll sources are relatively low/intermediate-luminosity LMXBs and are believed to possess low magnetic fields compared to the brightest LMXBs (the "Z sources"). The aperiodic variability of AMXPs includes phenomena at very high frequencies (the kHz QPOs) but also at lower frequency, such as the strong 1 Hz modulation observed in SAX J1808.4-3658 and NGC6440 X-2 (Sects. 4.3.1 and 4.3.11). Here we highlight the main aperiodic phenomena peculiar to AMXPs and refer to Chap. 1 of this book for a more detailed discussion.

The fastest variability in LMXBs is observed as kHz QPOs, which appear at frequencies of up to ~1300 Hz. Sometimes two kHz QPOs are observed simultaneously as "twins", with an "upper" QPO at frequency ν_u and a "lower" QPO at frequency ν_l. Such fast variability is observed only in LMXBs with NS accretors, and is considered to be a signature of the presence of a NS in the absence of pulsations or thermonuclear bursts. All models that try to explain kHz QPOs associate at least one of the frequencies with the orbital motion of plasma in the accretion disk. Such high frequency variability is expected given the short dynamical timescale (0.1–1 ms) characterizing the motion of matter near NSs. Such rapid variability is never observed in black hole LMXBs,[4] possibly because they have larger mass and hence a larger innermost stable circular orbit (ISCO). The ISCO is a limiting radius (at $6GM/c^2$ for non-rotating NSs) set by General Relativity that defines the smallest stable orbit in an accretion disk. The frequency of kHz QPOs is close to the value expected for stable Keplerian orbits around a NS, and can therefore be used to place constraints on the mass and radius of the NS (and hence the EoS of ultra-dense matter). The size of the orbit where kHz QPOs are produced must be larger than both NS radius and ISCO, both of which depend on NS mass (Fig. 4.10).

[4]Fast variability appears in black hole LMXBs below ~500 Hz as "high frequency QPOs".

The four AMXPs with twin kHz QPOs (SAX J1808.4-3658, XTE J1807-294, IGR J17511-305 and Aql X-1; Sect. 4.3) have been used to test kHz QPO formation models (e.g., the rejection of the beat-frequency model; Sect. 4.3.1) and have illuminated the relation with the NS spin. Taking into account also the NXPs with twin kHz QPOs, it is clear that the relation between kHz QPO separation $\Delta\nu = \nu_u - \nu_l$ and spin frequency ν_s is not as simple as originally thought: $\Delta\nu/\nu_s$ is not always equal to 1 or 0.5. $\Delta\nu$ and ν_s may even be unrelated: instead we see an average $\langle\Delta\nu\rangle \approx 300$ Hz for all sources *except* AMXPs [212, 366]. *Only* for the AMXPs, where the magnetic field is strong enough to channel accretion, is the ratio $\Delta\nu/\nu_s$ either 1 or 0.5. Some predictions of the sonic-point spin resonance and relativistic-resonance models [164, 175] are consistent with the AMXP results. In sonic-point spin resonance models, the magnetic and radiation fields rotating with the star excite a vertical motion of the disk gas at the "spin-resonance" radius r_{sr}. There, the vertical epicyclic frequency equals the difference between the orbital frequency and the spin. Depending on the "clumpiness" of the flow at r_{rs}, the X-ray flux exhibits $\Delta\nu$ equal to either ν_s (smooth flow) or $0.5\nu_s$ (clumpy flow). In relativistic resonance models, a non-linear 1:2 or 1:3 resonance between orbital and radial epicyclic motion emerges as a consequence of the deviation of the strong gravitational potential from the Newtonian $1/r$. A third model suggests that kHz QPOs are related to the three general relativistic epicyclic frequencies [318] or to the precession of frame dragging (Lense-Thirring precession [317]). So far none of these models satisfactorily accounts for all of the rich phenomenology of kHz QPOs.

Determining the origin of QPOs, in particular kHz QPOs is of fundamental importance as this is a phenomenon that is common in LMXBs. If a clear relation with the NS mass and/or radius is confirmed, it has the potential to place severe constraints on the EoS of ultra-dense matter. This field has however seen few advancements in the past few years, at least on the theoretical side. New attempts are now using numerical simulations to interpret the observations and test some of the original models [11, 80, 145, 146]. Most of these simulations use black hole accretors (but see [11]) which do not have kHz QPOs, but the results are promising in terms of understanding the generation of variability in LMXBs as a whole.

4.9 Open Problems and Final Remarks

Since the first discovery in 1998, the AMXPs have provided an incredibly body of observational data allowing us to understand NSs and their evolution. They have allowed the study of extreme phenomena on the surface of the NSs, in the strong gravity regime at relativistic rotational velocities, and have helped us to understand how radio pulsars form and evolve within the framework of the *recycling scenario*.

Proposed X-ray missions like the Indian *ASTROSAT*, the NASA *AXTAR* satellite and the ESA missions *LOFT* and *Athena+* will guarantee huge advances in our knowledge of these systems, building on the extraordinary results from *RXTE*,

XMM-Newton, *Chandra* and *Swift*. Long-term monitoring of future AMXP outbursts will allow measurement of long-term spin and orbital behaviour, something that has at present only been done for four systems. Increased collecting area will increase sensitivity to pulsations by orders of magnitude, leading (we hope) to the discovery of sub-millisecond pulsars or strengthening the 730 Hz spin distribution cutoff. In the former case this would place very tight constraints on the EoS of ultra-dense matter. Modeling of high signal to noise pulse profiles will reveal the subtle features imparted by strong gravity, enabling us to measure the NS compactness. We will be able to test models for kHz QPOs and burst oscillations with sufficient precision that we can finally determine the underpinning mechanism.

Of the many open problems, there are several that stand out. One big mystery is why do so very few LMXBs show pulsations? If magnetic field evolution and decay is involved, we may finally be able to understand the origin and the evolution of nature's strongest magnetic fields. What happens when accretion takes place during an outburst? Why we do not detect the effect of accretion torques in some AMXPs, contrary to expectations ? Are gravitational waves involved as a braking mechanism that counteracts the effect of accretion or are we again missing some fundamental ingredient? When do accreting pulsar switch on as radio pulsars ? Are all AMXPs turning on as radio millisecond pulsars during quiescence ? What is the relation between AMXPs and the radio pulsar "black-widows" and "red-backs" ? We look forward to discovering the answers to these questions in the years ahead.

Acknowledgments We would like to thank C. Heinke, J. Poutanen, R. Lovelace, M. Linares, T. Tauris, D. Altamirano, B. Haskell, P. D'Avanzo, D. Bhattacharya, J. Hessels, C. D'Angelo and N. Masetti for providing useful comments and suggestions. We acknowledge support from two NWO Vidi fellowships.

References

1. M.A. Alpar, A.F. Cheng, M.A. Ruderman, J. Shaham, Nature, **300**, 728 (1982)
2. D. Altamirano, P. Casella, A. Patruno, R. Wijnands, M. van der Klis, Astrophys. J. Lett. **674**, L45 (2008)
3. D. Altamirano, A. Patruno, C.O. Heinke, C. Markwardt, T.E. Strohmayer, M. Linares, R. Wijnands, M. van der Klis, J.H. Swank, Astrophys. J. Lett. **712**, L58 (2010)
4. D. Altamirano, A. Watts, M. Linares, C.B. Markwardt, T. Strohmayer, A. Patruno, Mon. Not. R. Astron. Soc. **409**, 1136 (2010)
5. D. Altamirano, R. Wijnands, M. van der Klis, A. Patruno, A. Watts, M. Armas Padilla, Y. Cavecchi, N. Degenaar, M. Kalamkar, R. Kaur, Y.J. Yang, P. Casella, M. Linares, P. Soleri, N. Rea, Astron. Tel. **2565**, 1 (2010)
6. D. Altamirano, Y. Cavecchi, A. Patruno, A. Watts, M. Linares, N. Degenaar, M. Kalamkar, M. van der Klis, N. Rea, P. Casella, M. Armas Padilla, R. Kaur, Y.J. Yang, P. Soleri, R. Wijnands, Astrophys. J. Lett. **727**, L18 (2011)
7. N. Andersson, K. Glampedakis, B. Haskell, A.L. Watts, Mon. Not. R. Astron. Soc. **361**, 1153 (2005)
8. M. Annala, J. Poutanen, Astron. Astrophys. **520**, A76 (2010)

9. A.M. Archibald, I.H. Stairs, S.M. Ransom, V.M. Kaspi, V.I. Kondratiev, D.R. Lorimer, M.A. McLaughlin, J. Boyles, J.W.T. Hessels, R. Lynch, J. van Leeuwen, M.S.E. Roberts, F. Jenet, D.J. Champion, R. Rosen, B.N. Barlow, B.H. Dunlap, R.A. Remillard, Science **324**, 1411 (2009)
10. J. Arons, S.M. Lea, Astrophys. J. **207**, 914 (1976)
11. M. Bachetti, M.M. Romanova, A. Kulkarni, L. Burderi, T. di Salvo, Mon. Not. R. Astron. Soc. **403**, 1193 (2010)
12. D.C. Backer, S.R. Kulkarni, C. Heiles, M.M. Davis, W.M. Goss, Nature **300**, 615 (1982)
13. M.C. Baglio, P. D'Avanzo, T. Muñoz-Darias, R.P. Breton, S. Campana, Astron. Astrophys. **559**, A42 (2013)
14. C. Baldovin, E. Kuulkers, C. Ferrigno, E. Bozzo, J. Chenevez, S. Brandt, V. Beckmann, A. Bird, A. Domingo, K. Ebisawa, P. Jonker, P. Kretschmar, C. Markwardt, T. Oosterbroek, A. Paizis, D. Risquez, C. Sanchez-Fernandez, S. Shaw, R. Wijnands, Astron. Tel. **2196**, 1 (2009)
15. D. Barret, M. Boutelier, M.C. Miller, Mon. Not. R. Astron. Soc. **384**, 1519 (2008)
16. C.G. Bassa, A. Patruno, J.W.T. Hessels, E.F. Keane, B. Monard, E.K. Mahony, S. Bogdanov, S. Corbel, P.G. Edwards, A.M. Archibald, G.H. Janssen, B.W. Stappers, S. Tendulkar, Mon. Not. R. Astron. Soc. **441**, 1825 (2014). https://doi.org/10.1093/mnras/stu708
17. M. Bejger, M. Fortin, P. Haensel, J.L. Zdunik, Astron. Astrophys. **536**, A87 (2011)
18. T. Belloni, L. Stella, E. Bozzo, G. Israel, S. Campana, Astron. Tel. **2568**, 1 (2010)
19. A.M. Beloborodov, Astrophys. J. Lett. **566**, L85 (2002)
20. C.G. Bernal, D. Page, W.H. Lee, Astrophys. J. **770**, 106 (2013)
21. S. Bhattacharyya, T.E. Strohmayer, Astrophys. J. Lett. **641**, L53 (2006)
22. S. Bhattacharyya, T.E. Strohmayer, Astrophys. J. **656**, 414 (2007)
23. D. Bhattacharya, E.P.J. van den Heuvel, Phys. Rep. **203**, 1 (1991)
24. S. Bhattacharyya, T.E. Strohmayer, M.C. Miller, C.B. Markwardt, Astrophys. J. **619**, 483 (2005)
25. L. Bildsten, Astrophys. J. Lett. **577**, L27 (2002)
26. L. Bildsten, E.F. Brown, Astrophys. J. **477**, 897 (1997)
27. L. Bildsten, D. Chakrabarty, Astrophys. J. **557**, 292 (2001)
28. G.S. Bisnovatyi-Kogan, B.V. Komberg, Sov. Astron. **18**, 217 (1974)
29. S. Bogdanov, A. Patruno, A.M. Archibald, C. Bassa, J.W.T. Hessels, G.H. Janssen, B.W. Stappers, Astrophys. J. **789**, 40 (2014). https://doi.org/10.1088/0004-637X/789/1/40
30. E. Bozzo, C. Ferrigno, M. Falanga, S. Campana, J.A. Kennea, A. Papitto, Astron. Astrophys. **509**, L3 (2010)
31. E. Bozzo, T. Belloni, G. Israel, L. Stella, Astron. Tel. **2567**, 1 (2010)
32. J. Brainerd, F.K. Lamb, Astrophys. J. Lett. **317**, L33 (1987)
33. E.F. Brown, L. Bildsten, Astrophys. J. **496**, 915 (1998)
34. T. Bulik, H. Riffert, P. Meszaros, K. Makishima, T. Mihara, B. Thomas, Astrophys. J. **444**, 405 (1995)
35. L. Burderi, T. Di Salvo, F. D'Antona, N.R. Robba, V. Testa, Astron. Astrophys. **404**, L43 (2003)
36. L. Burderi, T. Di Salvo, M.T. Menna, A. Riggio, A. Papitto, Astrophys. J. Lett. **653**, L133 (2006)
37. L. Burderi, T. Di Salvo, G. Lavagetto, M.T. Menna, A. Papitto, A. Riggio, R. Iaria, F. D'Antona, N.R. Robba, L. Stella, Astrophys. J. **657**, 961 (2007)
38. L. Burderi, V. Testa, M.T. Menna, G.L. Israel, T. di Salvo, A. Papitto, A. Riggio, A. Antonelli, E. Bozzo, M. Burgay, N. D'Amico, E. Gallo, R. Iaria, A. Possenti, Astron. Tel. **1132**, 1 (2007)
39. L. Burderi, A. Riggio, T. di Salvo, A. Papitto, M.T. Menna, A. D'Aì, R. Iaria, Astron. Astrophys. **496**, L17 (2009)
40. E.M. Cackett, R. Wijnands, C.O. Heinke, P.D. Edmonds, W.H.G. Lewin, D. Pooley, J.E. Grindlay, P.G. Jonker, J.M. Miller, Astrophys. J. **620**, 922 (2005)
41. E.M. Cackett, D. Altamirano, A. Patruno, J.M. Miller, M. Reynolds, M. Linares, R. Wijnands, Astrophys. J. Lett. **694**, L21 (2009)

42. C. Cadeau, S.M. Morsink, D. Leahy, S.S. Campbell, Astrophys. J. **654**, 458 (2007)
43. S. Campana, P. D'Avanzo, J. Casares, S. Covino, G. Israel, G. Marconi, R. Hynes, P. Charles, L. Stella, Astrophys. J. Lett. **614**, L49 (2004)
44. P. Casella, D. Altamirano, A. Patruno, R. Wijnands, M. van der Klis, Astrophys. J. Lett. **674**, L41 (2008)
45. A.J. Castro-Tirado, A. Caccianiga, J. Gorosabel, P. Kilmartin, P. Tristram, P. Yock, C. Sanchez-Fernandez, M.E. Alcoholado-Feltstrom, IAU Circ. **7895**, 1 (2002)
46. Y. Cavecchi, A. Patruno, B. Haskell, A.L. Watts, Y. Levin, M. Linares, D. Altamirano, R. Wijnands, M. van der Klis, Astrophys. J. Lett. **740**, L8 (2011)
47. D. Chakrabarty, in *A Decade of Accreting Millisecond X-ray Pulsars*, *AIPC Series*, vol. 1068, ed. by R. Wijnands, D. Altamirano, P. Soleri, N. Degenaar, N. Rea, P. Casella, A. Patruno, M. Linares (2008), pp. 67–74
48. M. Chakraborty, S. Bhattacharyya, Astrophys. J. Lett. **730**, L23 (2011)
49. M. Chakraborty, S. Bhattacharyya, Mon. Not. R. Astron. Soc. **422**, 2351 (2012)
50. D. Chakrabarty, E.H. Morgan, Nature **394**, 346 (1998)
51. D. Chakrabarty, E.H. Morgan, M.P. Muno, D.K. Galloway, R. Wijnands, M. van der Klis, C.B. Markwardt, Nature **424**, 42 (2003)
52. D. Chakrabarty, J.H. Swank, C.B. Markwardt, E. Smith, Astron. Tel. **1660**, 1 (2008)
53. D. Chakrabarty, C.B. Markwardt, M. Linares, P.G. Jonker, Astron. Tel. **3606**, 1 (2011)
54. M. Chakraborty, S. Bhattacharyya, A. Mukherjee, Mon. Not. R. Astron. Soc. **418**, 490 (2011)
55. J. Chenevez, E. Kuulkers, V. Beckmann, A. Bird, S. Brandt, A. Domingo, K. Ebisawa, P. Jonker, P. Kretschmar, C. Markwardt, T. Oosterbroek, A. Paizis, D. Risquez, C. Sanchez-Fernandez, S. Shaw, R. Wijnands, Astron. Tel. **2235**, 1 (2009)
56. J. Chenevez, E. Kuulkers, J. Alfonso-Garzón, V. Beckmann, T. Bird, P. Bordas, S. Brandt, T. Courvoisier, M. Del Santo, A. Domingo, K. Ebisawa, C. Ferrigno, P. Jonker, P. Kretschmar, C. Markwardt, T. Oosterbroek, A. Paizis, K. Pottschmidt, C. Sánchez-FernÁndez, R. Wijnands, Astron. Tel. **2924**, 1 (2010)
57. C. Chevalier, S.A. Ilovaisky, Astron. Astrophys. **251**, L11 (1991)
58. C. Chevalier, S.A. Ilovaisky, IAU Circ. **6806**, 2 (1998)
59. C. Chevalier, S.A. Ilovaisky, P. Leisy, F. Patat, Astron. Astrophys. **347**, L51 (1999)
60. Y. Chou, Y. Chung, C.P. Hu, T.C. Yang, Astrophys. J. **678**, 1316 (2008)
61. A.R. Choudhuri, S. Konar, Mon. Not. R. Astron. Soc. **332**, 933 (2002)
62. W. Coburn, W.A. Heindl, R.E. Rothschild, D.E. Gruber, I. Kreykenbohm, J. Wilms, P. Kretschmar, R. Staubert, Astrophys. J. **580**, 394 (2002)
63. H.N. Cohn, P.M. Lugger, S. Bogdanov, C.O. Heinke, M. Van Den Berg, G. Sivakoff, Astron. Tel. **5031**, 1 (2013)
64. R.H.D. Corbet, J.L. Sokoloski, K. Mukai, C.B. Markwardt, J. Tueller, Astrophys. J. **675**, 1424 (2008)
65. R. Cornelisse, J.J.M. in't Zand, F. Verbunt, E. Kuulkers, J. Heise, P.R. den Hartog, M. Cocchi, L. Natalucci, A. Bazzano, P. Ubertini, Astron. Astrophys. **405**, 1033 (2003)
66. W. Cui, Astrophys. J. Lett. **482**, L163 (1997)
67. W. Cui, E.H. Morgan, L.G. Titarchuk, Astrophys. J. Lett. **504**, L27+ (1998)
68. A. Cumming, E. Zweibel, L. Bildsten, Astrophys. J. **557**, 958 (2001)
69. M. Czerny, B. Czerny, J.E. Grindlay, Astrophys. J. **312**, 122 (1987)
70. C.R. D'Angelo, H.C. Spruit, Mon. Not. R. Astron. Soc. **406**, 1208 (2010)
71. P. D'Avanzo, S. Campana, S. Covino, G.L. Israel, L. Stella, G. Andreuzzi, Astron. Astrophys. **472**, 881 (2007)
72. P. D'Avanzo, S. Campana, J. Casares, S. Covino, G.L. Israel, L. Stella, Astron. Astrophys. **508**, 297 (2009)
73. P. D'Avanzo, S. Campana, T. Muñoz-Darias, T. Belloni, E. Bozzo, M. Falanga, L. Stella, Astron. Astrophys. **534**, A92 (2011)
74. J.E. Deeter, P.E. Boynton, F.K. Lamb, G. Zylstra, Astrophys. J. **336**, 376 (1989)
75. N. Degenaar, R. Wijnands, S. Campana, D. Galloway, W. Lewin, J. Homan, D. Chakrabarty, P. Jonker, E. Cackett, J. Miller, Astron. Tel. **1098**, 1 (2007)

76. C.J. Deloye, L. Bildsten, Astrophys. J. **598**, 1217 (2003)
77. C.J. Deloye, C.O. Heinke, R.E. Taam, P.G. Jonker, Mon. Not. R. Astron. Soc. **391**, 1619 (2008)
78. D. de Martino, M. Falanga, J.M. Bonnet-Bidaud, T. Belloni, M. Mouchet, N. Masetti, I. Andruchow, S.A. Cellone, K. Mukai, G. Matt, Astron. Astrophys. **515**, A25 (2010). https://doi.org/10.1051/0004-6361/200913802
79. D. de Martino, T. Belloni, M. Falanga, A. Papitto, S. Motta, A. Pellizzoni, Y. Evangelista, G. Piano, N. Masetti, J.M. Bonnet-Bidaud, M. Mouchet, K. Mukai, A. Possenti, Astron. Astrophys. **550**, A89 (2013). https://doi.org/10.1051/0004-6361/201220393
80. J. Dexter, P.C. Fragile, Astrophys. J. **730**, 36 (2011)
81. T. di Salvo, L. Burderi, A. Riggio, A. Papitto, M.T. Menna, Mon. Not. R. Astron. Soc. **389**, 1851 (2008)
82. D. Eckert, R. Walter, P. Kretschmar, M. Mas-Hesse, G.G.C. Palumbo, J.P. Roques, P. Ubertini, C. Winkler, Astron. Tel. **352**, 1 (2004)
83. D. Eckert, M. Del Santo, A. Bazzano, K. Watanabe, A. Paizis, E. Bozzo, C. Ferrigno, I. Caballero, L. Sidoli, L. Kuiper, Astron. Tel. **4925**, 1 (2013)
84. P. Elebert, P.J. Callanan, A.V. Filippenko, P.M. Garnavich, G. Mackie, J.M. Hill, V. Burwitz, Mon. Not. R. Astron. Soc. **383**, 1581 (2008)
85. R.F. Elsner, F.K. Lamb, Astrophys. J. **215**, 897 (1977)
86. M. Falanga, L. Titarchuk, Astrophys. J. **661**, 1084 (2007)
87. M. Falanga, J.M. Bonnet-Bidaud, J. Poutanen, R. Farinelli, A. Martocchia, P. Goldoni, J.L. Qu, L. Kuiper, A. Goldwurm, Astron. Astrophys. **436**, 647 (2005)
88. M. Falanga, L. Kuiper, J. Poutanen, E.W. Bonning, W. Hermsen, T. di Salvo, P. Goldoni, A. Goldwurm, S.E. Shaw, L. Stella, Astron. Astrophys. **444**, 15 (2005)
89. M. Falanga, L. Kuiper, J. Poutanen, D.K. Galloway, E.W. Bonning, E. Bozzo, A. Goldwurm, W. Hermsen, L. Stella, Astron. Astrophys. **529**, A68 (2011)
90. R. Fender, G. De Bruyn, G. Pooley, B. Stappers, Astron. Tel. **361**, 1 (2004)
91. C. Ferrigno, E. Bozzo, M. Falanga, L. Stella, S. Campana, T. Belloni, G.L. Israel, L. Pavan, E. Kuulkers, A. Papitto, Astron. Astrophys. **525**, A48 (2011)
92. C. Ferrigno, E. Bozzo, L.G.A.P.T.M. Belloni, Astron. Tel. **3560**, 1 (2011)
93. C. Ferrigno, E. Bozzo, A. Papitto, N. Rea, L. Pavan, S. Campana, M. Wieringa, M. Filipovic, M. Falanga, Astron. Astrophys. **576**, A77 (2014)
94. D.B. Fox, Astron. Tel. **526**, 1 (2005)
95. D.B. Fox, S.R. Kulkarni, Astron. Tel. **354**, 1 (2004)
96. M.Y. Fujimoto, T. Hanawa, S. Miyaji, Astrophys. J. **247**, 267 (1981)
97. D.K. Galloway, in *The Transient Milky Way: A Perspective for MIRAX*, *AIPC*, vol. 840, ed. by F. D'Amico, J. Braga, R.E. Rothschild (2006), pp. 50–54
98. D.K. Galloway, A. Cumming, Astrophys. J. **652**, 559 (2006)
99. D.K. Galloway, D. Chakrabarty, E.H. Morgan, R.A. Remillard, Astrophys. J. Lett. **576**, L137 (2002)
100. D.K. Galloway, C.B. Markwardt, E.H. Morgan, D. Chakrabarty, T.E. Strohmayer, Astrophys. J. Lett. **622**, L45 (2005)
101. D.K. Galloway, E.H. Morgan, M.I. Krauss, P. Kaaret, D. Chakrabarty, Astrophys. J. Lett. **654**, L73 (2007)
102. D.K. Galloway, E.H. Morgan, D. Chakrabarty, in *A Decade of Accreting Millisecond X-ray Pulsars*, *AIPC Series*, vol. 1068, ed. by R. Wijnands, D. Altamirano, P. Soleri, N. Degenaar, N. Rea, P. Casella, A. Patruno, M. Linares (2008), pp. 55–62
103. D.K. Galloway, M.P. Muno, J.M. Hartman, D. Psaltis, D. Chakrabarty, Astrophys. J. Suppl. Ser. **179**, 360 (2008)
104. F.P. Gavriil, T.E. Strohmayer, J.H. Swank, C.B. Markwardt, in *AAS/High Energy Astrophysics Division #9*, *Bulletin of the American Astronomical Society*, vol. 38 (2006), p. 336
105. F.P. Gavriil, T.E. Strohmayer, J.H. Swank, C.B. Markwardt, Astrophys. J. Lett. **669**, L29 (2007)
106. P. Ghosh, F.K. Lamb, Astrophys. J. Lett. **223**, L83 (1978)

107. P. Ghosh, F.K. Lamb, Astrophys. J. **232**, 259 (1979)
108. P. Ghosh, F.K. Lamb, Astrophys. J. **234**, 296 (1979)
109. L. Gibaud, A. Bazzano, E. Bozzo, M. Cadolle-Bel, J. Chenevez, S.P. Drave, C. Ferrigno, D. Gotz, M. Kadler, I. Kreykenbohm, G. Puehlhofer, J. Rodriguez, C. Sanchez-Fernandez, K. Watanabe, Astron. Tel. **3551**, 1 (2011)
110. M. Gierliński, C. Done, Mon. Not. R. Astron. Soc. **337**, 1373 (2002)
111. M. Gierliński, J. Poutanen, Mon. Not. R. Astron. Soc. **359**, 1261 (2005)
112. M. Gierliński, C. Done, D. Barret, Mon. Not. R. Astron. Soc. **331**, 141 (2002)
113. A.B. Giles, K.M. Hill, J.G. Greenhill, Mon. Not. R. Astron. Soc. **304**, 47 (1999)
114. A.B. Giles, J.G. Greenhill, K.M. Hill, E. Sanders, Mon. Not. R. Astron. Soc. **361**, 1180 (2005)
115. M. Gilfanov, M. Revnivtsev, R. Sunyaev, E. Churazov, Astron. Astrophys. **338**, L83 (1998)
116. E. Göğüş, M.A. Alpar, M. Gilfanov, Astrophys. J. **659**, 580 (2007)
117. S.A. Grebenev, S.V. Molkov, R.A. Sunyaev, Astron. Tel. **446**, 1 (2005)
118. J.G. Greenhill, A.B. Giles, K.M. Hill, IAU Circ. **7889**, 1 (2002)
119. J.G. Greenhill, A.B. Giles, C. Coutures, Mon. Not. R. Astron. Soc. **370**, 1303 (2006)
120. S. Greiss, D. Steeghs, T. Maccarone, P.G. Jonker, M.A.P. Torres, O. Gonzalez, N. Masetti, A. Rojas, Astron. Tel. **3562**, 1 (2011)
121. J. Halpern, GRB Coordinates Network **5210**, 1 (2006)
122. J.M. Hartman, D. Chakrabarty, D.K. Galloway, M.P. Muno, P. Savov, M. Mendez, S. van Straaten, T. Di Salvo, Bull. Am. Astron. Soc. **35**, 865 (2003)
123. J.M. Hartman, A. Patruno, D. Chakrabarty, D.L. Kaplan, C.B. Markwardt, E.H. Morgan, P.S. Ray, M. van der Klis, R. Wijnands, Astrophys. J. **675**, 1468 (2008)
124. J.M. Hartman, A.L. Watts, D. Chakrabarty, Astrophys. J. **697**, 2102 (2009)
125. J.M. Hartman, A. Patruno, D. Chakrabarty, C.B. Markwardt, E.H. Morgan, M. van der Klis, R. Wijnands, Astrophys. J. **702**, 1673 (2009)
126. J.M. Hartman, D.K. Galloway, D. Chakrabarty, Astrophys. J. **726**, 26 (2011)
127. B. Haskell, A. Patruno, Astrophys. J. Lett. **738**, L14 (2011)
128. A. Heger, A. Cumming, S.E. Woosley, Astrophys. J. **665**, 1311 (2007)
129. C.O. Heinke, S.A. Budac, Astron. Tel. **2139**, 1 (2009)
130. C.O. Heinke, P.G. Jonker, R. Wijnands, C.J. Deloye, R.E. Taam, Astrophys. J. **691**, 1035 (2009)
131. C.O. Heinke, D. Altamirano, H.N. Cohn, P.M. Lugger, S.A. Budac, M. Servillat, M. Linares, T.E. Strohmayer, C.B. Markwardt, R. Wijnands, J.H. Swank, C. Knigge, C. Bailyn, J.E. Grindlay, Astrophys. J. **714**, 894 (2010)
132. C.O. Heinke, A. Bahramian, R. Wijnands, D. Altamirano, Astron. Tel. **4927**, 1 (2013)
133. K. Heng, A. Spitkovsky, Astrophys. J. **703**, 1819 (2009)
134. J.W.T. Hessels, B.W. Stappers, Astron. Tel. **1129**, 1 (2007)
135. J.W.T. Hessels, S.M. Ransom, I.H. Stairs, P.C.C. Freire, V.M. Kaspi, F. Camilo, Science **311**, 1901 (2006)
136. A. Hewish, S.J. Bell, J.D. Pilkington, P.F. Scott, R.A. Collins, Nature **217**, 709 (1968)
137. J.S. Heyl, Astrophys. J. **600**, 939 (2004)
138. K.H. Hinkle, F.C. Fekel, R.R. Joyce, P.R. Wood, V.V. Smith, T. Lebzelter, Astrophys. J. **641**, 479 (2006)
139. R.M. Hjellming, X. Han, D. Roussel-Dupre, IAU Circ. **5112**, 1 (1990)
140. L. Homer, P.A. Charles, D. Chakrabarty, L. van Zyl, Mon. Not. R. Astron. Soc. **325**, 1471 (2001)
141. M.N. Iacolina, M. Burgay, L. Burderi, A. Possenti, T. di Salvo, Astron. Astrophys. **519**, A13 (2010)
142. A. Ibragimov, J. Poutanen, Mon. Not. R. Astron. Soc. **400**, 492 (2009)
143. A. Ibragimov, J.J.E. Kajava, J. Poutanen, Mon. Not. R. Astron. Soc. **415**, 1864 (2011)
144. A.F. Illarionov, R.A. Sunyaev, Astron. Astrophys. **39**, 185 (1975)
145. A. Ingram, C. Done, Mon. Not. R. Astron. Soc. **415**, 2323 (2011)
146. A. Ingram, C. Done, Mon. Not. R. Astron. Soc. **419**, 2369 (2012)

147. J.J.M. in 't Zand, J. Heise, J.M. Muller, A. Bazzano, M. Cocchi, L. Natalucci, P. Ubertini, Astron. Astrophys. **331**, L25 (1998)
148. J.J.M. in 't Zand, F. Verbunt, T.E. Strohmayer, A. Bazzano, M. Cocchi, J. Heise, M.H. van Kerkwijk, J.M. Muller, L. Natalucci, M.J.S. Smith, P. Ubertini, Astron. Astrophys. **345**, 100 (1999)
149. J.J.M. in't Zand, R. Cornelisse, E. Kuulkers, J. Heise, L. Kuiper, A. Bazzano, M. Cocchi, J.M. Muller, L. Natalucci, M.J.S. Smith, P. Ubertini, Astron. Astrophys. **372**, 916 (2001)
150. J.J.M. in't Zand, M.H. van Kerkwijk, D. Pooley, F. Verbunt, R. Wijnands, W.H.G. Lewin, Astrophys. J. Lett. **563**, L41 (2001)
151. J.J.M. in't Zand, P.G. Jonker, C.B. Markwardt, Astron. Astrophys. **465**, 953 (2007)
152. P.G. Jonker, M. van der Klis, Astrophys. J. Lett. **553**, L43 (2001)
153. P.G. Jonker, M. Méndez, M. van der Klis, Astrophys. J. Lett. **540**, L29 (2000)
154. P.G. Jonker, G. Nelemans, Z. Wang, A.K.H. Kong, D. Chakrabarty, M. Garcia, P.J. Groot, M. van der Klis, T. Kerr, B. Mobasher, M. Sullivan, T. Augusteijn, B.W. Stappers, P. Challis, R.P. Kirshner, J. Hjorth, A. Delsanti, Mon. Not. R. Astron. Soc. **344**, 201 (2003)
155. P.G. Jonker, M.A.P. Torres, D. Steeghs, Astrophys. J. **680**, 615 (2008)
156. A.M. Juett, D.K. Galloway, D. Chakrabarty, Astrophys. J. **587**, 754 (2003)
157. P. Kaaret, J.J.M. i. Zand, J. Heise, J.A. Tomsick, Astrophys. J. **598**, 481 (2003)
158. P. Kaaret, E.H. Morgan, R. Vanderspek, J.A. Tomsick, Astrophys. J. **638**, 963 (2006)
159. J.J.E. Kajava, A. Ibragimov, M. Annala, A. Patruno, J. Poutanen, Mon. Not. R. Astron. Soc. **417**, 1454 (2011)
160. M. Kalamkar, D. Altamirano, M. van der Klis, Astrophys. J. **729**, 9 (2011)
161. N. Kawai, M. Suzuki, Astron. Tel. **534**, 1 (2005)
162. L. Keek, D.K. Galloway, J.J.M. in't Zand, A. Heger, Astrophys. J. **718**, 292 (2010)
163. M.G.F. Kirsch, K. Mukerjee, M.G. Breitfellner, S. Djavidnia, M.J. Freyberg, E. Kendziorra, M.J.S. Smith, Astron. Astrophys. **423**, L9 (2004)
164. W. Kluźniak, M.A. Abramowicz, Astrophys. Space Sci. **300**, 143 (2005)
165. S. Konar, A.R. Choudhuri, Mon. Not. R. Astron. Soc. **348**, 661 (2004)
166. K. Koyama, H. Inoue, K. Makishima, M. Matsuoka, T. Murakami, M. Oda, Y. Osgawara, T. Ohashi, N. Shibazaki, Y. Tanaka, F.J. Marshall, I.S. Kondo, S. Hayakawa, H. Kunieda, F. Makino, K. Masai, F. Nagase, Y. Tawara, S. Miyamoto, H. Tsunemi, K. Yamashita, Astrophys. J. Lett. **247**, L27 (1981)
167. U. Kraus, S. Blum, J. Schulte, H. Ruder, P. Meszaros, Astrophys. J. **467**, 794 (1996)
168. M.I. Krauss, A. Dullighan, D. Chakrabarty, M.H. van Kerkwijk, C.B. Markwardt, IAU Circ. **8154**, 3 (2003)
169. M.I. Krauss, Z. Wang, A. Dullighan, A.M. Juett, D.L. Kaplan, D. Chakrabarty, M.H. van Kerkwijk, D. Steeghs, P.G. Jonker, C.B. Markwardt, Astrophys. J. **627**, 910 (2005)
170. H.A. Krimm, C.B. Markwardt, C.J. Deloye, P. Romano, D. Chakrabarty, S. Campana, J.R. Cummings, D.K. Galloway, N. Gehrels, J.M. Hartman, P. Kaaret, E.H. Morgan, J. Tueller, Astrophys. J. Lett. **668**, L147 (2007)
171. A.K. Kulkarni, M.M. Romanova, Astrophys. J. **633**, 349 (2005)
172. A.K. Kulkarni, M.M. Romanova, Mon. Not. R. Astron. Soc. **386**, 673 (2008)
173. E. Kuulkers, Nucl. Phys. B Proc. Suppl. **132**, 466 (2004)
174. E. Kuulkers, P.R. den Hartog, J.J.M. in't Zand, F.W.M. Verbunt, W.E. Harris, M. Cocchi, Astron. Astrophys. **399**, 663 (2003)
175. F.K. Lamb, M.C. Miller (2003). arXiv:0308179
176. F.K. Lamb, D. Pines, J. Shaham, Astrophys. J. **225**, 582 (1978)
177. F.K. Lamb, J. Shaham, D. Pines, Astrophys. J. **224**, 969 (1978)
178. F.K. Lamb, S. Boutloukos, S. Van Wassenhove, R.T. Chamberlain, K.H. Lo, A. Clare, W. Yu, M.C. Miller, Astrophys. J. **706**, 417 (2009)
179. F.K. Lamb, S. Boutloukos, S. Van Wassenhove, R.T. Chamberlain, K.H. Lo, M.C. Miller, Astrophys. J. Lett. **705**, L36 (2009)
180. C. Lange, F. Camilo, N. Wex, M. Kramer, D.C. Backer, A.G. Lyne, O. Doroshenko, Mon. Not. R. Astron. Soc. **326**, 274 (2001)

181. J.P. Lasota, New Astron. Rev. **45**, 449 (2001)
182. J.M. Lattimer, M. Prakash, Science **304**, 536 (2004)
183. J.M. Lattimer, M. Prakash, Phys. Rep. **442**, 109 (2007)
184. K. Lazaridis, J.P.W. Verbiest, T.M. Tauris, B.W. Stappers, M. Kramer, N. Wex, A. Jessner, I. Cognard, G. Desvignes, G.H. Janssen, M.B. Purver, G. Theureau, C.G. Bassa, R. Smits, Mon. Not. R. Astron. Soc. **414**, 3134 (2011)
185. D.A. Leahy, S.M. Morsink, Y.Y. Chung, Y. Chou, Astrophys. J. **691**, 1235 (2009)
186. D.A. Leahy, S.M. Morsink, Y. Chou, Astrophys. J. **742**, 17 (2011)
187. A. Levine, C.P. Ma, J. McClintock, S. Rappaport, M. van der Klis, F. Verbunt, Astrophys. J. **327**, 732 (1988)
188. W.H.G. Lewin, J. van Paradijs, R.E. Taam, Space Sci. Rev. **62**, 223 (1993)
189. F. Lewis, D.M. Russell, P.G. Jonker, M. Linares, V. Tudose, P. Roche, J.S. Clark, M.A.P. Torres, D. Maitra, C.G. Bassa, D. Steeghs, A. Patruno, S. Migliari, R. Wijnands, G. Nelemans, L.J. Kewley, V.E. Stroud, M. Modjaz, J.S. Bloom, C.H. Blake, D. Starr, Astron. Astrophys. **517**, A72 (2010)
190. M. Linares, Astron. Tel. **4960**, 1 (2013)
191. M. Linares, M. van der Klis, D. Altamirano, C.B. Markwardt, Astrophys. J. **634**, 1250 (2005)
192. M. Linares, V. Tudose, S. Migliari, Astron. Tel. **1667**, 1 (2008)
193. M. Linares, D. Chakrabarty, M. van der Klis, Astrophys. J. Lett. **733**, L17 (2011)
194. M. Linares, D. Altamirano, A. Watts, T. Strohmayer, D. Chakrabarty, A. Patruno, M. van der Klis, R. Wijnands, P. Casella, M. Armas-Padilla, Y. Cavecchi, N. Degenaar, M. Kalamkar, R. Kaur, Y. Yang, N. Rea, Astron. Tel. **3568**, 1 (2011)
195. M. Linares, E. Bozzo, D. Altamirano, N. Degenaar, R. Wijnands, P. Soleri, T. Belloni, T. Di Salvo, A. D'Ai, A. Papitto, A. Riggio, L. Burderi, Astron. Tel. **3661**, 1 (2011)
196. M. Linares, D. Altamirano, D. Chakrabarty, A. Cumming, L. Keek, Astrophys. J. **748**, 82 (2012)
197. M. Linares, A. Bahramian, C. Heinke, R. Wijnands, A. Patruno, D. Altamirano, J. Homan, S. Bogdanov, D. Pooley, Mon. Not. R. Astron. Soc. **438**, 251 (2014). https://doi.org/10.1093/mnras/stt2167
198. M. Long, M.M. Romanova, R.V.E. Lovelace, Mon. Not. R. Astron. Soc. **386**, 1274 (2008)
199. M. Long, M.M. Romanova, F.K. Lamb, New Astron. **17**, 232 (2012)
200. D.R. Lorimer, M. Kramer, *Handbook of Pulsar Astronomy* (Cambridge University Press, Cambridge, 2004)
201. C.B. Markwardt, T.E. Strohmayer, Astrophys. J. Lett. **717**, L149 (2010)
202. C.B. Markwardt, J.H. Swank, IAU Circ. **8144**, 1 (2003)
203. C.B. Markwardt, J.H. Swank, Astron. Tel. **1045**, 1 (2007)
204. C.B. Markwardt, J.H. Swank, T.E. Strohmayer, J.J.M.I. Zand, F.E. Marshall, Astrophys. J. Lett. **575**, L21 (2002)
205. C.B. Markwardt, E. Smith, J.H. Swank, IAU Circ. **8080**, 2–25 (2003)
206. C.B. Markwardt, J.H. Swank, T.E. Strohmayer, Astron. Tel. **353**, 1 (2004)
207. C.B. Markwardt, D.K. Galloway, D. Chakrabarty, E.H. Morgan, T.E. Strohmayer, Astron. Tel. **360**, 1 (2004)
208. C.B. Markwardt, D. Pereira, J.H. Swank, Astron. Tel. **1051**, 1 (2007)
209. C.B. Markwardt, H.A. Krimm, J.H. Swank, Astron. Tel. **1108**, 1 (2007)
210. C.B. Markwardt, D. Altamirano, T.E. Strohmayer, J.H. Swank, Astron. Tel. **2237**, 1 (2009)
211. C.B. Markwardt, D. Altamirano, J.H. Swank, T.E. Strohmayer, M. Linares, D. Pereira, Astron. Tel. **2197**, 1 (2009)
212. M. Méndez, T. Belloni, Mon. Not. R. Astron. Soc. **381**, 790 (2007)
213. M.C. Miller, F.K. Lamb, D. Psaltis, Astrophys. J. **508**, 791 (1998)
214. J.M. Miller, R. Wijnands, M. Méndez, E. Kendziorra, A. Tiengo, M. van der Klis, D. Chakrabarty, B.M. Gaensler, W.H.G. Lewin, Astrophys. J. Lett. **583**, L99 (2003)
215. J.C.A. Miller-Jones, D.M. Russell, S. Migliari, Astron. Tel. **2232**, 1 (2009)

216. J.C.A. Miller-Jones, G.R. Sivakoff, D. Altamirano, V. Tudose, S. Migliari, V. Dhawan, R.P. Fender, M.A. Garrett, S. Heinz, E.G. Körding, H.A. Krimm, M. Linares, D. Maitra, S. Markoff, Z. Paragi, R.A. Remillard, M.P. Rupen, A. Rushton, D.M. Russell, C.L. Sarazin, R.E. Spencer, Astrophys. J. Lett. **716**, L109 (2010)
217. J.C.A. Miller-Jones, C.O. Heinke, G.R. Sivakoff, D. Pooley, J. Homan, D. Altamirano, Astron. Tel. **2377**, 1 (2010)
218. M. Monelli, G. Fiorentino, L. Burderi, F. D'Antona, N. Robba, V. Testa, in *Interacting Binaries: Accretion, Evolution, and Outcomes*, *American Institute of Physics Conference Series*, vol. 797, ed. by L. Burderi, L.A. Antonelli, F. D'Antona, T. di Salvo, G.L. Israel, L. Piersanti, A. Tornambè, O. Straniero (2005), pp. 565–568
219. E. Morgan, P. Kaaret, R. Vanderspek, Astron. Tel. **523**, 1 (2005)
220. S.M. Morsink, D.A. Leahy, Astrophys. J. **726**, 56 (2011)
221. S. Motta, A. D'Aı, A. Papitto, A. Riggio, T. di Salvo, L. Burderi, T. Belloni, L. Stella, R. Iaria, Mon. Not. R. Astron. Soc. **414**, 1508 (2011)
222. D. Mukherjee, D. Bhattacharya, A. Mignone, Mon. Not. R. Astron. Soc. **435**, 718 (2013)
223. M.P. Muno, D. Chakrabarty, D.K. Galloway, P. Savov, Astrophys. J. Lett. **553**, L157 (2001)
224. M.P. Muno, D. Chakrabarty, D.K. Galloway, D. Psaltis, Astrophys. J. **580**, 1048 (2002)
225. M.P. Muno, F. Özel, D. Chakrabarty, Astrophys. J. **581**, 550 (2002)
226. M.P. Muno, F. Özel, D. Chakrabarty, Astrophys. J. **595**, 1066 (2003)
227. M.P. Muno, D.K. Galloway, D. Chakrabarty, Astrophys. J. **608**, 930 (2004)
228. R. Narayan, J.S. Heyl, Astrophys. J. **599**, 419 (2003)
229. G. Nelemans, P.G. Jonker, D. Steeghs, Mon. Not. R. Astron. Soc. **370**, 255 (2006)
230. L.A. Nelson, S. Rappaport, Astrophys. J. **598**, 431 (2003)
231. I.D. Novikov, K.S. Thorne, in *Black Holes (Les Astres Occlus)*, ed. by C. Dewitt, B.S. Dewitt (1973), pp. 343–450
232. M.A. Nowak, A. Paizis, J. Wilms, J. Rodriguez, S. Chaty, K. Ebisawa, M. Del Santo, R. Farinelli, P. Ubertini, T. Courvoisier, Astron. Tel. **2215**, 1 (2009)
233. S. Ortolani, B. Barbuy, E. Bica, Astron. Astrophys. **108**, 653 (1994)
234. F. Özel, Astrophys. J. **691**, 1678 (2009)
235. B. Paczyński, Acta. Astronom. **17**, 287 (1967)
236. A. Paizis, M.A. Nowak, J. Wilms, T. J-L. Courvoisier, K. Ebisawa, J. Rodriguez, P. Ubertini, Astron. Astrophys. **444**, 357 (2005)
237. C. Pallanca, E. Dalessandro, F.R. Ferraro, B. Lanzoni, G. Beccari, Astrophys. J. **773**, 122 (2013)
238. A. Papitto, M.T. Menna, L. Burderi, T. Di Salvo, F. D'Antona, N.R. Robba, Astrophys. J. Lett. **621**, L113 (2005)
239. A. Papitto, T. di Salvo, L. Burderi, M.T. Menna, G. Lavagetto, A. Riggio, Mon. Not. R. Astron. Soc. **375**, 971 (2007)
240. A. Papitto, T. di Salvo, L. Burderi, E. Gallo, M.T. Menna, A. Riggio, M.I. Krauss, D. Chakrabarty, A. Antonelli, E. Bozzo, M. Burgay, N. D'Amico, R.I.G.L. Israel, A. Possenti, V. Testa, Astron. Tel. **1133**, 1 (2007)
241. A. Papitto, M.T. Menna, L. Burderi, T. di Salvo, A. Riggio, Mon. Not. R. Astron. Soc. **383**, 411 (2008)
242. A. Papitto, T. di Salvo, A. D'Aı, R. Iaria, L. Burderi, A. Riggio, M.T. Menna, N.R. Robba, Astron. Astrophys. **493**, L39 (2009)
243. A. Papitto, A. Riggio, T. di Salvo, L. Burderi, A. D'Aì, R. Iaria, E. Bozzo, M.T. Menna, Mon. Not. R. Astron. Soc. **407**, 2575 (2010)
244. A. Papitto, A. D'Aì, S. Motta, A. Riggio, L. Burderi, T. di Salvo, T. Belloni, R. Iaria, Astron. Astrophys. **526**, L3 (2011)
245. A. Papitto, A. Riggio, L. Burderi, T. di Salvo, A. D'Aí, R. Iaria, Astron. Astrophys. **528**, A55 (2011)
246. A. Papitto, E. Bozzo, C. Ferrigno, L. Pavan, P. Romano, S. Campana, Astron. Tel. **4959**, 1 (2013)

247. A. Papitto, C. Ferrigno, E. Bozzo, N. Rea, L. Pavan, L. Burderi, M. Burgay, S. Campana, T. di Salvo, M. Falanga, M.D. Filipović, P.C.C. Freire, J.W.T. Hessels, A. Possenti, S.M. Ransom, A. Riggio, P. Romano, J.M. Sarkissian, I.H. Stairs, L. Stella, D.F. Torres, M.H. Wieringa, G.F. Wong, Nature **501**, 517 (2013)
248. A. Papitto, J.W.T. Hessels, M. Burgay, S. Ransom, N. Rea, A. Possenti, I. Stairs, C. Ferrigno, E. Bozz, Astron. Tel. **5069**, 1 (2013)
249. A. Patruno, Astrophys. J. **722**, 909 (2010)
250. A. Patruno, Astrophys. J. Lett. **753**, L12 (2012)
251. A. Patruno, Astron. Tel. **5068**, 1 (2013)
252. A. Patruno, C. D'Angelo, Astrophys. J. **771**, 94 (2013)
253. A. Patruno, D. Altamirano, J.W.T. Hessels, P. Casella, R. Wijnands, M. van der Klis, Astrophys. J. **690**, 1856 (2009)
254. A. Patruno, J.M. Hartman, R. Wijnands, D. Chakrabarty, M. van der Klis, Astrophys. J. **690**, 1856 (2009)
255. A. Patruno, A. Watts, M. Klein Wolt, R. Wijnands, M. van der Klis, Astrophys. J. **707**, 1296 (2009)
256. A. Patruno, C.B. Markwardt, T.E. Strohmayer, J.H. Swank, S.E. Smith, D. Pereira, Astron. Tel. **2130**, 1 (2009)
257. A. Patruno, N. Rea, D. Altamirano, M. Linares, R. Wijnands, M. van der Klis, Mon. Not. R. Astron. Soc. **396**, L51 (2009)
258. A. Patruno, R. Wijnands, M. van der Klis, Astrophys. J. Lett. **698**, L60 (2009)
259. A. Patruno, J.M. Hartman, R. Wijnands, D. Chakrabarty, M. van der Klis, Astrophys. J. **717**, 1253 (2010)
260. A. Patruno, D. Altamirano, C. Messenger, Mon. Not. R. Astron. Soc. **403**, 1426 (2010)
261. A. Patruno, D. Altamirano, A. Watts, M. Armas Padilla, Y. Cavecchi, N. Degenaar, M. Kalamkar, R. Kaur, P. Soleri, Y.J. Yang, M. van der Klis, R. Wijnands, P. Casella, M. Linares, N. Rea, C.O. Heinke, D. Pooley, Astron. Tel. **2407**, 1 (2010)
262. A. Patruno, P. Bult, A. Gopakumar, J.M. Hartman, R. Wijnands, M. van der Klis, D. Chakrabarty, Astrophys. J. Lett. **746**, L27 (2012)
263. A. Patruno, B. Haskell, C. D'Angelo, Astrophys. J. **746**, 9 (2012)
264. A. Patruno, M.A. Alpar, M. van der Klis, E.P.J. van den Heuvel, Astrophys. J. **752**, 33 (2012)
265. A. Patruno, A.M. Archibald, J.W.T. Hessels, S. Bogdanov, B.W. Stappers, C.G. Bassa, G.H. Janssen, V.M. Kaspi, S. Tendulkar, A.G. Lyne, Astrophys. J. Lett. **781**, L3 (2014). https://doi.org/10.1088/2041-8205/781/1/L3
266. L. Pavan, J. Chenevez, E. Bozzo, E. Kuulkers, J. Alfonso-Garzon, V. Beckmann, T. Bird, S. Brandt, T. Courvoisier, A. Domingo, K. Ebisawa, P. Jonker, P. Kretschmar, C. Markwardt, T. Oosterbroek, A. Paizis, C. Sanchez-Fernandez, R. Wijnands, Astron. Tel. **2548**, 1 (2010)
267. L. Pavan, G. Wong, M.H. Wieringa, N.F.H. Tothill, M.D. Filipovic, E. Bozzo, C. Ferrigno, A. Papitto, P. Romano, Astron. Tel. **4981**, 1 (2013)
268. D.J.B. Payne, A. Melatos, Mon. Not. R. Astron. Soc. **351**, 569 (2004)
269. D.J.B. Payne, A. Melatos, Mon. Not. R. Astron. Soc. **376**, 609 (2007)
270. A.L. Piro, L. Bildsten, Astrophys. J. **629**, 438 (2005)
271. G. Pooley, Astron. Tel. **355**, 1 (2004)
272. D. Pooley, W.H.G. Lewin, F. Verbunt, L. Homer, B. Margon, B.M. Gaensler, V.M. Kaspi, J.M. Miller, D.W. Fox, M. van der Klis, Astrophys. J. **573**, 184 (2002)
273. A. Possenti, M. Murgia, J. Reynolds, F. Camilo, L. Burderi, T. di Salvo, A. Antonelli, E. Bozzo, M. Burgay, N. D'Amico, E. Gallo, R.I.G.L. Israel, M.T. Menna, A. Papitto, A. Riggio, V. Testa, Astron. Tel. **1128**, 1 (2007)
274. K. Postnov, N. Shakura, A. González-Galán, E. Kuulkers, P. Kretschman, S. Larsson, M.H. Finger, A. Kochetkova, G. Lü, L. Yungelson, in *Proceedings of the 8th INTEGRAL Workshop "The Restless Gamma-ray Universe", September 27–30, Dublin* (2010)
275. J. Poutanen, Adv. Space Res. **38**, 2697 (2006)

276. J. Poutanen, in *A Decade of Accreting Millisecond X-ray Pulsars, AIPC Series*, vol. 1068, ed. by R. Wijnands, D. Altamirano, P. Soleri, N. Degenaar, N. Rea, P. Casella, A. Patruno, M. Linares (2008), pp. 77–86
277. J. Poutanen, A.M. Beloborodov, Mon. Not. R. Astron. Soc. **373**, 836 (2006)
278. J. Poutanen, M. Gierliński, Mon. Not. R. Astron. Soc. **343**, 1301 (2003)
279. J. Poutanen, A. Ibragimov, M. Annala, Astrophys. J. Lett. **706**, L129 (2009)
280. D. Psaltis, D. Chakrabarty, Astrophys. J. **521**, 332 (1999)
281. V. Radhakrishnan, G. Srinivasan, Curr. Sci. **51**, 1096 (1982)
282. S. Rappaport, P.C. Joss, R.F. Webbink, Astrophys. J. **254**, 616 (1982)
283. S.A. Rappaport, J.M. Fregeau, H. Spruit, Astrophys. J. **606**, 436 (2004)
284. R. Remillard, Astron. Tel. **357**, 1 (2004)
285. R.A. Remillard, J. Swank, T. Strohmayer, IAU Circ. **7893**, 1 (2002)
286. M. Revnivtsev, E. Churazov, M. Gilfanov, R. Sunyaev, Astron. Astrophys. **372**, 138 (2001)
287. A. Riggio, T. Di Salvo, L. Burderi, M.T. Menna, A. Papitto, R. Iaria, G. Lavagetto, Astrophys. J. **678**, 1273 (2008)
288. A. Riggio, L. Burderi, T. di Salvo, A. Papitto, A. D'Aì, R. Iaria, M.T. Menna, Astron. Astrophys. **531**, A140 (2011)
289. A. Riggio, A. Papitto, L. Burderi, T. di Salvo, M. Bachetti, R. Iaria, A. D'Aì, M.T. Menna, Astron. Astrophys. **526**, A95 (2011)
290. A. Riggio, L. Burderi, T. Di Salvo, A. D'Ai', A. Papitto, R. Iaria, N. Robba, A. Sanna, E. Bozzo, C. Ferrigno, N. Rea, L. Pavan, Astron. Tel. **5086**, 1 (2013)
291. P. Roche, D. Chakrabarty, L. Morales-Rueda, R. Hynes, S.M. Slivan, C. Simpson, P. Hewett, IAU Circ. **6885** (1998)
292. G. Roelofs, P.G. Jonker, D. Steeghs, M. Torres, G. Nelemans, Astron. Tel. **356**, 1 (2004)
293. M.M. Romanova, G.V. Ustyugova, A.V. Koldoba, R.V.E. Lovelace, Astrophys. J. **610**, 920 (2004)
294. M.M. Romanova, G.V. Ustyugova, A.V. Koldoba, R.V.E. Lovelace, Astrophys. J. Lett. **635**, L165 (2005)
295. M.M. Romanova, A.K. Kulkarni, R.V.E. Lovelace, Astrophys. J. Lett. **673**, L171 (2008)
296. R.W. Romani, Nature **347**, 741 (1990)
297. J. Roy, B. Bhattacharyya, P.S. Ray, Astron. Tel. **5890**, 1 (2014)
298. M. Ruderman, Astrophys. J. **366**, 261 (1991)
299. M.P. Rupen, V. Dhawan, A.J. Mioduszewski, IAU Circ. **7893**, 2 (2002)
300. M.P. Rupen, A.J. Mioduszewski, V. Dhawan, Astron. Tel. **286**, 1 (2004) (2002)
301. M.P. Rupen, V. Dhawan, A.J. Mioduszewski, Astron. Tel. **524** (2005)
302. M.P. Rupen, A.J. Mioduszewski, V. Dhawan, Astron. Tel. **530**, 1 (2005)
303. M.P. Rupen, A.J. Mioduszewski, V. Dhawan, Astron. Tel. **491**, 1 (2005)
304. D.M. Russell, F. Lewis, D. Altamirano, P. Roche, Astron. Tel. **3622**, 1 (2011)
305. P. Schady, A.P. Beardmore, F.E. Marshall, D.M. Palmer, E. Rol, G. Sato, GRB Coord. Netw. **5200**, 1 (2006)
306. M. Serino, T. Takagi, H. Negoro, S. Ueno, H. Tomida, S. Nakahira, M. Kimura, M. Ishikawa, T. Mihara, M. Sugizaki, M. Morii, T. Yamamoto, J. Sugimoto, M. Matsuoka, N. Kawai, R. Usui, K. Ishikawa, T. Yoshii, A. Yoshida, T. Sakamoto, Y. Nakano, H. Tsunemi, M. Sasaki, M. Nakajima, K. Fukushima, T. Onodera, K. Suzuki, Y. Ueda, M. Shidatsu, T. Kawamuro, Y. Tsuboi, M. Higa, M. Yamauchi, K. Yoshidome, Y. Ogawa, H. Yamada, K. Yamaoka, Astron. Tel. **4961**, 1 (2013)
307. N.I. Shakura, R.A. Syunyaev, Astron. Astrophys. **24**, 337 (1973)
308. M.M. Shara, Astrophys. J. **261**, 649 (1982)
309. A. Spitkovsky, Astrophys. J. Lett. **648**, L51 (2006)
310. A. Spitkovsky, Y. Levin, G. Ushomirsky, Astrophys. J. **566**, 1018 (2002)
311. H.C. Spruit, R.E. Taam, Astrophys. J. **402**, 593 (1993)
312. G. Srinivasan, Nucleic Acids Res. **54**, 93 (2010)
313. B.W. Stappers, A. Archibald, C. Bassa, J. Hessels, G. Janssen, V. Kaspi, A. Lyne, A. Patruno, A.B. Hill, Astron. Tel. **5513**, 1 (2013)

314. D. Steeghs, IAU Circ. **8155**, 2 (2003)
315. D. Steeghs, M.A.P. Torres, C. Blake, J.S. Bloom, Astron. Tel. **533**, 1 (2005)
316. D. Steeghs, M.A.P. Torres, M.R. Garcia, J.E. McClintock, J.M. Miller, P.G. Jonker, P.J. Callanan, P. Zhao, P. Berlind, R. Hutchins, C. Watson, Astron. Tel. **543**, 1 (2005)
317. L. Stella, M. Vietri, Astrophys. J. Lett. **492**, L59+ (1998)
318. L. Stella, M. Vietri, Phys. Rev. Lett. **82**, 17 (1999)
319. T. Strohmayer, L. Bildsten, New views of thermonuclear bursts, in *Compact Stellar X-ray Sources* (Cambridge University Press, Cambridge, 2006), pp. 113–156
320. T.E. Strohmayer, C.B. Markwardt, Astron. Tel. **2569**, 1 (2010)
321. T.E. Strohmayer, C.B. Markwardt, Astron. Tel. **2929**, 1 (2010)
322. T.E. Strohmayer, W. Zhang, J.H. Swank, A. Smale, L. Titarchuk, C. Day, U. Lee, Astrophys. J. Lett. **469**, L9+ (1996)
323. T.E. Strohmayer, W. Zhang, J.H. Swank, I. Lapidus, Astrophys. J. Lett. **503**, L147 (1998)
324. T.E. Strohmayer, C.B. Markwardt, J.H. Swank, J. in't Zand, Astrophys. J. Lett. **596**, L67 (2003)
325. M. Suzuki, N. Kawai, T. Tamagawa, A. Yoshida, Y.E. Nakagawa, K. Tanaka, Y. Shirasaki, M. Matsuoka, G.R. Ricker, R. Vanderspek, N. Butler, D.Q. Lamb, C. Graziani, G. Pizzichini, R. Sato, M. Arimoto, J. Kotoku, M. Maetou, M. Yamauchi, Publ. Astron. Soc. Jpn. **59**, 263 (2007)
326. J.H. Swank, C.B. Markwardt, E.A. Smith, Astron. Tel. **449**, 1 (2005)
327. T.M. Tauris, Science **335**, 561 (2012)
328. T.M. Tauris, E.P.J. van den Heuvel, Formation and evolution of compact stellar X-ray sources, in *Compact Stellar X-ray Sources* (Cambridge University Press, Cambridge, 2006), pp. 623–665
329. T. Tauris, N. Langer, M. Kramer, Mon. Not. R. Astron. Soc. **425**, 1601 (2012)
330. J. Thorstensen, P. Charles, S. Bowyer, Astrophys. J. Lett. **220**, L131 (1978)
331. L. Titarchuk, W. Cui, K. Wood, Astrophys. J. Lett. **576**, L49 (2002)
332. L. Titarchuk, S. Kuznetsov, N. Shaposhnikov, Astrophys. J. **667**, 404 (2007)
333. M.A.P. Torres, P. Rodriguez-Gil, D. Steeghs, J.M. Corral-Santana, J. Casares, P.G. Jonker, Astron. Tel. **1090**, 1 (2007)
334. M.A.P. Torres, P.G. Jonker, D. Steeghs, G.H.A. Roelofs, J.S. Bloom, J. Casares, E.E. Falco, M.R. Garcia, T.R. Marsh, M. Mendez, J.M. Miller, G. Nelemans, P. Rodríguez-Gil, Astrophys. J. **672**, 1079 (2008)
335. M.A.P. Torres, P.G. Jonker, D. Steeghs, I. Damjanov, E. Caris, K. Glazebrook, Astron. Tel. **2233**, 1 (2009)
336. M.A.P. Torres, O. Madej, P.G. Jonker, D. Steeghs, S. Greiss, N. Morrell, M. Roth, Astron. Tel. **3638**, 1 (2011)
337. V. Tudose, Z. Paragi, J. Yang, J.C.A. Miller-Jones, R. Fender, M. Garrett, A. Rushton, R. Spencer, Astron. Tel. **5158**, 1 (2013)
338. G.V. Ustyugova, A.V. Koldoba, M.M. Romanova, R.V.E. Lovelace, Astrophys. J. **646**, 304 (2006)
339. M. van den Berg, J. Grindlay, P. Zhao, J. Hong, M. Servillat, Astron. Tel. **3634**, 1 (2011)
340. M. van der Klis, Astrophys. J. **561**, 943 (2001)
341. M. van der Klis, D. Chakrabarty, J.C. Lee, E.H. Morgan, R. Wijnands, C.B. Markwardt, J.H. Swank, IAU Circ. **7358**, 3 (2000)
342. R. Vanderspek, E. Morgan, G. Crew, C. Graziani, M. Suzuki, Astron. Tel. **516**, 1 (2005)
343. F. Verbunt, M.H. van Kerkwijk, J.J.M. in't Zand, J. Heise, Astron. Astrophys. **359**, 960 (2000)
344. K. Viironen, J. Poutanen, Astron. Astrophys. **426**, 985 (2004)
345. Y.M. Wang, Astron. Astrophys. **183**, 257 (1987)
346. Z. Wang, D. Chakrabarty, P. Roche, P.A. Charles, E. Kuulkers, T. Shahbaz, C. Simpson, D.A. Forbes, S.F. Helsdon, Astrophys. J. Lett. **563**, L61 (2001)
347. Z. Wang, C. Bassa, A. Cumming, V.M. Kaspi, Astrophys. J. **694**, 1115 (2009)
348. Z. Wang, R.P. Breton, C.O. Heinke, C.J. Deloye, J. Zhong, Astrophys. J. **765**, 151 (2013)
349. A.L. Watts, Annu. Rev. Astron. Astrophys. **50**, 609 (2012)

350. A.L. Watts, T.E. Strohmayer, Mon. Not. R. Astron. Soc. **373**, 769 (2006)
351. A.L. Watts, T.E. Strohmayer, C.B. Markwardt, Astrophys. J. **634**, 547 (2005)
352. A.L. Watts, A. Patruno, M. van der Klis, Astrophys. J. Lett. **688**, L37 (2008)
353. A.L. Watts, D. Altamirano, C.B. Markwardt, T.E.S., Astron. Tel. **2199**, 1 (2009)
354. A.L. Watts, D. Altamirano, M. Linares, A. Patruno, P. Casella, Y. Cavecchi, N. Degenaar, N. Rea, P. Soleri, M. van der Klis, R. Wijnands, Astrophys. J. Lett. **698**, L174 (2009)
355. W.F. Welsh, E.L. Robinson, P. Young, Astron. J. **120**, 943 (2000)
356. R.A.M.J. Wijers, Mon. Not. R. Astron. Soc. **287**, 607 (1997)
357. R.A.M.J. Wijers, in *Pulsar Timing, General Relativity and the Internal Structure of Neutron Stars*, ed. by Z. Arzoumanian, F. Van der Hooft, E.P.J. van den Heuvel (Koninklijke Nederlandse Akademie van Wetenschappen, Amsterdam, 1999), p. 293
358. R. Wijnands, Astrophys. J. **588**, 425 (2003)
359. R. Wijnands, A. Reynolds, Astron. Tel. **166**, 1 (2003)
360. R. Wijnands, M. van der Klis, Nature **394**, 344 (1998)
361. R. Wijnands, M. Méndez, C. Markwardt, M. van der Klis, D. Chakrabarty, E. Morgan, Astrophys. J. **560**, 892 (2001)
362. R. Wijnands, M. van der Klis, J. Homan, D. Chakrabarty, C.B. Markwardt, E.H. Morgan, Nature **424**, 44 (2003)
363. R. Wijnands, E. Rol, E. Cackett, R.L.C. Starling, R.A. Remillard, Mon. Not. R. Astron. Soc. **393**, 126 (2009)
364. K.S. Wood, C. Ftaclas, M. Kearney, Astrophys. J. Lett. **324**, L63 (1988)
365. R.Z. Yahel, Astron. Astrophys. **90**, 26 (1980)
366. H.X. Yin, C.M. Zhang, Y.H. Zhao, Y.J. Lei, J.L. Qu, L.M. Song, F. Zhang, Astron. Astrophys. **471**, 381 (2007)
367. W. Zhang, E.H. Morgan, K. Jahoda, J.H. Swank, T.E. Strohmayer, G. Jernigan, R.I. Klein, Astrophys. J. Lett. **469**, L29 (1996)
368. W. Zhang, K. Jahoda, R.L. Kelley, T.E. Strohmayer, J.H. Swank, S.N. Zhang, Astrophys. J. Lett. **495**, L9+ (1998)

Chapter 5
Thermonuclear X-ray Bursts

Duncan K. Galloway and Laurens Keek

Contents

D. K. Galloway (✉)
School of Physics & Astronomy, Monash University, Clayton, VIC, Australia
e-mail: duncan.galloway@monash.edu

L. Keek
CRESST and X-ray Astrophysics Laboratory NASA/GSFC, Greenbelt, MD, USA

Department of Astronomy, University of Maryland, College Park, MD, USA
e-mail: laurens@xrb.space

© Springer-Verlag GmbH Germany, part of Springer Nature 2021

T. M. Belloni et al. (eds.), *Timing Neutron Stars: Pulsations, Oscillations and Explosions*, Astrophysics and Space Science Library 461,
https://doi.org/10.1007/978-3-662-62110-3_5

Abstract Type-I X-ray bursts arise from unstable thermonuclear burning of accreted fuel on the surface of neutron stars. In this chapter we review the fundamental physics of the burning processes, and summarise the observational, numerical, and nuclear experimental progress over the preceding decade. We describe the current understanding of the conditions that lead to burst ignition, and the influence of the burst fuel on the observational characteristics. We provide an overview of the processes which shape the burst X-ray spectrum, including the observationally elusive discrete spectral features. We report on the studies of timing behaviour related to nuclear burning, including burst oscillations and mHz quasi-periodic oscillations. We describe the increasing role of nuclear experimental physics in the interpretation of astrophysical data and models. We survey the simulation projects that have taken place to date, and chart the increasing dialogue between modellers, observers, and nuclear experimentalists. Finally, we identify some open problems with prospects of a resolution within the timescale of the next such review.

5.1 Overview

Thermonuclear (type-I) bursts are triggered by unstable ignition of accreted fuel on the surface of neutron stars (e.g. [68]). The first such events were detected in 1975 from the ultracompact binary 3A 1820−30 (in the globular cluster NGC 6624) with the *Small Astronomy Satellite 3* (*SAS-3*) and the *Astronomical Netherlands Satellite* (*ANS*; [26, 59]). Since then, a growing population of bursters (currently numbered at 115[1]) has been observed by almost every major X-ray satellite, sometimes for a substantial fraction of their mission duration. These events remain a high observational priority, as a key diagnostic of the nature of newly-discovered transient X-ray binaries. In addition, thermonuclear bursts have been increasingly exploited to constrain the fundamental properties of the host neutron stars, and to probe the accretion environment of the binary system.

In this chapter we focus on the research areas that have been of highest priority to both observers and theorists in the previous decade. These areas include the fundamentals of ignition of different types of bursts; the study of "long" bursts, both intermediate-duration (He) bursts and superbursts (the latter thought to burn carbon, rather than the H/He fuel for short bursts); the study of burst oscillations; the physics of the formation of the burst spectra, and the focus on deriving neutron-star parameters from them; the searches for, and theoretical predictions of, discrete

[1] http://burst.sci.monash.edu/sources.

features in the burst spectra; the phenomenon of mHz QPOs, and the link with steady burning; the role of nuclear experimental physics, and the growing dialogue between the different communities; progress in simulating bursts; and the growing role for bursts in probing the environment around the neutron star.

This chapter serves as an update to a series of thorough previous reviews, beginning in 1993 with [115] and the subsequent work developed from it [116]; the 1998 review [14], covering the current state of the theory and the unexpected results from the *European X-ray Observatory Satellite* (EXOSAT); the 2006 review (published initially in 2003), with its focus on new phenomena discovered in the preceding decade, including burst oscillations [171]; and the 2012 review of burst oscillations and related phenomena [193].

5.1.1 Theory of Burst Ignition and Nuclear Burning Regimes

Soon after their discovery, measurements of the ratio of burst to accretion energy (the so-called α-parameter) established that X-ray bursts are produced by thermonuclear burning of accreted material [127, 199]. Here we introduce the mechanism of runaway burning and present the classical picture of the different burning regimes.

5.1.1.1 Fuel Accretion from a Binary Companion Star

The fuel for X-ray bursts originates from the outer layers of a binary companion star. Typically, it is expected to have a composition similar to the Sun, predominantly hydrogen and helium, as well as small amounts of CNO and other metals. An exception are the ultra-compact X-ray binaries (UCXBs; e.g., [79]) where most of the accreted material is helium, with at most a small mass fraction of hydrogen ($\lesssim$10%; [31]). The material is transferred by Roche-lobe overflow to the neutron star via an accretion disk at rates up to the Eddington limit, of $\dot{M}_{\rm Edd} = 3.0 \times 10^{-8}\,{\rm M_{\odot}year^{-1}}\,(1+X)^{-1}(R/10\,{\rm km})$, with X the hydrogen mass fraction and R the neutron star radius (ignoring general relativistic corrections). $\dot{M}$ can vary by orders of magnitude on time scales of days to decades. In transient systems, accretion occurs episodically, and is effectively switched off ("quiescence") for long periods of time in between. We will see that the properties of X-ray bursts depend on $\dot{M}$, and therefore a wide range of bursting behaviour may be observed from a single accreting neutron star, as its $\dot{M}$ changes with time.

Material falling on the neutron star is expected to rapidly spread across its surface, due to the lateral pressure gradients imposed by the strong ($\approx 10^{14}$ cm s^{-2}) gravity. The presence of accretion-powered pulsations in some 16 low-mass binaries (of which 8 are also burst sources; [148, 173]) implies that the magnetic field strength is sometimes sufficient to confine the accreted fuel. For most bursting sources the magnetic field is thought to be dynamically unimportant.

5.1.1.2 Runaway Thermonuclear Burning in a Thin Shell

Accretion grows the fuel layer on top of the neutron star until the depth for thermonuclear ignition is reached. It is convenient to express the depth of the layer as the column depth at radius r: $y(r) = \int_r^\infty \rho(r')\mathrm{d}r'$, with ρ the mass density. The column depth y measures the mass above a unit area at radius r, making y a mass coordinate measured from the outside in. Hydrostatic equilibrium in the layer can be expressed as $\mathrm{d}P = -g\rho\mathrm{d}r$, with P the pressure and g the gravitational acceleration. Because the envelope is relatively thin, g can be approximated as being constant in the outer layers. Hydrostatic equilibrium is then simplified to $P = -gy$, which shows that y is also a pressure coordinate.

Compression increases the density and temperature in the fuel layer with depth. A few meters below the stellar surface (from $y \simeq 10^8\,\mathrm{g\,cm^{-2}}$ for hydrogen/helium), the conditions for thermonuclear fusion can be achieved, and the neutron star envelope is heated by nuclear burning. The thin-shell instability (aided by mild electron degeneracy) prevents the neutron star envelope from cooling by expansion, and allows the fuel layer to continue to heat up (see Sect. 5.2 for further discussion).

For several of the important nuclear processes, the thermonuclear burning rate increases sharply with temperature (Fig. 5.1). Therefore, any heating from nuclear burning further increases the burning rate. This leads to a *thermonuclear runaway*, where the burning rate increases quickly to the point where most fuel is locally consumed within a second.

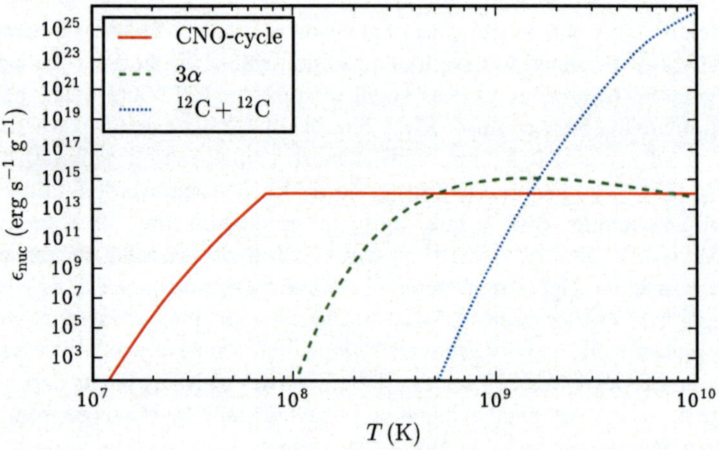

Fig. 5.1 The specific energy generation rate, ϵ_{nuc}, as a function of temperature, T for the three nuclear processes important for burst ignition. We calculate ϵ_{nuc} at a density of $\rho = 10^5\,\mathrm{g\,cm^{-3}}$ assuming a solar composition ($X = 0.73$, $Y = 0.25$, $Z = 0.02$) for the CNO cycle and 3α, and 0.2 of ^{12}C for carbon fusion. For CNO and 3α, ϵ_{nuc} increases sharply with T at lower temperatures, which leads to a thermonuclear runaway. At higher temperatures, the trend of ϵ_{nuc} as a function of T flattens, and burning is stable

5.1.1.3 Ignition Conditions and the Stability of Burning

The specific energy generation rate ϵ_{nuc} for the three nuclear processes important for burst ignition rises steeply at lower T and flattens at higher T (Fig. 5.1). For helium burning through the 3α process, ϵ_{nuc} depends weakly on T above $T \gtrsim 3 \times 10^8$ K, and stable helium burning occurs instead of a thermonuclear runaway. Similarly, hydrogen burning through the CNO cycle becomes stable at $T \gtrsim 0.7 \times 10^8$ K. These temperature thresholds are set by the depth of the Coulomb barrier which nuclei must overcome through quantum tunnelling. Because the H burning processes involve the conversion of protons into neutrons, they include weak decays, which delay the process overall. The timescale for burning is set by the β-decay of ^{14}O and ^{15}O, with half-lives of 70.620 s and 122.24 s, respectively. At high temperatures H burning is known as the "hot CNO cycle" or the "β-limited CNO cycle", where burning cannot run away and is, therefore, stable [192].

Nuclear burning is only efficient if it is stronger than radiative cooling, which is described by the specific cooling rate $\epsilon_{\mathrm{cool}} = -\frac{acT^4}{3\kappa y}$, with radiation constant a, speed of light c, and opacity κ (e.g., [14, 48]). Stable burning in a steady-state takes place at a depth where $\epsilon_{\mathrm{nuc}} = \epsilon_{\mathrm{cool}}$. For thermonuclear runaways the ignition condition is $\frac{\mathrm{d}\epsilon_{\mathrm{nuc}}}{\mathrm{d}T} > \frac{\mathrm{d}\epsilon_{\mathrm{cool}}}{\mathrm{d}T}$, which means that cooling cannot moderate small perturbations in the nuclear burning rate. Both ϵ_{nuc} and ϵ_{cool} depend on T and y, such that we can use these two conditions to map out the ignition conditions for stable and unstable burning (Fig. 5.2; e.g., [14, 48, 50, 95, 139]). For H and He, burning is unstable at lower T, and is stable at higher values. Carbon burning is unstable over the entire considered range of parameters.[2]

5.1.1.4 Burning Regimes as a Function of Accretion Rate

Which ignition conditions are reached on a neutron star, depends for a large part on $\dot{M}$. The amount of heating due to compression is proportional to $\dot{M}$. Therefore, T is higher for larger $\dot{M}$. Furthermore, steady-state burning requires that fuel is burned at the same rate as at which it is accreted. For stable burning, the temperature profile of the neutron star envelope adjusts to facilitate this equilibrium.

The simple ignition conditions presented in Fig. 5.2 are determined by considering hydrogen, helium, and carbon burning separately. Most bursters accrete a mixture of hydrogen and helium, and the (typically steady) burning of hydrogen influences the ignition of helium, via the heat that is contributed to the fuel layer. Steady H burning also affects the composition of the burst fuel at ignition, and hence the overall specific energy released by the burst, Q_{nuc}. Where H is present in the burst fuel, the burning can proceed to heavier species via (α, p) reactions and proton captures (the rapid-proton, or *rp*-process; e.g. [159]). Q_{nuc} can be inferred from measurements of the so-called α parameter, the ratio of the burst fluence to the peak flux (e.g. [53]).

[2] Multi-zone models find carbon burning to be stable in hot envelopes, depending on $\dot{M}$ [94].

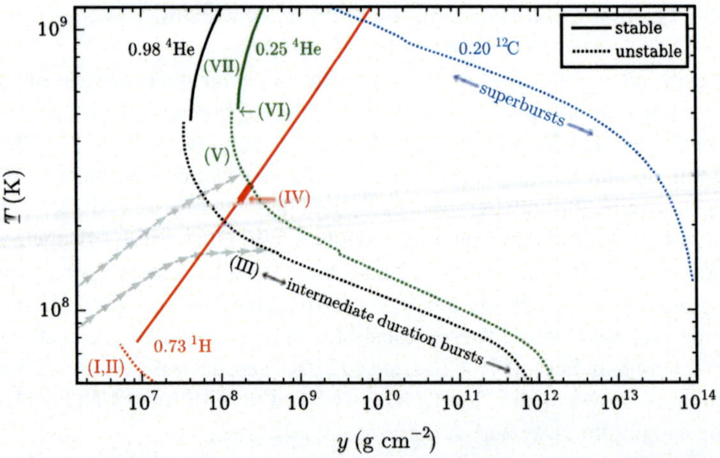

Fig. 5.2 Burning conditions as a function of column depth, y, and temperature, T, calculated with the one-zone model presented in [95]. Lines indicate the locations of stable burning and burst ignition (unstable burning) for three salient compositions: first, mixed hydrogen and helium at solar composition; second, a helium mass fraction of 0.98 (for the case where all the hydrogen has burned prior to ignition); and third, typical "superburst" fuel containing 20% carbon. Due to continual accretion, matter is compressed to higher y and T increases; lines of arrows (predicted by multi-zone models) exemplify two tracks that result in a mixed H/He burst (top) and a pure He burst (bottom). Labels indicate the burning regimes, with the roman numerals matching those in Table 5.1. The new stable H/He burning regime at sub-Eddington accretion rates (IV) is indicated by the thick red line

Table 5.1 The $\dot{m}/\dot{m}_{\rm End}$ values in column 1 represent the expected lower bound of the accretion rate range for the corresponding burst regime in column 2 [For solar accretion composition and base flux $Q_b = 0.1\,{\rm MeV\,u^{-1}}$ (see Sect. 5.1.1.5; [95])]

$\dot{m}/\dot{m}_{\rm Edd}$		Burning regime
	(I)	Deep H flash (burns He)
∼0.1%[a]	(II)	Shallow H flashes and deep He flash
0.4%	(III)	He flash (stable H burning)
8 %	(IV)	Stable H/He burning
11 %	(V)	Mixed H/He flash
∼100 %[b]	(VI)	Marginally stable burning of H/He
	(VII)	Stable H/He burning

[a] Peng et al. [149], including sedimentation
[b] Heger et al. [72]. See also [98, 103, 206]

The variation of ignition and burning conditions for H and He leads to the prediction of a range of burning regimes as a function of $\dot{M}$ (Table 5.1; [14, 48, 95]), some of which have been observed (Sect. 5.1.2):

I At $T \lesssim 7\times10^8$ K, hydrogen burns unstably. The hydrogen-ignited flash quickly raises T. If the ignition depth, $y_{\rm ign}$, is sufficiently large the ignition curve for unstable helium burning is crossed, and helium burns along with hydrogen in the burst.

II If $y_{\rm ign}$ is too shallow for runaway helium burning, the hydrogen flash does not ignite helium. Instead, helium continues to pile up until it reaches its ignition conditions at much larger y. This regime, therefore, exhibits brief hydrogen flashes and long helium bursts. (No observations matching case I or case II bursting have been identified.)

III At $T \gtrsim 7 \times 10^8$ K, βCNO cycle burning of hydrogen is stable. It heats the envelope and converts hydrogen into helium. At $\dot{M} \lesssim 0.08\, \dot{M}_{\rm Edd}$, the buildup of the fuel layer is sufficiently slow for all hydrogen to burn stably to helium. As most of the heating in the envelope comes from the hydrogen burning, T hardly increases with depth once hydrogen is exhausted (see the lower series of arrows in Fig. 5.2). Once the helium ignition curve is reached, a pure helium burst ignites.

IV In a narrow range around $0.1\, \dot{M}_{\rm Edd}$, heating by hydrogen burning creates the conditions for helium to burn stably before reaching the helium ignition curve [95]. This regime of stable hydrogen and helium burning produces pure carbon ashes, which may in turn ignite upon reaching the carbon ignition curve and power superbursts (see Sect. 5.7.2).

V At higher $\dot{M}$, there is insufficient time for βCNO burning to deplete hydrogen. Upon reaching helium ignition, hydrogen is still present in the fuel layer, and it burns along with helium in the burst. In this case burst ignition is more complicated because of the interplay between hydrogen and helium burning. As 3α burning of helium creates ^{12}C, the increase in the CNO abundance promotes βCNO cycle burning of hydrogen. In turn, hydrogen burning produces helium, which powers more 3α burning. This mode can be considered a combined runaway of hydrogen and helium burning. Furthermore, once $T \gtrsim 5 \times 10^8$ K, break-out reactions from the βCNO cycle such as ^{15}O (α, γ) ^{19}Ne boost $\epsilon_{\rm nuc}$, and have an important effect on the ignition conditions [35, 44, 103]. See [46] for a detailed description of the onset of a mixed hydrogen/helium burst.

VI Near $\dot{M}_{\rm Edd}$ helium burning is marginally stable due to competition between the burning and cooling processes. It results in an oscillatory burning mode (see Sect. 5.6).

VII For $\dot{M} \gtrsim \dot{M}_{\rm Edd}$ the envelope is sufficiently hot and fuel is accumulated fast enough for steady-state (stable) helium burning. Hydrogen also burns stably.

If the accreted fuel does not contain hydrogen, helium flashes are expected to occur at all mass accretion rates below the transition to stable helium burning (regime VII). Carbon burning is further discussed in Sect. 5.7.2.

5.1.1.5 Base Heating, Rotational Mixing, and Gravitational Separation

The above theoretical picture successfully describes many of the observed burning regimes as a function of mass accretion rate. The range of $\dot{M}$ where each regime occurs observationally are, however, different from the predicted values. For example, the transition from bursts to stable burning (around regime VI) is observed at $\sim$30% $\dot{M}_{\rm Edd}$ [28, 53, 187]. Furthermore, close to this transition the observed bursts have relatively long or irregular recurrence times with indications of stable burning in-between bursts (high α-values, indicative of low $Q_{\rm nuc}$). This bursting regime does not match any of those in Table 5.1 (see also Sect. 5.1.2). To improve the theoretical predictions, several processes have been investigated in recent years.

First, a "base heating" term is included for most models, which only describe the outer layers of the neutron star. This additional flux, usually specified by the parameter $Q_{\rm b}$ (in units of MeV per accreted nucleon), models the heat flowing from the underlying neutron star crust into the envelope. The corresponding base luminosity, $L_{\rm b} = \dot{M} Q_{\rm b}$, is thought to include contributions from pycnonuclear and electron-capture reactions in the crust. Most simulations use a low value of $Q_{\rm b} = 0.1$–0.15 MeV u^{-1}, stemming from an early prediction [66]. However, more recent work indicates that the generated heat may be larger, up to $Q_{\rm b} = 2$ MeV u^{-1} [60, 67]. Additionally, a yet-unknown shallow heat source may increase $Q_{\rm b}$ [18, 39].

The degree of heating from below the fuel layer $Q_{\rm b}$ may be reduced by the competing effect of neutrino cooling. Neutrino emissions from the core and crust strongly depend on temperature (e.g., [32]), and are therefore higher at larger $\dot{M}$: $Q_{\rm b} \simeq 1.0$ MeV u^{-1} at 0.01 $\dot{M}_{\rm Edd}$ and $Q_{\rm b} \simeq 0.1$ MeV u^{-1} at ≥ 0.1 $\dot{M}_{\rm Edd}$ [32]. Recently, a new Urca neutrino cooling process in the outer crust has been predicted, which further complicates estimates of $Q_{\rm b}$ [161]. This parameter can affect the accretion rate range in which the burst regimes of Sect. 5.1.1.4 occur; for example, larger value of $Q_{\rm b}$ places the transition between bursts and stable burning at a lower $\dot{M}$ [98, 206].

Second, rotational mixing may be induced through the accretion of matter onto the neutron star, which includes the transfer of angular momentum. This mixing is turbulent, and may involve a rotationally-induced magnetic field [98, 151]. Mixing transports the accreted fuel more quickly down to the ignition depth, where it can burn in a stable manner. Therefore, rotational mixing also reduces the $\dot{M}$ where the stability of nuclear burning is predicted to change.

Third, under the influence of the strong surface gravity, heavier isotopes sink faster than lighter ones, leading to gravitational separation. At low mass accretion rates ($\dot{M} \lesssim 0.01$ $\dot{M}_{\rm Edd}$) the burst recurrence times are sufficiently long for sedimentation of the fuel layer to occur. When He and CNO separate from H, this influences the ignition conditions of H-ignited bursts (regimes I and II in Table 5.1) [149]. Deeper in the envelope, chemical separation may occur of carbon and heavier isotopes, which influences the ignition conditions of superbursts [129, 130].

Additional factors that may be important include the ignition latitude (Sect. 5.2.3), multi-zone and multi-dimensional effects (Sect. 5.8), and the accuracy of nuclear physics data (Sect. 5.9). These additional effects lead to much larger parameter-space for different ignition regimes than the $\dot{M}$ dependence alone. Only a small part of this parameter space has been explored with numerical models at this time.

5.1.2 *Status of Burst Observations*

Alongside the developments in theory have been commensurate improvements in our understanding of the phenomenology of bursts, enabled in part by the assembly of large samples from long-duration X-ray missions.

A clear distinction has been established between sources that accrete mixed H/He, and those which accrete pure (or almost pure) He, as from an "ultracompact" companion (e.g. [53]). Sources accreting mixed H/He characteristically show regular, consistent bursts with long ($\approx$5 s) rise times and decays of a few minutes, understood to be powered by *rp*-process burning (e.g. [159]; Fig. 5.3). However, for most sources such bursts are only seen episodically, within the low/hard spectral state (see below; a notable exception is the so-called "Clocked Burster", GS 1826-24 [183]). At high accretion rates, mixed H/He accretors tend to show bursts with much shorter rise and decay times, often with burst oscillations (see Sect. 5.5) and photospheric radius expansion (PRE; [114, 182]). This latter phenomenon is detectable by a temporary rise in the apparent (blackbody) radius (Fig. 5.4; see also Sect. 5.3.1) around the peak of the burst, accompanied by a decrease in temperature sufficient to maintain a roughly constant luminosity. This behaviour is understood to arise when the burst luminosity reaches the (local) Eddington limit at the surface, resulting in a temporary expansion of the photosphere. When the energy input from the burning can no longer support the expanded photosphere, it falls back onto the star, resulting in a secondary increase in temperature (the "touchdown point") and followed by cooling at roughly constant radius.

Mixed H/He accretors are also the only sources that exhibit short recurrence time bursts (e.g. [99]), weak events occurring just a few minutes after a brighter event. These events are significant because the interval since the previous burst is insufficient to reach the critical conditions for burst ignition (see Sect. 5.2.2).

Burst sources where the accreted fuel is H-deficient, consistently show bursts characteristic of largely He-fuel, with short ($\lesssim$1 s) rise times and durations of 10–20 s (Fig. 5.3; see also Sect. 5.7). One of the best-studied examples is 3A 1820−30, which consists of a neutron star in an 11-min orbit with an evolved companion (e.g. [32]). PRE bursts are more common, except at high accretion rates.

A secondary distinction has emerged between bursts that occur in the low (hard) persistent spectral state, and those in the high (soft) spectral state. These states are defined both by their spectral shape, and their periodic (and aperiodic) timing features (e.g. [69]). The persistent spectral states are thought to indicate different characteristic accretion rate regimes; the low (hard) state, a truncated disk with the innermost accretion occurring through an optically thin, spherical flow; and the high (soft) state, with a disk extending to the NS surface and terminating in a boundary layer providing most of the persistent X-ray emission (e.g. [40]). Remarkably, these different accretion states also seem to give rise to markedly different burst behaviour.

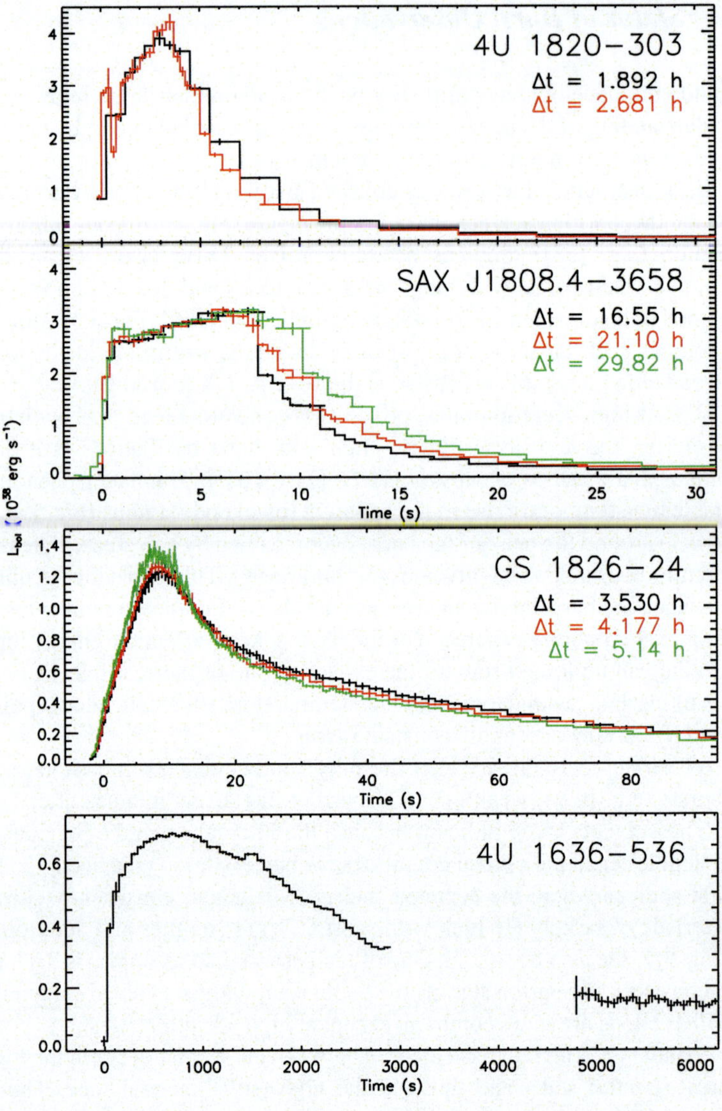

Fig. 5.3 Burst light curves observed by the *Rossi X-ray Timing Explorer* (*RXTE*) Proportional Counter Array (PCA) illustrating several different cases of ignition, labelled with the source name and the recurrence time Δt, where known. The top panel shows likely pure He bursts ignited on the neutron star in the ultracompact system 3A 1820−30. The second panel shows He-rich bursts ignited after the accreted H has been exhausted at the base prior to ignition (case III in Table 5.1), in the accretion-powered millisecond pulsar SAX J1808.4−3658. The third panel shows mixed H/He bursts from the "Clocked burster", GS 1826−24 (case V). The bottom panel shows an example of a superburst observed from 4U 1636−536. Adapted from [56]

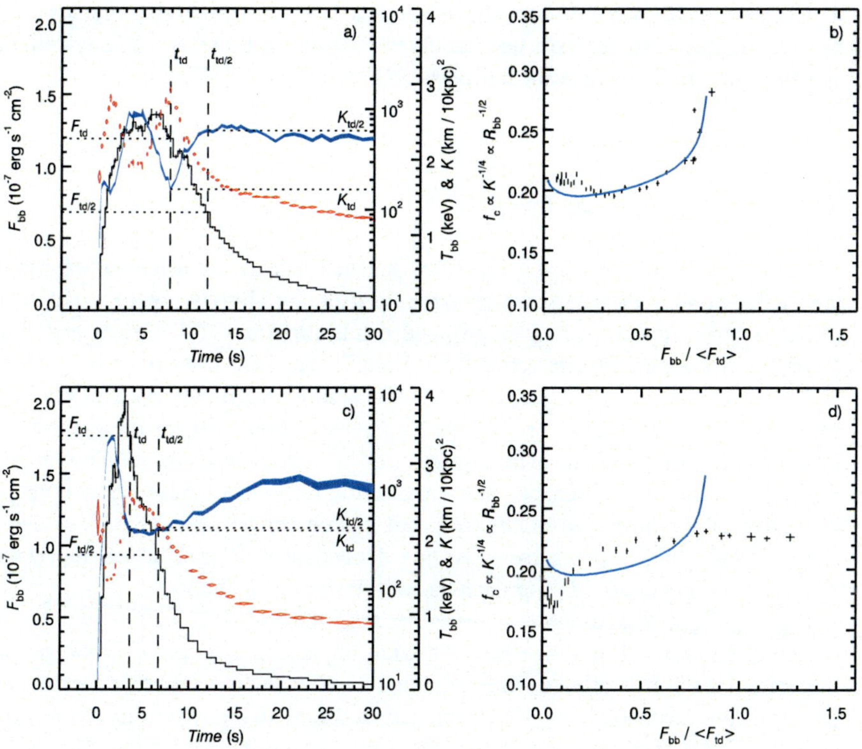

Fig. 5.4 Time-resolved spectroscopy of two X-ray bursts observed with *RXTE*/PCA from the transient 4U 1608−52 illustrating the differences between hard state (top panels) and soft state (bottom panels) bursts. The left-hand panels (**a** and **c**) show the time-resolved spectroscopic parameters, with bolometric flux F_{bb} (*black line*, left-hand y-axis); blackbody normalisation K_{bb} (*blue ribbon*, inner right-hand y-axis); and blackbody temperature T_{bb} (*red diamonds*, outer right-hand y-axis). Note the qualitative difference in the variation of K_{bb} during the two bursts (both events would be characterised as PRE). The right-hand panels (**b** and **d**) show the relationship between the inverse square root of the blackbody radius (proportional to the colour-correction factor f_c; see Sect. 5.3.1) and the blackbody flux F_{bb} (scaled using the mean touchdown flux F_{td} for the source). The blue line (identical in both panels) is a model prediction for a pure hydrogen NS atmosphere with a surface gravity of $\log g = 14.3$, taken from [179]. The agreement between the atmosphere model and the data is much better for the hard state burst. Adapted from [91]

5.1.2.1 Low (Hard) State Bursts

Thermonuclear bursts which occur when the persistent spectrum is hard (in colour-colour diagrams this state is referred to as the "island" state in *atoll* systems; e.g. [69]) tend to exhibit long *rp*-process tails, in systems that accrete mixed H/He. Radius-expansion bursts are relatively rare (e.g. [136]), as are detections of burst oscillations (in systems where they have been seen). The burst rises tend to have a convex shape, suggesting ignition near the equator (see Sect. 5.2.3). The spectral evolution during these bursts shows a characteristic inflection which roughly matches the expectation of atmosphere models (e.g. [178]; see also Fig. 5.4). For this reason, these bursts have been adopted by some authors as the only reliable events from which the neutron star mass and radius can be inferred (e.g. [91]; see also Sect. 5.3.1).

5.1.2.2 High (Soft) State Bursts

When the persistent spectrum is soft (the "banana" state in colour-colour diagrams) the characteristics of bursts are remarkably different. Bursts are irregular, with longer recurrence times (on average), and exhibiting rapid ($\lesssim$1 s) rises and short ($\approx$10–20 s) durations characteristic of primarily He fuel, even in those systems which accrete mixed H/He. Radius-expansion is more frequently seen, as are burst oscillations. The burst rises tend to have a concave shape, suggesting the ignition location has moved away from the equator. The blackbody normalisation is typically flat in the burst tail, which has motivated some observers to use the average as indicative of the projected radius [64]; but as this behaviour is at odds with the expected behaviour (arising from the flux-dependent distortion of the underlying blackbody component) this approach has raised some objections.

We highlight this dichotomy here because the physical mechanism is not presently understood, but clearly has a wide impact on a range of investigations involving bursts. It is often assumed that the burst behaviour is independent of the details of the mechanism by which the accreted fuel arrives onto the neutron star (see Sect. 5.1.1.1); the observations seem to suggest this cannot be true. That is, the observations suggest that the disk *geometry* (inferred from the persistent spectral state) influences the burst physics, in addition to the accretion *rate*. Alternatively, it is possible that the different accretion regimes give rise to different (local) conditions, perhaps allowing some ongoing level of steady He-burning. Such effects would certainly explain the weak bursts typically observed in the high (soft) state, as well as the decrease in burst rate at increasing accretion rates observed for many systems [28]. As we will see, one of the growing areas of research over the past decade has been the interaction between the bursts and the accretion disk and environment of the neutron star (Sect. 5.4).

5.1.2.3 The *Rossi X-ray Timing Explorer*

Although many X-ray instruments have contributed to our present understanding of thermonuclear bursts, a key instrument that has enabled much of the new results in burst phenomenology over the last decade is the Proportional Counter Array (PCA; [86]) on-board the *Rossi X-ray Timing Explorer* (*RXTE*). This instrument, with an effective area of $\approx$6500 cm^2 and fast timing down to 1 μs has allowed detailed studies of many prolific burst sources, with high signal-to-noise data permitting precise measurements of spectral and timing properties. The assembly of the accumulated $\approx$1100 bursts over more than a decade of observations from more than 50 sources, led to a much greater capacity for broad-scale investigation into the burst phenomenology [53]. Much of these investigations have focussed on the burst oscillation phenomenon, as described in Sect. 5.5. The *RXTE* burst sample, as with earlier compilations (e.g. [28]) from other instruments, revealed a much more complex picture for unstable burning than predicted by theory.

An example is the study of photospheric radius-expansion bursts, which are thought to reach the local Eddington luminosity, and have been used as "standard candles" to determine the distance to bursting sources [184]. Observations of bursts from globular cluster systems, which have an independently determined distance, allowed empirical measurement of the peak PRE burst luminosity, which is somewhat in excess of theoretical predictions for pure He fuel [109]. Such studies also showed that in some systems there are substantial variations in the peak flux of PRE bursts from the same source. In 4U 1636−536, a small subset of PRE bursts reach a peak flux a factor of 1.7 lower than the majority of events, explained as a variation in the electron scattering dominated opacity between pure He and mixed H/He at solar composition. This result, obtained for a sample of PRE bursts much larger than in any other source, further suggests that typical PRE bursts reach the Eddington flux for material devoid of hydrogen, even in systems accreting a mix of H and He [52]. In other systems, the typically 5–10% variation in the peak fluxes may be attributed to interactions with the accretion disk [51]. Eddington-limited bursts remain the primary source of information on the distances to burst sources, but more work is required to reduce the effect of systematic contributions to the peak flux distribution.

Independently of the studies that accompanied the catalogue data itself, the provision of an index (i.e. burst times and observation IDs) into an otherwise undifferentiated collection of burst observations enabled other researchers to perform their own analyses on the burst sample, further increasing its impact. The *RXTE* sample has been extended with further observations through to the end of the mission in 2012 January, and now totals more than 2000 bursts. These data will be incorporated into a new, and more comprehensive sample of bursts, the Multi-INstrument Burst ARchive (MINBAR[3]). Together with burst observations of the *BeppoSAX*/WFCs [28], and *INTEGRAL*/JEM-X, MINBAR contains over 7000 bursts.

5.2 X-ray Burst Ignition

In Sect. 5.1.1.2 we explained burst ignition as due to runaway thermonuclear burning. Here we discuss the other fundamental processes that facilitate the ignition of different types of bursts, including the so-called short recurrence time bursts, as well as the role of the latitude of the ignition location.

[3] http://burst.sci.monash.edu/minbar.

5.2.1 Thin-Shell Instability and Electron Degeneracy

The typical response of a star to heating is expansion, which reduces the pressure and thereby the temperature. For runaway nuclear burning to occur, this cooling mechanism must be circumvented, via a combination of the thin-shell instability and electron degeneracy. Some past reviews highlight the latter [115], whereas others present the former [171]. Here we review the two mechanisms.

X-ray bursts are an example of nuclear burning that happens in a thin shell [162, 205]. If the radial extent of the burning layer is small with respect to the stellar radius, expansion of the thin shell does not result in a sufficient change in pressure to reduce the temperature. This allows for the temperature to continue to increase, and is known as the *thin-shell instability*. For X-ray bursts and superbursts the ignition layer has a depth of a few meters up to $\sim$100 m, which is very thin compared to the $\approx$10 km neutron star radius. Therefore, the thin-shell instability allows for a growing temperature increase due to burning, which leads to the ignition of bursts.

Electron degeneracy, identified as the primary mechanism for the ignition of novae on white dwarfs [57], also plays a role in bursts on neutron stars. In degenerate material, the pressure is independent of the temperature, such that a reduction in pressure does not counter a temperature increase. On neutron stars, however, the electrons are only mildly degenerate near $y \simeq 10^8\,\mathrm{g\,cm^{-2}}$: the ratio of the electron Fermi energy and thermal energy prior to a burst is $\eta \simeq 3$. A small amount of heating at the burst onset lifts the degeneracy, but the runaway continues.

A simple numerical experiment elucidates the relative importance of the two mechanisms. Employ a one-zone model to simulate the burning behaviour with and without a pressure term for degenerate electrons.[4] When ignoring the pressure term from electron degeneracy, bursts are still produced. Of course, electrons do contribute to the pressure, and ignoring that term leads to a slightly larger ignition depth. Therefore, the primary mechanism for the thermal instability of bursts is the thin-shell instability, while the electron degeneracy is important for obtaining accurate ignition conditions.

Intermediate duration bursts and superbursts ignite deeper in the envelope (Table 5.2), where the electron degeneracy is larger, up to $\eta \simeq 10^2$. Nevertheless, their fuel layer is a thin shell, which implies that long duration bursts would also occur without the degeneracy effect.

5.2.2 Reignition After a Short Recurrence Time

In the classical picture, a burst burns nearly all hydrogen and helium in the neutron star envelope, and a completely new fuel layer must be accreted before the next burst

[4]For example, the one-zone helium ignition model available at https://github.com/andrewcumming/onezone.

Table 5.2 Typical properties of Type I bursts with different durations

	Normal	Intermediate	Superburst
Duration	10–100 s	Few minutes–hour	hours
Fluence (erg)	10^{39}	10^{40}–10^{41}	10^{42}
Recurrence time	hours–days	weeks–months	1 year
Ignition column depth (g cm^{-2})	10^8	10^9–10^{10}	10^{11}–10^{12}
Number observed	>7000 [a]	≈70 [b]	26 [c]
Number of sources	115	≈34	15
Fuel	H/He	He	C

[a] MINBAR, http://burst.sci.monash.edu/minbar
[b] The long burst catalog assembled by the International Space Science Institute team on thermonuclear bursts (PI: A. Cumming), http://www.issibern.ch/teams/ns_burster
[c] in't Zand [75]

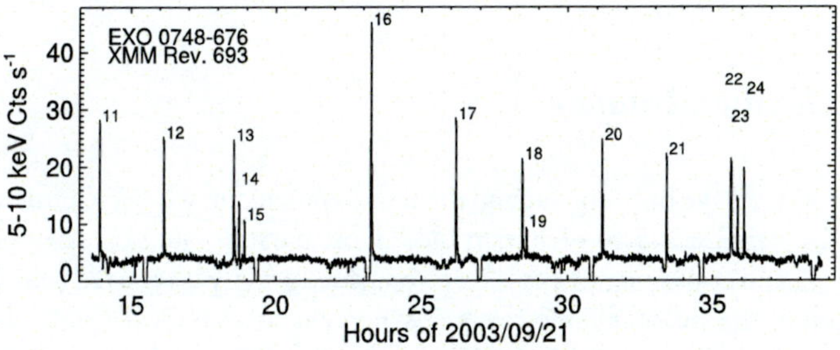

Fig. 5.5 Day-long light curve of EXO 0748−676 observed with *XMM-Newton* exhibits double and triple events, where the bursts are separated by short recurrence times of ~10 min. The second and third bursts are typically weaker than the first burst in an event, and the burst shape indicates that less hydrogen is present in their fuel. Eclipses by the companion star periodically obscure the source. Reproduced with permission from [16] © ESO

can ignite. Observations of bursts that recur within ~5–20 min suggest, however, a fraction of the fuel survives, as there is insufficient time to accrete a fresh layer [49, 58, 138]. Only recently have multi-zone simulations been able to show that a substantial fraction of the fuel may be left unburned after a burst [96]. Particularly if ignition occurs at a relatively shallow depth (resulting in a weaker burst), over ≈50% of the fuel remains unburned. Only the fuel near $y_{\rm ign}$ burns, whereas the material at smaller depths is left-over. The recurrence times of the bursts are reduced, since accretion needs only to replace the burned fraction of the fuel column.

It is possible for the left-over fuel to ignite and produce a new burst mere minutes after the previous burst (Fig. 5.5). If the fraction of unburned fuel is large, this material resides relatively close to $y_{\rm ign}$. Turbulent mixing can transport the unburned material down to this depth, where it ignites a new burst. This secondary burst is typically weak, with a lower peak flux and shorter duration than the previous burst, because it is powered by a diluted mixture of fuel and ashes. Different origins for the turbulent mixing have been proposed, such as rotationally induced shear mixing [49] and post-burst convection due to an inversion of the mean molecular weight

[200].[5] Recently, multi-zone simulations that include a full nuclear network show that the turbulence is convection that is driven by an anti-correlation between the temperature and the opacity in the ashes layer. As the ashes cool after a burst, the opacity increases until convection switches on. Convective mixing brings fresh fuel into the ashes layer, and the recurrence time of the resulting burst is the cooling time of the order of minutes. Because of the stochastic nature of convection, sufficient fuel for a follow-up burst is mixed down only ≈30% of the time. Furthermore, after one follow-up burst, the mixing continues and has the potential to produce another follow-up burst. This model reproduces many of the observed features of such bursts (Fig. 5.5): the follow-up bursts have a lower peak luminosity and shorter duration than the first burst, their recurrence time ranges from a few minutes to a few tens of minutes, and several follow-up bursts can occur in so-called double, triple, and quadruple events [16, 58, 99, 138].

5.2.3 *Ignition Latitude*

Up to now, we have only considered the radial dimension when locating the ignition point. The latitude is also of importance: most bursting neutron stars spin rapidly (Sect. 5.5), and consequently the effective surface gravity is reduced near the equator compared to the polar regions (by a few percent for a typical burster [166]). The surface of a spinning star follows an isobar (surface of constant pressure), and using the relation $y = -P/g$ from Sect. 5.1.1.1, we see that a smaller $g_{\rm eff}$ at the equator yields a larger y than at the poles. Furthermore, this implies that the *local* mass accretion rate is higher at the equator. Because $y_{\rm ign}$ depends only weakly on $g_{\rm eff}$ (e.g., Equation 32 in [14]), burst ignition preferentially occurs at the equator.

The spreading of the flame on the surface of the star was modelled by Spitkovsky et al. [166]. The difference in speed between the spreading in the latitudinal and longitudinal directions (with the latter significantly faster at low latitudes) is thought to give rise to distinct, observable differences in the shape of the burst rises, which is referred to as the "convexity" of the rise [128]. However, a study of a large sample of bursts observed by *RXTE* suggests that bursts occurring when the source is in the low (hard) spectral state typically ignite on the equator, while bursts observed in the high (soft) spectral state ignite at higher latitudes [128, 208]. The obvious question is, why do the latter bursts not also ignite on the equator?

An attractive explanation is that in the high (soft) state, both H and He fuel burns *steadily* on the equator, while unsteady burning (likely triggered by He) continues in regions at higher and lower latitudes. This explanation would also account for the infrequent, weak bursts that are typically observed in the high (soft) state (cf.

[5] Irregular bursting behaviour in early multi-zone models may have been an artefact of their reduced nuclear reaction networks [181, 200], as this behaviour is absent in later work with large networks [201].

with Sect. 5.1.2). At present this scenario remains highly speculative, and it is computationally challenging to test with multi-dimensional models (Sect. 5.8.3). Certainly, the balance in pressure resulting from the decrease in effective gravity at the equator gives a larger effective accretion rate per unit area there, but not sufficient to stabilise the burning [166]. Resolving this puzzle will likely require a substantial improvement in our understanding of the global burst properties and perhaps also the influence of the accretion flow.

5.3 The Burst Spectral Energy Distribution

Most of the energy from nuclear burning is generated at least several meters below the neutron star surface, and processed before exiting the star from a cm-thick photosphere. The burst radiation is, therefore, thermalized and leaves the burning layer as a blackbody spectrum. Scattering off electrons and ions in the photosphere, however, modifies the continuum spectrum and may introduce absorption lines and edges. Those departures from a pure blackbody spectrum offer opportunities to measure the neutron star mass and radius. In Sect. 5.4 we further discuss how some fraction of the burst spectrum may be reprocessed by the accretion environment.

5.3.1 The Continuum Spectrum

The majority of observed burst spectra are well described by a Planck (blackbody) distribution [180]. The rise of the burst flux, set by a combination of the time scales of nuclear burning and flame spreading, can be as fast as a millisecond, or as long as several seconds. The measured blackbody temperature increases to a maximum value of 2–3 keV. In the burst tail, the decrease of the temperature is determined by the time scales of cooling and of waiting points in the *rp*-process (Sect. 5.9), and follows that of the burst luminosity, typically lasting ~10–100 s. This cooling of the neutron star atmosphere in the burst tail is often cited as prime evidence of a thermonuclear flash (Type I burst) as opposed to an accretion flare (Type II burst; [6, 115]). A small subset of thermonuclear events exhibit only weak evidence of cooling. In an analysis of the remarkable burst source and 11 Hz pulsar IGR J17480−2446, it has been shown that cooling may not be detectable when the ratio of peak burst flux to persistent (accretion) flux is low, and the bursts correspondingly have a low peak temperature [117].

Because of the evolution of the spectral parameters, time resolved spectroscopy has to be performed, with each spectrum being collected in a short time interval. The short duration over which each spectrum is accumulated in typical bursts strongly limits the photon counts and the detail present in a single spectrum. Only when considering exceptionally long bursts [172] or large numbers of bursts [203] can deviations from a blackbody shape be detected. Some deviations are due to

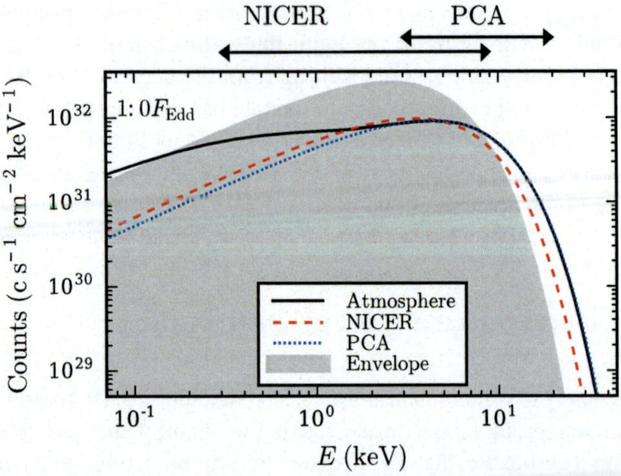

Fig. 5.6 Photon counts spectrum as a function of energy, E, for models of thermal burst emission (burst atmosphere model from [178]). The top of the grey area marks the Planck spectrum in the neutron star envelope during a burst. Scattering in the atmosphere modifies the spectrum (solid line). Historically the observed spectrum is fit with a Planck model. Two such fits are shown for *NICER* and *RXTE*/PCA in the energy bands indicated at the top. The ratio of the observationally inferred temperature and that of the envelope spectrum is the colour correction factor

changes in the persistent spectrum induced by the burst (Sect. 5.4), in addition to the modifications of the intrinsic spectrum discussed here.

Compton and free-free scattering of thermal emission in the neutron star atmosphere modifies the burst spectrum (Fig. 5.6). Specifically, up-scattering boosts the energy of a fraction of the soft photons, such that the "colour temperature" measured by fitting a blackbody is larger than the effective temperature of the photosphere [145]. This effect is normally accounted for using a "colour-correction" factor, f_c, giving the expected ratio of the (measured) colour temperature to the effective temperature. At fluxes close to Eddington, f_c is expected to be $\approx$2, dropping to $\approx$1.4 in the burst tail (e.g. [178]).

Radiative transfer simulations are employed to calculate all scattering processes in the atmosphere taking into account the composition, although initial models were not fully converged [122, 124]. Current state-of-the-art spectral models are one-dimensional, assume hydrostatic and radiative equilibrium, and have been calculated for a range of compositions and surface gravities [140, 178, 179]. The models have several limitations. They assume uniform emission across the entire stellar surface, whereas the detection of burst oscillations argues that this is not always the case (Sect. 5.5). Furthermore, the neutron star's rotation will cause Doppler broadening of the spectrum [12]. The impact of these effects is, however, likely small. More importantly, the assumed equilibria imply that the models are not representative of the PRE phase, which will require simulations that include both radiation transfer and hydrodynamics (cf. with [145]). In a new approach, time-dependent stochastic (Monte Carlo) models are being developed, which in time may overcome these limitations [131].

When observed in a restricted instrumental energy band, a neutron star atmosphere spectrum is still well-approximated by a Planck function (Fig. 5.6). The inferred temperature of the blackbody T_{bb} is expected to differ from the atmosphere's effective temperature T_{eff} by the colour-correction factor, $f_c \equiv T_{bb}/T_{eff}$, which is expected to vary as a function of the burst luminosity. f_c has been calculated for assumed model atmospheres for application to *RXTE*/PCA observations [178], and can similarly be obtained for other instruments using the model-predicted spectra.

Aside from the temperature, the blackbody flux depends on the size of the emitting area: the neutron star surface. It is typically assumed that the whole surface radiates uniformly.[6] At the onset of the brightest bursts, the flux may exceed the Eddington limit, causing photospheric radius expansion and possibly also driving a wind from the neutron star surface [41, 141, 153, 196]. During PRE the photospheric radius increases typically by up to a factor ~10, but even larger expansions have been observed ("superexpansion"; Sect. 5.7.1). Once the flux is reduced below the Eddington limit, the photosphere settles back at its original radius (the "touchdown" point). After touchdown, one expects the photospheric radius to remain constant, but substantial evolution in the blackbody normalisation, K_{bb}, can be observed [64, 207]. We discussed how deviations from a pure blackbody spectrum introduce a colour correction factor for T_{bb}. To preserve the total blackbody flux $F \propto K_{bb} T_{bb}^4$, this produces an equivalent correction to K_{bb}: the proportionality between the measured blackbody normalisation and the colour correction is $K_{bb}^{-1/4} \propto f_c$ (see Fig. 5.4).

The properties of the continuum burst spectrum have been employed to measure the neutron star mass and radius (e.g. [133, 134, 185, 186]; see also Miller, this volume). Two approaches have been used: the "touchdown" and the "cooling-tail" methods. The former method uses the flux at touchdown as a measure of the Eddington limit, which depends on the neutron star mass, in combination with measurements of the blackbody radius and (usually) the source distance [61, 62]. The radius is measured through the blackbody normalisation, $K_{bb} \propto 4\pi R^2$, assuming that f_c is constant. It is possible that the photosphere has not completely settled onto the neutron star at touchdown [167]. Additionally, an accurate measure of f_c and the Eddington limit requires knowledge of the composition of the photosphere as well as the source distance, which are generally known only approximately.

The cooling-tail method uses the flux-dependent colour corrections from model atmosphere spectra [177, 178]. Because these models are only valid at fluxes below the Eddington limit, only the tail of a burst is considered. By measuring $K_{bb}^{-1/4}$, one observes how f_c changes as the burst flux decreases. The atmosphere models

[6]Emission from only part of the surface has been inferred for the 1999 superburst from 4U 1820–30 [17]. This study ignored, however, disk reflection [9], which complicates the interpretation of superburst spectra [101, 102, 104].

predict f_c as a function of the Eddington ratio and the neutron star mass and radius. Fitting these f_c curves to the observed data allows for the neutron star parameters to be measured. This method does not depend on prior knowledge of the source distance or the atmosphere composition, as the required information is encoded in the f_c curves.

The radii obtained with the cooling-tail method are systematically larger than with the touchdown method. Each method has a set of assumptions that may be the source of this systematic uncertainty [63–65, 152, 167]. For example, the inclusion of stellar rotation in the atmosphere models may reduce the derived radius [152]. Furthermore, each method appears to work best for a different set of burst observations. Bursts in the low (hard) flux state conform to the predictions of the model atmospheres, whereas in the high (soft) flux state the bursts do not exhibit the expected $f_c - F$ relation [91]. In the high flux state the accretion disk is expected to extend to the neutron star surface, and a spreading layer may cover a substantial part of the star [156], which reprocesses the burst spectrum. Conversely, studies using the touchdown method have preferred bursts that show only small deviations from a blackbody, rather than those that follow the expected $f_c - F$ relation [64]. The source 4U 1608−522 exhibits bursts in both flux states, and an analysis with both methods illustrates the differences in the obtained results [152]. Further details on the measurement of mass and radius are presented in [135].

5.3.2 *Discrete Spectral Features*

Before the thermalized burst spectrum leaves the neutron star, the photons scatter off electrons in the photosphere. If the photosphere is incompletely ionized, discrete absorption features may appear in the spectrum. The photon's energy may take the electron from one bound state to another (bound-bound transition), which produces an absorption line at that energy. Alternatively, the photon energy may exceed the ionization energy required to unbind the electron (bound-free transition). This creates an absorption edge that starts at the ionization energy and continues towards higher energies. In the photosphere of an accreting neutron star, hydrogen and helium are fully ionized. Metals, however, may be only partially ionized and can produce lines and edges, particularly in the extended photosphere driven by a strong radius-expansion burst (see Sect. 5.4). A range of metals may produce these features, depending on the composition of the photosphere [196]. Iron is typically one of the most abundant heavy metals, and has relatively many transitions to create discrete absorption features.

Detection of surface features has been a high observational priority for decades, due to the prospects of constraining the neutron star compactness via the gravitational redshift, measurable from comparing the observed and rest-frame energies of such features. Although there have been several notable claims for detections of features (see below), many have been unverified in subsequent observations and/or been shown to be unlikely to arise from the NS surface. There are a number of mechanisms which may act to remove elements that might give rise to features, from the upper layers of the photosphere.

Gravitational settling is thought to rapidly separate accreted metals from the photosphere on a timescale of 10^{-3} s [19, 149]. The metals that can produce lines and edges may, therefore, usually be absent from the neutron star surface. Just a few meters below the surface, a plethora of metals is created by nuclear burning during bursts. These elements may be revealed at the neutron star surface—only during the most powerful bursts—because of two processes. First, convection at the onset of the burst transports the burning ashes towards the surface. Convection typically does not reach the surface, but the more powerful the initial burning, the closer to the surface the metal-rich ashes are convected. Secondly, if the burst has PRE, a wind during the Eddington-limited phase may blow off the upper atmosphere, facilitating the exposure of heavy metals [196]. The appearance of metals in the photosphere can also give rise to a changing colour correction factor [92]. In bursts where the fuel is helium rich, most of the burning happens at the burst onset, and the Eddington limit is reached. Powerful helium bursts are, therefore, the prime candidates for observing absorption features during the PRE phase. For example, a weak absorption line was detected during the PRE phase of an intermediate duration burst from GRS 1741.9−2853 [10].

The energy at which a line or edge is observed is reduced by the gravitational redshift of the neutron star, that also provides a measure of the star's compactness. Discrete features are also expected to be affected by relativistic Doppler broadening, since most bursting neutron stars rotate rapidly (e.g. [11]). These effects reduces the detectability of such features against the continuum, increasing the difficulty of constraining the equation of state.

The studies of EXO 0748−676 provide an illustrative example. The source was a calibration target for *XMM-Newton*, and consequently a large number of bursts were observed with the Reflection Grating Spectrometer. Stacking the spectra of a series of bursts revealed a tentative detection of an iron absorption line [29, 154]. The inferred gravitational redshift of the line was employed to constrain the mass and radius of this neutron star [143], favoring a soft equation of state. These analyses assumed a low neutron star spin frequency, supported by the detection of a weak burst oscillation at 45 Hz [190]. Later, a much stronger burst oscillation was detected in two bursts, at a significantly higher frequency: 552 Hz [54]. If the later detection instead corresponded to the neutron star spin in this system, the expected Doppler broadening was, therefore, substantially larger than assumed, casting doubt on the line origin. Furthermore, a larger sample of bursts observed with *XMM-Newton* [30] and *Chandra* [99] do not exhibit the line feature. The status of the line detection from EXO 0748−676 remains, therefore, uncertain. However, the potential for absorption features from the neutron star surface to constrain the neutron star equation of state remains high.

Because a wider range of energies is absorbed to produce an edge, edges typically have a larger equivalent width than lines [196], and consequently should be easier to detect. The strongest absorption features associated with the surface of neutron stars are nickel edges observed in the PRE phase of three superexpansion bursts from 4U 0614+091 and 4U 1722−30 [76]. These detections were made with the *RXTE*/PCA, which has modest spectral resolution ($\Delta E/E \approx 0.17$ at 6 keV, where ΔE is the full-width at half maximum; [87]). Improved constraints on the neutron star equation of state will result from future observations of edges in helium-rich bursts performed with high-resolution spectrometers, but it has proven challenging to schedule such observations of relatively infrequent helium flashes.

5.4 Interaction with the Accretion Environment

The dynamical influence of thermonuclear bursts on the accretion disk has generally been studied in the context of radius-expansion bursts, which exceed the Eddington limit at the neutron star surface, and may be expected to drive winds into and across the disk. More broadly, three principal effects have been discussed in the context of interactions between bursts and the accretion disk (e.g. [8]): structural changes to the disk (and/or corona) resulting from burst-induced X-ray heating (or cooling); radiatively (or thermally) driven outflows; and inflow due to Poynting-Robertson drag. In addition, the observed spectrum may be modified by interaction with, and reflection from, the accretion disk.

The evidence for structural changes comes from some observations of long-duration bursts (see also Sect. 5.7). In some of these events we observe strong flux variability late in the tail of the burst [37, 81] (Fig. 5.10). This variability has been suggested to arise from Thompson scattering in the settling "fragments" of disk material, following disruption by an extremely energetic radius-expansion burst. Decreases in the intensity of high-energy (30–50 keV) photons beginning a few seconds after the burst onset have also been observed in a few sources [88, 121]. These photons are thought to arise from Compton scattering in a hot corona, and it is expected that the sudden rise in (relatively) cool photons will result in a transient change in the coronal structure. While the principal effect predicted is additional cooling of the corona, the estimated change in the cooling rates is in the other direction, i.e. predicting *reduced* cooling [89]. Such explanations also seem at odds with the evidence supporting the role of magnetic fields energizing coronae in other accreting sources (e.g. [132]). Other possible mechanisms which might instead give the observed behaviour are related to the attenuation of the inner disk in response to the burst (see below), possibly mediated via the magnetic field threading the disk [88].

Irradiation of the accretion disk may drive a wind from its surface. The prospects of this phenomenon have been discussed in the context of spectral features observed from an intermediate duration burst of IGR J17062−6143 [37]. Additionally, such powerful bursts may eject a shell from the atmosphere of the neutron star in a superexpansion phase, and enrich the surroundings in metal-rich burst ashes [196]. This mechanism has been argued to be the source of discrete spectral features detected in intermediate duration bursts ([76]; see also Sect. 5.3.2).

The effects of Poynting-Robertson drag arising from the increase in X-ray flux during the burst have been predicted to lead to an increase in the accretion rate (and hence the persistent flux component) during the burst ([191]; see also [111]). This effect is potentially at odds with the conventional approach for time-resolved spectroscopic analysis, which involves subtracting off the pre-burst emission (usually incorporating both the persistent emission and instrumental background). This approach implicitly assumes that the burst and persistent component are independent, even late in the tail of the burst, which cannot be completely true (e.g. [108]). Nevertheless, the net burst spectrum is then usually well-fit with a blackbody (e.g. [180]).

The burst sample accumulated by *RXTE* over its mission offers the most stringent test of this conventional analysis approach. A study of 332 PRE bursts observed from 40 sources found significant evidence for an increase in the persistent contribution to the flux early in the bursts [203]. This increase, up to a factor of 20, was interpreted as the expected increase in the accretion rate in response to Poynting-Robertson drag. It is somewhat surprising that this effect is seen in PRE bursts, since the luminosity of those bursts should exceed the Eddington limit at the peak, temporarily *halting* accretion. The effect has also been confirmed in an even larger sample of 1759 non-PRE bursts from 56 sources [204], although at with smaller increase factors. Additionally, a bright burst from SAX J1808.4−3658 observed simultaneously with *Chandra* and *RXTE* finds even more compelling evidence for increases in the accretion component during the burst [82], from the much broader band-pass ($\approx$0.5–30 keV) available for the two instruments combined (Fig. 5.7).

Unfortunately the available data is insufficient to test for possible changes in the *shape* of the persistent spectrum accompanying the burst-induced increase in accretion rate, which might be expected in parallel to the spectral shape variations that are seen on longer timescales (see e.g. Sect. 5.1.2). Furthermore, while the modified analysis approach improves the combined distribution of fit statistics for the ensemble, it does not yet achieve the expected distribution for a well-fitting model, suggesting that the net burst spectrum still deviates significantly from a blackbody. As we have discussed, these deviations include the recombination edge features observed in very energetic bursts (e.g. [76]; see also Sects. 5.7 and 5.3.2). The residuals also appear inconsistent with the more modest deviations predicted by spectral models (see Sect. 5.3.1). The question of the true shape of the burst spectrum over the full range of burst types remains open, with the available data not sufficiently sensitive to clearly measure deviations from a blackbody. However, with the analysis of the *RXTE* sample, the key contributions of variations in the persistent contribution, discrete features, as well as the effect of reflection from the accretion disk (Sect. 5.4.1) are now well-established.

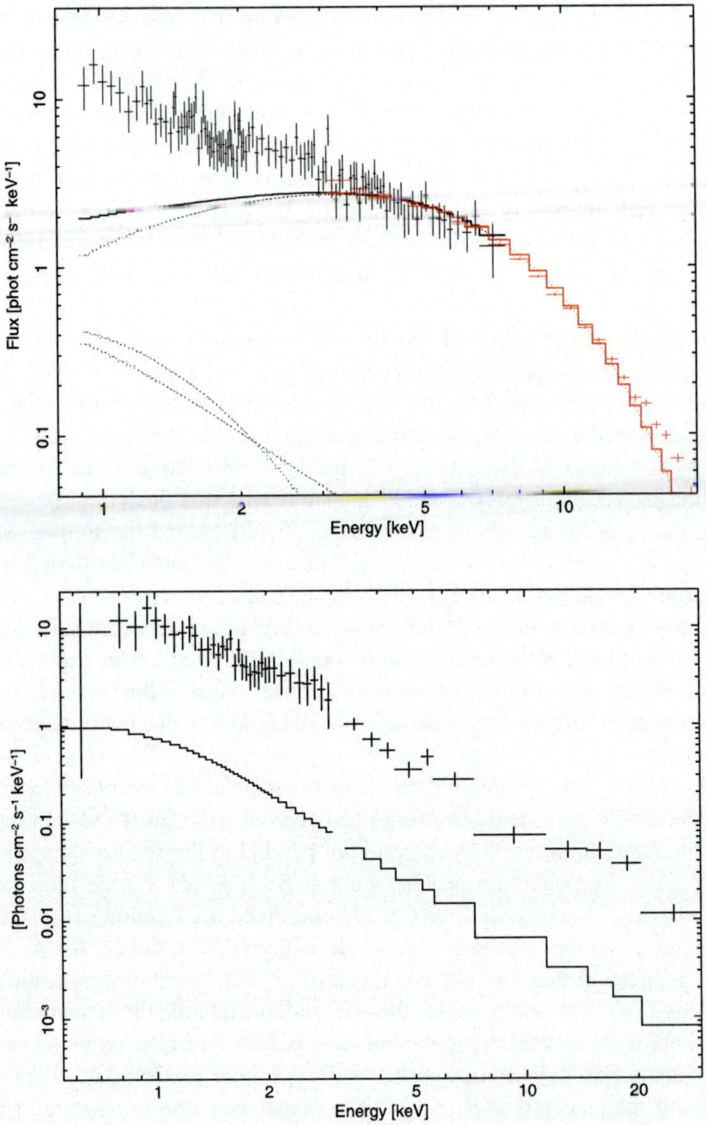

Fig. 5.7 X-ray spectra during a bright burst from the millisecond pulsar SAX J1808.4−3658, observed simultaneously with *RXTE* and *Chandra*, illustrating the apparent increase in the persistent flux level during the burst. The top panel shows the *Chandra* (black crosses) and *RXTE* (red crosses) spectrum, integrated over an interval 2–5 s after burst onset. The solid curve shows the best-fit blackbody spectrum, including the model components for the pre-burst (persistent) radiation. The model is a poor fit (reduced $\chi^2_\nu = 3.505$ for $\nu = 268$ degrees of freedom), but is dramatically improved if the normalization of the persistent component is also free to vary ($\chi^2_\nu = 0.963$ for $\nu = 267$). In the bottom panel the residual spectrum is plotted after subtraction of the fitted black body model (crosses). The model for the non-burst spectrum is plotted as a histogram. For clarity *Chandra* data is plotted up to 3 keV and *RXTE* data beyond 3 keV. Note the similarity in shape of the residual and the persistent spectrum. Reproduced with permission from [82] © ESO

5.4.1 *Reflection by the Accretion Disk*

A substantial fraction of the X-rays emitted by the bursting neutron star fall on the accretion disk [47, 113]. For a thin disk this fraction is about 25% of the total burst emission, and most flux falls on the inner few tens of kilometers closest to the star. The radiation undergoes diffusive scattering off the disk's photosphere, which is referred to as X-ray "reflection." Depending on the inclination angle of the disk with respect to the line of sight, a substantial part of the observed burst flux may have arisen from photons reflected by the disk. The reflection fraction is defined as the ratio of the reflected and the directly observed burst flux components. For a thin disk, reflection fractions of up to approximately 0.5 are expected.

If the disk has a different shape, such as a steeply increasing height, the reflection fraction can potentially exceed unity: most of the burst signal is observed in reflection [70]. A simple analogy is a candle in a teacup: when viewed under certain angles, the flame of the candle is not directly visible, but one still sees its light scatter off the inner sides of the cup.

When the burst scatters off the disk, the X-rays are reprocessed, and their spectrum is modified [7]. Above 3 keV, the reflection spectrum is shaped like the illuminating burst spectrum with the addition of a fluorescent iron emission line near 6.4 keV and an iron absorption edge near 9 keV (Fig. 5.8). The shape of the

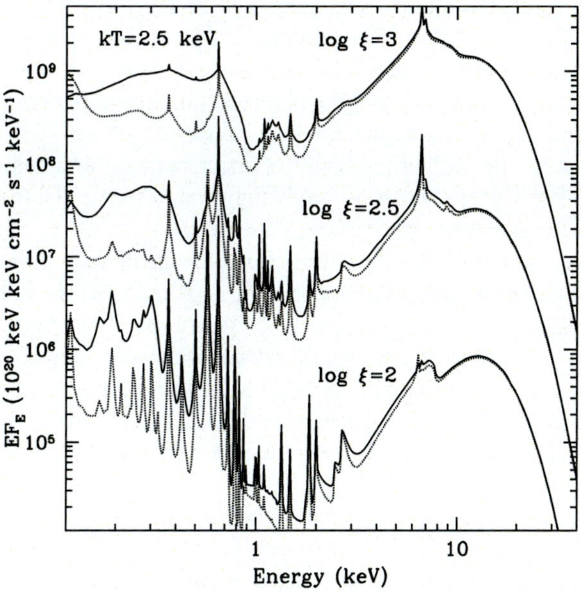

Fig. 5.8 Models of burst reflection spectra as a function of energy (from [7]). For an irradiating blackbody spectrum with $kT = 2.5$ keV, the resulting reflection spectra are shown for three values of the disk ionization parameter ξ. Especially below $\approx$3 keV, the reflection spectra depend on the density of the inner disk ($n = 10^{18}$ cm^{-3} for solid lines and $n = 10^{15}$ cm^{-3} for dotted lines). Observed reflection spectra are typically modified by relativistic Doppler broadening

line depends on the ionization state of iron in the disk, quantified by the ionization parameter $\xi = 4\pi F n^{-1}$, with F the illuminating flux measured at the disk and n its number density. Reflection further contributes a number of emission features below 3 keV as well as a free-free continuum. The strength of these features depends on the density and composition of the reflecting material (Fig. 5.8).

The reflection spectrum undergoes relativistic Doppler broadening from the motion of gas in the disk, which is stronger if the inner disk is closer to the neutron star (cf. with Sect. 5.1.2). A reflection spectrum, therefore, contains information on the location of the inner disk, its ionization state, composition, and density, and the geometry of the disk. Current reflection models cover only part of this large parameter space, making accurate interpretation of the spectra challenging. Thus far, reflection has been detected in two superbursts [9, 101, 103, 104, 172] and one intermediate duration burst [105]. The reflection signals hint at a strong impact of these bursts on their environment, showing the evolution of the disk inner radius and its ionization parameter.

5.4.2 *Anisotropic Emission*

An interesting consequence of reflection combined with obscuration of the neutron star by the disk is that the observed burst flux depends on the inclination angle of the disk with respect to the line of sight. Both the burst flux from the neutron star and the flux from the disk are anisotropic. If the inclination angle is known, the anisotropy factors ξ can be calculated, which are defined as $L = 4\pi d^2 \xi F$, with L and F the intrinsic luminosity and the observed flux, respectively, of either the star or the disk, and with source distance d [47, 70, 113].

Unfortunately, the inclination angle of most bursting systems is poorly constrained (e.g. [55]). Moreover, most predictions for ξ assume that the accretion disk is thin and extends down to the surface of the neutron star [47, 113]. At lower mass accretion rates, however, the disk is truncated at some distance from the star. Also, different disk shapes will alter the anisotropy ξ [70]. Lack of information on the disk geometry hampers accurate predictions of ξ. The observation of reflection fractions in excess of $\gtrsim$0.5 (the maximum for thin disks) indicates a concave geometry. For the two superbursts observed with *RXTE*/PCA large reflection fractions in excess of unity were inferred [9, 103], suggesting such a disk shape. However, because of the limited data quality alternate interpretations are possible that are consistent with a thin disk [104].

The anisotropy has important consequences for a range of measured quantities, from the burst peak flux and fluence, to the mass accretion rate derived from the persistent flux. It impacts distance measurements that use the peak flux of Eddington-limited bursts. Different values of the Eddington flux have been derived from bursts of the same source [52], but the effect of variations in ξ in different accretion states has not been investigated yet. Radius measurements are impacted as well, as they are often derived from the normalization of the spectrum.

Unfortunately, the effects of anisotropy are often neglected, because the system inclination is difficult to measure accurately.

5.5 Burst Oscillations and the Neutron Star Spin

The evolutionary role for accretion-powered neutron stars in low-mass binaries as precursors to rotation-powered millisecond pulsars [1] has been well established with the discovery of several types of coherent millisecond oscillations, by which the neutron-star spin frequencies may be measured. Most compelling has been the discovery of persistent X-ray pulsations at frequencies typically in the range 200–600 Hz (e.g. [148] and Patruno and Watts, this volume). Those sources also confirmed the earlier discoveries of burst oscillations as an alternative tracer of the neutron star spin [24].

Burst oscillations are coherent, periodic variations in the X-ray intensity of bursting sources that are observed during a thermonuclear burst (see [193] for a review). These oscillations were first detected in the well-studied burst source 4U 1728−34 by *RXTE* shortly after its launch, in 1996 [175]. To date oscillations have been detected in 17 sources, with tentative (low significance, or single-burst detections, or both) claims for a further 9 sources. The oscillations are detected at a typical fractional amplitude of 0.05–0.2, and are within a few Hz of a frequency value characteristic for each source. The oscillation frequency tends to drift upwards by 1–2 Hz while it is present. For those sources which also show pulsations in the persistent (accretion-powered) X-ray emission, the frequencies of the persistent pulsations and burst oscillations are within a few Hz of each other, confirming the burst oscillation frequency as (roughly) the neutron star spin [24]. Oscillations are not observed in every burst, but instead are preferentially seen at high accretion rates [137, 142], when the spectrum is in the "soft" state (see Sect. 5.1.2). Even in those bursts where they are present, the oscillations are not found throughout the burst, instead tending to occur most often in the burst rise, and less frequently through to the burst tail [53].

The mechanism giving rise to the oscillations remains unknown; candidates include the spreading hotspot model, the related "cooling wake" that follows, and models involving surface modes of various kinds. The rise in fractional burst oscillation amplitude in the early phases of the burst generally supports the hotspot model [25, 176], although once the burning has spread to the entire surface it is difficult to understand how oscillations can persist. If the cooling timescale is similar to the spreading timescale, it seems possible that the temperature contrast can also produce oscillations in the tail [123], although such models have difficulty achieving the high amplitudes that are observed. Models invoking rotation modes [73] suffer from difficulties reproducing the observed frequency drifts [13].

In the absence of certainty as to the mechanism, much effort has been focussed on understanding why oscillations are not seen in every burst. Studies of the assembled *RXTE* sample have confirmed that oscillations preferentially occur at higher accretion rates [142]. There is some evidence [207, 208] linking the presence of oscillations in the burst tails with the characteristic evolution of the blackbody normalisation (proportional to the emitting area, up to a factor of the colour correction; cf. with Fig. 5.4). However, since the spectral evolution is also correlated with the persistent spectral state (e.g. [91, 152]), the causal relationship between these phenomena remains unclear.

At the same time, theoretical studies have revisited the microphysics of the burning taking place in bursts. Simulations of flame propagation for helium detonations, with possible application to intermediate-duration bursts (see Sect. 5.7.1) have been performed [165, 210]; however, most bursts are likely deflagrations rather than detonations. Other simulations have modelled the propagation of the burning front (e.g. [21]), and the influence of the latitude of the ignition point and the neutron star spin. Spreading is understood to occur at different speeds in the latitudinal and longitudinal directions, and even at different speeds towards and away from the equator [22, 166]. The discovery of Terzan 5 X-2, a burster with a spin rate of 11 Hz (more than an order of magnitude slower than the next fastest burster) has provided an additional avenue to such studies [20].

Along with accretion-powered pulsations, burst oscillations have held the promise for constraining neutron star mass and radius, via measurements of the Doppler shift of the burst spectrum. However, the most recent studies remain inconclusive [5], suggesting that significantly higher signal-to-noise is required, as might be provided by a next-generation mission such as the Enhanced X-ray Timing and Polarization mission (*eXTP*; [209]), currently in phase A study with a launch expected from 2025 or beyond. Another proposed mission offering effective area significantly in excess of the *RXTE*/PCA is the Spectroscopic Time-Resolving Observatory for Broadband Energy X-rays (*STROBE-X*; [197])

5.6 mHz Oscillations and Marginally Stable Burning

Quasi-periodic oscillations (QPOs) at mHz frequencies have been detected from $\approx 5\%$ of bursting sources. The behaviour of the mHz QPOs is related to the occurrence of bursts, and thus the QPOs are thought to result from an oscillatory burning mode, as expected from marginally stable burning (regime VI in Table 5.1). Numerical simulations of this burning regime broadly reproduce the observed properties of mHz QPOs, although some of the details do not match. The discrepancies are principally related to the difficulty in matching theoretically-predicted burning regimes to the observed mass accretion rates, and mHz QPOs may provide important insight in how to improve theoretical expectations. Here we review the observations of the mHz QPOs and the theory of marginally stable burning, and discuss to what extent they match.

5.6.1 Observations of mHz QPOs

QPOs at frequencies of 5–10 mHz with fractional amplitudes of $\approx 2\%$ have been detected in observations of 5 bursting sources: 4U 1636−536, Aql X-1, 4U 1608−52 [155] (Fig. 5.9), Terzan 5 X-2 [118], and 4U 1323−619 [170]. All sources likely accrete hydrogen-rich material, judging from the properties of the Type I bursts that they exhibited. The power spectrum reveals an oscillation mode with a period of a few minutes, and the oscillations can often be seen by eye in the X-ray light curve. The QPOs occur in a narrow range of persistent flux (or mass accretion rate) for each source [118, 120]. For 4U 1636–536 it was shown that the frequency evolves with time, and when it drops below a certain value, an X-ray burst occurs [2, 119, 120]. This correlation is a strong indication that the oscillatory behaviour is related to the nuclear burning processes in the neutron star envelope. After the burst, the mHz QPO is absent for a while, and it returns to repeat the frequency evolution before the occurrence of the next burst. There is no apparent correlation between the frequency evolution and the persistent flux. Furthermore, no (strong) correlation has been found between the frequency and the photospheric temperature [119, 120]. However, the limits are still consistent with predictions from multi-zone models. Phase-resolved spectroscopy has been performed for one case, and it suggests that the oscillations are produced by a hot spot of constant temperature that is periodically changing in size [169]. This is at odds with the

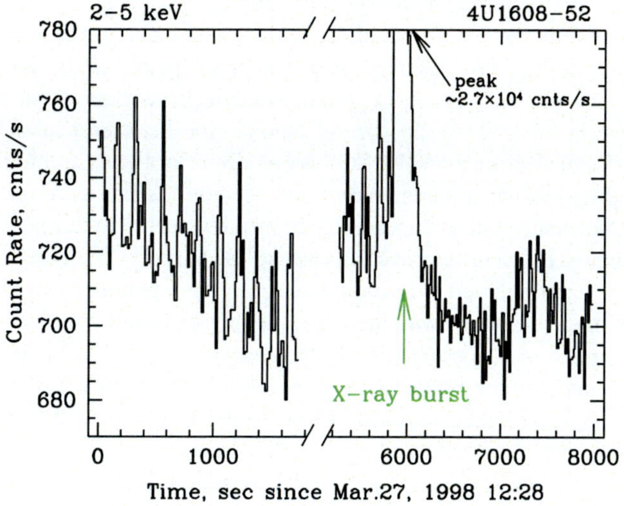

Fig. 5.9 mHz QPOs are observable in the light curve of 4U 1608−522 up to the X-ray burst around 6000 s. The amplitude of the oscillations is much smaller than the burst, and the burst peak is not visible in this zoomed-in figure. After the burst the oscillations are no longer detected, and the visible variability is Poisson noise. It suggests that the mHz oscillations originate from the same location as the burst, and are likewise produced by nuclear burning in the neutron star envelope (regime VI in Table 5.1). Reproduced with permission from [155] © ESO

common theoretical interpretation, but it remains unclear whether it signals a failing of the theory or is the result of the limited quality of the data.

The behaviour that we described was found for all mentioned sources, except Terzan 5 X-2. The behaviour of the bursts and mHz QPOs of this source has been unique among the known bursters. This transient source went into a month-long outburst in 2011, in which the accretion rate steadily increased and subsequently decreased until the source was once again quiescent. During the outburst rise, the burst rate steadily increased until the recurrence time was a mere 3 min, and the light curve exhibited oscillations. This interval was followed by an absence of bursts or oscillations at the highest persistent flux. As the flux gradually returned to quiescence, the reverse was observed: oscillations reappeared, followed by bursts, whose recurrence time increased as the flux decreased. The behaviour of Terzan 5 X-2 is different from the other four sources, in which bursts and QPOs alternated while the persistent flux remained approximately constant at a value corresponding to a much lower accretion rate. Furthermore, for those sources there is a large difference between the burst recurrence time of several hours and the oscillation period of a few minutes. For Terzan 5 X-2 there was no clear transition at which bursts ceased and mHz QPOs appeared.

5.6.2 *Theoretical Interpretation: Marginally Stable Burning*

From the apparent interaction between the mHz QPOs and Type I bursts, as well as from the energetics of the observed oscillations compared to the persistent flux, it was argued that the mHz QPOs could be produced by a nuclear burning mode [2, 155]. Oscillatory burning modes are predicted to be present at the transition of stable and unstable burning [144], because of competition between heating and cooling processes. Stable burning results from an equilibrium between cooling processes and heating by nuclear burning, taking into account the rate at which fresh fuel is accreted. Conversely, bursts appear after a period of accretion during which cooling was sufficiently efficient to prevent most nuclear burning. Once the bottom of the fuel column is compressed to reach a sufficiently high temperature and density, the nuclear burning rate runs away unstably, quickly overtaking the cooling rate and producing a burst. At the transition between the stable and unstable burning modes, nuclear burning is "marginally" stable: the nuclear burning rate starts to rise, but the cooling rate quickly catches up and prevents a runaway. This alternates to produce an oscillatory burning mode with a period, P, that is the geometric mean of the cooling (thermal) timescale, t_{therm}, and the accretion timescale (time to replenish the burned fuel), t_{acc}: $P \simeq \sqrt{t_{\mathrm{therm}} t_{\mathrm{acc}}}$ [72]. This picture is confirmed by multi-zone numerical simulations. The marginally stable burning mode is reproduced in a narrow range of mass accretion rates at the stability transition. For current standard theory, this transition takes place at the Eddington limit, whereas mHz QPOs are typically observed at a ten times lower mass accretion rate.

The mass accretion rate of the stability transition ($\dot{M}_{st}$) depends on the conditions in the neutron star atmosphere and potentially the burning latitude. These conditions include the effective gravity in the neutron star envelope and the composition of the accreted material [72, 103]. Equally important are the details of the processes that take place in the atmosphere. In particular, an increased heat flux into the atmosphere substantially lowers $\dot{M}_{st}$ [98]. Similarly, turbulent mixing induced by rotation or a magnetic field lowers $\dot{M}_{st}$ by reducing the time it takes to bring fresh fuel to the burning depth [98, 151]. Furthermore, several nuclear reactions have been identified that play a key role in setting the stability of the burning processes. The CNO breakout reaction $^{15}O(\alpha, \gamma)^{19}Ne$ is one of the most important in this respect [35, 44, 45], but its reaction rate is poorly constrained by nuclear experiments, leading to a substantial uncertainty in $\dot{M}_{st}$ [103]. Competition of this reaction with another breakout reaction, $^{18}Ne(\alpha, p)^{21}Na$, has been shown to increase the range of $\dot{M}$ where burning is marginally stable [103].

The oscillation period depends on t_{acc}, and is therefore expected to be a function of $\dot{M}_{st}$. Similarly, the waveform and the amplitude of the oscillations depend on $\dot{M}_{st}$. Additionally, changes are expected as a function of the "distance" from the stability transition. For example, at a mass accretion rate close to $\dot{M}_{st}$, the amplitude is smaller, the waveform is more symmetric, and the frequency is higher [103].

Marginally stable burning is an elegant explanation for mHz QPOs, as it is expected to occur in a narrow range of mass accretion rate, broadly matching observations. The predicted periods are consistent with the observed values of a few minutes. However, if the observed accretion timescale is considered, the predictions for the period are actually larger. Furthermore, marginally stable burning takes place at the transition from bursts to stable burning; this appears to be the behaviour observed from Terzan 5 X-2, but not for the other sources, where bursts and mHz QPOs alternate, and the transition to stable burning takes place at higher persistent flux [97]. Instead, these sources more closely resemble models that include a slowly decreasing heat flux into the neutron star envelope. Those simulations display oscillations with an evolving frequency up to the ignition of a burst [98]. This suggests that the observed mHz QPOs are indicative of cooling of the atmosphere, possibly heated by a preceding burst [120]. However, multi-zone models that self-consistently include the heating and cooling of deeper layers, do not exhibit alternating oscillations and bursts. Therefore, while marginally stable burning is the most likely process to power mHz QPOs, the circumstances under which it is observed are challenging to explain with current theory.

5.7 Burst Duration and Fuel Composition

The majority of observed bursts have durations of ~10–100 s. A typical burst ignites at a depth of $y \simeq 10^8\,\mathrm{g\,cm^{-2}}$, where the thermal timescale is of the order of 10 s. If all fuel burns quickly after ignition, this cooling timescale sets the burst duration. If the fuel is hydrogen-rich, however, nuclear burning via the *rp*-process can prolong the burst for up to $\approx$100 s [e.g., 159].

Historically the burst timescale has been measured in a number of ways that sometimes make it difficult to compare different analyses. Measurements that rely on the length of the interval during which the burst emission exceeds some factor of the persistent noise level, are subject to variations due to different instrumental sensitivities. Burst decays have been fitted with single (or sometimes multiple) exponential decay segments, with the decay timescale quoted as a measure of the burst duration (e.g., [53]). In high signal-to-noise lightcurves, such models generally offer poor fits to the decay, and in fact more recent analysis favours power-law or more complex combinations [83, 85]. A third, oft-quoted measure is the ratio of the burst fluence to the peak flux, i.e. $\tau = E_b/F_{\mathrm{peak}}$ (e.g., [187]). However, this quantity does not offer an unambiguous mapping to the burst fuel composition. For example, mixed H/He bursts with long *rp*-process tails may have similar τ-values as long, intense PRE bursts arising from ignition of a deep pile of pure He.

In rare cases bursts have much longer durations, of minutes or tens of minutes (intermediate duration bursts) to many hours (so-called "superbursts"). These durations cannot be explained by prolonged burning, but are comparable to the cooling timescales of deeper layers. As it takes a longer time to accumulate a larger fuel layer, these bursts have long recurrence times of months to years. Observations are, therefore, rare, and have mostly been performed with wide-field or all-sky instruments, which yield data of modest quality.

The long bursts reach similar peak fluxes and photospheric temperatures as regular bursts. The intermediate duration bursts all reach the Eddington limit, whereas most superbursts have a lower peak flux. From an observer's perspective, the main discriminating characteristic is the long duration and correspondingly large fluence of these events (Table 5.2).

Long bursts are of particular interest, because they probe deeper regions of the neutron star envelope down to the outer crust. Furthermore, long durations allow for more detailed spectra to be obtained, displaying interesting behaviour such as interaction with the accretion environment (Sect. 5.4).

5.7.1 *Intermediate Duration Bursts*

Intermediate burst durations have been observed with exponential decay timescales of several minutes up to $\sim$40 min. These events are thought to be powered by deep helium ignition. At low mass accretion rates of $\approx$0.001–0.01 $\dot{M}_{\mathrm{Edd}}$, the neutron star envelope is relatively cool, allowing a large helium pile to accumulate prior to ignition. Furthermore, at low temperatures, the ignition depth strongly increases with temperature (Fig. 5.2), such that small temperature variations produce bursts with a large range of ignition depths and durations. The durations can be anywhere between a few minutes to tens of minutes, and may even rival the hours-long superbursts (Sect. 5.7.2). In a few cases of very low persistent flux, the burst decay could be observed for several hours [36, 105]. Observationally, the start of these longer events often falls in a data gap, so that the burst duration and fluence are only known approximately, making it difficult to accurately classify them. These events are observed both from sources where the accretion composition is helium-rich [110] and hydrogen-rich [42]. In the latter case, hydrogen burns at a shallow depth either stably (regime III in Table 5.1) or in weak flashes (regime II), producing a deep layer of helium.

As in all energetic helium-rich bursts, the burst flux reaches the Eddington limit, causing PRE (Fig. 5.10). Often the radius expansion is particularly strong, producing so-called "superexpansion" where the photospheric radius increases by a factor $\sim 10^2$. For two intermediate duration bursts, expansion velocities of up to 30% of the speed of light were inferred [84]. In the tail of some intermediate duration bursts strong flux variability is observed (Fig. 5.10), which may indicate interaction between the burst and the accretion environment (see also Sect. 5.4). Finally, convection at the onset of these powerful bursts can potentially mix heavy ashes into the photosphere. Combined with their relatively long duration, this

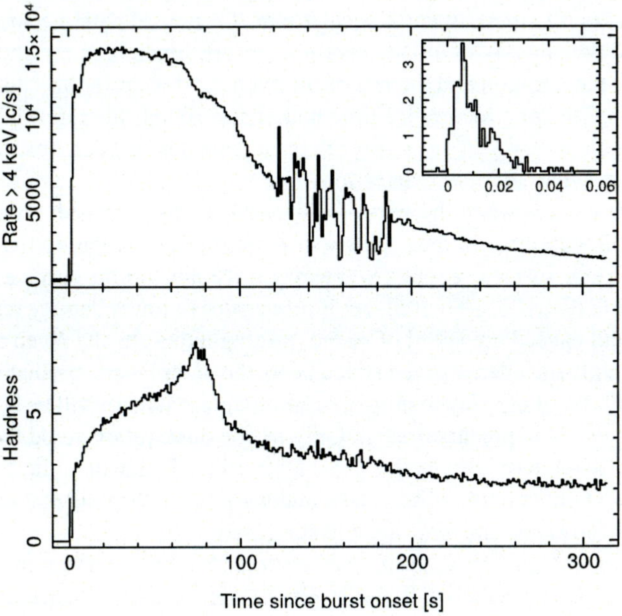

Fig. 5.10 Light curve of an intermediate duration burst from 2S 0918−549 observed with *RXTE*/PCA. The observed count rate in the top panel exhibits strong variability in the burst tail, which is achromatic in the PCA band pass, as it is not apparent in the hardness (the ratio of the counts exceeding 4 keV to those below; bottom panel). The inset in the top panel zooms in on a short "precursor" at the burst onset, marking the start of a brief period of superexpansion, during which the burst spectrum temporarily drops below the PCA band. Although superexpansion lasts mere seconds, photospheric radius expansion continues up to the peak in the hardness ("touch down"). Adapted with permission from [81] © ESO

makes intermediate duration bursts the prime candidates for detecting redshifted spectral features from the neutron star surface to constrain the dense matter EOS (Sect. 5.3.2).

5.7.2 Superbursts

X-ray bursts that last between several hours and a day are termed "superbursts". Their duration corresponds to the cooling timescale at a column depth of 10^{11}–$10^{12}\,\mathrm{g\,cm^{-2}}s$. At a mass accretion rate of $0.1\,\dot{M}_{\mathrm{Edd}}$, it takes roughly a year to accumulate such a layer. During that time all superbursting sources exhibit short bursts. Therefore, the freshly accreted hydrogen and helium burns at a smaller depth, and superbursts ignite in the carbon-rich ashes (Fig. 5.2). Most superbursts are observed from sources with accretion rates of the order of $0.1\,\dot{M}_{\mathrm{Edd}}$, but exceptional cases are known: 4U 0614+09 with $\sim 0.01\,\dot{M}_{\mathrm{Edd}}$ exhibited two superbursts [110, 164], and GX 17+2 at $\sim 1\,M_{\mathrm{Edd}}$ showed four [78]. Because their duration is often much longer than the typical satellite (low-Earth) orbit of ≈90 min, the start is in many cases obscured by Earth occultations. Combined with the sparse sampling and limited data quality of all-sky instruments, this limits in many cases our ability to measure the duration and fluence of the event, and we refer instead to *candidate* superbursts. The long recurrence time makes superbursts the rarest burst category: since their discovery [27, 172] only 25 (candidates) have been observed (see [75] for a recent review and list of detections).

In the rare cases when the start of a superburst was observed, a short precursor is visible directly prior to the superburst. After the carbon ignition ∼100 m below the neutron star surface, a sound wave may travel toward the surface ahead of the carbon flame [94, 100, 194, 195]. As it propagates to lower density regions, it may speed up and exceed the speed of sound, turning into a shock. When it reaches the surface, the kinetic energy of the shock heats the outer layers, producing a short X-ray burst. Furthermore, any hydrogen or helium present there will be burned, adding to the energy of the precursor burst. Only in one observation could the precursor's fluence be accurately determined: it was 40–100% larger than the fluence of the regular short helium bursts observed from this source, indicating that another source of energy contributed, such as shock heating [93].

Following the precursor, the rise of a superburst light curve typically is rather slow: in 2001, the superburst flux from 4U 1636−536 reached its peak only after ∼900 s (Fig. 5.3 bottom). With the exception of the 1999 superburst from 4U 1820−20, which reached the Eddington limit [172], most superbursts likely have such a slow rise. The shape of the light curve is attributed to the cooling of an envelope with a particular temperature profile [104]. This profile is left behind by the radially outward moving carbon flame. Fits with cooling models measure a profile close to $T \propto y^{1/4}$. This suggests that as the carbon flame moved into lower density regions near the surface, the fraction of carbon that burned decreased, and the flame likely stalled before reaching the photosphere.

After the superburst has decayed, it continues to have an effect on the neutron star envelope by quenching the occurrence of regular bursts for several weeks [107]. Heat deposited by the superburst in the envelope causes freshly accreted hydrogen and helium to burn in a stable manner (regime VII in Table 5.1). Even after the superburst has decayed, the stable burning can sustain itself for several weeks. Once the envelope has cooled down sufficiently, the burning mode transitions back to producing bursts [100].

Superbursts ignite close to the outer crust, and are sensitive to the thermal properties at that depth. This region is rather poorly understood, with observations of cooling quiescent sources suggest the presence of an unknown heat source [18, 38], and where Urca neutrino cooling may play an important role in regulating the temperature [161]. In recent years superburst candidates have been detected from transient sources [97, 164]. These observations pose several challenges. First, the persistent flux may evolve on similar timescales to a superburst, and it is difficult to make the definite determination of the thermonuclear nature of the event. Second, the superbursts occur within months or sometimes days of the start of an accretion outburst [3, 163]. Current theory predicts that the outer crust of the neutron star is not heated sufficiently at that time to ignite a carbon flash. Yet, superbursts are observed, suggesting the presence of an unknown heat source near the superburst ignition depth. This may be related to the shallow heat source inferred for cooling quiescent sources [39].

It is challenging to explain the origin of the carbon fuel. Comparison of the X-ray light curves to cooling models finds that the carbon mass fraction of the superburst fuel is 15–30% [32]. The rest of the fuel may be iron and heavier elements produced during the *rp*-process. If a substantial fraction is heavy isotopes near the end point of the *rp*-process, their photodisintegration could account for as much as 50% of the superburst energetics [160]. However, simulations typically predict most heavy isotopes to be near the iron-group [90, 201]. Carbon is typically not accreted in substantial quantities, and therefore has to be produced by nuclear burning of the accreted hydrogen and helium. 3α burning of helium during "normal" (mixed H/He) bursts produces carbon, but the high temperatures reached during these events enable α- and proton-captures also to destroy carbon. Detailed simulations predict that the net carbon production by normal bursting activity is at most a mass fraction of $\approx 5\%$ [201]. Stable burning of hydrogen and helium can produce large quantities of carbon [168], but this burning has also been thought to only take place at high accretion rates, where the temperature in the envelope again is high enough to allow for the reactions that destroy carbon. A new stable regime has recently been identified in simulations [95], where stable burning produces copious quantities of carbon at lower mass accretion rates near $0.1\,\dot{M}_{\rm Edd}$, similar to where most superbursts are observed. All superbursting sources also exhibit short hydrogen or helium flashes, but they have a high value of the α-parameter [77]. This suggests that a substantial part of hydrogen and helium are burned in a stable manner in between the short bursts. Future studies will need to determine if this involves the newly suggested stable regime as a source of the carbon fuel for the superbursts.

In the absence of spectroscopic information, superbursts provide the only observational constraints on the composition of the ashes of hydrogen/helium

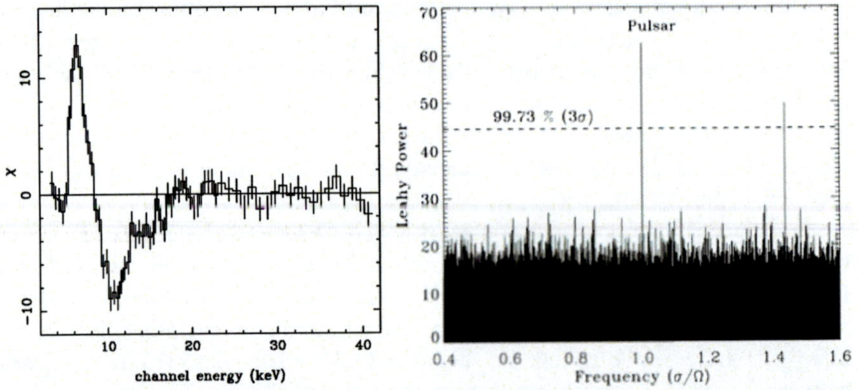

Fig. 5.11 Detailed spectroscopy and timing studies of the superburst from 4U 1636−536 observed in 2001 with *RXTE*/PCA (see also Fig. 5.3 bottom). Left: after subtraction of the thermal continuum spectrum, the iron emission line and absorption edge are visible as signatures of reflection off the disk (from [172]). Right: power spectrum during the burst showing pulsations detected at the neutron star spin frequency, Ω, as well as another mode of unknown origin at higher frequency (from [174])

burning in neutron star envelopes. Furthermore, superburst ashes are compressed and form the crust. These ashes consist predominantly of ^{56}Fe created in nuclear statistical equilibrium, where any heavy *rp*-process isotopes have been destroyed by photodisintegration.

The long durations of superbursts allow for detailed spectra to be obtained compared to short bursts. Most detections are, however, with instruments that have a relatively small effective area and a high background. The highest quality spectra were obtained for two superbursts with the *RXTE*/PCA: 4U 1820−30 in 1999 and 4U 1636−536 in 2001. The spectra exhibited photoionized reflection of the burst off the accretion disk (Fig. 5.11, left) and a variable persistent component (see also Sect. 5.4). A detailed timing study also revealed pulsations at the neutron star spin frequency, modulated by Doppler shifts from the binary orbit, as well as a (presumed) global oscillation mode at higher frequency (Fig. 5.11 right).

5.8 Thermonuclear Burst Simulations

Simulating X-ray bursts is a problem with extreme scales. The strong dependence of the nuclear reaction rates on temperature demands time steps of less than a nanosecond, whereas the recurrence time of bursts is hours and that of superbursts is typically a year or longer. Resolving convective mixing at the onset of a burst requires a spatial resolution of half a centimeter [125], whereas a flame spreading around the star travels over 30 km. Furthermore, for each zone of the model thousands of nuclear reactions have to be evaluated among hundreds of isotopic species. Numerical simulations of X-ray bursts, therefore, need a large number of time steps and zones as well as an extensive nuclear network, a computational task which is not yet feasible on present-day hardware. All current simulations reduce the problem. 2D and 3D models explicitly model turbulent mixing, but typically simulate only part of a burst with a small selection of nuclear reactions to limit the computational expense. On the other hand, one-dimensional simulations include an approximation of turbulence, but this allows them to use an implicit scheme and calculate series of multiple complete bursts. Moreover, the number of zones in such a model is small enough to use a full nuclear network. Here we discuss the progress in recent years for the different categories of models.

5.8.1 Single-Zone Models

The largest simplification of any simulation is the one-zone model, where the entire neutron star envelope is reduced to a single zone. Time evolution can be implemented by considering the changes in the the fuel column due to accretion and nuclear burning. Such simplification is helpful when illustrating the different regimes of nuclear burning ([48, 72]; see also Fig. 5.2). Single-zone models, however, miss several important effects from nuclear burning taking place simultaneously at different depths or from the mixing of the fuel and ashes layers. That behaviour can only be resolved with multi-zone models.

By considering only a single zone, most computational resources can be applied to large nuclear networks. Such models are used to evaluate the impact on X-ray burst light curves and burning ashes of uncertainties in a large number of reaction rates [34, 74, 146]. Whereas some one-zone models include the thermodynamic response of the neutron atmosphere to the nuclear burning, others rely on the ignition conditions or even full thermodynamic trajectories from multi-zone models. The latter is not a good approach for studying X-ray bursts, because changes to any rate potentially produce a significant change in the temperature, invalidating the trajectory. Such changes to the trajectory are difficult to predict a priori, as the nuclear flow may progress through different branches, and a change in one reaction may cause the nuclear flow to be rerouted through another path. The collective effect of the reactions along the altered path can significantly affect the ashes composition and the observed light curve. Therefore, an important aspect of evaluating a nuclear reaction rate is its impact on *changing* the thermodynamic trajectory. Given that multi-zone models can now be run relatively fast on current hardware, the use of trajectories should be avoided, and the only remaining role for one-zone models may be in exploratory studies of large parameter spaces, to be followed up with multi-zone models [34].

5.8.2 One-Dimensional Multi-Zone Models

One-dimensional models resolve the neutron star envelope into multiple zones in the radial direction, usually on a Lagrangian grid which employs mass coordinates.

Most simulations use an implicit hydrodynamics scheme, which assumes hydrostatic equilibrium and uses mixing length theory to approximate convection and other turbulent mixing processes. This allows for the model to be evolved over long times, simulating multiple complete bursts. The past decade saw the introduction of full nuclear reaction networks covering up to ≈1300 isotopes [44, 90, 201] (Fig. 5.14). These models are being used to study a wide range of bursts and burning phenomena [112]. A major success of these models is their good agreement with the observed light curve of bursts from the "Clocked Burster", GS 1826-24 [71, 80] (Fig. 5.12).

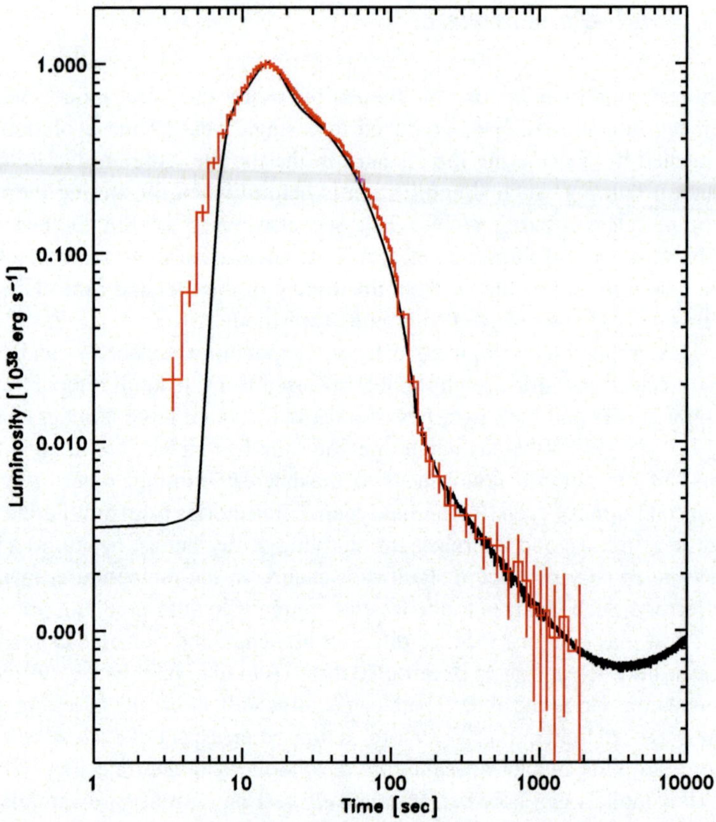

Fig. 5.12 Comparison of a burst observed from GS 1826−24 with *RXTE*/PCA (red histogram with error bars) and a one-dimensional multi-zone simulation with KEPLER (black line). The model of a mixed hydrogen/helium burst (regime V in Table 5.1) reproduces the burst shape well. The light curve from the peak to ≈120 s is shaped by the *rp*-process, which is modeled in full detail. Afterwards, the flux decay follows the cooling of the neutron star envelope, which can be traced in the observation for over 1000 s. The observed rise is slower than predicted: this may be the effect of flame spreading, requiring multi-dimensional models. Adapted with permission from [80] © ESO

The implementation of large nuclear networks confirmed that approximated networks were responsible for irregular bursts in past models, especially with respect to approximations of the *rp*-process [181]: full reaction networks are required to simulate behaviour consistent with observations [201]. Moreover, interactions between different zones influence the nuclear burning. Ashes of the previous burst are mixed with fresh fuel, and typically shorten the waiting time for the next burst to ignite. This effect is referred to as "compositional inertia" [90, 181, 201]. Bursts with a smaller ignition column depth are less powerful, reach lower peak temperatures, and the nuclear flow does not extend as far along the *rp*-process path. Therefore, compositional inertia changes the composition of the ashes that fuel superbursts and forms the outer crust. Although it is typically not taken into account self-consistently in burst simulations, nuclear heating and cooling processes in the crust depend very sensitively on the ash composition [39, 161]. Furthermore, as many of the required nuclear reaction rates are highly uncertain, extensive studies have been undertaken to find the most influential reactions and to quantify their effect on burst observables [34, 74, 146]. The CNO breakout reaction $^{15}O(\alpha, \gamma)^{19}Ne$ is found to be particularly important for the stability of the burning processes ([34, 35, 44, 45, 103]; see also Sects. 5.1.1.4 and 5.6.2).

The burning processes powering pure helium bursts were thought to consist of 3α and a straightforward series of α-captures. Simulations with full networks find, however, that (α, p) reactions can produce a small number of protons. The protons act as a catalyst and substantially speed up the burning process by by-passing $^{12}C(\alpha, \gamma)^{16}O$ through the faster reactions $^{12}C(p, \gamma)^{13}N(\alpha, p)^{16}O$.

In recent years multi-zone models of deep carbon burning have been created to model superbursts [100, 194, 195]. These models directly accrete carbon-rich material to avoid the computational expense of simulating $\sim 10^3$ hydrogen/helium flashes, as well as to avoid the problem that models of hydrogen/helium typically do not produce sufficient carbon. In some studies a hydrogen/helium atmosphere is added to the model close to the moment of ignition, such that the effect of the deep carbon flash on the outer atmosphere can be investigated. Hydrogen/helium burning at the superburst onset contributes to the precursor burst [100, 194]. Furthermore, the hot ashes of the superburst cause freshly accreted hydrogen and helium to burn stably, quenching short X-ray bursts for days to weeks [100, 107]. An unsolved issue is that the carbon burning propagates radially as a convective flame, but convection is not properly modeled by implicit codes. The one-dimensional simulations find the flame to spread supersonically as a detonation [195], whereas multi-dimensional models of comparable carbon flames in Type Ia supernovae produce deflagrations [202].

Multi-zone models have been very successful at simulating a wide range of observed bursts and other burning behaviour (e.g. [112]). It has proven a challenge, however, to reproduce the observed conditions where the different burning regimes occur. In part this is due to the large parameter space that needs to be investigated because of uncertainties in, for example, the accretion composition and key nuclear reaction rates. Furthermore, heating and cooling processes in the crust have a strong effect on the burning behaviour [94, 98, 206], but their influence has been explored in a limited way thus far. The same holds true for rotationally induces turbulent mixing [98, 151].

5.8.3 Multi-Dimensional Models

Multi-dimensional models of X-ray bursts either simulate a small box in the neutron star envelope aiming to resolve convection, or they simulate flame spreading across the entire envelope with a cruder implementation of turbulent mixing. The earliest efforts in the former category are simulations of deep helium ignition [165, 210], which may be applicable to the most powerful intermediate duration bursts. For pure helium bursts, a small nuclear network may suffice. Furthermore, these simulations find that at large depths, the fast helium flame travels as a detonation, such that only a short simulation time is required. Fast flame propagation could be a requirement for explaining exceptionally fast rise times observed in the most powerful helium bursts [84]. Recognizing that most observed bursts ignite at a shallower depth and thus propagate more slowly as a deflagration, a new "low Mach number" code has been developed to make simulations more efficient. The first simulations with the MAESTRO code have explored the resolution requirements, and find that a zone size of 0.5 cm is required to resolve convective mixing during the burst [125, 126, 211] (Fig. 5.13). Potentially, such simulations can be used to calibrate the approximation for mixing in the one-dimensional codes.

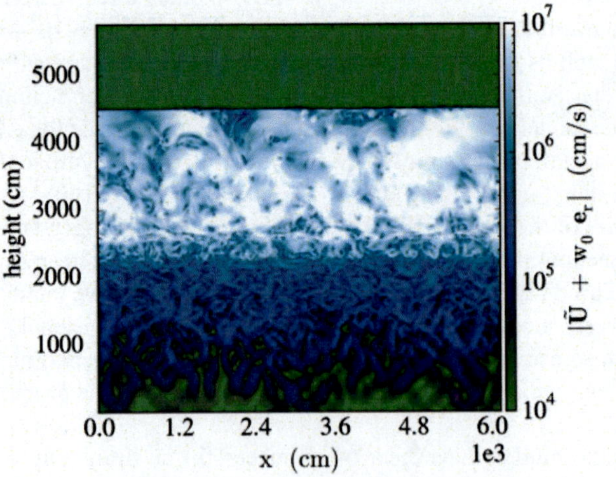

Fig. 5.13 Magnitude of the velocity field in a two-dimensional hydrodynamics simulation with MAESTRO. Due to the large amount of heat generated by the thermonuclear runaway, convection is initiated. It dominates the energy transport shortly after the burst onset, and mixing of the composition may deposit burst ashes close to the surface. To correctly model convective mixing during an X-ray burst, a spatial resolution of 0.5 cm is required [125]. Adapted from [211]

At the onset of a burst, the flame travels a longer path than what can be fully resolved by current simulations. Two-dimensional simulations that include the radial and latitudinal (or longitudinal) directions, adopting zones that are elongated in the latter dimension. The zones are not fine enough to resolve the small-scale turbulence during the convective stage at the burst onset, but they are sufficient for modeling the larger scale mechanisms that drive the flame propagation. These simulations find that initially, the flame must be confined by the Coriolis force and/or magnetic tension in order to prevent it from dying out [21–23, 166]. Subsequently, a combination of the Coriolis force, magnetic tension-induced friction in the sheared burning front, and conduction causes the flame to spread across the surface.

5.9 Nuclear Experimental Physics

A significant fraction of the research effort directed at thermonuclear bursts over the last few decades has been in the area of nuclear experimental physics. The nuclear reactions that may take place in bursts, where the temperature can reach 10^8–10^9 K, were first explored by Wallace and Woosley [192]. The most important reactions are the β-limited CNO cycle, in systems that accrete hydrogen, and the triple-α reaction, which fuses helium to carbon.

The triple-α reaction first produces carbon, which (via the CNO cycle) catalyses the burning of hydrogen. CNO breakout reactions including $^{15}\mathrm{O}(\alpha, \gamma)^{19}\mathrm{Ne}$ then lead to a series of (α, p) and (p, γ) reactions (the αp-process), feeding into the rapid-proton, or *rp*-process. The *rp*-process was identified as a key reaction chain that could produce elements far heavier than any of the precursor reactions. However, experimental data for the neutron-deficient nuclei involved was lacking for many years, forcing modelers to rely on theoretical calculations for masses and rates [158]. At the same time, it was initially unclear how far the *rp*-process reactions extended. Early simulations were based on limited reaction networks, under the assumption that further proton captures beyond $^{56}\mathrm{Ni}$ could be neglected. Such simulations also tended to find non-negligible reaction products accumulating at the limits of the reaction rate networks. A simulation study using a one-zone model determined the ultimate limit of *rp*-process burning in the Sn-Sb-Te cycle [159] (see also [106]). This result implies that *rp*-process reaction products are limited to atomic number $Z \leq 54$, and that the ashes forming the crust are made up of nuclei lighter than mass number $A \approx 107$.

With the maximum extent of the *rp*-process burning now constrained, the goals of experimenters became to improve the mass measurements for proton-rich nuclei (e.g. [157]), as well as determining the key reactions that might influence the properties of the bursts. The experimental studies informed, and were further motivated by, development of more accurate numerical models of bursts, as described in Sect. 5.8. Simulations enabled a very detailed picture of the nuclear burning processes during a burst to emerge (e.g. [46]), although the fidelity of this picture remains uncertain.

Any individual reaction will have little impact on the burst profile unless a significant fraction of the fuel mass is processed through that reaction, and it's timescale is unusually slow [188]. Substantial experimental work has established several such "waiting points", including ^{60}Zn, ^{64}Ge, ^{68}Se and ^{72}Kr (e.g. [15, 150, 198]). The effect of the reactions which produce these nuclei is illustrated in Fig. 5.14, with a substantial fraction of the mass flow "piling up" in these isotopes. Further proposed waiting points, of ^{22}Mg, ^{26}Si, ^{30}S, and ^{34}Ar were suggested as the explanation for seldom-observed bolometrically double-peaked bursts [43].

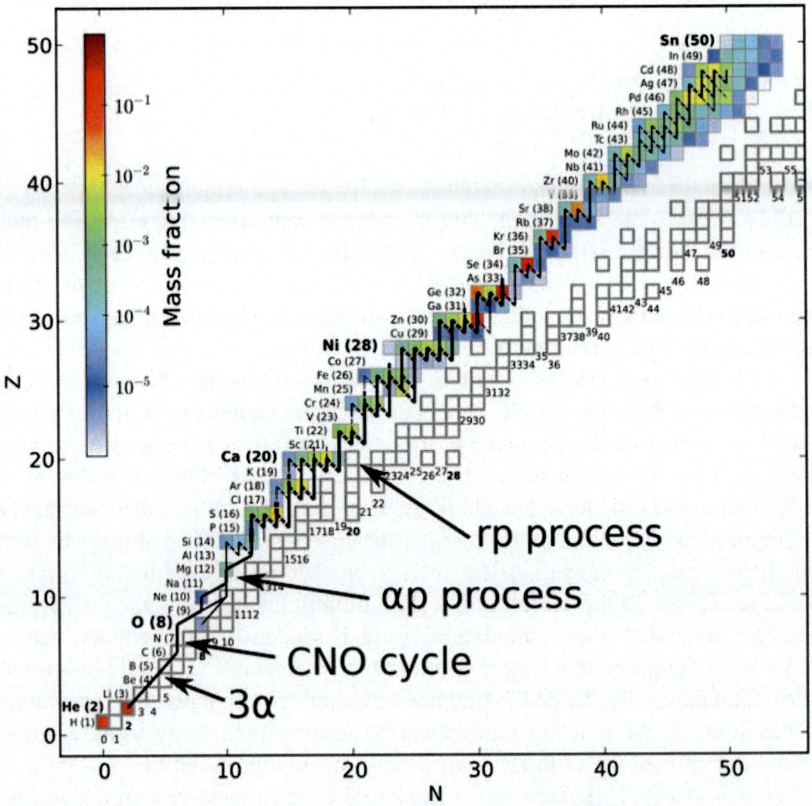

Fig. 5.14 Flow of nuclear reactions between isotopes with neutron number N and proton number Z, where thicker lines represent larger net flows; from a KEPLER simulation with a solar accretion composition [100] approximately a millisecond into the burst. Colours indicate the mass fraction of the isotopes, and a thick border marks the stable isotopes. Helium burns via the 3α process, followed by hydrogen burning in the βCNO cycle. Break-out from the cycle flows into the αp process and subsequently the *rp*-process, which extends up to tin (Sn). At this stage of the burst, most isotopes are unstable, meaning that after the burst they β-decay to form stable isotopes. The most abundant isotopes along the *rp*-process path are at "waiting points", such as ^{64}Ge, ^{68}Se and ^{72}Kr, where reaction products accumulate due to slow subsequent β-decays

However, an explanation for these events in terms of nuclear physics must also necessarily explain why the double-peaked behaviour is not observed in every burst.

More systematic investigations using a combination of single- and multi-zone models attempted to identify all of the most significant reactions that might be expected to affect the burst lightcurves [33]. These investigations were integrated into the ongoing REACLIB[7] project, which seeks to collect, curate, and distribute data on nuclear reaction rates and masses to broadly impact the astrophysical modelling community. The most recent results from this project have provided a clear list of priority reactions for investigation via nuclear experiment [34]. However, it is worth noting that the sensitivity studies are based on a single burst model, and it is possible that different reactions may have significant impacts on the lightcurves of bursts in different regimes (see Sect. 5.1.1.4).

A detailed description of the nuclear experiments focussed on reactions or masses of relevance for thermonuclear bursts is beyond the scope of this review (see instead [147]). However, we comment that in the past few years an increasing level of interaction has taken place between the experimental community, and burst observers and modellers, enabled in a large part by the activities of the Joint Institute for Nuclear Astrophysics: Centre for the Evolution of the Elements (JINA-CEE[8]). This growing dialogue, in combination with new experimental techniques, and upcoming hardware including the Facility for Rare Isotope Beams (from 2021 onwards), suggest that the most exciting time for studying the nuclear reactions in bursts is yet to come.

5.10 Summary and Outlook

In the last decade, studies of thermonuclear bursts have entered an exciting new phase, where the interplay of observations, numerical modelling, and nuclear experiments are making substantial progress on understanding the burst phenomenology. At the same time, the interpretation of observational data (motivated in part by constraining the mass and radius; see Miller chapter, this volume) has driven new investigations into the spectral formation processes.

The diversity of burst behaviour, although still not completely understood, presents a continuing challenge to these investigations. The computational challenges in simulating the entire neutron star surface remain, but even 1-D models are achieving success in reproducing an increasing range of observed phenomena. Assembly and analysis of large databases of observations and simulations appear to offer the best chance for understanding all the possible ignition cases. More comprehensive sensitivity studies have the prospects of conclusively identifying the nuclear reactions most important to the full range of burst types, and hence

[7] https://groups.nscl.msu.edu/jina/reaclib/db.

[8] http://jinaweb.org.

directly motivating future experimental efforts. Thermonuclear burst physics has been identified as a substantial priority for future nuclear physics experiments [4].

New observations are continually being sought, with the available fleet of long-duration X-ray missions in addition to a handful of recently-launched instruments. The current prospects for detection of new examples of burst oscillations appear excellent, with the 2015 launch of India's *ASTROSAT* mission, featuring a large-area proportional counter with similar capabilities to *RXTE*'s PCA. Early observations of 4U 1728−34 have detected high-frequency variability as seen earlier with *RXTE* [189]. The 2017 deployment of the *NICER* instrument to the International Space Station offered another avenue for detection of high-frequency variability, and the "first-light" observations have detected burst oscillations in 4U 1608−52.[9]

These new data also offer the prospect of new insights, particularly where the observational capabilities exceed those to date. In particular, there are exciting prospects of more detailed burst timing and spectral information from recently-launched large-area instruments including *NICER* and LAXPC, aboard *ASTROSAT*; and in the more distant future, *eXTP* and *Strobe-X*. The data from these instruments may more clearly distinguish the reflection spectrum, and perhaps also finally realise the promise of burst oscillations for precisely constraining neutron star mass and radius.

Acknowledgments The authors are grateful for helpful comments from Y. Cavecchi, M.C. Miller, and H. Schatz. This work was supported in part by the National Science Foundation under Grant No. PHY-1430152 (JINA Center for the Evolution of the Elements). The authors are grateful for support received as part of the International Team on Nuclear Reactions in Superdense Matter by the International Space Science Institute in Bern, Switzerland.

References

1. M.A. Alpar, A.F. Cheng, M.A. Ruderman, J. Shaham, A new class of radio pulsars. Nature **300**, 728–730 (1982)
2. D. Altamirano, M. van der Klis, R. Wijnands, A. Cumming, Millihertz oscillation frequency drift predicts the occurrence of type I X-ray bursts. Astrophys. J. Lett. **673**, L35–L38 (2008)
3. D. Altamirano, L. Keek, A. Cumming, G.R. Sivakoff, C.O. Heinke, R. Wijnands, N. Degenaar, J. Homan, D. Pooley, A superburst candidate in EXO 1745-248 as a challenge to thermonuclear ignition models. Mon. Not. R. Astron. Soc. **426**, 927–934 (2012)
4. A. Arcones, D.W. Bardayan, T.C. Beers, L.A. Berstein, J.C. Blackmon, B. Messer, B.A. Brown, E.F. Brown, C.R. Brune, A.E. Champagne, A. Chieffi, A.J. Couture, P. Danielewicz, R. Diehl, M. El-Eid, J. Escher, B.D. Fields, C. Fröhlich, F. Herwig, W.R. Hix, C. Iliadis, W.G. Lynch, G.C. McLaughlin, B.S. Meyer, A. Mezzacappa, F. Nunes, B.W. O'Shea, M. Prakash, B. Pritychenko, S. Reddy, E. Rehm, G. Rogachev, R.E. Rutledge, H. Schatz, M.S. Smith, I.H. Stairs, A.W. Steiner, T.E. Strohmayer, F.X. Timmes, D.M. Townsley, M. Wiescher, R.G.T. Zegers, M. Zingale, White paper on nuclear astrophysics. ArXiv e-prints, March 2016

[9] http://www.nasa.gov/press-release/goddard/2017/nasa-neutron-star-mission-begins-science-operations.

5. R. Artigue, D. Barret, F.K. Lamb, K.H. Lo, M.C. Miller, Testing the rotating hotspot model using X-ray burst oscillations from 4U 1636-536. Mon. Not. R. Astron. Soc. **433**, L64–L68 (2013)
6. T. Bagnoli, J.J.M. in't Zand, C.R. D'Angelo, D.K. Galloway, A population study of type II bursts in the rapid burster. Mon. Not. R. Astron. Soc. **449**, 268–287 (2015)
7. D.R. Ballantyne, Reflection spectra from an accretion disc illuminated by a neutron star X-ray burst. Mon. Not. R. Astron. Soc. **351**, 57–62 (2004)
8. D.R. Ballantyne, J.E. Everett, On the dynamics of suddenly heated accretion disks around neutron stars. Astrophys. J. **626**, 364–372 (2005)
9. D.R. Ballantyne, T.E. Strohmayer, The evolution of the accretion disk around 4U 1820-30 during a superburst. Astrophys. J. Lett. **602**, L105–L108 (2004)
10. N.M. Barrière, R. Krivonos, J.A. Tomsick, M. Bachetti, S.E. Boggs, D. Chakrabarty, F.E. Christensen, W.W. Craig, C.J. Hailey, F.A. Harrison, J. Hong, K. Mori, D. Stern, W.W. Zhang, NuSTAR observation of a type I X-ray burst from GRS 1741.9-2853. Astrophys. J. **799**, 123 (2015)
11. M. Bauböck, E. Berti, D. Psaltis, F. Özel, Relations between neutron-star parameters in the Hartle-Thorne approximation. Astrophys. J. **777**, 68 (2013)
12. M. Bauböck, F. Özel, D. Psaltis, S.M. Morsink, Rotational corrections to neutron-star radius measurements from thermal spectra. Astrophys. J. **799**, 22 (2015)
13. R.G. Berkhout, Y. Levin, Evolution of the bursting-layer wave during a type I X-ray burst. Mon. Not. R. Astron. Soc. **385**, 1029–1035 (2008)
14. L. Bildsten, Thermonuclear burning on rapidly accreting neutron stars, in *The Many Faces of Neutron Stars*, ed. by R. Buccheri, J. van Paradijs, A. Alpar (Kluwer, Dordrecht, 1998), p. 419
15. B. Blank, S. Andriamonje, S. Czajkowski, F. Davi, R. del Moral, J.P. Dufour, A. Fleury, A. Musquère, M.S. Pravikoff, R. Grzywacz, Z. Janas, M. Pfützner, A. Grewe, A. Heinz, A. Junghans, M. Lewitowicz, J.-E. Sauvestre, C. Donzaud, New isotopes from ^{78}Kr fragmentation and the ending point of the astrophysical rapid-proton-capture process. Phys. Rev. Lett. **74**, 4611–4614 (1995)
16. L. Boirin, L. Keek, M. Méndez, A. Cumming, J.J.M. in 't Zand, J. Cottam, F. Paerels, W.H.G. Lewin, Discovery of X-ray burst triplets in EXO 0748-676. Astron. Astrophys. **465**, 559–573 (2007)
17. S. Boutloukos, M.C. Miller, F.K. Lamb, Super-Eddington fluxes during thermonuclear X-ray bursts. Astrophys. J. Lett. **720**, L15–L19 (2010)
18. E.F. Brown, A. Cumming, Mapping crustal heating with the cooling light curves of quasi-persistent transients. Astrophys. J. **698**, 1020–1032 (2009)
19. E.F. Brown, L. Bildsten, P. Chang, Variability in the thermal emission from accreting neutron star transients. Astrophys. J. **574**, 920–929 (2002)
20. Y. Cavecchi, A. Patruno, B. Haskell, A.L. Watts, Y. Levin, M. Linares, D. Altamirano, R. Wijnands, M. van der Klis, Implications of burst oscillations from the slowly rotating accreting pulsar IGR J17480-2446 in the Globular Cluster Terzan 5. Astrophys. J. Lett. **740**, L8 (2011)
21. Y. Cavecchi, A.L. Watts, J. Braithwaite, Y. Levin, Flame propagation on the surfaces of rapidly rotating neutron stars during Type I X-ray bursts. Mon. Not. R. Astron. Soc. **434**, 3526–3541 (2013)
22. Y. Cavecchi, A.L. Watts, Y. Levin, J. Braithwaite, Rotational effects in thermonuclear type I bursts: equatorial crossing and directionality of flame spreading. Mon. Not. R. Astron. Soc. **448**, 445–455 (2015)
23. Y. Cavecchi, Y. Levin, A.L. Watts, J. Braithwaite, Fast and slow magnetic deflagration fronts in type I X-ray bursts. Mon. Not. R. Astron. Soc. **459**, 1259–1275 (2016)
24. D. Chakrabarty, E.H. Morgan, M.P. Muno, D.K. Galloway, R. Wijnands, M. van der Klis, C.B. Markwardt, Nuclear-powered millisecond pulsars and the maximum spin frequency of neutron stars. Nature **424**, 42–44 (2003)
25. M. Chakraborty, S. Bhattacharyya, Evidence of thermonuclear flame spreading on neutron stars from burst rise oscillations. Astrophys. J. **792**, 4 (2014)

26. G.W. Clark, J.G. Jernigan, H. Bradt, C. Canizares, W.H.G. Lewin, F.K. Li, W. Mayer, J. McClintock, H. Schnopper, Recurrent brief x-ray bursts from the globular cluster NGC 6624. Astrophys. J. Lett. **207**, L105–L108 (1976)
27. R. Cornelisse, J. Heise, E. Kuulkers, F. Verbunt, J.J.M. in 't Zand, The longest thermonuclear X-ray burst ever observed?. A BeppoSAX Wide Field Camera observation of 4U 1735-44. Astron. Astrophys. **357**, L21–L24 (2000)
28. R. Cornelisse, J.J.M. in 't Zand, F. Verbunt, E. Kuulkers, J. Heise, P.R. den Hartog, M. Cocchi, L. Natalucci, A. Bazzano, P. Ubertini, Six years of BeppoSAX wide field cameras observations of nine galactic type I X-ray bursters. Astron. Astrophys. **405**, 1033–1042 (2003)
29. J. Cottam, F. Paerels, M. Méndez, Gravitationally redshifted absorption lines in the X-ray burst spectra of a neutron star. Nature **420**, 51–54 (2002)
30. J. Cottam, F. Paerels, M. Méndez, L. Boirin, W.H.G. Lewin, E. Kuulkers, J.M. Miller, The burst spectra of EXO 0748-676 during a Long 2003 XMM-Newton observation. Astrophys. J. **672**, 504–509 (2008)
31. A. Cumming, Models of type I X-ray bursts from 4U 1820-30. Astrophys. J. **595**, 1077–1085 (2003)
32. A. Cumming, J. Macbeth, J.J.M.i. Zand, D. Page, Long type I X-ray bursts and neutron star interior physics. Astrophys. J. **646**, 429–451 (2006)
33. R.H. Cyburt, A.M. Amthor, R. Ferguson, Z. Meisel, K. Smith, S. Warren, A. Heger, R.D. Hoffman, T. Rauscher, A. Sakharuk, H. Schatz, F.K. Thielemann, M. Wiescher, The JINA REACLIB database: its recent updates and impact on type-I X-ray bursts. Astrophys. J. Suppl. Ser. **189**, 240–252 (2010)
34. R.H. Cyburt, A.M. Amthor, A. Heger, E. Johnson, L. Keek, Z. Meisel, H. Schatz, K. Smith, Dependence of X-ray burst models on nuclear reaction rates. Astrophys. J. **830**, 55 (2016)
35. B. Davids, R.H. Cyburt, J. José, S. Mythili, The influence of uncertainties in the $^{15}O(\alpha,\gamma)^{19}Ne$ reaction rate on models of type I X-ray bursts. Astrophys. J. **735**, 40 (2011)
36. N. Degenaar, R. Wijnands, R. Kaur, Swift detection of an intermediately long X-ray burst from the very faint X-ray binary XMMU J174716.1-281048. Mon. Not. R. Astron. Soc. **414**, L104–L108 (2011)
37. N. Degenaar, J.M. Miller, R. Wijnands, D. Altamirano, A.C. Fabian, X-ray emission and absorption features during an energetic thermonuclear X-ray burst from IGR J17062-6143. Astrophys. J. Lett. **767**, L37 (2013)
38. A. Deibel, A. Cumming, E.F. Brown, D. Page, A strong shallow heat source in the accreting neutron Star MAXI J0556-332. Astrophys. J. Lett. **809**, L31 (2015)
39. A. Deibel, Z. Meisel, H. Schatz, E.F. Brown, A. Cumming, Urca cooling pairs in the neutron star ocean and their effect on superbursts. Astrophys. J. **831**, 13 (2016)
40. C. Done, M. Gierliński, A. Kubota, Modelling the behaviour of accretion flows in X-ray binaries. Everything you always wanted to know about accretion but were afraid to ask. Astron. Astrophys. Rev. **15**, 1–66 (2007)
41. T. Ebisuzaki, T. Hanawa, D. Sugimoto, Mass loss from the neutron stars associated with X-ray bursts. Publ. Astron. Soc. Jpn. **35**, 17–32, 1983.
42. M. Falanga, A. Cumming, E. Bozzo, J. Chenevez, The new intermediate long-bursting source XTE J1701-407. Astron. Astrophys. **496**, 333–342 (2009)
43. J.L. Fisker, F.-K. Thielemann, M. Wiescher, The nuclear reaction waiting points: ^{22}Mg, ^{26}Si, ^{30}S, and ^{34}Ar and bolometrically double-peaked type I X-ray bursts. Astrophys. J. Lett. **608**, L61–L64 (2004)
44. J.L. Fisker, J. Görres, M. Wiescher, B. Davids, The importance of $^{15}O(\alpha,\gamma)^{19}Ne$ to X-ray bursts and superbursts. Astrophys. J. **650**, 332–337 (2006)
45. J.L. Fisker, W. Tan, J. Görres, M. Wiescher, R.L. Cooper, The $^{15}O(\alpha,\gamma)^{19}Ne$ reaction rate and the stability of thermonuclear burning on accreting neutron stars. Astrophys. J. **665**, 637–641 (2007)
46. J.L. Fisker, H. Schatz, F.-K. Thielemann, Explosive hydrogen burning during type I X-ray bursts. Astrophys. J. Suppl. Ser. **174**, 261–276 (2008)

47. M.Y. Fujimoto, Angular distribution of radiation from low-mass X-ray binaries. Astrophys. J. **324**, 995–1000 (1988)
48. M.Y. Fujimoto, T. Hanawa, S. Miyaji, Shell flashes on accreting neutron stars and X-ray bursts. Astrophys. J. **247**, 267–278 (1981)
49. M.Y. Fujimoto, M. Sztajno, W.H.G. Lewin, J. van Paradijs, On the theory of type I X-ray bursts - the energetics of bursts and the nuclear fuel reservoir in the envelope. Astrophys. J. **319**, 902–915 (1987)
50. I. Fushiki, D.Q. Lamb, New insights from a global view of X-ray bursts. Astrophys. J. Lett. **323**, L55–L60 (1987)
51. D.K. Galloway, D. Psaltis, D. Chakrabarty, M.P. Muno, Eddington-limited x-ray bursts as distance indicators. I. Systematic trends and spherical symmetry in bursts from 4U 1728-34. Astrophys. J. **590**, 999–1007 (2003)
52. D.K. Galloway, D. Psaltis, M.P. Muno, D. Chakrabarty,. Eddington-limited X-ray bursts as distance indicators. II. Possible compositional effects in bursts from 4U 1636-536. Astrophys. J. **639**, 1033 (2006)
53. D.K. Galloway, M.P. Muno, J.M. Hartman, D. Psaltis, D. Chakrabarty, Thermonuclear (Type I) X-ray bursts observed by the Rossi X-ray timing explorer. Astrophys. J. Suppl. Ser. **179**, 360–422 (2008)
54. D.K. Galloway, J. Lin, D. Chakrabarty, J.M. Hartman, Discovery of a 552 Hz burst oscillation in the low-mass X-ray binary EXO 0748-676. Astrophys. J. Lett. **711**, L148–L151 (2010)
55. D.K. Galloway, A.N. Ajamyan, J. Upjohn, M. Stuart, Intermittent dipping in a low-mass X-ray binary. Mon. Not. R. Astron. Soc. **461**, 3847–3853 (2016)
56. D.K. Galloway, A.J. Goodwin, L. Keek, Thermonuclear burst observations for model comparisons: a reference sample. Publ. Astron. Soc. Aust. **34**, e019 (2017)
57. P. Giannone, A. Weigert, Entwicklung in engen Doppelsternsystemen III. Z. Astrophys. **67**, 41, 1967.
58. M. Gottwald, F. Haberl, A.N. Parmar, N.E. White, The properties of bursts with short recurrence times from the transient X-ray source EXO 0748-676. Astrophys. J. **323**, 575–581 (1987)
59. J. Grindlay, H. Gursky, H. Schnopper, D.R. Parsignault, J. Heise, A.C. Brinkman, J. Schrijver, Discovery of intense X-ray bursts from the globular cluster NGC 6624. Astrophys. J. Lett. **205**, L127–L130 (1976)
60. S. Gupta, E.F. Brown, H. Schatz, P. Möller, K.-L. Kratz, Heating in the accreted neutron star ocean: implications for superburst ignition. Astrophys. J. **662**, 1188–1197 (2007)
61. T. Güver, F. Özel, A. Cabrera-Lavers, P. Wroblewski, The distance, mass, and radius of the neutron star in 4U 1608-52. Astrophys. J. **712**, 964–973 (2010)
62. T. Güver, P. Wroblewski, L. Camarota, F. Özel, The mass and radius of the neutron star in 4U 1820-30. Astrophys. J. **719**, 1807–1812 (2010)
63. T. Güver, F. Özel, D. Psaltis, Systematic uncertainties in the spectroscopic measurements of neutron-star masses and radii from thermonuclear X-ray bursts. II. Eddington limit. Astrophys. J. **747**, 77 (2012)
64. T. Güver, D. Psaltis, F. Özel, Systematic uncertainties in the spectroscopic measurements of neutron-star masses and radii from thermonuclear X-ray bursts. I. Apparent radii. Astrophys. J. **747**, 76 (2012)
65. T. Güver, F. Özel, H. Marshall, D. Psaltis, M. Guainazzi, M. Díaz-Trigo, Systematic uncertainties in the spectroscopic measurements of neutron-star masses and radii from thermonuclear X-ray bursts. III. Absolute flux calibration. Astrophys. J. **829**, 48 (2016)
66. P. Haensel, J.L. Zdunik, Non-equilibrium processes in the crust of an accreting neutron star. Astron. Astrophys. **227**, 431–436 (1990)
67. P. Haensel, J.L. Zdunik, Nuclear composition and heating in accreting neutron-star crusts. Astron. Astrophys. **404**, L33–L36 (2003)
68. C.J. Hansen, H.M. van Horn, Steady-state nuclear fusion in accreting neutron-star envelopes. Astrophys. J. **195**, 735–741 (1975)

69. G. Hasinger, M. van der Klis, Two patterns of correlated X-ray timing and spectral behaviour in low-mass X-ray binaries. Astron. Astrophys. **225**, 79–96 (1989)
70. C.-C. He, L. Keek, Anisotropy of X-ray bursts from neutron stars with concave accretion disks. Astrophys. J. **819**, 47 (2016)
71. A. Heger, A. Cumming, D.K. Galloway, S.E. Woosley, Models of type I X-ray bursts from GS 1826-24: a probe of RP-process hydrogen burning. Astrophys. J. Lett. **671**, L141–L144 (2007)
72. A. Heger, A. Cumming, S.E. Woosley, Millihertz quasi-periodic oscillations from marginally stable nuclear burning on an accreting neutron star. Astrophys. J. **665**, 1311–1320 (2007)
73. J.S. Heyl, R-modes on rapidly rotating, relativistic stars. I. Do type I bursts excite modes in the neutron star ocean? Astrophys. J. **600**, 939–945 (2004)
74. C. Iliadis, P.M. Endt, N. Prantzos, W.J. Thompson, Explosive hydrogen burning of ^{27}Si, ^{31}S, ^{35}Ar, and ^{39}Ca in novae and X-ray bursts. Astrophys. J. **524**, 434–453 (1999)
75. J. in't Zand, Understanding superbursts, in *7 years of MAXI: Monitoring X-ray Transients, held 5–7 December 2016 at RIKEN*, ed. by M. Serino, M. Shidatsu, W. Iwakiri, T. Mihara (2017), p. 121
76. J.J.M. in 't Zand, N.N. Weinberg, Evidence of heavy-element ashes in thermonuclear X-ray bursts with photospheric superexpansion. Astron. Astrophys. **520**, A81 (2010)
77. J.J.M. in 't Zand, E. Kuulkers, F. Verbunt, J. Heise, R. Cornelisse, A superburst from 4U 1254-69. Astron. Astrophys. **411**, L487–L491 (2003)
78. J.J.M. in 't Zand, R. Cornelisse, A. Cumming, Superbursts at near-Eddington mass accretion rates. Astron. Astrophys. **426**, 257–265 (2004)
79. J.J.M. in 't Zand, P.G. Jonker, C.B. Markwardt, Six new candidate ultracompact X-ray binaries. Astron. Astrophys. **465**, 953–963 (2007)
80. J.J.M. in 't Zand, L. Keek, A. Cumming, A. Heger, J. Homan, M. Méndez, Long tails on thermonuclear X-ray bursts from neutron stars: a signature of inward heating? Astron. Astrophys. **497**, 469–480 (2009)
81. J.J.M. in 't Zand, D.K. Galloway, D.R. Ballantyne, Achromatic late-time variability in thermonuclear X-ray bursts. An accretion disk disrupted by a nova-like shell? Astron. Astrophys. **525**, A111+ (2011)
82. J.J.M. in 't Zand, D.K. Galloway, H.L. Marshall, D.R. Ballantyne, P.G. Jonker, F.B.S. Paerels, D.M. Palmer, A. Patruno, N.N. Weinberg, A bright thermonuclear X-ray burst simultaneously observed with Chandra and RXTE. Astron. Astrophys. **553**, A83 (2013)
83. J.J.M. in 't Zand, A. Cumming, T.L. Triemstra, R.A.D.A. Mateijsen, T. Bagnoli, The cooling rate of neutron stars after thermonuclear shell flashes. Astron. Astrophys. **562**, A16 (2014)
84. J.J.M. in't Zand, L. Keek, Y. Cavecchi, Relativistic outflow from two thermonuclear shell flashes on neutron stars. Astron. Astrophys. **568**, A69 (2014)
85. J.J.M. in 't Zand, M.E.B. Visser, D.K. Galloway, J. Chenevez, L. Keek, E. Kuulkers, C. Sánchez-Fernández, H. Wörpel, Neutron star cooling and the RP process in thermonuclear X-ray bursts. Astron. Astrophys. **606**, A130 (2017)
86. K. Jahoda, J.H. Swank, A.B. Giles, M.J. Stark, T. Strohmayer, W. Zhang, E.H. Morgan, In-orbit performance and calibration of the Rossi X-ray Timing Explorer (*RXTE*) Proportional Counter Array (PCA), in *Proceedings of SPIE*, vol. 2808, October 1996, pp. 59–70
87. K. Jahoda, C.B. Markwardt, Y. Radeva, A.H. Rots, M.J. Stark, J.H. Swank, T.E. Strohmayer, W. Zhang, Calibration of the Rossi X-ray timing explorer proportional counter array. Astrophys. J. Suppl. Ser. **163**, 401–423 (2006)
88. L. Ji, S. Zhang, Y. Chen, S.-N. Zhang, D.F. Torres, P. Kretschmar, J. Li, The hard X-ray shortages prompted by the clock bursts in GS 1826-238. Astrophys. J. **782**, 40 (2014)
89. L. Ji, S. Zhang, Y. Chen, S.-N. Zhang, D.F. Torres, P. Kretschmar, E. Kuulkers, J. Li, Z. Chang, Diagnosing the burst influence on accretion in the clocked burster GS 1826-238. Astrophys. J. **806**, 89 (2015)
90. J. José, F. Moreno, A. Parikh, C. Iliadis, Hydrodynamic models of type I X-ray bursts: metallicity effects. Astrophys. J. Suppl. Ser. **189**, 204–239 (2010)

91. J.J.E. Kajava, J. Nättilä, O.-M. Latvala, M. Pursiainen, J. Poutanen, V.F. Suleimanov, M.G. Revnivtsev, E. Kuulkers, D.K. Galloway, The influence of accretion geometry on the spectral evolution during thermonuclear (type I) X-ray bursts. Mon. Not. R. Astron. Soc. **445**, 4218–4234 (2014)
92. J.J.E. Kajava, J. Nättilä, J. Poutanen, A. Cumming, V. Suleimanov, E. Kuulkers, Detection of burning ashes from thermonuclear X-ray bursts. Mon. Not. R. Astron. Soc. **464**, L6–L10 (2017)
93. L. Keek, Photospheric radius expansion in superburst precursors from neutron stars. Astrophys. J. **756**, 130 (2012)
94. L. Keek, A. Heger, Multi-zone models of superbursts from accreting neutron stars. Astrophys. J. **743**, 189 (2011)
95. L. Keek, A. Heger, Carbon production on accreting neutron stars in a new regime of stable nuclear burning. Mon. Not. R. Astron. Soc. **456**, L11–L15 (2016)
96. L. Keek, A. Heger, Thermonuclear bursts with short recurrence times from neutron stars explained by opacity-driven convection. Astrophys. J. **842**, 113 (2017)
97. L. Keek, J.J.M. in 't Zand, E. Kuulkers, A. Cumming, E.F. Brown, M. Suzuki, First superburst from a classical low-mass X-ray binary transient. Astron. Astrophys. **479**, 177–188 (2008)
98. L. Keek, N. Langer, J.J.M. in 't Zand, The effect of rotation on the stability of nuclear burning in accreting neutron stars. Astron. Astrophys. **502**, 871–881 (2009)
99. L. Keek, D.K. Galloway, J.J.M. in 't Zand, A. Heger, Multi-instrument X-ray observations of thermonuclear bursts with short recurrence times. Astrophys. J. **718**, 292–305 (2010)
100. L. Keek, A. Heger, J.J.M. in 't Zand, Superburst models for neutron stars with hydrogen- and helium-rich atmospheres. Astrophys. J. **752**, 150 (2012)
101. L. Keek, D.R. Ballantyne, E. Kuulkers, T.E. Strohmayer, Characterizing the evolving X-ray spectral features during a superburst from 4U 1636-536. Astrophys. J. **789**, 121 (2014)
102. L. Keek, D.R. Ballantyne, E. Kuulkers, T.E. Strohmayer, X-raying an accretion disk in realtime: the evolution of ionized reflection during a superburst from 4U 1636-536. Astrophys. J. Lett. **797**, L23 (2014)
103. L. Keek, R.H. Cyburt, A. Heger, Reaction rate and composition dependence of the stability of thermonuclear burning on accreting neutron stars. Astrophys. J. **787**, 101 (2014)
104. L. Keek, A. Cumming, Z. Wolf, D.R. Ballantyne, V.F. Suleimanov, E. Kuulkers, T.E. Strohmayer, The imprint of carbon combustion on a superburst from the accreting neutron star 4U 1636-536. Mon. Not. R. Astron. Soc. **454**, 3559–3566 (2015)
105. L. Keek, W. Iwakiri, M. Serino, D.R. Ballantyne, J.J.M. in't Zand, T.E. Strohmayer, X-ray reflection and an exceptionally long thermonuclear helium burst from IGR J17062-6143. Astrophys. J. **836**, 111 (2017)
106. O. Koike, M.-A. Hashimoto, R. Kuromizu, S.-I. Fujimoto, Final products of the RP-process on accreting neutron stars. Astrophys. J. **603**, 242–251 (2004)
107. E. Kuulkers, The observers' view of (very) long X-ray bursts: they are super! Nucl. Phys. B Proc. Suppl. **132**, 466–475 (2004)
108. E. Kuulkers, J. Homan, M. van der Klis, W.H.G. Lewin, M. Méndez, X-ray bursts at extreme mass accretion rates from GX 17+2. Astron. Astrophys. **382**, 947–973 (2002)
109. E. Kuulkers, P.R. den Hartog, J.J.M. in 't Zand, F.W.M. Verbunt, W.E. Harris, M. Cocchi, Photospheric radius expansion X-ray bursts as standard candles. Astron. Astrophys. **399**, 663–680 (2003)
110. E. Kuulkers, J.J.M. in 't Zand, J.-L. Atteia, A.M. Levine, S. Brandt, D.A. Smith, M. Linares, M. Falanga, C. Sánchez-Fernández, C.B. Markwardt, T.E. Strohmayer, A. Cumming, M. Suzuki, What ignites on the neutron star of 4U 0614+091? Astron. Astrophys. **514**, A65+ (2010)
111. F.K. Lamb, M.C. Miller, Critical radiation fluxes and luminosities of black holes and relativistic stars. Astrophys. J. **439**, 828–845 (1995)
112. N. Lampe, A. Heger, D.K. Galloway, The influence of accretion rate and metallicity on thermonuclear bursts: predictions from KEPLER models. Astrophys. J. **819**, 46 (2016)

113. I.I. Lapidus, R.A. Sunyaev, Angular distribution and polarization of X-ray-burster radiation (during stationary and flash phases). Mon. Not. R. Astron. Soc. **217**, 291–303 (1985)
114. W.H.G. Lewin, W.D. Vacca, E.M. Basinska, Precursors to X-ray bursts - the result of expansion and subsequent contraction of the neutron star's photosphere. Astrophys. J. Lett. **277**, L57–L60 (1984)
115. W.H.G. Lewin, J. van Paradijs, R.E. Taam, X-ray bursts. Space Sci. Rev. **62**, 223–389 (1993)
116. W.H.G. Lewin, J. van Paradijs, R.E. Taam, X-ray bursts, *X-ray Binaries*, ed. by W.H.G. Lewin, J. van Paradijs, E.P.J. van den Heuvel (Cambridge University Press, Cambridge, 1995), p. 175
117. M. Linares, D. Chakrabarty, M. van der Klis, On the cooling tails of thermonuclear X-ray bursts: the IGR J17480-2446 link. Astrophys. J. Lett. **733**, L17 (2011)
118. M. Linares, D. Altamirano, D. Chakrabarty, A. Cumming, L. Keek, Millihertz quasi-periodic oscillations and thermonuclear bursts from Terzan 5: a showcase of burning regimes. Astrophys. J. **748**, 82 (2012)
119. M. Lyu, M. Méndez, A. Sanna, J. Homan, T. Belloni, B. Hiemstra, Iron-line and continuum variations in the XMM-Newton and Suzaku spectra of the neutron-star low-mass X-ray binary 4U 1636-53. Mon. Not. R. Astron. Soc. **440**, 1165–1178 (2014)
120. M. Lyu, M. Méndez, G. Zhang, L. Keek, Spectral and timing analysis of the mHz QPOs in the neutron-star low-mass X-ray binary 4U 1636-53. Mon. Not. R. Astron. Soc. **454**, 541–549 (2015)
121. T.J. Maccarone, P.S. Coppi, Spectral fits to the 1999 Aql X-1 outburst data. Astron. Astrophys. **399**, 1151–1157 (2003)
122. J. Madej, P.C. Joss, A. Różańska, Model atmospheres and X-ray spectra of bursting neutron stars: hydrogen-helium comptonized spectra. Astrophys. J. **602**, 904–912 (2004)
123. S. Mahmoodifar, T. Strohmayer, X-ray burst oscillations: from flame spreading to the cooling wake. Astrophys. J. **818**, 93 (2016)
124. A. Majczyna, J. Madej, P.C. Joss, A. Różańska, Model atmospheres and X-ray spectra of bursting neutron stars. II. Iron rich comptonized spectra. Astron. Astrophys. **430**, 643–654 (2005)
125. C.M. Malone, A. Nonaka, A.S. Almgren, J.B. Bell, M. Zingale, Multidimensional modeling of type I X-ray bursts. I. Two-dimensional convection prior to the outburst of a pure ^{4}He accretor. Astrophys. J. **728**, 118 (2011)
126. C.M. Malone, M. Zingale, A. Nonaka, A.S. Almgren, J.B. Bell, Multidimensional modeling of type I X-ray bursts. II. Two-dimensional convection in a mixed H/He accretor. Astrophys. J. **788**, 115 (2014)
127. L. Maraschi, A. Cavaliere, Gamma-ray bursts from thermonuclear explosions on neutron stars, in *Highlights in Astronomy*, vol. 4, ed. by E.A. Müller (Reidel, Dordrecht, 1977), p. 127
128. I. Maurer, A.L. Watts, Ignition latitude and the shape of Type I X-ray bursts. Mon. Not. R. Astron. Soc. **383**, 387–398 (2008)
129. Z. Medin, A. Cumming, Compositionally driven convection in the oceans of accreting neutron stars. Astrophys. J. **730**, 97 (2011)
130. Z. Medin, A. Cumming, A signature of chemical separation in the cooling light curves of transiently accreting neutron stars. Astrophys. J. Lett. **783**, L3 (2014)
131. Z. Medin, M. von Steinkirch, A.C. Calder, C.J. Fontes, C.L. Fryer, A.L. Hungerford, Model atmospheres for X-ray bursting neutron stars. Astrophys. J. **832**, 102 (2016)
132. A. Merloni, A.C. Fabian, Accretion disc coronae as magnetic reservoirs. Mon. Not. R. Astron. Soc. **321**, 549–552 (2001)
133. M.C. Miller, Astrophysical constraints on dense matter in neutron stars. ArXiv e-prints, November 2013
134. M.C. Miller, F.K. Lamb, Observational constraints on neutron star masses and radii. Eur. Phys. J. A, **52**, 63 (2016)
135. J.M. Miller, M.L. Parker, F. Fuerst, M. Bachetti, D. Barret, B.W. Grefenstette, S. Tendulkar, F.A. Harrison, S.E. Boggs, D. Chakrabarty, F.E. Christensen, W.W. Craig, A.C. Fabian, C.J. Hailey, L. Natalucci, F. Paerels, V. Rana, D.K. Stern, J.A. Tomsick, W.W. Zhang, Constraints

on the neutron star and inner accretion flow in Serpens X-1 Using NuSTAR. Astrophys. J. Lett. **779**, L2 (2013)
136. M.P. Muno, D. Chakrabarty, D.K. Galloway, P. Savov, Millisecond oscillations and photospheric radius expansion in thermonuclear X-ray bursts. Astrophys. J. Lett. **553**, L157–L160 (2001)
137. M.P. Muno, D.K. Galloway, D. Chakrabarty, The effect of neutron star rotation on the properties of thermonuclear X-ray bursts. Astrophys. J. **608**, 930–934 (2004)
138. T. Murakami, H. Inoue, K. Koyama, K. Makishima, M. Matsuoka, M. Oda, Y. Ogawara, T. Ohashi, N. Shibazaki, Y. Tanaka, S. Hayakawa, H. Kunieda, F. Makino, K. Masai, F. Nagase, Y. Tawara, S. Miyamoto, H. Tsunemi, K. Yamashita, I. Kondo, Two successive X-ray bursts at ten-minute interval from 1608-522. Publ. Astron. Soc. Jpn. **32**, 543 (1980)
139. R. Narayan, J.S. Heyl, Thermonuclear stability of material accreting onto a neutron star. Astrophys. J. **599**, 419–449 (2003)
140. J. Nättilä, V.F. Suleimanov, J.J.E. Kajava, J. Poutanen, Models of neutron star atmospheres enriched with nuclear burning ashes. Astron. Astrophys. **581**, A83 (2015)
141. L. Nobili, R. Turolla, I. Lapidus, Winds from neutron stars and strong type I X-ray bursts. Astrophys. J. **433**, 276–286 (1994)
142. L.S. Ootes, A.L. Watts, D.K. Galloway, R. Wijnands, The accretion rate dependence of burst oscillation amplitude. Astrophys. J. **834**, 21 (2017)
143. F. Özel, Soft equations of state for neutron-star matter ruled out by EXO 0748 - 676. Nature **441**, 1115–1117 (2006)
144. B. Paczynski, A one-zone model for shell flashes on accreting compact stars. Astrophys. J. **264**, 282–295 (1983)
145. B. Paczynski, N. Anderson, Models of extended relativistic envelopes of neutron stars. Astrophys. J. **302**, 1–10 (1986)
146. A. Parikh, J. José, F. Moreno, C. Iliadis, The effects of variations in nuclear processes on type I X-ray burst nucleosynthesis. Astrophys. J. Suppl. Ser. **178**, 110–136 (2008)
147. A. Parikh, J. José, G. Sala, C. Iliadis, Nucleosynthesis in type I X-ray bursts. Prog. Part. Nucl. Phys. **69**, 225–253 (2013)
148. A. Patruno, A.L. Watts, Accreting millisecond X-ray pulsars. ArXiv e-prints, June 2012
149. F. Peng, E.F. Brown, J.W. Truran, Sedimentation and type I X-ray bursts at low accretion rates. Astrophys. J. **654**, 1022–1035 (2007)
150. R. Pfaff, D.J. Morrissey, W. Benenson, M. Fauerbach, M. Hellström, C.F. Powell, B.M. Sherrill, M. Steiner, J.A. Winger, Fragmentation of ^{78}Kr projectiles. Phys. Rev. C **53**, 1753–1758 (1996)
151. A.L. Piro, L. Bildsten, Turbulent mixing in the surface layers of accreting neutron stars. Astrophys. J. **663**, 1252–1268 (2007)
152. J. Poutanen, J. Nättilä, J.J.E. Kajava, O.-M. Latvala, D.K. Galloway, E. Kuulkers, V.F. Suleimanov, The effect of accretion on the measurement of neutron star mass and radius in the low-mass X-ray binary 4U 1608-52. Mon. Not. R. Astron. Soc. **442**, 3777–3790 (2014)
153. T. Quinn, B. Paczynski, Stellar winds driven by super-Eddington luminosities. Astrophys. J. **289**, 634–643 (1985)
154. T. Rauch, V. Suleimanov, K. Werner, Absorption features in the spectra of X-ray bursting neutron stars. Astron. Astrophys. **490**, 1127–1134 (2008)
155. M. Revnivtsev, E. Churazov, M. Gilfanov, R. Sunyaev, New class of low frequency QPOs: signature of nuclear burning or accretion disk instabilities? Astron. Astrophys. **372**, 138–144 (2001)
156. M.G. Revnivtsev, V.F. Suleimanov, J. Poutanen, On the spreading layer emission in luminous accreting neutron stars. Mon. Not. R. Astron. Soc. **434**, 2355–2361 (2013)
157. H. Schatz, W.-J. Ong, Dependence of X-ray burst models on nuclear masses. Astrophys. J. **844**, 139 (2017)
158. H. Schatz, A. Aprahamian, J. Goerres, M. Wiescher, T. Rauscher, J.F. Rembges, F.-K. Thielemann, B. Pfeiffer, P. Moeller, K.-L. Kratz, H. Herndl, B.A. Brown, H. Rebel, RP-process nucleosynthesis at extreme temperature and density conditions. Phys. Rep. **294**, 167 (1998)

159. H. Schatz, A. Aprahamian, V. Barnard, L. Bildsten, A. Cumming, M. Ouellette, T. Rauscher, F.-K. Thielemann, M. Wiescher, End point of the RP process on accreting neutron stars. Phys. Rev. Lett. **86**, 3471–3474 (2001)
160. H. Schatz, L. Bildsten, A. Cumming, Photodisintegration-triggered nuclear energy release in superbursts. Astrophys. J. Lett. **583**, L87–L90 (2003)
161. H. Schatz, S. Gupta, P. Möller, M. Beard, E.F. Brown, A.T. Deibel, L.R. Gasques, W.R. Hix, L. Keek, R. Lau, A.W. Steiner, M. Wiescher, Strong neutrino cooling by cycles of electron capture and β^- decay in neutron star crusts. Nature **505**, 62–65 (2014)
162. M. Schwarzschild, R. Härm, Thermal instability in non-degenerate stars. Astrophys. J. **142**, 855 (1965)
163. M. Serino, T. Mihara, M. Matsuoka, S. Nakahira, M. Sugizaki, Y. Ueda, N. Kawai, S. Ueno, Superburst with outburst from EXO 1745-248 in Terzan 5 with MAXI. Publ. Astron. Soc. Jpn. **64**, 91 (2012)
164. M. Serino, W. Iwakiri, T. Tamagawa, T. Sakamoto, S. Nakahira, M. Matsuoka, K. Yamaoka, H. Negoro, MAXI observations of long X-ray bursts. Publ. Astron. Soc. Jpn. (2016). https://doi.org/10.1093/pasj/psw086
165. V.A. Simonenko, D.A. Gryaznykh, I.A. Litvinenko, V.A. Lykov, A.N. Shushlebin, Mechanism of thermonuclear burning propagation in a helium layer on a neutron star surface: a refined model with heat conduction and subgrid turbulence. Astron. Lett. **38**, 305–320 (2012)
166. A. Spitkovsky, Y. Levin, G. Ushomirsky, Propagation of thermonuclear flames on rapidly rotating neutron stars: extreme weather during type I X-ray bursts. Astrophys. J. **566**, 1018–1038 (2002)
167. A.W. Steiner, J.M. Lattimer, E.F. Brown, The equation of state from observed masses and radii of neutron stars. Astrophys. J. **722**, 33–54 (2010)
168. J. Stevens, E.F. Brown, A. Cumming, R. Cyburt, H. Schatz, Carbon synthesis in steady-state hydrogen and helium burning on accreting neutron stars. Astrophys. J. **791**, 106 (2014)
169. H. Stiele, W. Yu, A.K.H. Kong, Millihertz quasi-periodic oscillations in 4U 1636-536: putting possible constraints on the neutron star size. Astrophys. J. **831**, 34 (2016)
170. T.E. Strohmayer, D. Altamirano, Marginally stable nuclear burning, in *American Astronomical Society Meeting Abstracts*, vol. 219, January 2012, p. 249.03
171. T. Strohmayer, L. Bildsten, New views of thermonuclear bursts, in *Compact Stellar X-Ray Sources*, ed. by W.H.G. Lewin, M. van der Klis (Cambridge University Press, Cambridge, 2006), astro–ph/0301544
172. T.E. Strohmayer, E.F. Brown, A remarkable 3 hour thermonuclear burst from 4U 1820-30. Astrophys. J. **566**, 1045–1059 (2002)
173. T. Strohmayer, L. Keek, IGR J17062-6143 is an accreting millisecond X-ray pulsar. Astrophys. J. Lett. **836**, L23 (2017)
174. T. Strohmayer, S. Mahmoodifar, Discovery of a neutron star oscillation mode during a superburst. Astrophys. J. Lett. **793**, L38 (2014)
175. T.E. Strohmayer, W. Zhang, J.H. Swank, A. Smale, L. Titarchuk, C. Day, U. Lee, Millisecond X-ray variability from an accreting neutron star system. Astrophys. J. Lett. **469**, L9–12 (1996)
176. T.E. Strohmayer, W. Zhang, J.H. Swank, 363 Hz oscillations during the rising phase of bursts from 4U 1728-34: evidence for rotational modulation. Astrophys. J. Lett. **487**, L77–L80 (1997)
177. V. Suleimanov, J. Poutanen, M. Revnivtsev, K. Werner, A neutron star stiff equation of state derived from cooling phases of the X-ray burster 4U 1724-307. Astrophys. J. **742**, 122 (2011)
178. V. Suleimanov, J. Poutanen, K. Werner, X-ray bursting neutron star atmosphere models: spectra and color corrections. Astron. Astrophys. **527**, A139+ (2011)
179. V. Suleimanov, J. Poutanen, K. Werner, X-ray bursting neutron star atmosphere models using an exact relativistic kinetic equation for Compton scattering. Astron. Astrophys. **545**, A120 (2012)
180. J.H. Swank, R.H. Becker, E.A. Boldt, S.S. Holt, S.H. Pravdo, P.J. Serlemitsos, Spectral evolution of a long X-ray burst. Astrophys. J. Lett. **212**, L73–LL76 (1977)
181. R.E. Taam, S.E. Woosley, T.A. Weaver, D.Q. Lamb, Successive x-ray bursts from accreting neutron stars. Astrophys. J. **413**, 324–332 (1993)

182. Y. Tawara, T. Kii, S. Hayakawa, H. Kunieda, K. Masai, F. Nagase, H. Inoue, K. Koyama, F. Makino, K. Makishima, M. Matsuoka, T. Murakami, M. Oda, Y. Ogawara, T. Ohashi, N. Shibazaki, Y. Tanaka, S. Miyamoto, H. Tsunemi, K. Yamashita, I. Kondo, A very long X-ray burst with a precursor from XB 1715-321. Astrophys. J. Lett. **276**, L41–L44 (1984)
183. P. Ubertini, A. Bazzano, M. Cocchi, L. Natalucci, J. Heise, J.M. Muller, J.J.M. in 't Zand, Bursts from GS 1826−238: a clocked thermonuclear flashes generator. Astrophys. J. Lett. **514**, L27–L30 (1999)
184. J. van Paradijs, Average properties of X-ray burst sources. Nature **274**, 650–653 (1978)
185. J. van Paradijs, Possible observational constraints on the mass-radius relation of neutron stars. Astrophys. J. **234**, 609–611 (1979)
186. J. van Paradijs, H.G. Lewin, The interpretation of blackbody spectra and radii during X-ray bursts. Astron. Astrophys. **157**, L10–L12 (1986)
187. J. van Paradijs, W. Penninx, W.H.G. Lewin, On the relation between x-ray burst properties and the persistent X-ray luminosity. Mon. Not. R. Astron. Soc. **233**, 437–450 (1988)
188. L. van Wormer, J. Görres, C. Iliadis, M. Wiescher, F.-K. Thielemann, Reaction rates and reaction sequences in the RP-process. Astrophys. J. **432**, 326–350 (1994)
189. J. Verdhan Chauhan, J.S. Yadav, R. Misra, P.C. Agrawal, H.M. Antia, M. Pahari, N. Sridhar, D. Dedhia, T. Katoch, P. Madhwani, R.K. Manchanda, B. Paul, P. Shah, AstroSat/LAXPC detection of millisecond phenomena in 4U 1728-34. Astrophys. J. **841**, 41 (2017)
190. A.R. Villarreal, T.E. Strohmayer, Discovery of the neutron star spin frequency in EXO 0748-676. Astrophys. J. Lett. **614**, L121–L124 (2004)
191. M.A. Walker, Radiation dynamics in X-ray binaries. I - Type 1 bursts. II - Type 2 bursts. III - Extremely compact objects. Astrophys. J. **385**, 642–664 (1992)
192. R.K. Wallace, S.E. Woosley, Explosive hydrogen burning. Astrophys. J. Suppl. Ser. **45**, 389–420 (1981)
193. A.L. Watts, Thermonuclear burst oscillations. Annu. Rev. Astron. Astrophys. **50**, 609–640 (2012)
194. N.N. Weinberg, L. Bildsten, Carbon detonation and shock-triggered helium burning in neutron star superbursts. Astrophys. J. **670**, 1291–1300 (2007)
195. N.N. Weinberg, L. Bildsten, E.F. Brown, Hydrodynamic thermonuclear runaways in superbursts. Astrophys. J. Lett. **650**, L119–L122 (2006)
196. N.N. Weinberg, L. Bildsten, H. Schatz, Exposing the nuclear burning ashes of radius expansion type I X-ray bursts. Astrophys. J. **639**, 1018–1032 (2006)
197. C.A. Wilson-Hodge, P.S. Ray, K. Gendreau, D. Chakrabarty, M. Feroci, T. Maccarone, Z. Arzoumanian, R.A. Remillard, K. Wood, C. Griffith, STROBE-X Collaboration, STROBE-X: X-ray timing spectroscopy on dynamical timescales from microseconds to years, in *American Astronomical Society Meeting Abstracts*, vol. 229, January 2017, p. 309.04
198. J.A. Winger, D.P. Bazin, W. Benenson, G.M. Crawley, D.J. Morrissey, N.A. Orr, R. Pfaff, B.M. Sherrill, M. Steiner, M. Thoennessen, S.J. Yennello, B.M. Young, Halflife measurements of the RP-process nuclei 61GA, 63GE, and 65AS. Phys. Lett. B **299**(3), 214–218 (1993)
199. S.E. Woosley, R.E. Taam, Gamma-ray bursts from thermonuclear explosions on neutron stars. Nature **263**, 101–103 (1976)
200. S.E. Woosley, T.A. Weaver, Repeated thermonuclear flashes on an accreting neutron star, in *High Energy Transients in Astrophysics* (1984), p. 273
201. S.E. Woosley, A. Heger, A. Cumming, R.D. Hoffman, J. Pruet, T. Rauscher, J.L. Fisker, H. Schatz, B.A. Brown, M. Wiescher. Models for type I X-ray bursts with improved nuclear physics. Astrophys. J. Suppl. Ser. **151**, 75–102 (2004)
202. S.E. Woosley, A.R. Kerstein, A.J. Aspden, Flames in type Ia Supernova: deflagration-detonation transition in the oxygen-burning flame. Astrophys. J. **734**, 37 (2011)
203. H. Worpel, D.K. Galloway, D.J. Price, Evidence for accretion rate change during type I X-ray bursts. Astrophys. J. **772**, 94 (2013)
204. H. Worpel, D.K. Galloway, D.J. Price, Evidence for enhanced persistent emission during sub-Eddington thermonuclear bursts. Astrophys. J. **801**, 60 (2015)

205. S.-C. Yoon, N. Langer, M. van der Sluys, On the stability of thermonuclear shell sources in stars. Astron. Astrophys. **425**, 207–216 (2004)
206. M. Zamfir, A. Cumming, C. Niquette, The thermal stability of helium burning on accreting neutron stars. Mon. Not. R. Astron. Soc. **445**, 3278–3288 (2014)
207. G. Zhang, M. Méndez, T.M. Belloni, J. Homan, Coherent oscillations and the evolution of the apparent emission area in the decaying phase of radius-expansion bursts from 4U 1636-53. Mon. Not. R. Astron. Soc. **436**, 2276–2285 (2013)
208. G. Zhang, M. Méndez, M. Zamfir, A. Cumming, The link between coherent burst oscillations, burst spectral evolution and accretion state in 4U 1728-34. Mon. Not. R. Astron. Soc. **455**, 2004–2017 (2016)
209. S.N. Zhang, M. Feroci, A. Santangelo, Y.W. Dong, H. Feng, F.J. Lu, K. Nandra, Z.S. Wang, S. Zhang, E. Bozzo, S. Brandt, A. De Rosa, L.J. Gou, M. Hernanz, M. van der Klis, X.D. Li, Y. Liu, P. Orleanski, G. Pareschi, M. Pohl, J. Poutanen, J.L. Qu, S. Schanne, L. Stella, P. Uttley, A. Watts, R.X. Xu, W.F. Yu, J.J.M. in 't Zand, S. Zane, L. Alvarez, L. Amati, L. Baldini, C. Bambi, S. Basso, Bhattacharyya S., B. , R., T. Belloni, P. Bellutti, S. Bianchi, A. Brez, M. Bursa, V. Burwitz, C. Budtz-Jørgensen, I. Caiazzo, R. Campana, X. Cao, P. Casella, C.Y. Chen, L. Chen, T. Chen, Y. Chen, Y. Chen, Y.P. Chen, M. Civitani, F. Coti Zelati, W. Cui, W.W. Cui, Z.G. Dai, E. Del Monte, D. de Martino, S. Di Cosimo, S. Diebold, M. Dovciak, I. Donnarumma, V. Doroshenko, P. Esposito, Y. Evangelista, Y. Favre, P. Friedrich, F. Fuschino, J.L. Galvez, Z.L. Gao, M.Y. Ge, O. Gevin, D. Goetz, D.W. Han, J. Heyl, J. Horak, W. Hu, F. Huang, Q.S. Huang, R. Hudec, D. Huppenkothen, G.L. Israel, A. Ingram, V. Karas, D. Karelin, P.A. Jenke, L. Ji, S. Korpela, D. Kunneriath, C. Labanti, G. Li, X. Li, Z.S. Li, E.W. Liang, O. Limousin, L. Lin, Z.X. Ling, H.B. Liu, H.W. Liu, Z. Liu, B. Lu, N. Lund, D. Lai, B. Luo, T. Luo, B. Ma, S. Mahmoodifar, M. Marisaldi, A. Martindale, N. Meidinger, Y. Men, M. Michalska, R. Mignani, M. Minuti, S. Motta, F. Muleri, J. Neilsen, M. Orlandini, A.T. Pan, A. Patruno, E. Perinati, A. Picciotto, C. Piemonte, M. Pinchera, Rachevski A., M. Rapisarda, N. Rea, E.M.R. Rossi, A. Rubini, G. Sala, X.W. Shu, C. Sgro, Z.X. Shen, P. Soffitta, L. Song, G. Spandre, G. Stratta, T.E. Strohmayer, L. Sun, J. Svoboda, G. Tagliaferri, G. Tenzer, T. Hong, R. Taverna, G. Torok, R. Turolla, S. Vacchi, J. Wang, D. Walton, K. Wang, J.F. Wang, R.J. Wang, Y.F. Wang, S.S. Weng, J. Wilms, B. Winter, X. Wu, X.F. Wu, S.L. Xiong, Y.P. Xu, Y.Q. Xue, Z. Yan, S. Yang, X. Yang, Y.J. Yang, F. Yuan, W.M. Yuan, Y.F. Yuan, G. Zampa, N. Zampa, A. Zdziarski, C. Zhang, C.L. Zhang, L. Zhang, X. Zhang, Z. Zhang, W.D. Zhang, S.J. Zheng, P. Zhou, X.L. Zhou, eXTP: enhanced X-ray timing and polarization mission, in *Space Telescopes and Instrumentation 2016: Ultraviolet to Gamma Ray, Proceedings of SPIE*, vol. 9905, July 2016, p. 99051Q
210. M. Zingale, F.X. Timmes, B. Fryxell, D.Q. Lamb, K. Olson, A.C. Calder, L.J. Dursi, P. Ricker, R. Rosner, P. MacNeice, H.M. Tufo, Helium detonations on neutron stars. Astrophys. J. Suppl. Ser. **133**, 195–220 (2001)
211. M. Zingale, C.M. Malone, A. Nonaka, A.S. Almgren, J.B. Bell, Comparisons of two- and three-dimensional convection in type I X-ray bursts. Astrophys. J. **807**, 60 (2015)

Chapter 6
High-Frequency Variability in Neutron-Star Low-Mass X-ray Binaries

Mariano Méndez and Tomaso M. Belloni

Contents

Abstract Binary systems with a neutron-star primary accreting from a companion star display variability in the X-ray band on time scales ranging from years to milliseconds. With frequencies of up to ∼1300 Hz, the kilohertz quasi-periodic oscillations (kHz QPOs) represent the fastest variability observed from any astronomical object. The sub-millisecond time scale of this variability implies that the kHz QPOs are produced in the accretion flow very close to the surface of the neutron star, providing a unique view of the dynamics of matter under the influence of some of the strongest gravitational fields in the Universe. This offers

M. Méndez
Kapteyn Astronomical Institute, University of Groningen, Groningen, The Netherlands
e-mail: mariano@astro.rug.nl

T. M. Belloni (✉)
INAF - Osservatorio Astronomico di Brera, Merate, Italy
e-mail: tomaso.belloni@inaf.it

© Springer-Verlag GmbH Germany, part of Springer Nature 2021

T. M. Belloni et al. (eds.), *Timing Neutron Stars: Pulsations, Oscillations and Explosions*, Astrophysics and Space Science Library 461,
https://doi.org/10.1007/978-3-662-62110-3_6

the possibility to probe some of the most extreme predictions of General Relativity, such as dragging of inertial frames and periastron precession at rates that are 16 orders of magnitude faster than those observed in the solar system and, ultimately, the existence of a minimum distance at which a stable orbit around a compact object is possible. Here we review the last 20 years of research on kHz QPOs, and we discuss the prospects for future developments in this field.

6.1 Introduction

Fast time variability from accreting X-ray binaries has become in the past decades a very important tool for our understanding of the process of accretion onto compact objects. As emission properties change on time scales well below a second, it is impossible to ignore variability while concentrating solely on spectral analysis. It was the Rossi X-ray Timing Explorer (RXTE) satellite with its large-area PCA instrument that allowed us to probe very fast time scales, below 10 ms. In this regime, we are exploring the accreting flow very close to the compact object, whether it is a black hole or a neutron star. So deep into the potential well the effects of General Relativity in the strong-field regime can be observable and timing analysis is a very direct way to explore them.

In neutron-star binaries, the phenomenology is particularly rich and complex, with the presence of Quasi-Periodic Oscillations (QPOs) at frequencies of hundreds of Hz, and even faster than 1 kHz. The frequencies of these oscillations are linked to fundamental frequencies in a gravitational field, allowing us to probe General Relativity in extreme gravitational fields. Most of what we learned comes from the RXTE satellite, which ended in 2012, although new information is provided by current missions such as Astrosat and NICER. New, much more sensitive, instruments are being planned, like the Chinese-European satellite eXTP, and when they become operative we expect a real explosion of new results. This will take several years; here we review the current state of research for neutron-star binaries, concentrating on high-frequency oscillations.

6.2 History

The history of aperiodic variability from X-Ray Binaries began in the early days of X-ray astronomy. After several detections of spurious pulse periods from Cygnus X-1 were reported, the idea that the observed variability was the result of an incoherent process (shot noise) was put forward [155]. Further observations with more advanced instrumentation led to the production of the first statistically significant Power Density Spectra (PDS) from which the strong noise from this black hole candidate was defined and found to be more complex than a simple shot noise (see e.g. [123]).

For Neutron-Star Low-Mass X-ray Binaries (NS LMXB), the first important result was the discovery of Quasi-Periodic Oscillations (QPO) in GX 5-1 in 1985 with the EXOSAT satellite [162], obtained serendipitously when looking for X-ray pulsations (see the chapter by Patruno and Watts, this volume). From this first detection, QPOs were soon observed in many other sources of the same class. The detection of a signal with a very defined frequency provided the first precise time scale measurement of the accretion flow. The first interpretation of QPOs was in terms of the spin frequency of the neutron star. The values of their frequencies (typically ~1 Hz) and the fact that they were not coherent or constant in time excluded that they could be a direct observation of the spin. However, models that interpreted them as a beat between the rotation of the neutron star and the orbital motion at the inner radii of the accretion flow were proposed and were able to explain the observations (see e.g. [6, 88]). For this to be the case, the presence of a non-negligible magnetic field is necessary. Additional types of QPOs with different properties were also found, the origin of which was even more difficult to explain.

A few years later, similar QPOs started being observed from black-hole binaries (BHB), thanks to the new all-sky monitors that were able to discover transient systems (since most of the BHB are transient). Their frequency was lower (1–10 Hz), but they appeared to be rather similar to those in NS LMXB (see e.g. [118]). In addition, the broad-band noise component connected to QPOs was found, at least in some source states, to be extremely similar between the two classes. Since black holes do not have a solid surface nor a magnetic field, the NS models that depend on either could not be applied to black holes.

The launch of RXTE at the end of 1995 opened the way to the detection of high-frequency (>100 Hz) features in the PDS of accreting binaries. For NS systems, new quasi-periodic peaks at frequencies of hundreds of Hertz, called kilohertz QPOs (kHz QPOs), were discovered first in the brightest source, Sco X-1 [164], then soon in many other NS LMXBs. The model involving a beat with the neutron star spin was adapted to interpret these high frequencies, as the peaks often appear in pairs with roughly the same separation (e.g. [50, 151]), but new data presented problems for the model, which had to be abandoned. RXTE also allowed to bring the BHB QPOs into a phenomenological scheme that appears to be connected to that of NS LMXB low-frequency QPOs [34]. The kHz phenomenon appears to be very common in bright NS LMXBs. RXTE also discovered high-frequency QPOs (HFQPOs) from BHBs, but they are extremely rare to the extent that, excluding one peculiar source that had many detections [17, 119], only a handful of them were found in the 16 years of operation of the satellite [22]. RXTE also led to the discovery of other fast-timing phenomena from NS LMXBs that have completely changed our knowledge of these systems: burst oscillations, accreting millisecond pulsars and intermittent pulsars, all of which are dealt with in other chapters of this book.

After the end of the RXTE mission it has become much more difficult to detect fast-timing aperiodic phenomena, because missions like XMM-Newton, Chandra or Swift are not optimised for timing studies and do not yield the high count rates that are needed. In the recent years, the launch of the Indian satellite Astrosat, which contains an instrument similar to the main one on board RXTE [3] and of the NICER experiment on board the International Space Station [53] have opened a new window onto these phenomena, while future missions like eXTP are being studied.

6.3 Basic Frequencies Close to a Neutron Star

The accretion flow around a neutron star is a very complex physical system. In order to study the time variability of the emitted flux, it is important to consider the expected characteristic time scales that might be observed, leaving aside the issue of the mechanism that will give rise to flux variability.

- Neutron stars in LMXBs are expected to be rapidly rotating, based on evolutionary scenarios [154]. An obvious characteristic time to consider is the rotational period of the central object, which would manifest itself in the form of a coherent signal. Rotational frequencies higher than 100 Hz are known for 26 systems, with the fastest being currently 620 Hz (see chapter by Patruno and Watts and [179]).
- A particle orbiting a compact object defines an obvious time scale, that of the period of its orbit (hereafter dynamical time scale t_K). In the vicinity of a neutron star, the space time is affected by the presence of the compact object and an expression from General Relativity has to be used.
- The accretion flow around a compact object is made of different components whose physical nature and emission properties are very varied, more for a neutron star than for a black hole (see e.g. [94], and references therein). Depending on the model and on the source state, we have: (a) the surface of the neutron star, onto which the accreting matter is deposited; (b) a boundary layer between the star and the accretion flow, where the speed of the material in the disc needs to drop rather quickly to adjust to the slower rotation speed of the neutron-star surface; (c) a geometrically thin accretion disc; (d) a Comptonising medium whose spatial location is not yet firmly established; (e) a relativistic jet where matter is ejected from the system at a speed close to that of light. In addition, although the magnetic field of the neutron star in a LMXB is expected to be low, of the order of 10^8G, nevertheless the presence of a magnetosphere has influence onto the accretion flow.

Close the surface of the neutron star surface, matter orbits with a speed close to half the speed of light, c. A number of fundamental time scales can be identified. The light-crossing time, t_{LC}, is shorter than a millisecond and is potentially detectable in time delays between signals. For sub-Keplerian flows, the free-fall time scale t_{ff} can become important. In an optically thick and geometrically thin disc, in addition to the shortest characteristic time scale corresponding to the *dynamical* timescale t_K (see above), other important time scales are the *viscous* time scale, t_{disc}, on which matter diffuses through the disc due to viscosity, the *vertical* time scale, t_z, on which vertical deviations from the hydrostatic equilibrium are damped, and the *thermal* time scale, t_{th}. on which deviations from thermal equilibrium are damped (see [52]).

In the innermost regions of an accretion disc around a neutron star, t_K and t_z are of the order of milliseconds, t_{th} is higher by a factor of a few and t_{visc} is much higher. Moreover, all these time scales increase moving away from the neutron star, and have similar functional dependences on the orbital radius.

With the exception of the neutron star spin, all other timescales apply also to the case of black holes. Since the inner orbits of the accretion flow are comparable in radius between the two objects, similar frequencies are expected, although the mass difference will yield faster time scales for neutron stars. Notice that General Relativity predicts the presence of an innermost stable orbit around a compact object, which naturally imposes a lower limit on all these time scales.

6.4 Timing Phenomenology: QPOs 101

In this section we provide a brief introduction to the study of variability using Fourier power density spectra, we explain the concept of variability components in the Fourier power spectrum of accreting LMXBs, and we discuss the properties of one of those components, the high-frequency quasi-periodic oscillations in neutron-star LMXBs, the so-called kHz QPOs. In subsequent sections we expand on some of the properties of the kHz QPOs in more detail. Because this is meant to be a very general introduction to the topics discussed later in this chapter, and to improve the readability, in this section we try to keep the references to the minimum necessary. We give the appropriate references when we discuss the topics introduced here in more detail in the rest of the chapter.

A useful way to characterise the variability of a source is to use the Fourier power density spectrum (PDS) of the source light curve. The PDS gives the square of the amplitude, called power, of the variability in the light curve at each frequency over a range of frequencies (see [161], for a full explanation). The great advantage of using the Fourier PDS instead of studying the light curves directly is that, while in a light curve one is bound to study the variability over a single, broad range of time scales, from the longest time scale equal to the length of the observation to the shortest time scale equal to the time resolution of the light curve (more precisely, twice the time resolution), in the PDS one can isolate a certain range of frequencies (or equivalently time scales) to study those separately. For instance, it would be very difficult (to say the least) to study a weak, short-period, quasi-periodic signal (e.g., a truly periodic signal with a period that changes randomly during the observation time) in a light curve when that signal is superimposed to another signal that changes stochastically over a long time scale. The reason for this complication is that the two signals would be mixed up in the light curve; one would only be able to study the amplitude of the variability over the total range of time scales combined, and hence only see the combined effect of the two processes. On a PDS, however, one can isolate certain time scales to study the phenomena independently. Perhaps the best example is the case of a strictly periodic signal, e.g., from a pulsar; even if the pulsations appear on top of a very noisy light curve, the signal of the pulsar can be easily identified in the

PDS. This is so because the amplitude of the variability of the pulsar signal is spread over all time bins in the light curve, but the power is concentrated in a few frequency bins (ideally one) in a PDS. The same applies to signals that are not strictly periodic; the advantage in these cases is, again, that in the PDS one can isolate, and study separately, the properties of different variability components that are present in the light curve, but span only a limited range of frequencies, whereas this is impossible using the light curve directly. Also because of this, one final advantage of using the PDS is that, at each frequency, one can easily separate and subtract the part of the variability in the light curve due to the Poisson nature of the signal. The power per unit frequency of a constant, Poisson dominated, signal is also a constant that, when the units of the PDS are chosen conveniently [90], is equal to 2. In the remainder of this chapter we will use the PDS to characterise the variability components observed in accreting X-ray sources.

Without entering into too much details, a PDS gives the power per unit frequency of a signal as a function of frequency. The units of the power can be chosen arbitrarily, but the important point we want to make here is that this power is per unit frequency (therefore the word density in the name power density spectrum; as is common in the literature, here we use loosely the word power to refer to power density). The total power in a light curve over a certain range of time scales is the integral of the PDS with respect to frequency over the corresponding range of frequencies; this quantity is no longer a density (per unit frequency) and, because of Parseval's theorem, this integral is equal to the total variance in the light curve in that particular frequency range. By choosing the appropriate PDS normalisation, this quantity can represent the fractional root-mean square variability, also known as fractional rms, in the light curve over a range of frequencies (see [161], for details).

After producing the PDS of a light curve, the power can be fitted as a function of frequency with (a combination of) all kinds of mathematical functions, and use the parameters of those functions to characterise the properties of the components that those functions represent. Ideally those functions would have some underlying theoretical meaning but, even if they do not, one can still deduce interesting properties of the processes that produce that variability, and eventually about the sources themselves, from the parameters of those functions. For instance, a mathematical function that is commonly used to fit the PDS is a Lorentzian, or Cauchy, function:

$$P(\nu) = \frac{\Delta}{2\pi} \frac{N}{(\nu - \nu_0)^2 + (\frac{\Delta}{2})^2}. \tag{6.1}$$

This function has three parameters: The centroid frequency, ν_0, measures the frequency at which this variability component peaks in the PDS. When we fit a Lorentzian to a QPO, the centroid frequency of the Lorentzian provides information about the dynamics of the process that produces the QPO, e.g. an orbital frequency in the disc, or the frequency of a standing wave in the accretion disc. The next parameter is the full-width at half-maximum (FWHM), Δ, which measures the range of frequencies over which the power of this component contributes significantly to the variability. Instead of the width, some authors use the quality factor, Q, (sometimes also called the coherence, but we will reserve the name coherence for another property of the QPO signal), defined as the ratio of the centroid frequency and the FWHM of the QPO, $Q = \nu_0/\Delta$, to characterise the width of the Lorentzian. A narrow Lorentzian would then have a high quality factor. The width or, equivalently, the quality factor, can provide information about the lifetime of the process that produces the QPO, or how much the frequency of the QPO changes over the time interval that was used to produce the PDS. On the other hand, an initially very narrow QPO could be broadened if the oscillations are damped in an intervening medium between the source and the observer, e.g., an X-ray corona very close to the accreting object. Finally, the normalisation, N, equal to the integral of the Lorentzian from $-\infty$ to ∞, measures the total power of that variability component. As mentioned earlier, the integral of the power density over a certain frequency range gives the power contributed by, in this case, the Lorentzian component that represents the QPO and, because of Parseval's theorem, this is the part of the variance in the light curve that is produced by the QPO. The amplitude of the QPO is the square root of N, and is usually expressed as the rms variability of the signal that produces the QPO divided by the average intensity of the source (and normally given in percent), the so-called rms fractional amplitude, or rms amplitude for short [18, 118]. When it is not normalised by the average intensity, this amplitude is called the absolute rms variability [158]. Both the fractional and the absolute rms amplitudes provide a measure of the variability of the light curve of the source over the range of frequencies (or, equivalently, times scales) where the QPO dominates the power spectrum. The rms amplitude as a function of energy provides information about the radiative process that produces the QPO.

A narrow component, with a Q factor larger than 2, is usually called a QPO. The definition is a bit vague (should a component with a Q factor just a bit smaller or bigger than 2 be also called a QPO?), but it has been useful, and hence it sticked. Components that have $Q < 2$ are usually called bumps and, if the central frequency of this component is at $\nu_0 = 0$, they are called zero-centred Lorentzians. In general, all components that produce power over a broad frequency range are called broad-band noise components. (Notice that, in this case, the word noise refers to variability from the source.) Sometimes a broad-band noise component can be fitted by a combination of several, relatively broad and weak, Lorentzians. Since a Lorentzian is the Fourier transform of a sine (or cosine) function whose amplitude drops exponentially with time, a so-called shot, there have been many attempts to understand the variability in these sources in terms of a combination of shot noise components, with different periods, amplitudes and decay times, that add up together to produce the observed light curve. (The decay time of a shot in the light curve is inversely proportional to the FWHM of the Lorentzian in the Fourier PDS.) In recent years, however, it has been shown that the variability in these sources is inconsistent with additive shots, but it is rather a multiplicative process [159]. This raises the question of whether the bumps and broad-band noise components in the PDS of these sources are in reality a combination of several narrow Lorentzians. This standpoint is attractive because a broad-band noise component is complex, and it is difficult to assign a characteristic frequency (time scale) to it, whereas relatively narrow QPOs give good frequencies which are easier to extract and follow over time, and can be treated in a model-independent way. This approach has been tried in a few cases, and interesting correlations among the properties of those (sometimes weak) Lorentzians have emerged. We will mention some of those in the coming sections.

The Rossi RXTE mission yielded thousands of high sensitivity observations of X-ray binaries and revolutionised our knowledge of these objects. RXTE observations covered a large range of states of dozens of X-ray binaries, unveiling details of the variability of these objects that helped us understand them more deeply. One of the discoveries of RXTE was the existence of very-high frequency quasi-periodic variability components, up to ~1200 Hz, in several NS LMXBs. These variability components are what we call the kHz QPOs. Other variability components were also studied with RXTE, including low-frequency QPOs and broad-band noise components. We will mention some of those in passing when necessary, but here we will concentrate mainly on the properties of the kHz QPOs.

At a very basic level, NS LMXBs can be subdivided into three classes: (i) persistent sources at high luminosity, historically called "Z" sources, that can reach luminosities close to the Eddington limit for a neutron star, (ii) persistent and transient sources that can become rather bright but, with top luminosities of 0.1–0.2 Eddington, do not reach the same high luminosities as the Z sources, historically called "atoll" sources, and (iii) faint sources which, even when transient, remain at low luminosities, below 0.01 Eddington.

The power spectrum of the three classes of sources mentioned above show several differences. For example, the amplitude of the variability components in Z sources is generally (but not for all variability components) lower than in atoll and low-luminosity sources. In Fig. 6.1 we show examples of the PDS of a low-luminosity (Fig. 6.1a; the NS LMXB 1E 1724−3045 in the globular cluster Terzan 2) and an atoll source (Fig. 6.1b; the NS LMXB 4U 1636−53). As it is apparent in that Figure, all the PDS show a broad-band noise component extending up to ~10–20 Hz; above that frequency the power drops as the frequency increases, except for a few relatively narrow features peaking at some specific frequencies. The narrow peaks appearing above ~400 Hz are the kHz QPOs. Notice also that the scales in the *y* axis of the two Figures are different, and that the PDS on the Figure on the right approaches a power level of 2 at high frequencies, whereas the one the left drops to 0 (the *y* axis on the left panel is in a log scale). The difference is that the powers in Fig. 6.1a are in units of fractional rms^2 per Hz (see above), whereas in Fig. 6.1b the power has not been converted to rms units. Furthermore, in Fig. 6.1a the contribution of the constant level due to the Poisson nature of the counting process was subtracted from the PDS, leaving only the signal from the source.

The first two sources to show kHz QPOs were the Z source Sco X-1 [164] and the atoll source 4U 1728−34 [152]. Both sources displayed (sometimes) two QPOs appearing simultaneously in the PDS at frequencies between ~700 and ~1100 Hz. The two QPOs were then labeled "lower" and "upper" kHz QPO according to their frequencies, such that $\nu_{\rm upper} > \nu_{\rm lower}$. If observed frequently enough, most sources

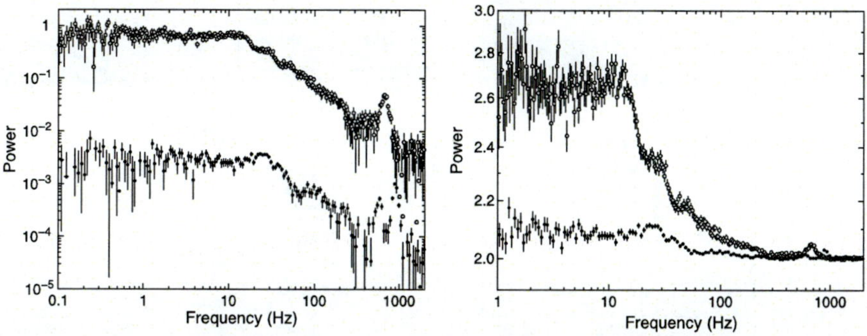

(a) (b)

Fig. 6.1 Left (**a**): Two power spectra of the low-luminosity source and X-ray burster 1E 1724−3045 in the globular cluster Terzan 2 [9, 126]. Right (**b**): Two power spectra of the atoll source 4U 1636−53. The power spectra of 4U 1636−53 were computed on the basis of data published in [20, 145]. In both panels the power spectra show the broad-band noise component, extending up to ∼20–30 Hz in 1E1724−3045 and up to ∼10–20 Hz in 4U 1636−53, some low-frequency QPOs at ∼30 Hz, and one or two simultaneous kHz QPOs above ∼500–600 Hz and up to ∼1000 Hz. The power spectra on the left have the Poisson noise component subtracted and the power is plotted in units of rms^2 per Hz. The power spectra on the right still contain the Poisson noise component and the power is given in units such that the Poisson level is 2

with kHz QPOs show two simultaneous QPOs in the PDS, but some (few cases) have so far only showed one. We will come to that below.

Given that the frequency of the upper kHz QPO is consistent with the Keplerian orbital frequency of a test particle at ∼10–20 km around a ∼1.5-$M_\odot$ neutron star (see Sect. 6.3), the kHz QPOs were immediately associated to motion of matter at the inner parts of the accretion disc. Because of this, and because the bolometric luminosity, and hence the observed flux and intensity, of the source is expected to be proportional to mass accretion rate, $\dot{M}$, while the inner radius of the accretion disc, R_{in}, is expected to decrease as $\dot{M}$ increases (and vice versa), the expectation was that the QPO frequency would increase with X-ray intensity. While this was the case over short time intervals (a day or less), the long term relation was more complex, with the QPO frequency tracing several, more or less parallel, tracks in a plot of QPO frequency vs. X-ray intensity (Fig. 6.2a). This kind of plots were then, indeed, called parallel tracks.

When enough observations of a single source are collected, a pattern of the detection of the kHz QPOs emerges. In an atoll source, the lower kHz QPO appears in a relatively narrow part of the colour-colour diagram, at an intermediate state, the transitional part of this diagram, between the low-luminosity hard state (called the island state) and the high-luminosity soft state (called the banana; we will try not to use these names here to avoid too much jargon, and we will call these low or hard and high or soft states). The X-ray colours of the source do not change much in the observations in which the lower kHz QPO is present, but the frequency of the QPO appears to correlate with the position of the source in this diagram, with the frequency increasing as the inferred mass accretion rate increases.

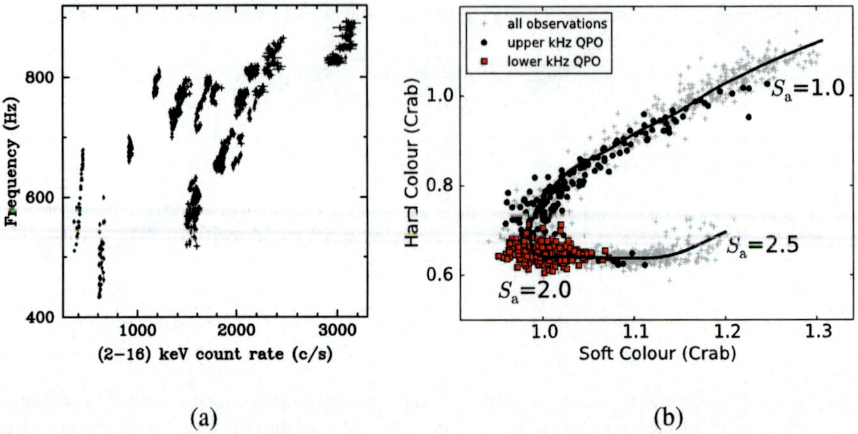

Fig. 6.2 Left panel (**a**): Frequency of the lower kHz QPO in 4U 1608−52 vs. the count rate of the source in the 2–16 keV band. Each point corresponds to a measurement over intervals of ∼64–128 s. The frequency and the source count rate are significantly correlated over intervals of a few thousand seconds, producing each of the tracks in the plot, but different observations, made a few days or weeks apart, produce different tracks. This plot shows the so-called parallel tracks (originally published as Figure 2 in [109]). Right panel (**b**): the distribution of the observations with kHz QPOs on the colour-colour diagram of 4U 1636−53 [188]. Black points correspond to observations in which only the upper kHz QPO was detected, while red points correspond to observations that showed only the lower, or both the lower and the upper, kHz QPO. The grey points indicate observations where no kHz QPO were detected. The parameter S_a gives the length along the solid line, which parametrises the position of the source in this diagram. High values of S_a correspond to high values of inferred $\dot{M}$

Figure 6.2b shows the colour-colour diagram of 4U 1636−53. The red and black points indicate, respectively, the observations in which the lower and the upper kHz QPOs were detected. The solid line parameterises the position of the source in this diagram, through the variable S_a. High values of S_a correspond to high values of inferred $\dot{M}$. The frequency of the lower kHz QPO increases as S_a increases. The upper kHz QPO, on the contrary, covers a broader range in the colour-colour diagram, with the frequency of the QPO increasing as the source moves from the low-luminosity hard state, via the transitional intermediate state to the high-luminosity soft state (increasing value of S_a). This is the sense in which mass accretion rate is inferred to increase in these sources. For completeness, the grey points mark observations in which no kHz QPO was detected. The observations with no kHz QPOs are at the extremes of the C-shaped figure traced by the source in the colour-colour diagram; at the top right the source is in the low-luminosity hard state, where the inferred $\dot{M}$ is the lowest, while at the bottom right it is in the high-luminosity soft state, where the inferred $\dot{M}$ is the highest.

Although not apparent from the plot, there are several observations in which both QPOs were detected simultaneously. This can be seen in Fig. 6.3; the top panel of that Figure shows the frequency of both kHz QPOs as a function of the hard colour,

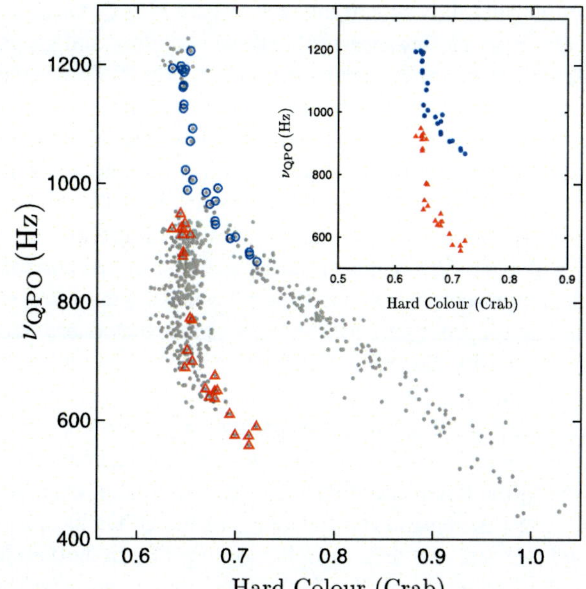

(a)

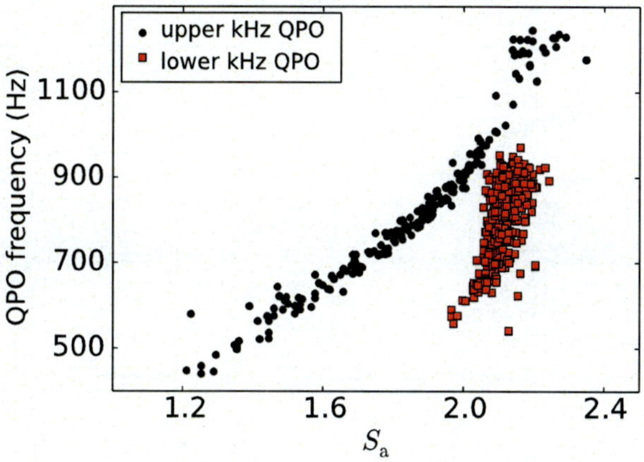

(b)

Fig. 6.3 Top (**a**): Frequency of the kHz QPOs in 4U 1636−53 vs. the source hard colour (see Fig. 6.2b). Grey symbols correspond to observations in which only one of the kHz QPOs was observed; red and blue symbols correspond to QPOs identified, respectively, as the lower and upper kHz QPO. The inset shows only the QPO frequencies of the observations in which two simultaneous kHz QPOs were detected [143]. Bottom (**b**): Frequency of the lower (red symbols) and upper (black symbols) kHz QPOs in 4U 1636−53 vs. S_a [188]. The quantity S_a measures the position of the source along the track in the colour-colour diagram in Fig. 6.2b while the bottom panel shows the frequency of both kHz QPOs as a function of S_a. The two kHz QPOs are clearly separated in these two plots, and the parallel tracks of Fig. 6.2a collapse into a single track (one for each kHz QPO) when the frequencies of the QPOs are plotted against the hard colour or S_a. By the way, since the position of the source in the colour-colour diagram, and therefore hard colour and the value of S_a, is driven by changes of the source spectrum, it should be no surprise that the relation of the QPO frequency with the parameters of the models used to fit the energy spectrum also consists of a single track. We will discuss this in Sect. 6.7.

In Z sources, the kHz QPOs appear mostly in the so-called horizontal and normal branches in the colour-colour diagram (roughly speaking these are, respectively, the top horizontal and diagonal parts of the letter Z that the source traces as it moves in the hardness-intensity diagram), and the QPOs disappear when the source is in the flaring branch (the bottom horizontal part of the letter Z in the hardness-intensity diagram). In Fig. 6.4a we show the hardness-intensity diagram of the Z source GX 5-1 with the branches indicated. As for the atoll sources, the parameter S_Z measures the position along the track traced by the source in this diagram, with inferred $\dot{M}$ increasing when S_Z increases. The frequency of the QPOs increases as the source moves from left to right and then from the top right to the bottom left along the Z shape in the colour-colour diagram, which is the same direction in which, according to work that preceded the discovery of kHz QPOs, $\dot{M}$ increases in these sources. In

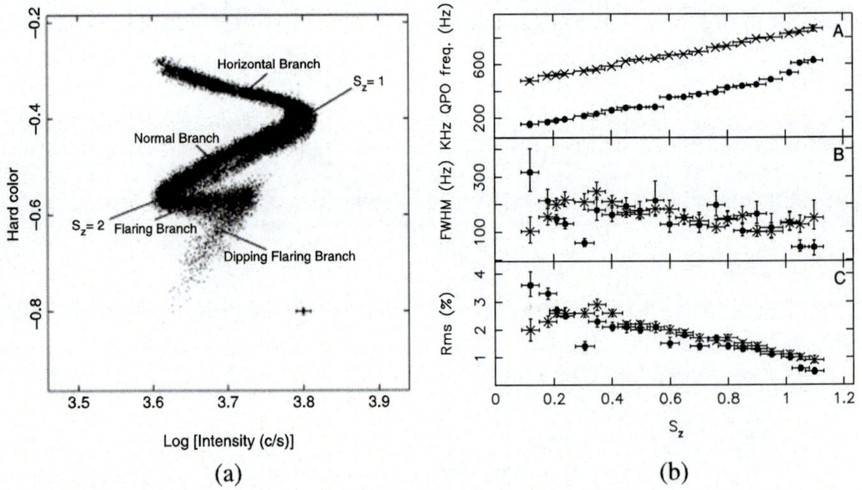

Fig. 6.4 Left panel (**a**): Hardness intensity diagram of the Z source GX 5-1 with the different branches of the Z shape indicated (originally published as Figure 1 in [69]). The position of the source along the Z-shaped track is parameterised through the quantity S_Z which is set to $S_Z = 1$ at the vertex between the Horizontal and the Normal branches, and to $S_Z = 2$ at the vertex between the Normal and the Flaring branches. Right panel (**b**): Frequency (top), FWHM (middle) and rms amplitude (bottom) of the kHz QPOs vs. S_Z for GX 5-1 (originally published as Figure 9 in [69]). From the values of S_Z in this plot it is apparent that the QPOs are detected in the Horizontal and the initial part of the Normal branches

Fig. 6.4b we show the frequency, FWHM and fractional rms amplitude of both kHz QPOs in GX 5-1 as a function of S_Z. The fact that QPO frequency increases with inferred mass accretion rate (increasing S_Z) made the identification of the upper kHz QPO with the Keplerian frequency at the inner disc radius plausible.

The frequency range that makes a QPO a kHz QPO is roughly 400–1200 Hz. The fact that kHz QPOs are not detected outside this frequency range can be understood from the dependence of the other properties of the kHz QPO upon QPO frequency. The rms amplitude and the Q factor of both kHz QPOs depend upon QPO frequency in a systematic way. The rms amplitude and the Q factor of the lower kHz QPO are maximum when the frequency of the QPO is around 700–800 Hz, and both the rms and Q decrease when the QPO frequency either increases or decreases. The typical range of fractional rms amplitudes of the lower kHz QPO is 3–15% considering photons in the full band covered by RXTE/PCA, nominally from 2 to 60 keV. At the same time, the Q factor of the lower kHz QPO ranges from ~10 to ~50, and can be as high as ~200–250 in some sources. For the upper kHz QPO, when the QPO frequency is low the rms amplitude is maximum and remains roughly constant and then drops more or less continuously as the QPO frequency increases whereas, at the same time, the Q factor remains constant or increases slightly. The rms amplitude of the upper kHz QPO is 2–20% in the 2–60-keV band, while Q is usually around 10 or less. In Fig. 6.5a and b we show, respectively, the rms amplitude and the Q factor, of the lower and upper kHz QPOs in 4U 1636−53.

The drop of both the rms amplitude and the Q factor limits the detectability of the lower kHz QPO at low and high QPO frequencies below ~400 Hz and above ~950 Hz. Similarly, the drop of the rms amplitude of the upper kHz QPO limits its detectability at frequencies above ~1200 Hz, whereas at low QPO frequencies the detectability of the upper kHz QPO is limited by the relatively low Q value and the fact that, when the frequency of the upper kHz QPO goes down to ~400 Hz the broad-band noise extends up to comparable frequencies such that the upper kHz QPO starts to appear on top of the broad-band noise, and hence it is difficult to detect. All in all, there is a range of frequencies at which the QPOs are the narrowest and the strongest, and hence the most significantly (and hence most often) detected. In the sources in which a single kHz QPO was detected, either the source was not observed for long enough to sample the range of states in which the QPOs are detected, or the source was relatively weak such that sensitivity to detecting kHz QPOs was not sufficient. The fact that the kHz QPOs most often appear in pairs is then a characteristic that needs to be explained.

The frequency of the two QPOs change when other source properties, e.g. the source intensity or colours, change; but an interesting fact is that, as the frequency of the QPOs changes, the difference of the centroid frequency of the two QPO peaks remains more or less constant. When burst oscillations and two simultaneous kHz QPOs were detected in 4U 1728−34, with the frequency separation between the two QPOs consistent with being equal to the frequency of the burst oscillations, a beat-frequency mechanism [115] was proposed to explain the double kHz QPOs. In the original model, the upper kHz QPO was identified with the Keplerian frequency at the inner edge of the disc, which is truncated at the sonic radius, the radius at

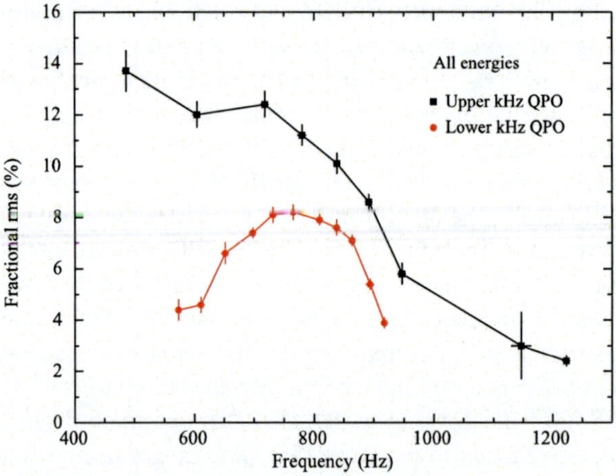

(a)

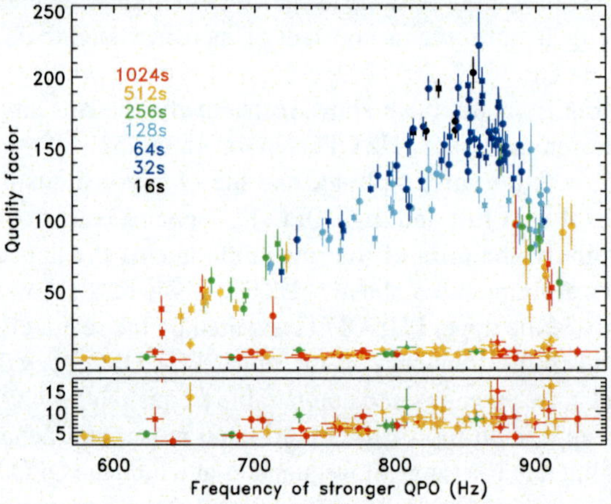

(b)

Fig. 6.5 Top (**a**): Fractional rms amplitude of the kHz QPOs in 4U 1636−53 vs. the frequency of that same QPO (adapted from [139]). Bottom (**b**): Quality factor of the kHz QPOs vs. the frequency of the strongest kHz QPO in the power spectrum of 4U 1636−52 [14]

which the radial component of the velocity of the material falling onto the neutron star goes from subsonic to supersonic. The lower kHz QPO was then interpreted as a beat between the oscillation at the Keplerian frequency and the neutron-star spin. Under those conditions, the frequency separation between the two QPOs, which is equal to the neutron-star spin, should remain constant as the frequencies of the QPO move. We will return to this below.

The initial observations of the kHz QPOs in Sco X-1 had already shown that the frequency separation between the two QPO peaks was not always the same, but decreased systematically, and significantly, by a few percent as the frequencies of the two simultaneous kHz QPOs increased. The beat-frequency model could still explain this behaviour if the material at the inner radius of the disc, where the beating took place, suffered from radiation drag and the beating took place as that material spiralled in towards the neutron star. Being a very luminous source, this effect could be strong in the case of Sco X-1. But soon after several other less luminous sources, starting with the atoll source 4U 1608−52, showed the same effect.

The situation got more complicated for this model when burst oscillations and two simultaneous kHz QPOs were detected in 4U 1636−53, with the frequency of the burst oscillations being twice the difference in frequency between the kHz QPOs. The original beat-frequency model could not explain this. The model had to be made more complex, by adding a possible excitation of vertical modes in the accretion disc at a radial distance where the difference between the Keplerian frequency at the inner edge of the disc (that causes the QPO at $\nu_{\rm upp}$) and the neutron-star spin frequency is equal to the vertical epicyclic frequency in the disc. Depending on whether the material in the disc is smooth or clumped, the excited frequency, which produces the lower kHz QPO, would be at $\nu_{\rm upp} - \nu_{\rm spin}$ or $\nu_{\rm upp} - \nu_{\rm spin}/2$.

These complications for the beat-frequency model triggered other proposals to explain the QPO frequencies. One of them, that could explain the phenomenology rather naturally, was the idea of periastron precession of the innermost parts of the accretion disc. Under this hypothesis, the upper kHz QPO was still identified as the epicyclic azimuthal (Keplerian) frequency of a test particle at the inner edge of the disc, under the influence of the general relativistic (GR) potential of the neutron star. In this model, however, the lower kHz QPO would be the difference between this azimuthal and the epicyclic radial frequency, the so-called periastron precession frequency, at the same spot in the disc. The difference between the azimuthal and the periastron precession frequency in the model changes generally in the same way as in the observations [148], although the calculations do not fit the exact trend of the observations.

One strong prediction of this model was that the difference between the frequency of the two QPOs should not only decrease at high, but also at low QPO frequencies, something that was later on observed in the system Cir X-1 (Fig. 6.6). To be fair to history, the relativistic-precession model, as this model was called, came about as an extension of the Lense-Thirring model [147] that was proposed a year earlier to explain the correlation between the frequency of the upper kHz QPO and a low-frequency QPO in neutron-star LMXBs. The Lense-Thirring and the relativistic-precession models became one consistent model for both the low- and the high-frequency variability. Notice, also, that in this model there is no relation between the frequencies of the kHz QPOs and the spin of the neutron star, therefore this model was also applicable to QPOs in black-hole systems.

If the kHz QPOs and the low-frequency QPOs are all GR frequencies in the disc (but notice that the calculations assume test particles, so no disc hydrodynamics), another prediction of this model is that the frequency of the low-frequency QPO

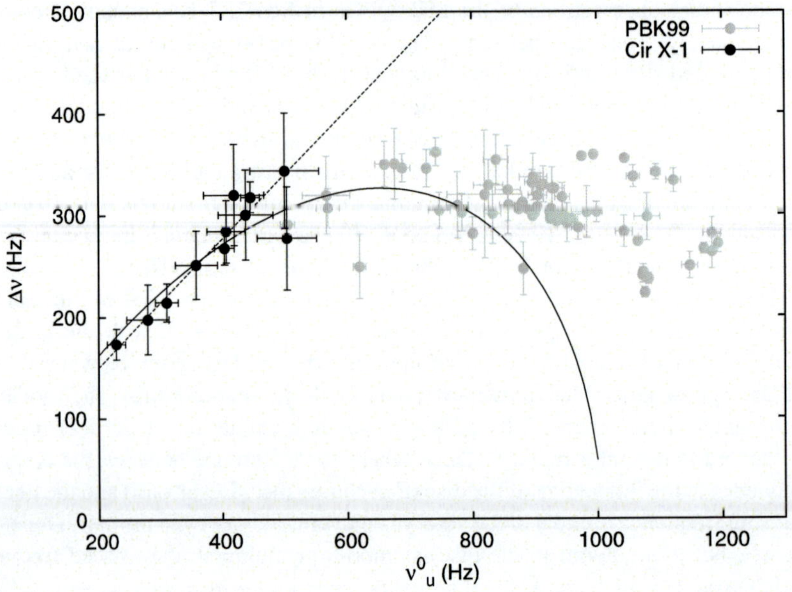

Fig. 6.6 Difference between the centroid frequencies of the kHz QPOs, $\Delta\nu = \nu_{\rm upp} - \nu_{\rm low}$, as a function of the frequency of the upper kHz QPO in Cir X-1 (originally published as Figure 11 in [28])

should be proportional to the square of the frequency of the upper kHz QPO. Figure 6.7a shows the PDS of three separate observations of 4U 1728−34 in which the low-frequency and upper kHz QPOs are marked with vertical lines. Figure 6.7b, on the other hand, shows the relation between the frequency of the low-frequency QPO and that of the upper kHz QPO in this same source, with the line corresponding to the best-fitting power to the data with index of 2.11 ± 0.11. We will expand on models in Sect. 6.5

One can extract useful information about the mechanism that causes the QPOs from the Q factor, if one has a model to explain its behaviour with QPO frequency. As shown in Fig. 6.5b, in 4U 1636−53 the Q factor of the lower kHz QPO first increases as the QPO frequency increases, it reaches a maximum at $\nu_{\rm low} \sim 800$–850 Hz, and drops rather abruptly as the frequency of the QPO continues to increase. This same behaviour was observed in all sources for which enough data were available. The rapid drop at high frequencies was interpreted as the inner radius of the accretion disc reaching closer and closer to the ISCO (see Sect. 6.3), where the faster and faster radial drift of the material in the disc towards the neutron star causes a drift of the QPO frequency over the lifetime of the process that produces that QPO, hence broadening the observed QPO peak, and reducing Q. We will discuss this effect, and other alternatives, in Sect. 6.8.

The other parameter of the Lorentzian function in Eq. (6.1) is the rms amplitude, equal to $\sqrt{N}$ in that equation. Already the initial observations showed that the

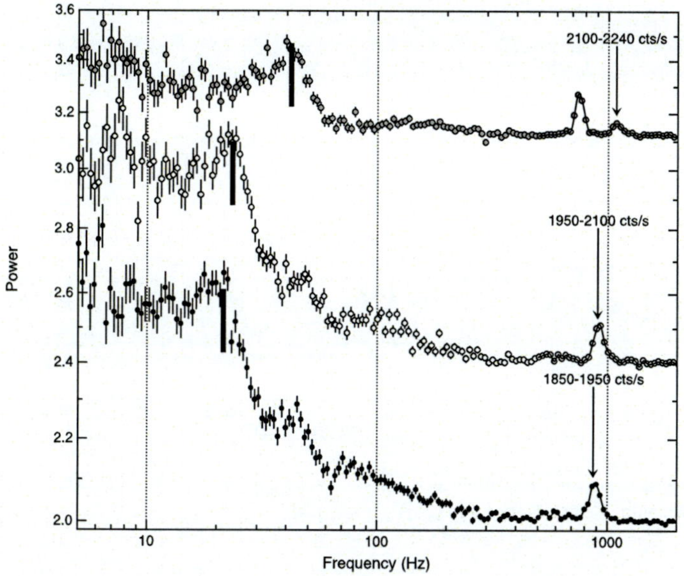

(a)

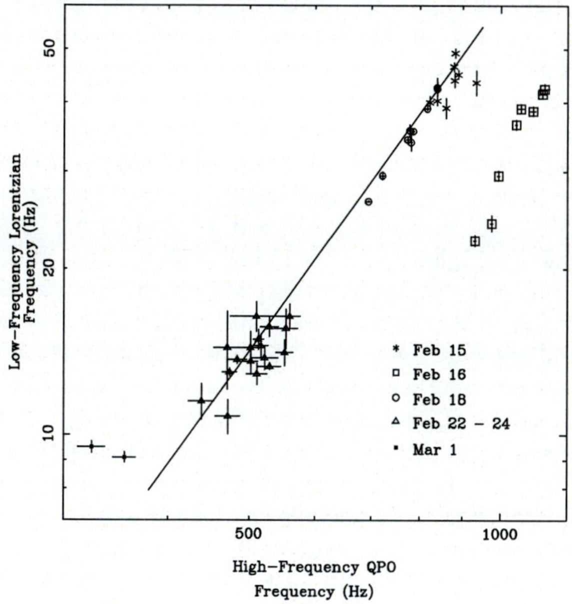

(b)

Fig. 6.7 Top (**a**): Power spectra of three separate observations of 4U 1728−34 with the low-frequency and upper kHz QPO indicated (the plot is based on the data published in [152]). Bottom (**b**): Plot of the frequency of the low-frequency QPO vs. that of the upper kHz QPO in 4U 1728−34. The solid line is the best-fitting power-law relation to the data (originally published as Figure 2 in [51])

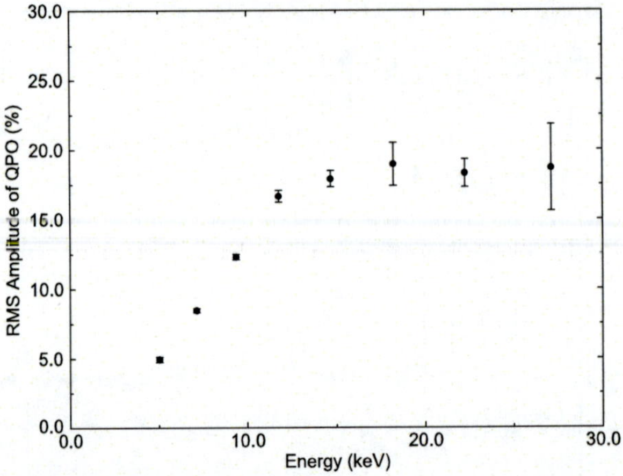

Fig. 6.8 Fractional rms amplitude as a function of energy for the lower kHz QPO in 4U 1608−52 (originally published as Figure 4 in [24])

spectrum of the variability is hard. In other words, the fractional rms amplitude of both kHz QPO increases with energy. For instance, in 4U 1608−52 and 4U 1636−53, the rms amplitude of the lower kHz QPO at ~25 keV is ~20%, while the rms of the upper kHz QPO in these two sources increases a bit less steeply with energy, reaching ~12% at ~20 keV. In Fig. 6.8 we show the rms spectrum of the lower kHz QPO in 4U 1608−52.

The soft thermal component, which is the combined emission from the neutron-star surface and the accretion disc, in the time-averaged X-ray energy spectrum of these sources peaks at ~3–6 keV and drops very rapidly as the energy increases. Therefore, the contribution of the disc and the neutron-star surface to the total emission at energies higher than ~10–15 keV is always negligible and, even if the disc or the neutron-star surface were oscillating with an rms amplitude of 100%, their contribution to the observed fractional rms amplitude at and above those energies would be totally negligible. This shows that, while the *dynamical* process that determines the frequency of the QPOs could take place in the disc, like in the models described above, the *radiative* process that modulates the source emission at the QPO frequency cannot come from either the neutron star or the disc. At those energies, the dominant spectral component is the corona, in which highly energetic electrons transfer energy to the soft photons emitted form the neutron star and the disc via inverse Compton scattering, redistributing those photons into a power-law shaped component in the energy spectrum. We will discuss this further in Sect. 6.8.1.

Finally, a property of the kHz QPOs (and any other variable signal) that is not represented in Eq. (6.1) is the energy-dependent phase lag (or, equivalently, time lag) of the signal. To understand the phase lag one needs to go back to the Fourier analysis of a signal. The power spectrum that we described at the beginning of this section, is the modulus square of the complex Fourier transform of the light curve of the source as a function of frequency or, equivalently, the product of the Fourier transform of the signal by its complex conjugate. If, instead, one multiplies the Fourier transform of a signal by the complex conjugate of another signal, both functions of frequency, the result is the cross-spectrum. In the same way that the power spectrum measures the variance of the signal per unit frequency (through Parseval's theorem), the modulus and the argument of the cross-spectrum measure, respectively, the covariance per unit frequency and the phase difference, also called phase lag, $\Delta\phi$, between the two signals as a function of Fourier frequency. In the same way that the power spectrum gives the degree of correlation of a light curve with itself, the autocorrelation of the light curve, the cross-spectrum gives the degree of correlation of one light curve with the other, the cross-correlation between the two light curves. If the signals are uncorrelated, at each Fourier frequency the covariance and the phase lag will be on average 0. (Notice that uncorrelated signals give a 0 phase lag, but a 0 phase lag does not imply that the signals are uncorrelated.) At any given frequency, ν, the phase lag can be converted into a time lag, $\Delta t(\nu) = \frac{\Delta\phi(\nu)}{2\pi\nu}$. The phase lags are defined[1] from $-\pi$ to π, while the time lags run between $-1/(2\nu)$ and $1/(2\nu)$. Since both quantities are related, depending on the context, we will either use the term phase or time lags to refer to the delay between the two light curves in the Fourier space.

If the two light curves used to compute the cross-spectrum come from two different energy bands, the time lag at each Fourier frequency represents the time delay between the light curves in those two energy bands at each Fourier frequency. For a QPO (and any other somewhat broad component) with a centroid frequency ν_0 and a FWHM Δ, we call the phase (or time) lag of the QPO to the average of the phase (or time) lags over a frequency range around the centroid frequency of the QPO, e.g. from $\nu_0 - \Delta$ to $\nu_0 + \Delta$. It is customary to take the light curve at the lowest energy band as the *reference band* and to measure the phase lag, with respect to reference band, of the light curve in the bands, called *subject bands*, at energies above the energy of the reference band. Under this convention, a positive phase/time lag, also called hard lag, indicates that the hard light curve lags (follows after) the soft one, whereas a negative phase/time lag, when the soft light curve leads (comes before) the hard light curve, is called soft lag. Alternatively, one can use the full band as the reference band, and narrow bands within the full band to measure the lags, provided that one corrects for the correlation introduced by the part of the signal that is both in the subject and the reference bands. In the end one obtains the energy dependent phase lags, of the subject bands with respect to the reference band, over the frequency range in which the QPOs dominate the variability of the source.

Because the lower kHz QPO is usually narrower and, therefore usually more significantly detected, than the upper, the first measurements of lags where obtained

[1]Phase lags equal to $\Delta\phi \pm 2n\pi$, with n any integer number, cannot be distinguished from a phase lag $\Delta\phi$.

for the lower kHz QPO. The magnitude of the time lags of the lower kHz QPO in 4U 1608−52 [174, 175] and 4U 1636−53 [73] was $\Delta t \sim 20$–25 μs, constraining the size of the region where the lags are produced to $c\Delta t \lesssim 10$ km. A remarkable fact of those detections was that the lags of the lower kHz QPO in these two sources were soft, contrary to the expectation if the lags were produced by inverse Compton scattering in the corona, since in that case the low-energy photons escape from the system before the photons that are up-scattered in the corona to high energies.

Fifteen years passed before new measurements of the lags of the kHz QPOs were published. In that period the number of sources with kHz QPOs, and the number of detections of QPOs in individual sources, covering a broad range of frequencies, allowed for more detailed studies of the lags as a function of energy and QPO frequency [11, 37]. At the same time, this also allowed to measure, for the first time, the lags of the upper kHz QPO [11, 37, 38, 130, 157].

Figure 6.9a shows the lags of the lower and the upper kHz QPO in 4U 1608−52 and 4U 1636−53 as a function of energy. The lags of the lower kHz QPO in both sources are soft and become softer as the energy increases, whereas the lags of the upper kHz QPO are either consistent with zero or increase slightly with energy. Figure 6.9b shows the lags measured between two broad energy bands for the lower and the upper kHz QPO in the same two sources as a function of frequency. The magnitude of the lags of the lower kHz QPO in 4U 1636−53 first increases and then decreases as the frequency of the QPO increases, whereas the lags of the upper kHz QPO remain more or less constant at zero. The lags of the lower and upper kHz QPOs in 4U 1608−52 have larger error bars, but their dependence upon QPO frequency is consistent with that of 4U 1636−53. These two plots show that two different radiative mechanisms operate to produce the lags (and, as we saw, the rms amplitude) of the lower and the upper kHz QPO, and this, in turn, provides valuable information for models that try and explain these phenomena. We will come back to this in Sect. 6.8.3.

6.5 Linking Observed Frequencies with Theoretical Expectations

The initial reports of the detections of the first kHz QPOs in Sco X-1 and 4U 1728−34 already put forward the suggestion that the observed frequency could be the Keplerian frequency at the inner edge of the disc. The IAU Circulars with those reports stated: "The high-QPO frequency, and its increase with mass-transfer rate, suggest that we may be seeing the keplerian frequency at the inner edge of the disk near the magnetospheric boundary, or its beat frequency with a slower (about 100 Hz) pulsar." [163], and "Explanations in terms of either keplerian frequencies or a beat-frequency model cannot yet be ruled out, although no evidence has yet been seen for a coherent pulsar frequency in the same data." [151].

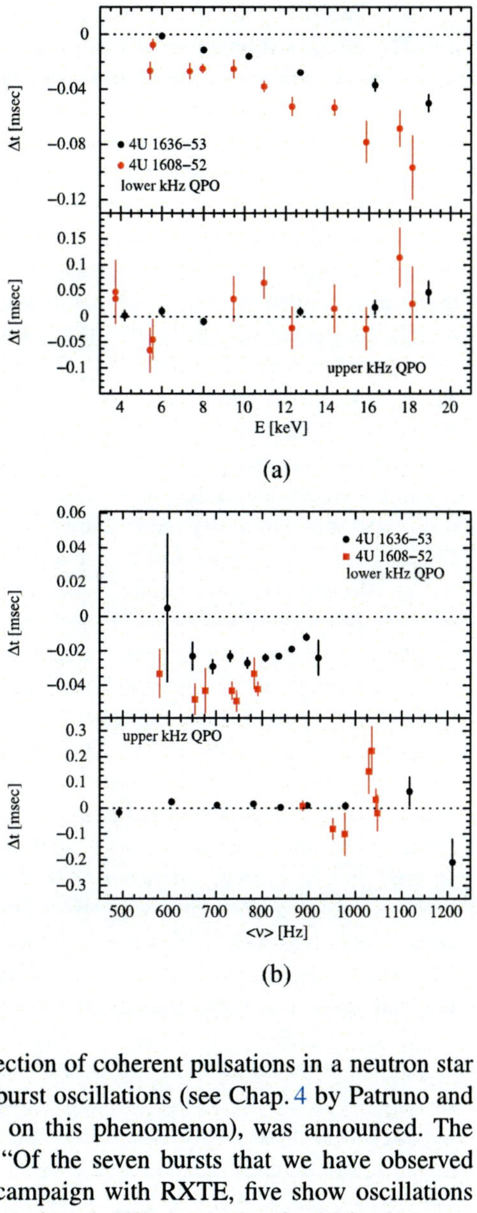

Fig. 6.9 Time lags as a function of energy (top, **a**) and frequency (bottom, **b**) for the lower (top panels) and the upper (bottom panels) kHz QPOs in 4U 1636−53 and 4U 1608-52 (adapted from [37])

About 2 months later the first detection of coherent pulsations in a neutron star during an X-ray burst, the so-called burst oscillations (see Chap. 4 by Patruno and Watts in this book for more details on this phenomenon), was announced. The IAU Circular with that report said: "Of the seven bursts that we have observed from 4U 1728−34 during a recent campaign with RXTE, five show oscillations with a frequency of 363 Hz.", and continued: "We have also found in the same data set two simultaneously-present kHz quasiperiodic oscillations (QPOs), one of which has been reported earlier (IAUC 6320). The centroid frequencies of the two QPOs change with intensity and time, but their difference appears to be always near 363 Hz. These observations are consistent with a neutron star spin period of 2.75 ms." [150].

These results set up the stage for the idea [115] of a beat-frequency model of the kHz QPOs. (Although published in 1998, the idea was first presented in detail by the same authors in a preprint in 1996, arXiv:astro-ph/9609157.) In this model, called the sonic-point beat-frequency model or, for short, the sonic-point model, the upper kHz QPO is a beaming oscillation produced by a hot spot on the neutron-star surface. This spot is the footprint of a stream of matter falling from material at the sonic radius (see Sect. 6.4) onto the neutron star. When this stream hits the star it heats a small area producing a footprint that rotates around the surface of the star at the same frequency as that of the material in the disc, at the sonic radius, where the stream originates. When the neutron star rotates, radiation from the pole(s) illuminates periodically the part in the disc where the stream starts and, because of radiation drag, increases momentarily the rate of mass that is injected into the stream and falls onto the neutron star. When this extra amount of material hits the neutron-star surface at the footprint of the stream, the temperature of the spot increases. The emission from the footprint is therefore modulated at a frequency that is equal to the Keplerian frequency at the sonic radius minus the neutron-star spin frequency. In this model, the luminosity modulation of the hot spot produces the lower kHz QPO (Please read [115], for the full explanation of the model).

One obvious conclusion of this scenario is that the frequency difference between the kHz QPOs, which is equal to the neutron-star spin frequency,[2] has to remain constant when the QPO frequencies move (Sect. 6.4). This was the case for most of the sources in which kHz QPO had been detected, except for Sco X-1 [165], in which the frequency difference decreased systematically as the QPO frequencies increased together. While this result posed a problem to the sonic-point model, the situation could be explained if the clumps in the disc, where the stream originates, spiralled in due to the strong radiation drag in this bright source [86]. In this case, the frequency difference could be less than the neutron-star spin frequency and decrease as the QPO frequencies increased. The situation became even more difficult for the sonic-point model when this effect was observed in more sources, all much weaker than Sco X-1 [105, 107, 108] and, especially, when the difference of the QPO frequencies in some observations of 4U 1636−53 [70] turned out to be larger than half the neutron-star spin frequency in this source, something that could not be explained in the sonic model and its subsequent extensions.

Almost at the same time, a different model that could explain the dependence of $\nu_{\rm upp} - \nu_{\rm low}$ vs. the frequency of the QPO was proposed. As in the sonic-point model, this relativistic-precession model [147, 148] considered that the frequency of the upper kHz QPO is the Keplerian frequency at the inner edge of the disc; but differently from the previous model, in this case the lower kHz QPO would be the periastron precession frequency, equal to the difference between the Keplerian and epicyclic radial frequencies at the inner edge of the disc. In this model the frequency difference between the kHz QPOs is independent of the neutron-star spin

[2] As explained in Sect. 6.4, in a modified version of the sonic-point model the frequency difference between the kHz QPOs can also be equal to half the neutron-star spin frequency [87].

and, as can be readily seen from the identification of the lower kHz QPO, should be equal to the epicyclic radial frequency at the inner radius of the disc. This epicyclic frequency is 0 at the ISCO, first increases as the radial distance in the disc increases, and then decreases again as the radial distance continues increasing. This implies that the frequency difference between the kHz QPOs should decrease both at high and low QPO frequencies, corresponding to small and large radial distances in the disc. As indicated, this explained the observed decrease of $\nu_{\rm upp} - \nu_{\rm low}$ with QPO frequencies as the QPO frequencies increase in Sco X-1 [106, 165] and other sources [70, 105, 107, 108], but also predicted a trend at low kHz QPO frequencies for which there were no data at the time. A few years later, the neutron-star LMXB Cir X-1 [28] showed exactly that (Fig. 6.6 in Sect. 6.4) and, since other predictions of the model for low-frequency variability had already been validated (Fig. 6.7b in Sect. 6.4), all this lent support to this model. Notice, however, that the model relies on frequencies of test particles around the neutron star, and therefore does not consider the hydrodynamical effects in the disc that may affect those frequencies. We will come to this again in Sect. 6.8.4.

A third class of models considers wave patterns in the disc as the cause of the kHz QPOs. These models also rely upon the three basic GR epicyclic frequencies discussed in the previous models (and sometimes also upon the neutron-star spin), but in this case those are not the frequencies of test particles orbiting the neutron star, but characteristic frequencies in a hydrodynamical flow that determine how pressure and gravity waves travel in the disc and, sometimes, lead to other frequencies that are resonances of the basic ones (some examples of those ideas can be found in [76–78, 80, 85, 124, 176], but the list is much longer). Among these models, one that received some attention [1, 2] argued that a resonance in the disc appears when the ratio of two of the epicyclic frequencies discussed above is the ratio of two small integer numbers, e.g. 2:3. Such a preferred frequency ratio was reported for the kHz QPOs in Sco X-1 [2], and the model gained popularity because, in two cases in which two simultaneous high-frequency QPOs were observed in black-hole systems, those QPOs appear at frequencies that are in a 2:3 ratio (e.g., at 300 and 450 Hz in the black-hole LMXB GRO J1655-44; see [136, 149]). In essence, this resonance model is equivalent to the example of a double pendulum discussed in books of Mechanics (e.g. [89]) with, in this case, a mechanism that couples two oscillating phenomena in the disc. The report of a 2:3 frequency ratio of the kHz QPOs in Sco X-1 has been subsequently disputed [19, 21, 113], but the model is still considered for high-frequency QPOs in black-hole LMXBs.

The models described in this section, and most of the models of the kHz QPOs that appeared in the last 20 years, aim at explaining the frequencies of the oscillations and, in that sense, are dynamical models of the QPO phenomenon. Very few models have attempted to give an explanation of the other, radiative, properties of the QPOs. We will discuss those radiative properties of the kHz QPOs in Sect. 6.8, and we will also mention some of the latest attempts to try and explain those properties.

6.6 QPO Frequency Correlations

Although the beat frequency model is unable to explain all observations, it is reasonable to think that the kHz QPOs could in some way be connected to the rotation of the neutron star. When the first source for which both burst oscillations and kHz QPOs were discovered, 4U 1728−34, it was realised that $\Delta\nu$ was around the same value as the burst oscillation frequency [150]. However, the next source with a double detection was 4U 1636−63, where $\Delta\nu \sim 270$ Hz and the burst oscillation frequency was 581 Hz, close to twice that value. After then, every time a new source showed kHz QPOs and had an estimate of the spin period either through burst oscillations or through a direct detection in the case of accreting millisecond pulsars, it turned out that the latter were close to $\Delta\nu$ or half of it. More specifically, if the ν_{spin} was slower than ~400 Hz, $\Delta\nu \sim \nu_{\rm spin}$, if it was faster $\Delta\nu \sim \nu_{\rm spin}/2$. The symbol $\sim$ here is to be intended as "close to", since $\Delta\nu$ is not constant for any particular source, but varies over a range. However, it was later realised that the data are also compatible with $\Delta\nu$ being essentially constant around 305 Hz [104], especially after multiplying the kHz QPO frequencies of accreting millisecond pulsars by 1.5, as suggested by an offset in the correlation with the low-frequency QPO frequencies [96, 172]. The situation can be seen in Fig. 6.10. Notice that the spin period of 4U 0614+09 was discovered after the original version of this plot was published and its $\Delta\nu$ values fall on the constant-$\Delta\nu$ track rather than the $\Delta\nu \sim \nu_{\rm spin}/2$ one.

It is interesting to compare the distribution of all $\Delta\nu$ values available in the literature and the distribution of detected (or derived from burst oscillations) spin periods (see chapter by Patruno and Watts, this book) as shown in Fig. 6.11. The distribution of $\Delta\nu$ values, coming from a large number of sources, peaks around 300 Hz and is well approximated by a Gaussian with centroid 305 Hz. The distribution of pulse periods, obviously less populated, is rather flat between 200 and 600 Hz. From these data, it appears that the kHz QPOs are not related to the spin period of the neutron star, although in a number of sources $\Delta\nu$ does increase towards $\nu_{\rm spin}$ with decreasing $\nu_{\rm upp}$ (e.g. [105, 107, 108], but see [70]).

Going back to the correlations between kHz QPO frequencies and theoretical models, it is interesting to produce an updated version of the plot shown in Fig. 6.6 (originally shown for a few sources in [148]), which gives $\Delta\nu$ vs. $\nu_{\rm upp}$, where all published values from RXTE are shown (the same values used for the top panel of Fig. 6.11). They can be seen in Fig. 6.12. Notice that a prediction of the relativistic-precession model is that, for these masses, $\Delta\nu$ should not exceed ~400 Hz, which indeed is what is observed.

However, when dealing with pairs of values, in this case $\nu_{\rm low}$ and $\nu_{\rm upp}$, it is best to plot them one versus the other. This was done in [104]; in Fig. 6.13 we show a new version of that plot with all published values included (the same values used for Fig. 6.12). The predictions of the relativistic-precession model for a neutron-star mass of 1.8, 2.0 and 2.2 solar masses (dashed lines) fit rather well the distribution of points at low frequencies, but diverge slightly at high frequencies, as can also be

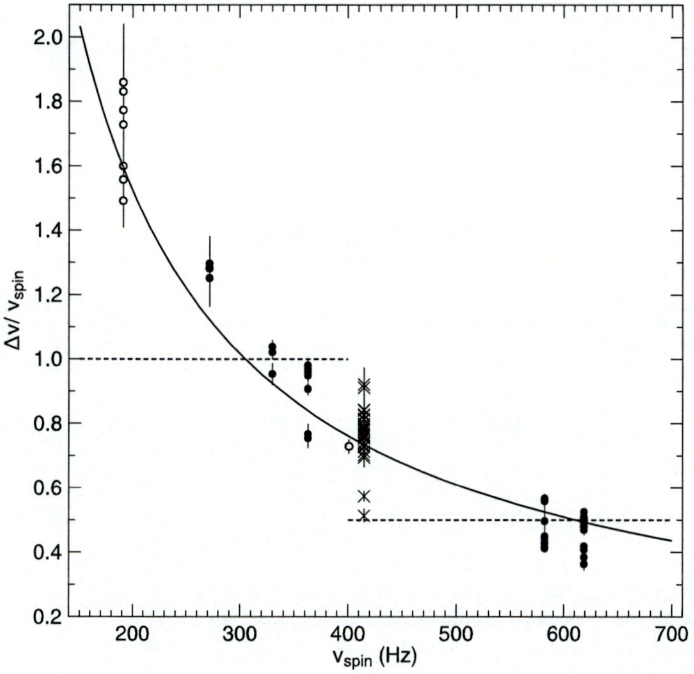

Fig. 6.10 $\Delta\nu/\nu_{\rm spin}$ vs.$\nu_{\rm spin}$ for all sources for which both kHz QPOs and pulse periods are available (adapted from [104]). The $\Delta\nu$ values for accreting millisecond pulsars (empty circles) have been multiplied by 1.5 (see text). The X symbols correspond to 4U 0614+09. The solid line corresponds to a constant $\Delta\nu = 305$ Hz, the dashed line is 1 until 400 Hz, then 0.5

seen from Fig. 6.12. Moreover, a constant 3:2 ratio, shown by the dotted line, fails to represent the data. What is important to note is that, despite the fact that the plot contains points from a number of different sources, the overall correlation is rather good. This suggests that the process that gives rise to the signal at these frequencies is not strongly dependent on other parameters of the sources, like the spin period.

6.7 Relation Between Properties of the kHz QPOs and Parameters of the Energy Spectrum

If kHz QPOs are produced in the accretion flow close to the neutron star, one would expect that properties of that accretion flow will affect the properties of the QPOs. For instance, the frequencies would be related to the radius of the disc, obtained from spectral fits, if the QPOs reflect the Keplerian frequency at that radius while, if the photons that oscillate at the QPO frequency are up-scattered in the corona, the fractional rms amplitude and the phase lags of the QPO would depend on the optical

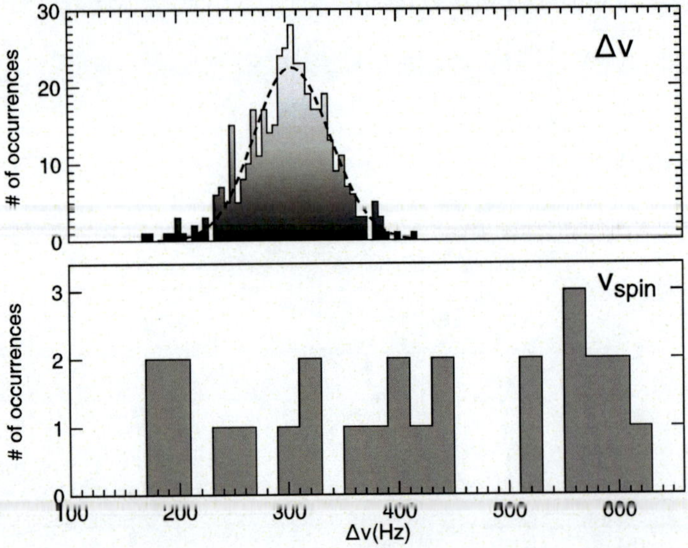

Fig. 6.11 Top: distribution of all available $\Delta\nu$ tabulated in the literature. The dashed line is a Gaussian fit, which yields a centroid of 305 Hz. Bottom: distribution of pulse periods for neutron-star LMXBs, both from accreting millisecond pulsars and burst oscillations. An earlier version of this figure can be found in [104]

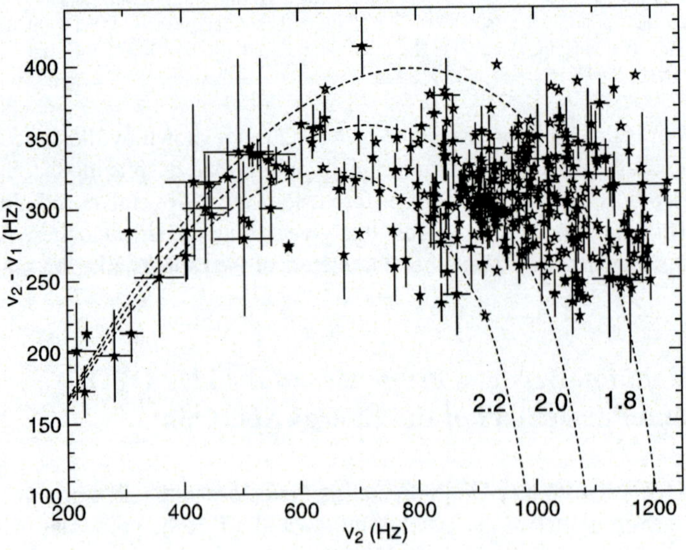

Fig. 6.12 $\Delta\nu$ vs. ν_{upp} for all values published in the literature. The three curves correspond to the prediction of the relativistic-precession model for a mass of 1.8, 2.0 and 2.2 solar masses

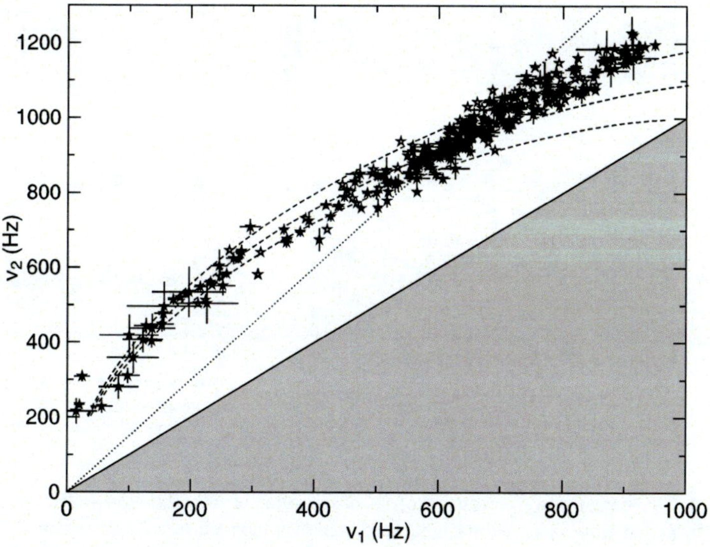

Fig. 6.13 The frequency of the upper kHz QPO, $\nu_{\rm upp}$, vs. that of the lower kHz QPO, $\nu_{\rm low}$, for all values published in the literature. The gray area is obviously not allowed. The dashed lines are the predictions of the relativistic-precession model for a neutron-star mass of 1.8, 2.0 and 2.2 solar masses. The dashed line is $\nu_{\rm upp} = 1.5\nu_{\rm low}$

depth and the electron temperature of the corona. One should keep in mind that, to recover a potential relation between timing and spectral parameters, one needs to study both the energy and the power spectrum of a source over time scales that are comparable to (or preferably shorter than) the time scales over which the properties of the accretion flow change.

Following the discussion in Sect. 6.3, a perturbation in the accretion disc travels through the disc over the viscous time scale, which in these systems is of the order of hundreds of seconds (e.g. [24]). This is also the time scale over which the frequency of the kHz QPOs was observed to change in the power spectrum of some of this sources. For instance, the left panel of Fig. 6.14 shows the dynamical power spectrum of an observation of 4U 1728−34 [109]. In a dynamical power spectrum one plots time in the x axis (in this case $t = 0$ corresponds to the start of the observation), Fourier frequency in the y axis (the plot shows only the frequencies above ~400 Hz to focus on the kHz QPOs), and the power density in the z coordinate (plotted with colours). The dark track in the dynamical power spectrum is the lower kHz QPO in this source. As it is apparent in the plot, the frequency of the QPO changes by ~100 Hz over time scales of a few thousand seconds. The right panel of Fig. 6.14 shows power spectra of six contiguous time intervals within that same observation, with the changes of the QPO frequency, going from ~700 to ~900 Hz over the period of the observation, visible in the individual power spectra. If one wants to compare, for instance, the frequency of the QPO with the inner radius

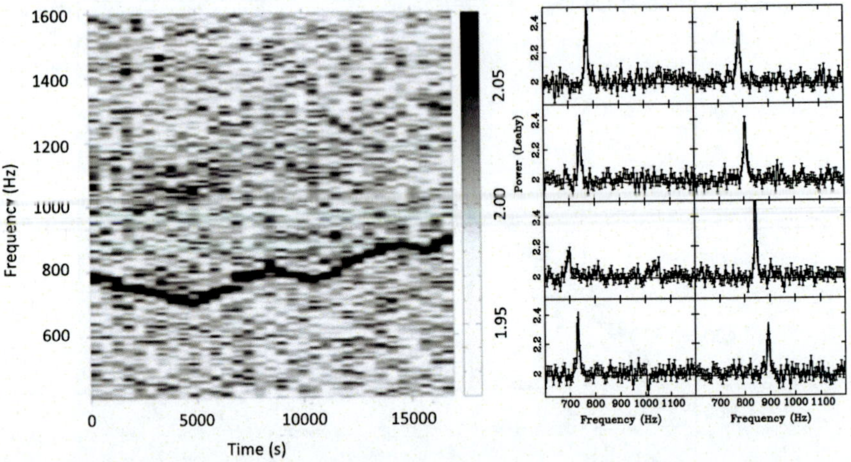

Fig. 6.14 Dynamical power spectrum of an observation of 4U 1728−34 (left), and six power spectra from that same observation (right) computed over intervals of ~2000 s (adapted from [109])

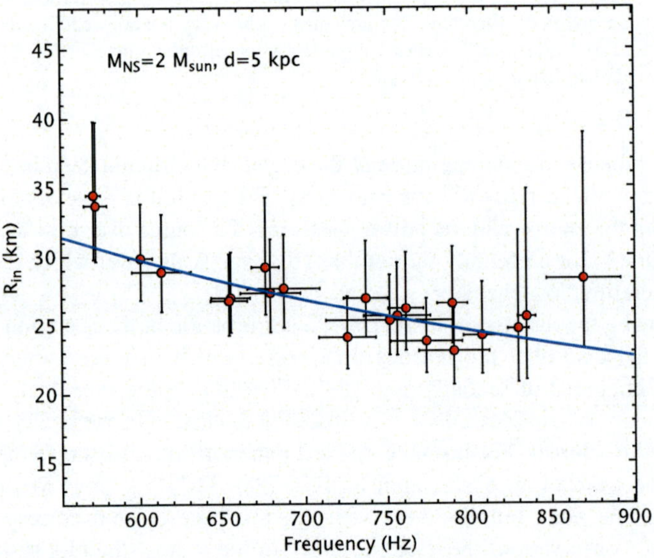

Fig. 6.15 Inner radius of the accretion disc, from fits to the energy spectra, as a function of the frequency of the lower kHz QPO, from fits to the power spectra, in 4U 1608−52 (originally published as Figure 5 in [11])

of the accretion disc, one needs to match the length of the observations used for the comparison with intervals over which the QPO frequency is more or less constant.

Figure 6.15 shows the inner radius of the accretion disc as a function of the frequency of the lower kHz QPO in 4U 1608−52 [11]. The solid line in the plot is the radius as a function of the Keplerian orbital frequency around a 2-$M_{\odot}$ neutron star. As expected, if the QPO frequency is equal to the orbital frequency at that radius, the radius decreases as the QPO frequency increases. Notice, however, that the match of the orbital frequency as a function of the radius with the QPO frequency in that Figure would imply that, contrary to what most models propose (see Sect. 6.4), the lower, not the upper, kHz QPO would reflect the Keplerian frequency at the inner disc radius.

Figure 6.16 shows the spectral parameters of the X-ray corona as a function of the kHz QPO frequencies in 4U 1636−53 ([139], see also [72]). The fits to the energy spectra yield Γ and kT_e, the power-law index and the electron temperature of the Comptonised component, respectively, whereas the optical depth, τ, of the corona is a function of the other two parameters [153]. In the left panel the red and black points correspond to, respectively, the lower and the upper kHz QPO. The right panel shows the same parameters but with the frequency of the lower kHz QPO shifted up by 300 Hz. From this Figure it is apparent that there is a smooth relation of the frequency of the QPOs and the parameters of the corona. Given that the corona is driven by the soft photons in the disc, it is no surprise that both the inner disc radius [11] and the corona parameters [72, 139] change with QPO frequency in a systematic way. The dependence of the rms amplitude of the lower and upper kHz QPOs upon the spectral parameters of the corona (see plots in [138]), however, do not match in the same way; in other words, one cannot apply a shift to the relation of the rms amplitude of one of the kHz QPOs vs. any of the spectral parameters and make it match the same plot of the other kHz QPO [138]. As we discuss below, the same applies to the quality factor and phase lags. The fact that, except for a frequency shift, the relation of the parameters of the corona vs. the QPO frequency is the same for both QPOs, whereas the relation of the rms amplitude is different, indicates that the dynamical mechanism that drives the frequency of both kHz QPOs can be the same, whereas the radiative mechanisms that modulate the QPO signals must be different.

6.8 Beyond QPO Frequencies

6.8.1 The Fractional rms Amplitude of the kHz QPOs

For both kHz QPOs, the spectrum of the fractional rms amplitude of the variability is hard. For instance, in 4U 1608−52 the fractional rms amplitude of the lower kHz QPO increases from ∼5% at ∼3 keV up to ∼20% at 20–25 keV [24, 56, 110]. A similar trend is seen for the lower kHz QPOs of 4U 1728−34 and Aql X-1 [110, 121], 4U 1636−53 [139], and the only kHz QPO in EXO 0748−676 (see [57], and Fig. 6.17a). For the upper kHz QPO the trend is similar, although the increase of the fractional rms amplitude with energy is less steep (Fig. 6.17a, b; but notice that, as we show below, the total rms amplitude and the slope of the rms spectrum of both

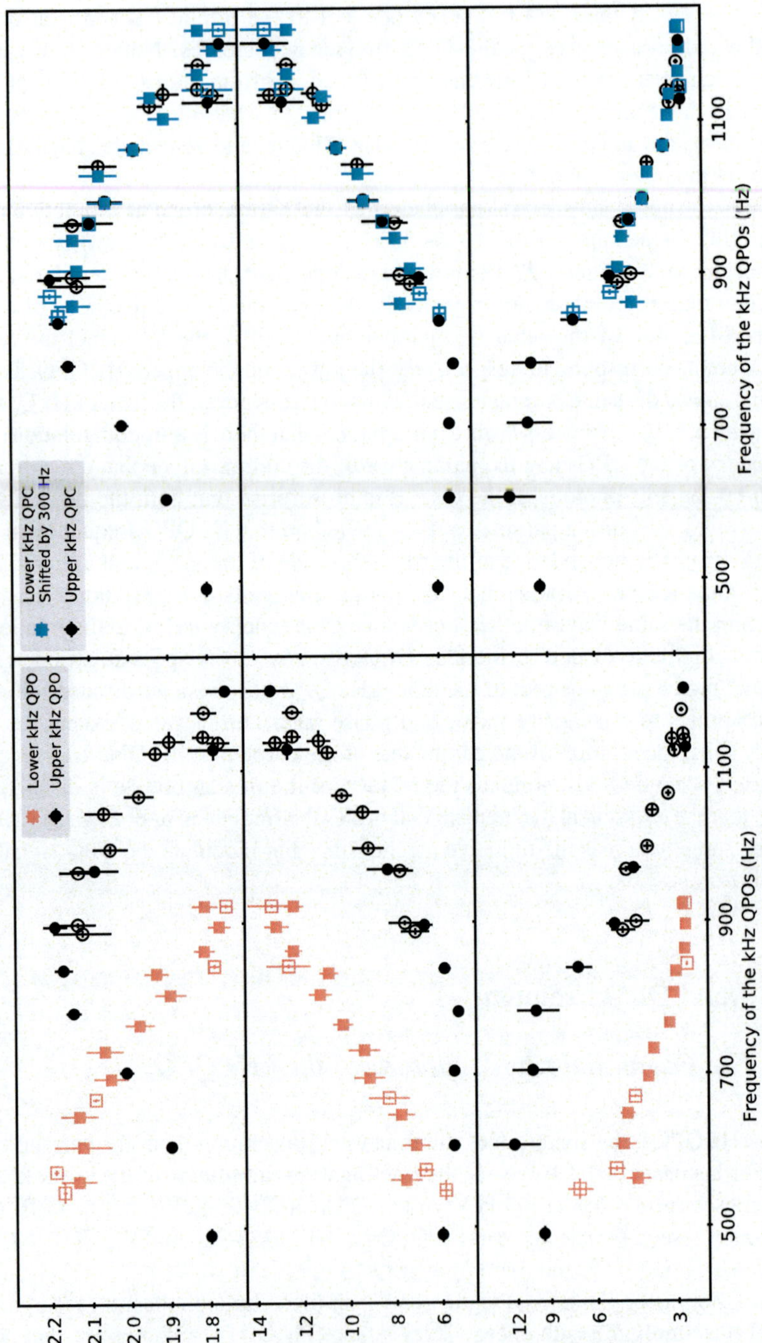

Fig. 6.16 Left: Spectral parameters of the lower and the upper kHz QPOs in 4U 1636−53 as a function of the QPO frequency. Right: Same as left panel, but with the frequency of the lower kHz QPO shifted up by 300 Hz [138]

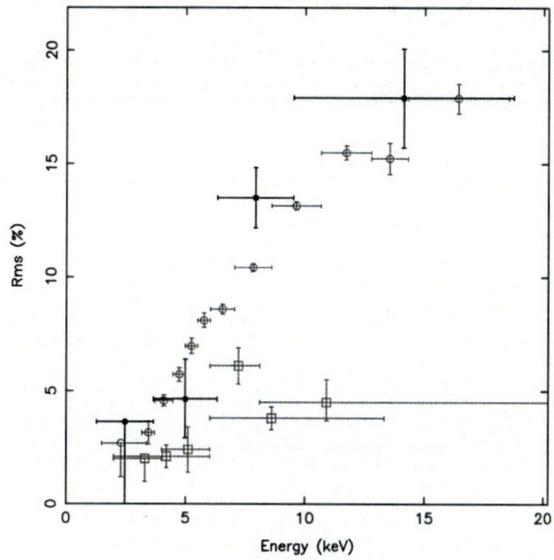

(a)

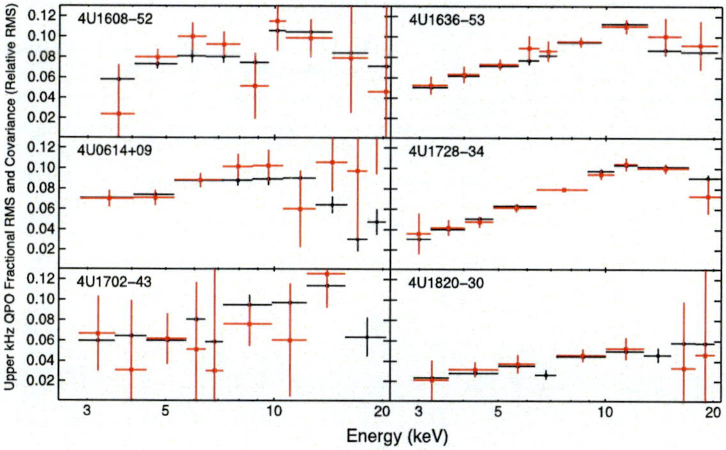

(b)

Fig. 6.17 Top (**a**): rms spectrum of the single kHz QPO in EXO 0748−676 (originally published as Figure 4 in [57]), and the lower and upper kHz QPOs in 4U 1608−52 [24, 110]. Bottom (**b**): rms (red) and covariance (black) spectrum of the upper kHz QPO in six sources (originally published as Figure 5 in [157])

QPOs depend upon QPO frequency, and therefore one has to consider that to draw conclusions from the comparisons). For instance, for 4U 1608−52 the amplitude of the upper kHz QPO increases from ~2–4% at ~3 keV to ~20% at ~20 keV [24, 110]. On the other hand, there is no evidence that the frequency or the width of either of the kHz QPOs change with energy [121].

Figure 6.18 shows the rms spectrum of the lower and the upper kHz QPO plotted in different colours for different QPO frequencies. For both kHz QPOs the

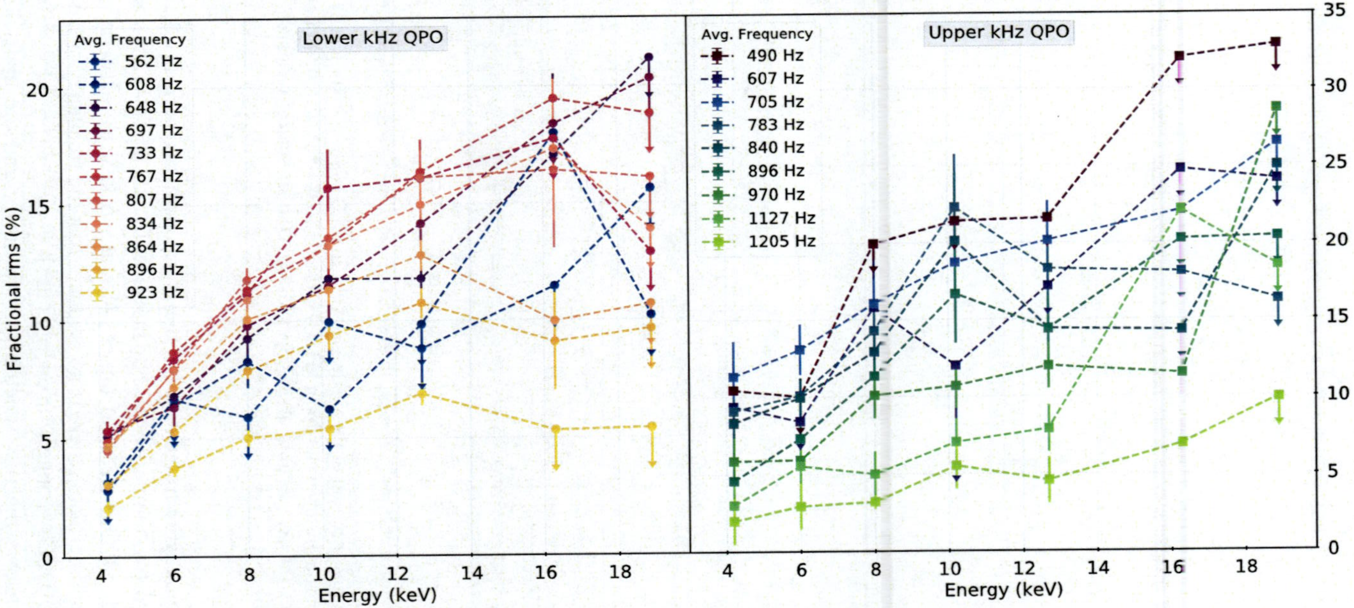

Fig. 6.18 Fractional rms amplitude of the lower (left) and upper (right) kHz QPOs as a function of energy for different frequencies of the QPOs. From the plots it is apparent that the slope of the rms-energy relation changes in a systematic way as a function of the frequency of the QPO [139]

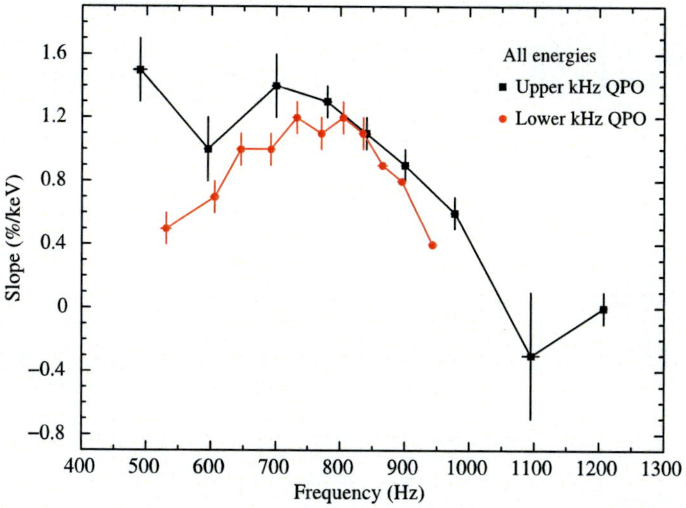

Fig. 6.19 Slope of the rms vs. energy relation of the lower (red) and upper (black) kHz QPO in 4U 1636−53 as a function of the QPO frequency. The slopes were obtained from fits to the data in Fig. 6.18. Compare this Figure with Fig. 6.5a that shows, for the same data, the total rms amplitude vs. QPO frequency (adapted from [139])

rms increases with energy (remember Fig. 6.8 showing the rms spectrum of the lower kHz QPO in 4U 1608−52) with, at least for the lower kHz QPO, the rate of increase being faster at low than at high energies. It is also apparent from this Figure that, for the lower kHz QPO, as the QPO frequency increases the slope of the rms spectrum first increases, and then decreases. For the upper kHz QPO the slope of the rms spectrum decreases more or less steadily as the frequency of the QPO increases.

This is seen more clearly in Fig. 6.19, which shows the slope of the rising part of the rms spectrum of both QPOs as a function of the frequency of each QPO. The dependence of the slope of the rms spectrum with QPO frequency is almost exactly the same as that of the total fractional rms (for all energies combined) as a function of QPO frequency shown in Fig. 6.5a. From this comparison one can conclude that the change of the rms amplitude of the QPOs with frequency is driven by a non-monochromatic change of the rms spectrum, rather than by an energy-independent shift of the rms amplitude of the QPOs at all energies (see [139], for more details).

On the other hand, the fractional rms amplitude of both kHz QPO changes in a systematic way with the frequency of the QPO. Figure 6.20 shows the behaviour of the total (for all energies combined) rms amplitude of both kHz QPOs as a function of the frequency of the upper kHz QPO in the atoll sources 4U 0614+09 and 4U 1728−34 [170]. Figure 6.5a showed the same for another atoll source, 4U 1636−53. (Notice that in that Figure the rms amplitude of each kHz QPO is plotted as a function of the frequency of the corresponding QPO itself, whereas in

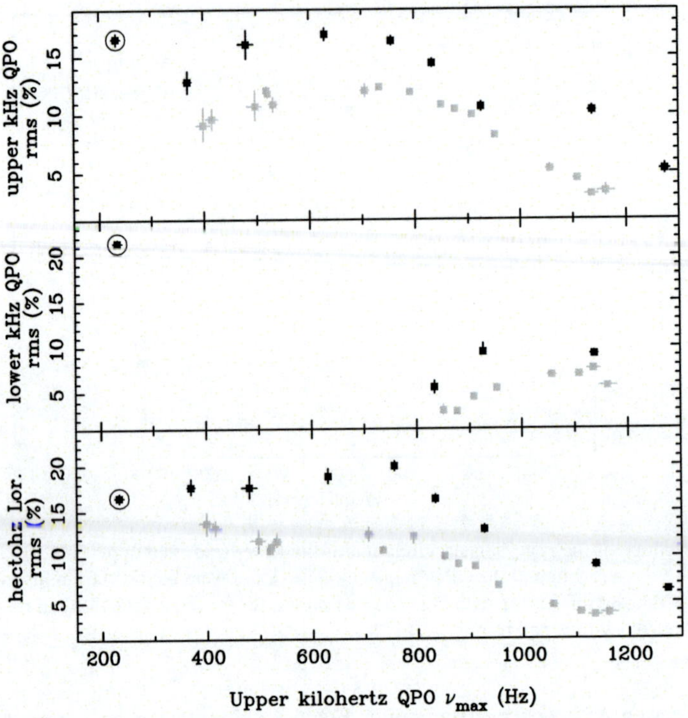

Fig. 6.20 Fractional rms amplitude of the lower (upper panel) and lower (middle panel) kHz QPOs as a function of the frequency of the upper kHz QPO in 4U 0614+09 (black points) and 4U 1728−34 (grey points). The bottom panel shows the rms amplitude of the hectohertz (hHz) QPO as a function of the frequency of the upper kHz QPO in these two sources (originally published as Figure 4 in [170])

Fig. 6.20 the rms amplitude of both QPOs is plotted as a function of the frequency of the upper kHz QPO.)

As shown in Fig. 6.4b, the case of Z sources is similar, albeit in those cases the custom is to plot all the QPO parameters as a function of S_Z, the variable that measures the position of the source along the branches in a colour-colour or hardness-intensity diagram. But since the QPO frequencies are correlated with S_Z (see upper panel of Fig. 6.4b), it follows that the rms amplitude of the QPOs in Z sources follows a similar trend as in atoll sources.

At first, at low QPO frequencies, the rms amplitude of the upper QPO increases slightly or stays more or less constant as the QPO frequency increases, and then decreases more or less steadily as the frequency increases further. For the lower kHz QPO the trend is similar, but the rising part when the frequency of the QPO increases, at low QPO frequencies, is steeper than that of the upper kHz QPO.

In fact, a similar behaviour has been observed in all sources for which enough measurements of the QPOs are available [7, 8, 14–16, 27, 38, 40, 42, 50, 51, 58, 68, 69, 103, 110, 138, 139, 142, 165–167, 169–172, 182–184]. Figure 6.21 shows the

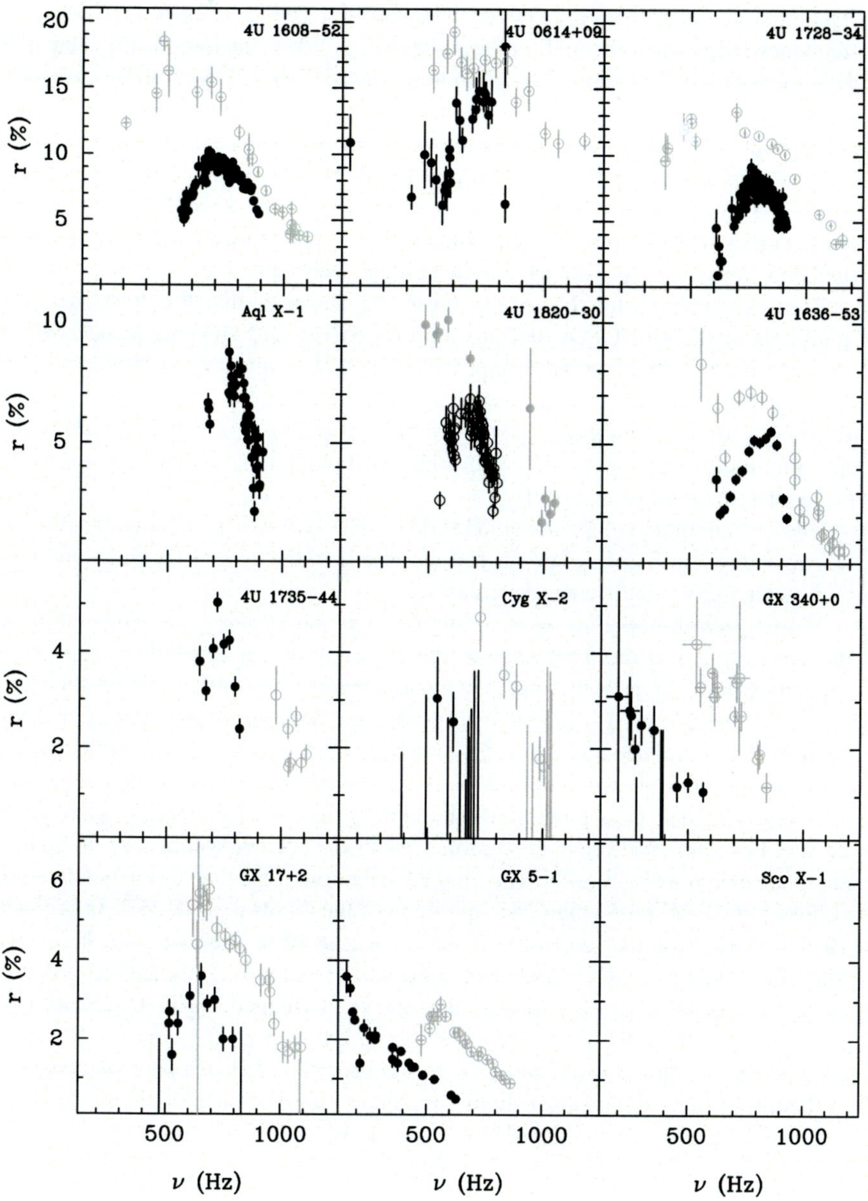

Fig. 6.21 Fractional rms amplitude of the lower (filled symbols) and the upper (open symbols) kHz QPOs as a function of the QPO frequency for seven atoll sources, 4U 1608−52, 4U 0614+09, 4U 1728−34, Aql X-1, 4U 1820−30, 4U 1636−53, and 4U 1735−44, and four Z sources, Cyg X-2, GX 340+0, GX 17+2 and GX 5-1 [103]. Because Sco X-1 is so bright, the detector's dead time made it impossible to measure the fractional rms amplitude of the kHz QPOs in this source. Notice that the scale in the y axis is different for the atoll and the Z sources, because the maximum fractional rms amplitude of the QPOs is generally larger in the former than in the latter

fractional rms amplitude of the lower and upper kHz QPO as a function of the QPO frequency for seven atoll and four Z sources [103]. While the trend is the same, the data are noisier in the case of the Z sources, because the QPOs in those cases are weaker (have lower fractional rms amplitude; notice the scale of the y axis in the different panels) and generally broader (see Sect. 6.8.2) than in the atoll sources. Since the spectrum of the Z sources is in general softer than that of the atoll sources (e.g. [35]), the difference between the rms amplitude of the kHz QPO in the Z and atoll sources suggests that the same mechanism that modulates the oscillations at the QPO frequency sets the shape of the emitted spectrum.

The conclusion from the results presented above is that the fractional rms amplitude of the kHz QPOs depends both on energy and QPO frequency. So far we have shown either the rms amplitude vs. QPO frequency, marginalised over energy, the rms amplitude vs. energy, marginalised over QPO frequency, or the rms vs. energy for a given frequency (the conditional plots). In Fig. 6.22 we show the rms amplitude of the lower and upper kHz QPO plotted vs. both energy and QPO frequency (the joint plots; [139]).

The rms amplitude of the lower kHz QPO in 4U 1636−53 is maximum at $\nu_{\rm low} \approx$ 800 Hz and $E \approx 13$ keV, while the rms amplitude of the upper kHz QPO increases both as the QPO and the energy increase.

From spectral modelling of accreting neutron-star systems, the temperature of the accreting gas at the inner edge of the accretion disc is typically ∼0.3–2 keV, depending on the state of the source, while the temperature of the neutron-star itself is ∼1–2 keV [54, 99, 144]. This implies that the emission from the disc and the neutron star components peaks at $\lesssim$1–6 keV, and drops quickly at energies higher than that, such that at energies above ∼10–15 keV the spectrum of accreting neutron stars is dominated by a power-law like component which is usually ascribed to inverse Compton scattering in a corona (with unspecified geometry) of highly-energetic electrons [153, 181]. The total contribution of the disc or the neutron-star surface at ∼20–25 keV, where the amplitude of the QPOs is ∼10–25%, is between 10^{-3} and 10^{-6} of the total flux of the source at those energies (e.g. [12, 41]). Therefore, even if the kHz QPOs may represent variations of a dynamical property of the accretion disc, e.g., one of the epicyclic frequencies in the relativistic precession model [147, 148], a beat between the Keplerian frequency at the inner edge of the accretion disc and the neutron-star spin [86, 115], or a perturbation wave in the disc [2, 93], the mechanism that modulates the QPO signal cannot be at the disc itself, but must be connected to the corona. We will return to this below.

6.8.2 *The Width of the kHz QPOs*

The quality factor, or equivalently the FWHM, of the kHz QPOs depends upon the QPO frequency [8, 13–16, 27, 40, 42, 50, 51, 58, 68, 69, 103, 110, 142, 165, 169–172, 182–184]. For the lower kHz QPO, the quality factor first increases slowly as the frequency of the QPO increases, and after reaching the maximum value it drops

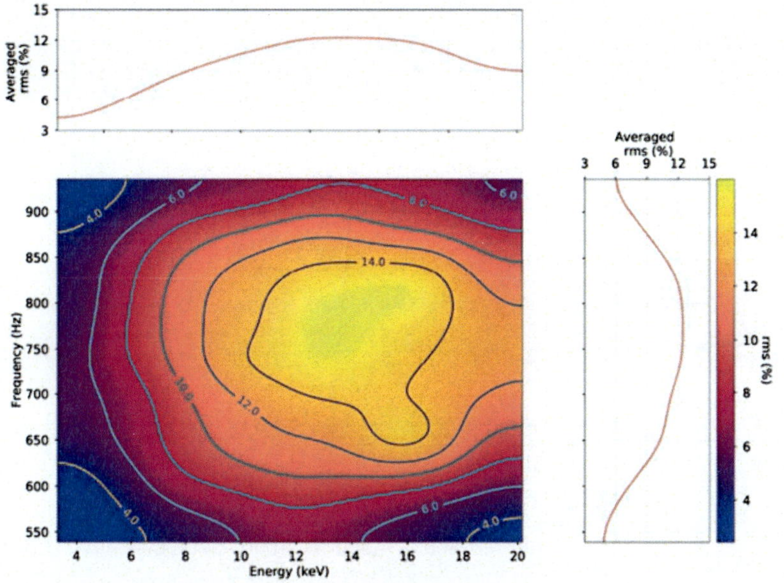

(a)

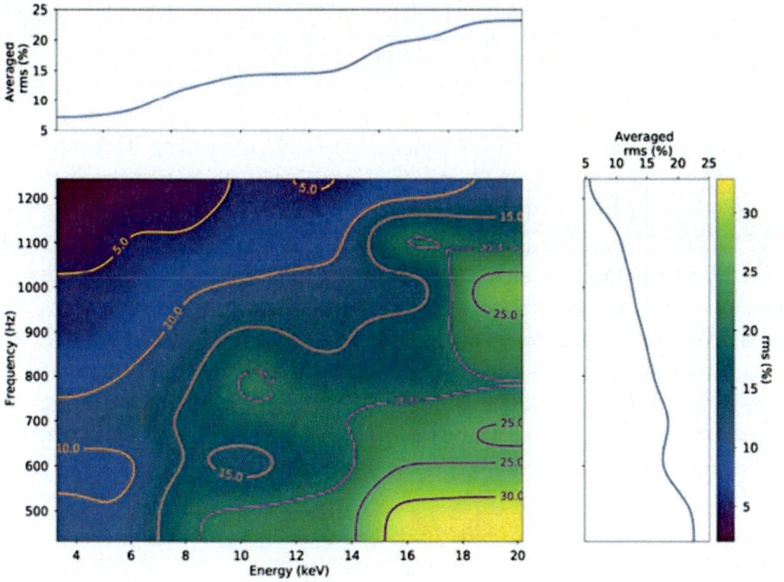

(b)

Fig. 6.22 Fractional rms amplitude of the lower (top, **a**) and upper (bottom, **b**) kHz QPO in 4U 1636−53 as a function of both energy (x axis) and QPO frequency (y axis). The colours (z coordinate) represent the rms amplitude of the QPO. The top and right panels next to each Figure show the rms amplitude of the QPOs marginalised over, respectively, QPO frequency and energy [139]

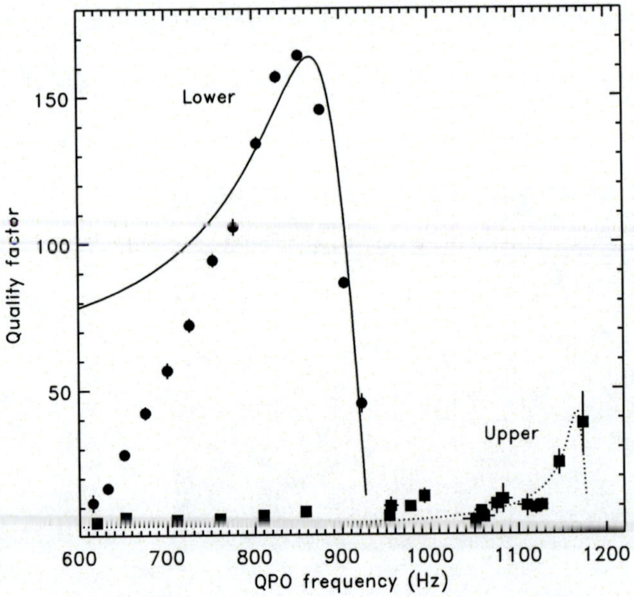

Fig. 6.23 Quality factor of the lower (filled circles) and upper (filled squares) kHz QPOs in 4U 1636−53. The solid line shows the trend expected from a simplified calculation of the effect of broadening of the profile of the QPO by: (a) radial drift of the inner parts of the disc as the gas falls onto the neutron star, (b) the range of radii in the disc at which the QPO is produced, and (c) the lifetime of the blobs that produce the oscillations [15]

more or less abruptly as the QPO frequency continues increasing. This can be seen in Fig. 6.23 for the case of 4U 1636−53 ([15], see also [40]). Notice also in this Figure the difference of the quality factors of the lower and the upper kHz QPOs: The lower kHz QPO is almost always narrower than the upper one, and can be as narrow as ~5 Hz. To reduce the errors in the measurements, this plot shows the quality factor averaged over intervals of ~20 Hz in frequency for the lower kHz QPO, and combines measurements for many separate observations, taken at different epochs, of the source. When measured over short intervals in single observations, the quality factor in 4U 1636−53 can be as large as 200–220 (see Fig. 6.5b) which, at ~850 Hz means that the lower kHz QPO can be as narrow as ~4 Hz FWHM (see also [120], for the source EXO 1745−248). The quality factor of the upper kHz QPO, on the other hand, is much less than that of the lower kHz QPO, and it remains more or less constant or increases slowly as the frequency of the QPO increases. Translated into a width, the FWHM of the upper kHz QPO goes from $\Delta \sim 40$ Hz at $\nu_{\rm upp} \sim 1150$ Hz to $\Delta \sim 120$ Hz at $\nu_{\rm upp} \sim 600$ Hz.

The more or less abrupt drop of the quality factor of the lower kHz QPO has been interpreted [13–15] as an indication that the inner radius of the accretion disc where, according to most models, the QPOs are formed, starts to reach the ISCO (Sect. 6.3). Close to the ISCO, three effects contribute to the width of the QPO peak: First, because of the rapid increase of the radial component of the velocity of a particle orbiting the neutron star when it approaches the ISCO, the gas will move a certain radial distance $\Delta r_{\rm drift}$ during the lifetime of the oscillation, leading to a change $\Delta \nu_{\rm drift} \approx \frac{3}{2}\frac{\Delta r_{\rm drift}}{r_{\rm orb}} \nu_{\rm orb}$ of the frequency of the oscillations during this time. Second, the region in the disc where the QPO is formed is stretched in the radial direction, leading to a range of frequencies of the oscillations $\Delta \nu_{\rm orb} \approx \frac{3}{2}\frac{\Delta r_{\rm orb}}{r_{\rm orb}} \nu_{\rm orb}$. Finally, the finite lifetime of the oscillations adds to the width of the QPO profile.

The solid curve in Fig. 6.23 shows a simplified calculation of these effects combined, for a set of parameters that roughly reproduce the data, and that yield a neutron-star mass of $\sim 2 M_{\odot}$ or higher. In this model, at frequencies above $\nu_{\rm low} \approx 800$ Hz the main contribution to the width of the QPO is from the drift of the QPO frequency, $\Delta \nu_{\rm drift}$. If this interpretation is correct, the drop of the quality factor of the lower kHz QPO in this and other sources would provide a constrain to the mass of the neutron star, and a direct evidence of the existence of the ISCO around the neutron star in these systems.

Measurements of the rate at which the QPO frequency changes as a function of the QPO frequency, however, appear to question this interpretation. As shown in Fig. 6.24 [143], the rate of change of the QPO frequency is the largest at low QPO frequencies, decreases as the QPO frequency increases, and is the smallest at the highest QPO frequency. On the contrary, under the interpretation that the drop of Q is due to the ISCO, at high QPO frequencies, when the radius of the disc is closest to the ISCO, the effect of $\Delta \nu_{\rm drift}$ and the rate of change of the frequency of the lower kHz QPO should be the largest.

In Fig. 6.21 we saw that the rms amplitude of the kHz QPOs is lower in Z than in atoll sources. In fact, also the quality factor of the kHz QPO, and in particular that of the lower kHz QPO, is in general lower in Z [58, 68, 127, 165, 184, 187]

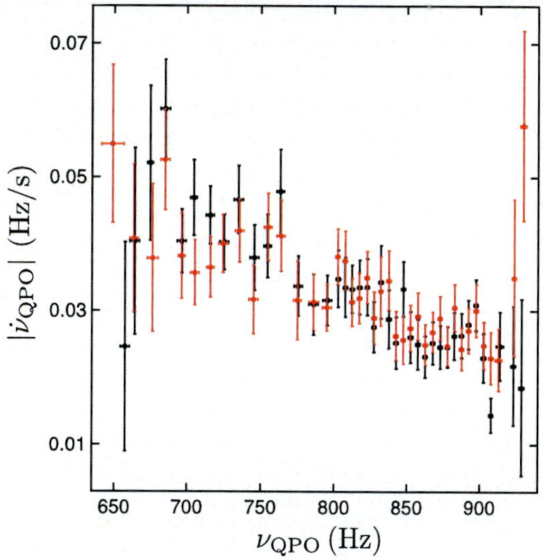

Fig. 6.24 Absolute values of the positive (red points) and negative frequency derivative (black points) for the lower kHz QPO in 4U 1636−53 as a function of the QPO frequency measured on time-scales of 64 s or shorter [143]

than in atoll sources [7, 8, 14–16, 27, 38, 40, 42, 50, 51, 103, 110, 138, 139, 142, 166, 167, 169–172, 182, 183]. This is shown in Fig. 6.25 in which the quality factor, both of the lower and the upper kHz QPO, in seven atoll and 5 Z sources is plotted as a function of the frequency of the QPO [103]. From this Figure it is apparent that the lower kHz QPO is narrower in the atoll than in the Z sources and that, within the atoll sources, the minimum width that the QPO can attain is different for different sources. The situation is less clear for the upper kHz QPO because the measurements, especially of the Z sources, have larger errors. The Z sources are more luminous and have a softer spectrum than the atoll sources, and this differences are likely due to the difference of the total mass accretion rate in these two classes of sources (see above in this section). It is, therefore, possible that the rms amplitude and quality factor of the kHz QPOs depend upon mass accretion rate, reflecting properties of the accretion flow that produces the X-ray spectrum in these sources.

Figure 6.26a, b summarise these points. The black symbols in those Figures show the maximum quality factor and maximum fractional rms amplitude of, respectively, the lower and the upper kHz QPO as a function of the source luminosity (in Eddington units), for the seven atoll and five Z sources in Figs. 6.21 and 6.25. (We will discuss the red points below.) The relation between the maximum quality factor of the lower kHz QPO and the luminosity of the source in this Figure resembles the relation between the quality factor and the frequency of the lower kHz QPO in 4U 1636−53 [15] and other sources (see Fig. 6.23). The same is true for the dependence of the maximum rms amplitude of the lower kHz QPO with luminosity in the set of sources seen in this Figure, and the relation between rms amplitude and QPO frequency for the lower kHz QPO in individual sources (e.g., Figs. 6.5a and 6.20). Since in individual sources the QPO frequency generally increases with luminosity (see Fig. 6.2a; despite being partially affected by the parallel-track phenomenon, this is generally the case), this suggests that the same mechanism is responsible for the drop of quality factor and rms amplitude of the lower kHz QPO with QPO frequency in 4U 1636−53 and other sources, as well as for the drop of the maximum QPO quality factor and maximum QPO rms amplitude with luminosity in the set of sources. This would imply that the drop of the quality factor in 4U 1636−53 is not driven by the inner edge of the disc approaching the ISCO, but to changes in the properties of the accretion flow, e.g. optical depth and temperature of the boundary layer [56] or the corona [103], where the signal of the QPO is likely modulated (see below).

The fact that the maximum rms amplitude and quality factor of the lower kHz QPO in the set of sources are lower in the Z than in the atoll sources offers the possibility to test these ideas. If the scenario in which the drop of the quality factor and rms amplitude of the lower kHz QPO in 4U 1636−53 and other sources is driven only by the inner edge of the disc approaching the ISCO was correct, one would expect that, if a source ever switched from atoll to Z, or vice versa, and continued showing kHz QPOs both in the atoll and Z phases, at the same QPO frequency, hence the same inner disc radius, the quality factor and rms amplitude of the lower kHz QPO would be the same. The reason for this is that the radius of the ISCO depends only on the mass, spin and equation of state of the neutron

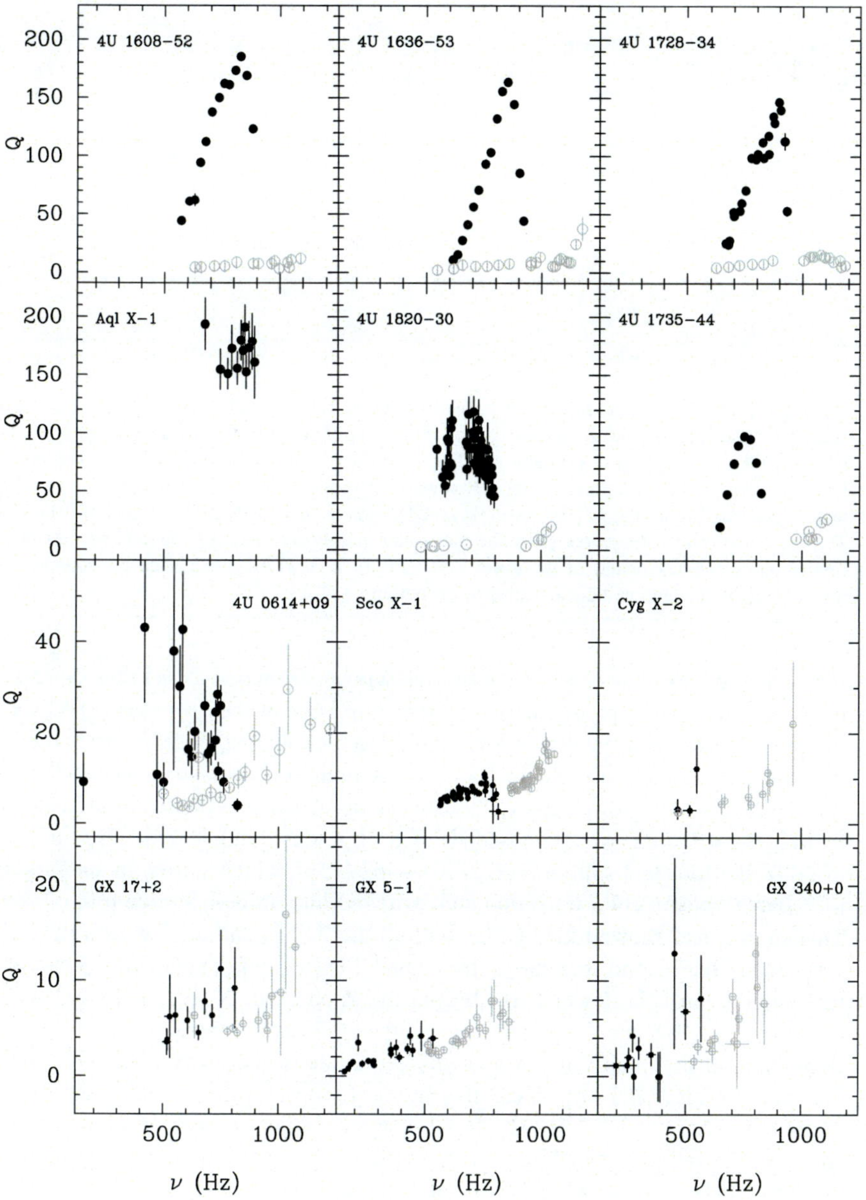

Fig. 6.25 Quality factor of the lower (filled symbols) and the upper (open symbols) kHz QPOs as a function of the QPO frequency for seven atoll sources, 4U 1608−52, 4U 1636−53, 4U 1728−34, Aql X-1, 4U 1820−30, 4U 1735−44, and 4U 0614+09, and five Z sources, Sco X-1, Cyg X-2, GX 17+2, GX 5-1, and GX 340+0 [103]. Notice that, as with the fractional rms (Fig. 6.21), the scale of the y axis is different for atoll and Z sources, reflecting the fact that the maximum value of the quality factor is larger in the former than in the latter

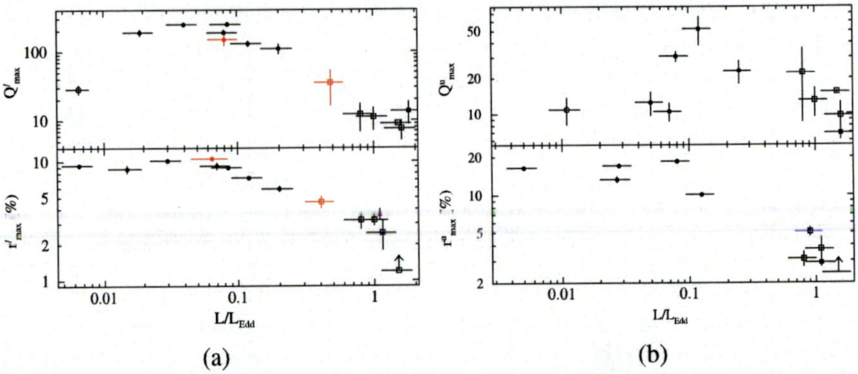

Fig. 6.26 Maximum quality factor (top panels), obtained from Fig. 6.25, and maximum fractional rms amplitude (lower panels), obtained from Fig. 6.21, of the lower (left, **a**) and upper (right, **b**) kHz QPOs of seven atoll (circles) and five Z sources (squares) as a function of the source luminosity [103]. The red points in the left panel correspond to the measurements of the quality factor and rms amplitude of the lower kHz QPO in the transient LMXB XTE J1701−462. The red circles and red squares show the maximum quality factor (upper panel) and maximum rms amplitude (lower panel) of the lower kHz QPO in XTE J1701−462 when the source was, respectively, in the atoll and the Z phases of the outburst [142]

star, which do not change when the source switches from one class to the other. On the other hand, if the quality factor and rms amplitude of the lower kHz QPO were driven (at least in part) by the properties of the accretion flow, since the properties of the accretion flow are different in Z and atoll sources, the average quality factor and rms amplitude of the lower kHz QPO between the Z and atoll phases would change.

In 2006, when this was proposed [103], such a source did not exist. But in April of 2007, the transient source XTE J1701−462 [135], which started its outburst as a Z source, underwent a transition and switched into an atoll source [60, 62, 95]. This source also showed kHz QPOs both in the Z [59] and atoll phases [61]. As proposed in the second scenario, at the same QPO frequency, the quality factor and rms amplitude of the lower kHz QPO in the Z phase of XTE J1701−462 were, respectively, ∼7–8 and ∼3 times larger in the atoll than in the Z phase [142]. Not only that, but the quality factor and rms amplitude of the lower kHz QPO in the Z phase of XTE J1701−462 were also similar to those of the other Z sources, and in the atoll phase of XTE J1701−462 they were similar to those of the other atoll sources. The red points in Fig. 6.25 show the quality factor and rms amplitude of the lower kHz QPO in XTE J1701−462 in the Z (open square) and atoll (filled circle) phase, perfectly in line with the other sources in that plot. This shows that in XTE J1701−462, and other sources, the quality factor and rms amplitude of the kHz QPOs are, at least in part, driven by the properties of the accretion flow. This questions the suggestion that the drop of the quality factor of the lower kHz QPO at high QPO frequencies provided evidence of the ISCO in these systems. This, on the other hand, offers an avenue to develop models to explain the radiative properties of the QPO.

6.8.3 The Energy-Dependent Lags and Coherence of the kHz QPOs

As we described in Sect. 6.4, one can study the phase/time lags of the QPOs using light curves in two different energy bands. Besides the lags, one can also study the coherence function between the two light curves [65, 125, 173]. The coherence function [23], $\gamma^2(\nu)$, measures the degree of linear correlation between two noiseless signals (light curves) as a function of the Fourier frequency. More precisely, $\gamma^2(\nu)$ should be called $\gamma^2(\nu; E_1, E_2)$ to indicate that it is the coherence function between two light curves at energies E_1 and E_2; when the lags and the coherence function are given as a function of energy, one means that those are the quantities measured at $E = E_2$ with respect to the reference band, in this case E_1. However, since the observed light curves are not noiseless, but are affected by Poisson counting noise (see Sect. 6.4), to study the degree of correlation between two X-ray light curves one uses the intrinsic coherence function, usually denoted as $\gamma_{\mathrm{I}}^2(\nu)$, that corrects for this (see [173], for an explanation). For simplicity, here we will use the term coherence function, and will write $\gamma^2(\nu)$, to refer to the intrinsic coherence function.

Figure 6.27 shows the time lags (top panel) and coherence function (bottom panel) around the frequency of the lower kHz QPO in 4U 1608−53, using light curves with energies around 3 and 8 keV [37]. This Figure shows that the two light curves used to compute the lags and coherence function are perfectly correlated (coherence function equal to 1 with small errors) in the frequency range at which the QPO signal dominates, and uncorrelated (coherence function consistent with 0

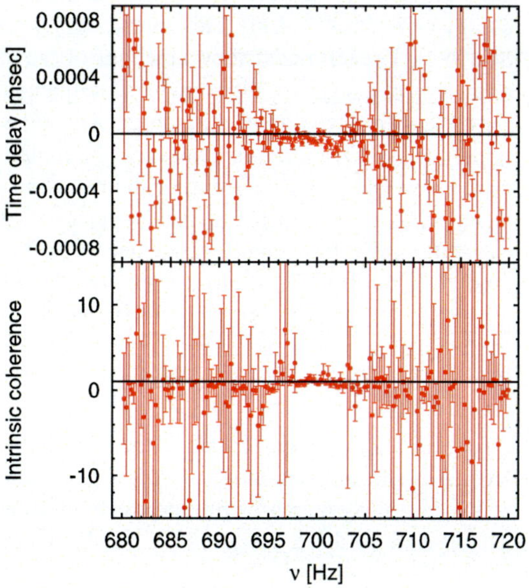

Fig. 6.27 Time lags (top panel) and coherence function (bottom panel) as a function of Fourier frequency, at frequencies around that of the lower kHz QPO in 4U 1608−52. The horizontal line in the top panel is at the zero time lag, while the horizontal line in the bottom panel shows the perfect coherence of 1 [37]

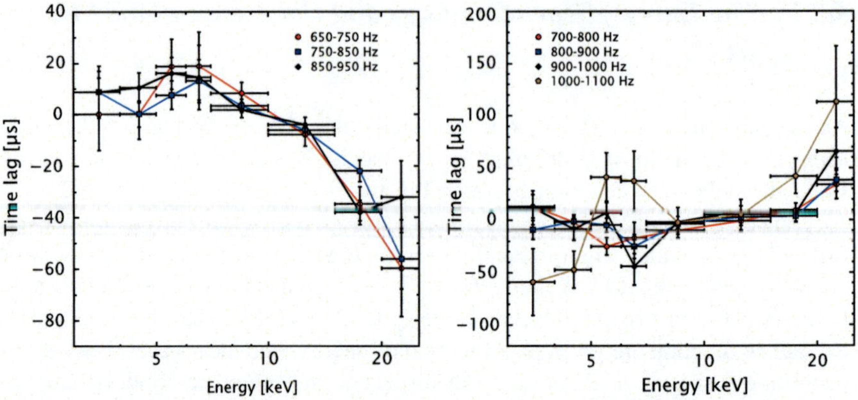

(a) (b)

Fig. 6.28 Time lags as a function of energy for the lower (left, **a**) and upper (right, **b**) kHz QPO in 4U 1728−34 for different QPO frequencies (originally published as Figure 3 in [130])

and large errors) outside that frequency range. This Figure also shows that the lags of the lower kHz QPO in this source are soft (negative), meaning that the low-energy photons lag the high-energy ones.

In Sect. 6.4 we showed that the lags of the lower kHz QPO are significantly different from those of the upper kHz QPO [37]. Specifically, the lags of the lower kHz QPO in 4U 1608−52, 4U 1636−53, Aql X-1 and 4U 1728−34 are soft and become softer as the energy increases [11, 37, 38, 130, 156]. On the contrary, the lags of the upper kHz QPO in 4U 1608−52, 4U 1636−53 and 4U 1728−34 are either consistent with zero or hard and, if anything, they become harder as the energy increases [37, 130]. This can be clearly seen in Fig. 6.28 (see also Fig. 6.9 in Sect. 6.4). The left panel shows the lags of the lower kHz QPO in 4U 1728−34 as a function of energy for different QPO frequencies. The right panel shows the same for the upper kHz QPO in this source. This difference in the lags of the lower and upper kHz QPOs indicates that either the radiative mechanisms that generate each of the kHz QPOs are different, or the mechanism is the same but each of the two signals that are modulated at the frequency of the lower and the upper kHz QPO travel through different parts of the accretion flow before reaching the observer.

In Sect. 6.4 (see Fig. 6.9b) we showed that the magnitude of the soft lags of the lower kHz QPO in 4U 1636−53 first increases and then decreases as the frequency of the QPO increases. Since the QPO frequency is a function of S_a (Fig. 6.3b), the dependence of the lags on S_a is similar to that on $\nu_{\rm low}$ (see Fig. 7 in [38]). Figure 6.29 shows the time lags of the lower kHz QPO in 4U 1608−52 as a function of QPO frequency [11]. The magnitude of the lags shows a significant increase as the QPO frequency increases from $\nu_{\rm low} \approx 550$ Hz to $\nu_{\rm low} \approx 700$, and a significant decrease as the frequency increases further. (Notice that in this Figure the convention is that soft lags are positive, therefore the trend appears be the opposite of what is shown

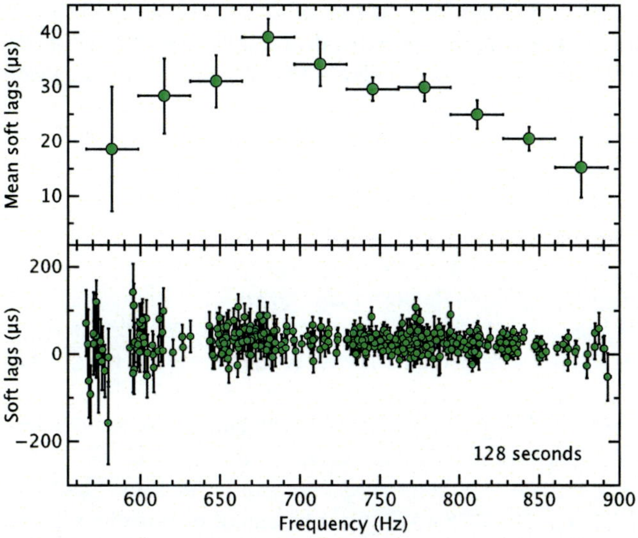

Fig. 6.29 Time lags as a function of frequency for the lower kHz QPO in 4U 1608−52 (originally published as Figure 2 in [11]). The lower panel shows the lags measured over 128-s time intervals, while the upper panel shows the same data binned into ten adjacent QPO-frequency intervals. Notice that by convention, in this Figure negative values correspond to soft lags, which is the opposite convention to the one used in Figs. 6.9 and 6.27

in Fig. 6.9b.) In other sources, e.g., 4U 1728−34 [130], Aql X-1 ([156], see also [157]), the trend is not as clear as in 4U 1608−52 and 4U 1636−53.

The magnitude of the lags of the lower kHz QPO in 4U 1608−52 is between 15 and 40 μs. These values translate into light-travel distances between 4.5 and 12 km, commensurate with the expected distance between the inner edge of the accretion disc and the neutron-star surface (e.g. [115]). It is therefore tempting to identify the lags with the light travel time in that environment. However, from the trend of the lags of the lower kHz QPO in 4U 1608−52 with QPO frequency in Fig. 6.29, and that of the frequency of the lower kHz QPO vs. the inner radius of the accretion disc for the same source in Fig. 6.15, it is apparent that the relation between the lags of the lower kHz QPO and the inner radius of the accretion disc is not monotonic. If the model used to fit the energy spectra of 4U 1606−52 [11] is correct (remember, it is just a model), this non-monotonic relation between the lags and the inner-disc radius implies that the lags cannot be just a delay due only to the light travel-time of the photons from the disc to the neutron star, or vice versa. A similar conclusion can be drawn about the lags of the upper kHz QPO. In this case the lags do not appear to depend upon QPO frequency (but the errors of the lags of the upper kHz QPO are larger than those of the lags of the lower kHz QPO); at the same time, the frequency of the upper kHz QPO changes by a factor of ~2 (see Figs. 6.9b and 6.28) and, comparing the range of frequencies spanned by the lower and the upper kHz QPOs and the range of inner-disc radii in Fig. 6.15, the inner-disc radius changes also by a factor of at least ~2. Therefore, the lags of the upper kHz QPO cannot be due only to a light travel-time delay between the disc and the neutron star either.

Since the lags of the lower kHz QPO are soft, whereas inverse Compton scattering in a corona should produce hard lags (Sect. 6.4; but see below), a different mechanism was required to model the lags of the lower kHz QPO. Reverberation due to reflection of hard photons from the corona off the accretion disc had been used to explain the soft lags of the broad-band noise component in the power spectrum of active galactic nuclei [49, 189, 190] and galactic black-hole candidates ([39, 160], see also [74]). Reverberation was therefore a promising mechanism to explain the lag spectrum of the lower kHz QPOs.

Figure 6.30a shows the lag spectrum of the lower kHz QPO in 4U 1608−52 [30] with the best-fitting reverberation model. While the lags of the QPO decrease more or less monotonically as the energy increases, the best-fitting model predicts that above $E \sim 8$ keV the lags should increase with energy, contrary to the observations. This led to the conclusion [30] that the lags of the lower kHz QPO in 4U 1608−52 cannot be due only to reverberation. Subsequently, the same idea was tested on the upper kHz QPO in 4U 1728−34 [36]; the result of the fits of a reverberation model to the lags of this QPO is shown in Fig. 6.30b. From this Figure it is apparent that, while the lags of the QPO first decrease with energy up to $E \sim$ 5–6 keV, and increase again above that energy, the model predicts that the lags should follow the opposite behaviour. This led to the conclusion that, as with the case of the lags of the lower kHz QPO in 4U 1608−52, the lags of the upper kHz QPO in 4U 1728−34 cannot be solely due to reverberation. Notice that, with few exceptions [66, 101], so far reverberation models have been used only to fit and explain the time/phase lags, and not the amplitude, of the observed light curves of accreting supermassive black holes in active galactic nuclei and stellar-mass black holes in galactic X-ray binaries, whereas a consistent model of the radiative mechanism that fits the data should be able to explain both the rms and lag spectrum of the variability in these sources.

Figure 6.31 shows some of the results of the most extensive study yet of the lags and rms spectra of the kHz QPOs in LMXBs [157]. That work presents the energy-dependent rms amplitude, time lags and intrinsic coherence function, plus the time lags as a function of frequency of the lower kHz QPOs in 14 sources, and the same information for the upper kHz QPO in six out of those 14 sources. In Fig. 6.31 we show only the rms and lag spectra of the lower kHz QPO in those 14 sources (we already showed the rms spectra of the upper kHz QPO in six of those 14 sources in Fig. 6.17b). The results shown in [157] reinforce the importance of considering both the rms and lag spectra to try and model the radiative properties of the kHz QPOs.

Models that involve inverse Compton scattering did not seem appropriate to explain the lags of the lower kHz QPO, given that those lags are soft, whereas Comptonisation would only produce hard lags (see Sect. 6.4). Contrary to those expectations, inverse Compton scattering could work, and produce soft lags, if there is feedback from the corona to the disc ([92], see also [91]). In this scenario, soft photons from the disc are up-scattered in the corona; part of those up-scattered photons go to the observer, and produce the power-law like component in the

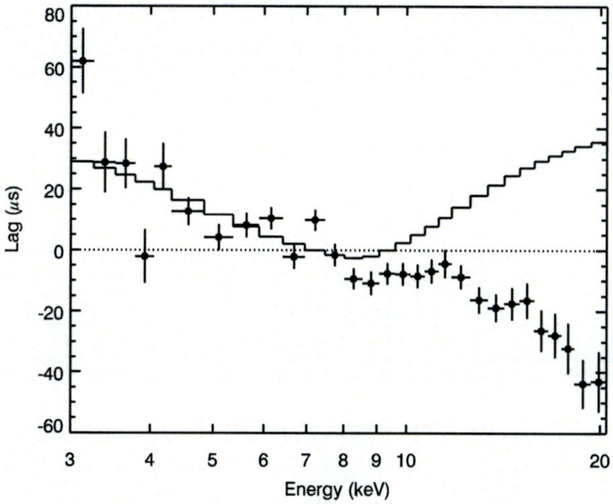

(a)

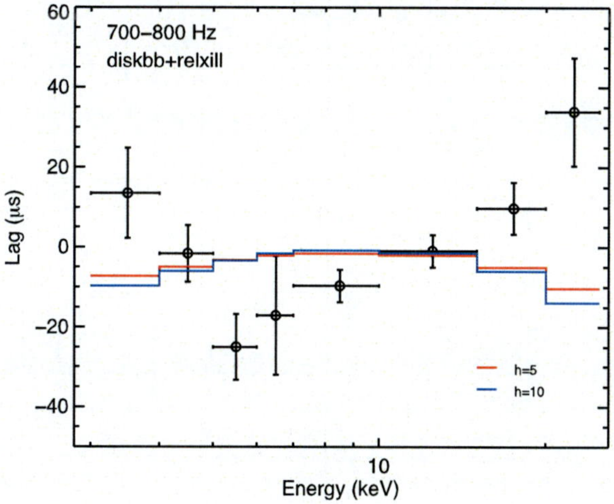

(b)

Fig. 6.30 Time lags of the lower kHz QPO in 4U 1608−52 (top (**a**); originally published as Figure 6 in [30]) and the upper kHz QPO in 4U 1728−34 (bottom (**b**); originally published as Figure 6 in [36]) as a function of energy, with the best-fitting reverberation model

spectrum, but a fraction of those corona photons will illuminate back the disc, increasing the disc temperature [55]. If the photon flux that is originally produced in the disc is modulated (e.g., at the QPO frequency), the flux that returns to the disc after being scattered in the corona will also be modulated and, as those photons reheat the disc, the disc temperature will vary at the frequency of that modulation. In this feedback loop, the photons that reach the observer from the corona through the

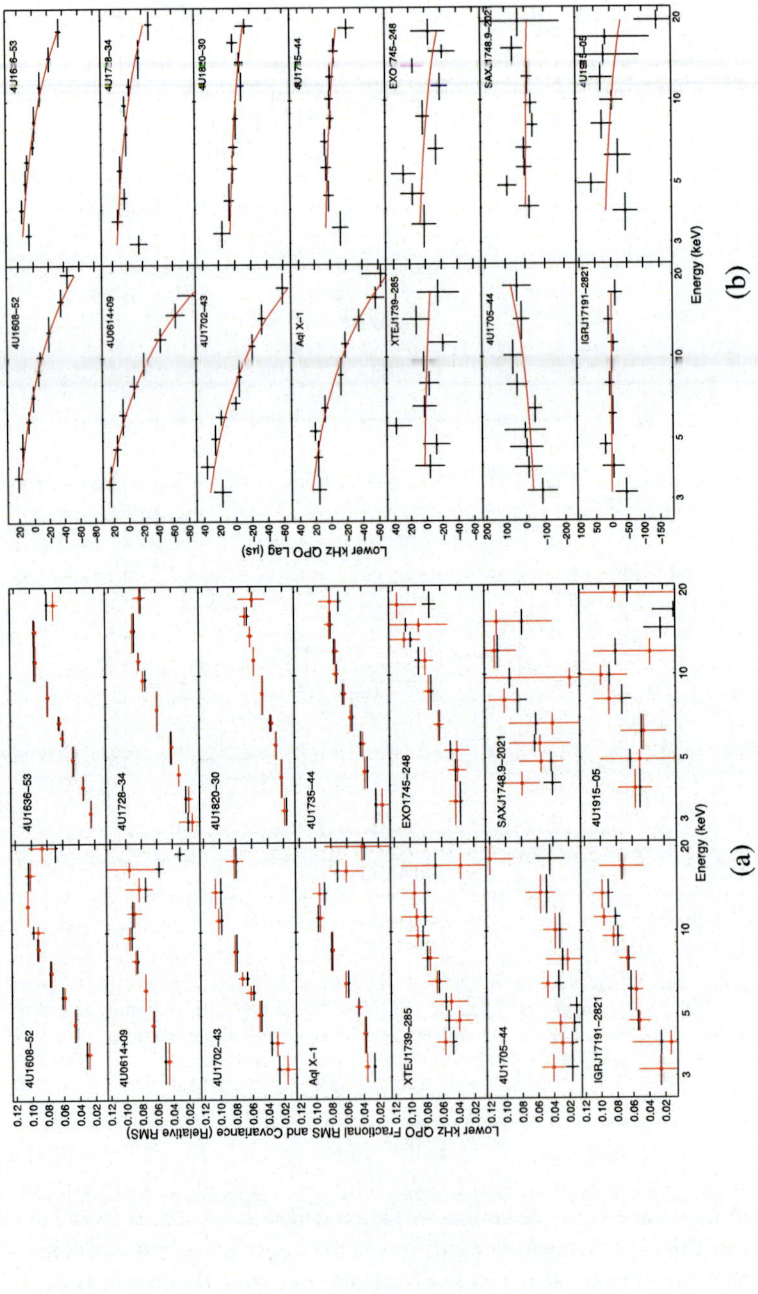

Fig. 6.31 Fractional rms (left, **a**) and time lags (right, **b**) as a function of energy for the lower kHz QPO in 14 atoll sources (originally published as Figures 4 and 8 in [157])

disc will be delayed with respect to those photons from the corona that are directly emitted towards the observer. Since the disc temperature is lower than that of the corona, the process results in a delay of the soft with respect to the hard photons, as observed.

This process can be modelled by solving the time-dependent version of the Kompaneets equation [81], assuming a geometry of the corona [75, 82, 83, 91, 92]. Since the inverse Compton scattering process cools down the electrons in the corona, there has to be an external source of heating [100, 111] to explain long-lived coronas in accreting LMXBs. If the photon flux from the disc that cools the corona is variable, the power provided by this external heating source must also be variable to maintain the equilibrium in the system. An interesting aspect of this approach is that the solution of the equation that describes the variability of the flux received by the observer provides both the energy-dependent amplitude and lags of the signal, and hence one can fit, with the same model, both the rms and lag spectrum of the QPOs.

The two panels in Fig. 6.32 show the rms and lag spectrum of the lower kHz QPO in 4U 1608−52 together with the results of the model that describes, simultaneously, the energy-dependent amplitude and lag of the variability produced by a feedback loop between the accretion disc and the corona [82]. The models reproduce the rms and lag spectrum of the lower kHz QPO in 4U 1608−52 fairly well when the corona is a few km thick and for feedback fractions between ~0.1 and ~0.5.

The plots in Fig. 6.32 correspond to data obtained [11, 24] when the frequency of the lower kHz QPO in 4U 1608−52 was ~800 Hz. We saw, however, that the rms spectrum of the lower kHz QPO changes significantly as the frequency of the QPO changes (see Sect. 6.8.1 and [139]); while it is likely that the lag spectrum of

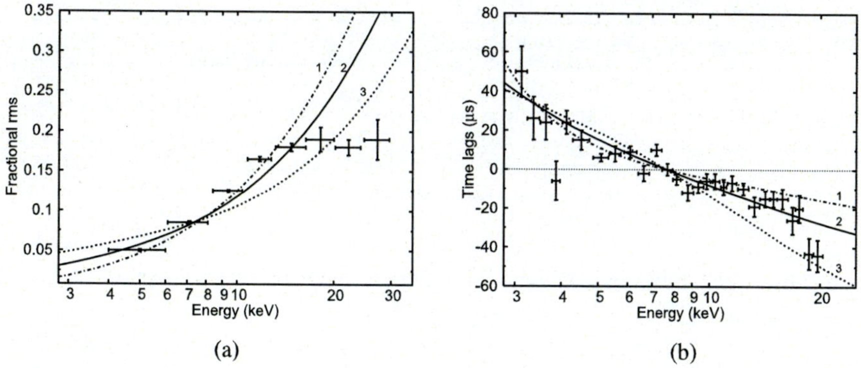

Fig. 6.32 Fractional rms amplitude (left, **a**) and time lags (right, **b**) of the lower kHz QPO in 4U 1608−52 as a function of energy. The three lines labeled 1, 2 and 3 correspond to calculations of the variability produced by a feedback loop between the accretion disc and a 1-km thick corona [82] with values of η, the fraction of photons of the corona that impinge back onto the disc, of 0.3, 0.4 and 0.5, respectively

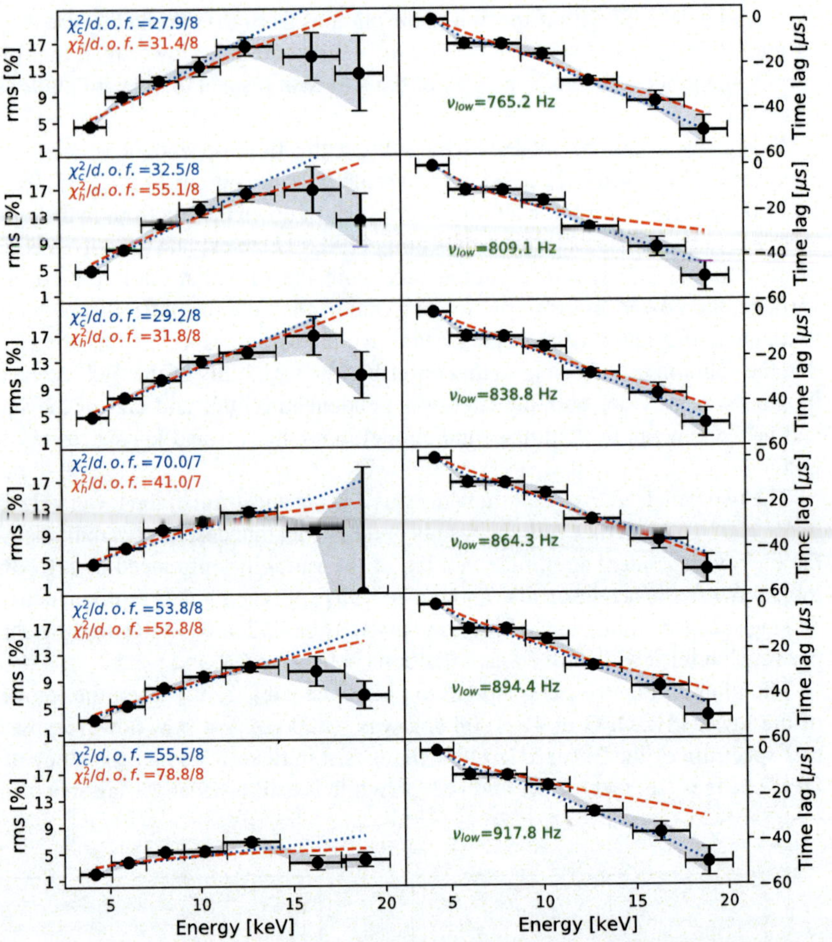

Fig. 6.33 Fractional rms (left) and phase lag (right) spectra of the lower kHz QPO in 4U 1636−53 for six values of the QPO frequency as indicated. [75]. The red and blue lines correspond to the best-fitting model of the variability model that includes feedback between the accretion disc and the corona for, respectively, a hot and a cold accretion disc (see [75], for details)

the lower kHz QPO also changes with QPO frequency, the error bars of the current measurements [37, 130] are too large to tell.

Figure 6.33 shows the fits to the rms and lag spectra of the lower kHz QPO in 4U 1636−53 for six different values of the QPO frequency, while Fig. 6.34 shows the full-band fractional rms amplitude vs. QPO frequency that results from these fits [75]. These fits provide (under the assumptions made in the model; see [75], for details) the evolution of the parameters of the corona, size, kT_e and τ, as a function of the QPO frequency that matches reasonably well the observations (e.g., Fig. 6.16; see [75], for details).

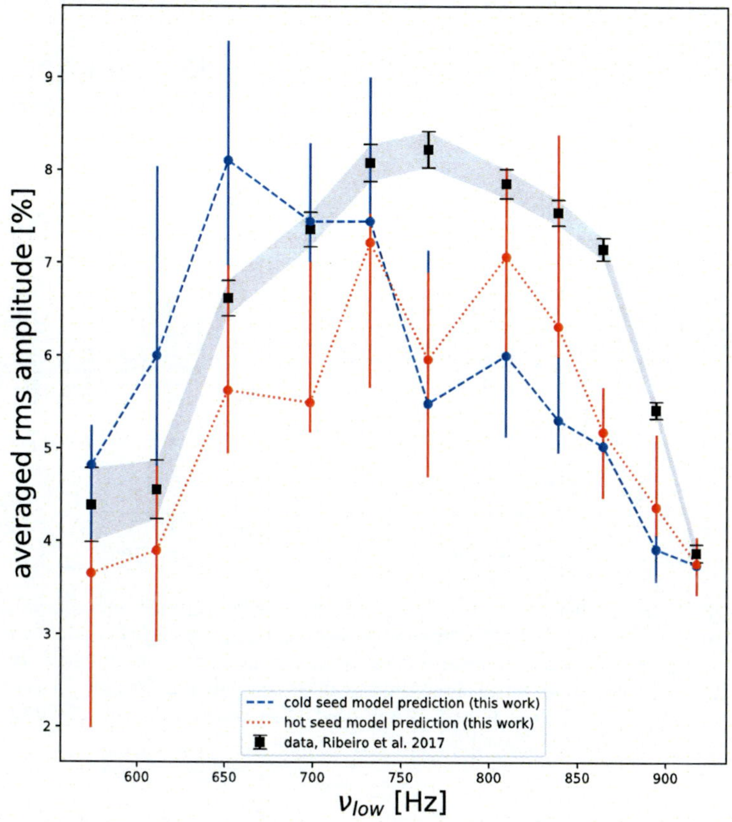

Fig. 6.34 Full-band fractional rms amplitude of the lower kHz QPO in 4U 1636−53 as a function of QPO frequency. The grey shaded area represents the same measurements [139] of the rms amplitude of the QPO shown in Fig. 6.5a, while the red and blue lines and symbols correspond to the best-fitting model of the variability model for, respectively, a hot and a cold accretion disc [75]

While the model appears to work reasonably well, it does not reproduce the data completely. As Figs. 6.32 and 6.33 show, the models fit the lags over the full energy range and the rms spectrum below ~12 keV quite accurately, but fail to reproduce the rms spectrum at energies above ~12–13 keV. Above those energies the rms spectrum stops increasing and levels off (see also Figs. 6.8 and 6.18), whereas the second derivative of the model is always positive (Fig. 6.32a). This is the energy range at which emission from the Compton hump due to reflection of corona photons off the accretion disc is expected (e.g., [49, 140, 141]). If the reflected signal is not (strongly) modulated, that extra emission component at those energies will reduce the rms variability without affecting the lags. This needs to be explored further using time-dependent reflection models.

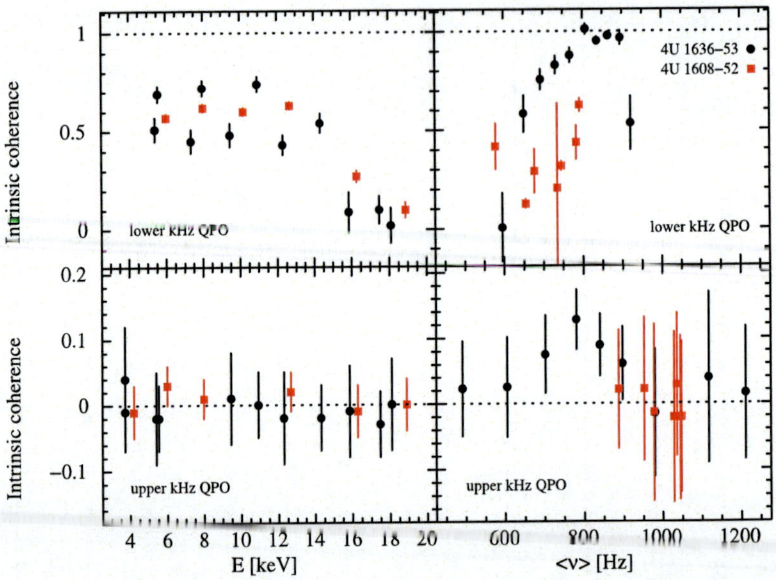

Fig. 6.35 Coherence function of the lower (top) and upper (bottom) kHz QPOs in 4U 1636−53 and 4U1608−52 as a function of energy (left) and QPO frequency (right). The plots on the left show the coherence between the photons with energies given by the value on the *x* axis relative those in the reference bands, for all observations with QPOs combined. The plots on the right show the coherence between photons in the 4–12 keV band relative to photons in the 12–20 keV band at different QPO frequencies (adapted from [37])

To end this part, we will now discuss briefly some results of the coherence function of the kHz QPOs. In Fig. 6.35 we show the coherence function of the lower and upper kHz QPOs in 4U 1636−53 and 4U1608−52 as a function of energy and QPO frequency (the plots are from [37], see that reference for details of the way the coherence function was calculated using all the QPO data for those two sources). There are four things that we would like to mention about the plots shown in this Figure: (i) The coherence function of the lower kHz QPO in both sources is independent of energy below $E \approx 12$ keV, and drops as the energy increases beyond that value. This is the same energy at which the rate of increase of the rms spectrum of the lower kHz QPO starts to flatten (Figs. 6.8, 6.18, 6.31a and 6.33). This suggests that at those energies another signal, independent of the one that produces the variability of the signal at the frequency of the lower kHz QPO, starts to become important in the light curves.

As mentioned previously, this extra component could be the Compton hump, which is part of the reflection component coming from the accretion disc and dominates the reflected spectrum at those energies [140]. (ii) In 4U 1636−53, the degree of linear correlation at the frequency of the lower kHz QPO between the light curves in the two energy bands used in this study (see caption in Fig. 6.35) increases from $\gamma^2 \approx 0$ to $\gamma^2 \approx 1$ as the QPO frequency increases from $\nu_{\rm low} \approx 600$ Hz to

$\nu_{\rm low} \approx 800$ Hz, and then drops to $\gamma^2 \approx 0.5$ as the QPO frequency increases further to $\nu_{\rm low} \approx 900$ Hz. The behaviour of the coherence function of the lower kHz QPO in 4U 1636−53 with QPO frequency is very similar to that of the quality factor and the rms amplitude of the lower kHz QPO in this source (Figs. 6.5a and 6.23). These similarities provide a strong indication that these three phenomena are related, and suggest that the amplitude and width (or the quality factor) of the lower kHz QPO in 4U 1636−53 (and, by extension, of the lower kHz QPO in all the other sources[3]) reflect (at least in part) the degree of linear correlation of the signals in different energies bands at the QPO frequency. In the model of the rms amplitude and the lags in which the corona and the disc are connected through a feedback loop [75, 82, 92], the temperatures of the corona and the accretion disc oscillate coherently; therefore, the soft and hard light curves produced in this scenario are linearly correlated and the intrinsic coherence is high. (iii) The degree of linear correlation between the light curves of the source in the two energy bands used to calculate the coherence function is much lower at the frequency of the upper than at the frequency of the lower kHz QPO. From the discussion in the previous point, this is probably the reason why the upper kHz QPO is much broader (lower quality factor) than the upper kHz QPO [15]. (iv) The coherence of the upper kHz QPO shows a small increase at $\nu_{\rm upp} \approx$ 700–800 Hz. This is the same frequency at which the rms amplitude of the upper kHz QPO (Fig. 6.5a) and the slope of the rms spectrum of the QPO (Fig. 6.19) as a function of QPO frequency show a local maximum. In the context of the model in which a feedback mechanism connects the corona and the disc [75, 82, 92], this could be interpreted as the source of soft photons and the Comptonising medium not oscillating coherently for the largest part of the range of QPO frequencies spanned by the upper kHz QPO, and the disc and the corona becoming resonant when the frequency of the upper kHz QPO is at around 800 Hz.

6.8.4 Other Phenomenology of the kHz QPOs

In this subsection we will briefly discuss two additional phenomena that are interesting to try and understand the mechanisms that produce the kHz QPOs, but do not fit thematically in the previous subsections: Harmonics of the kHz QPOs and frequency and amplitude modulation of the QPO signal. These topics have not yet been fully explored, partly because the analysis required is not standard, and partly because the data available do not allow us to go beyond what has been done so far. For instance, there is only one paper published exploring the harmonic content of the kHz QPOs, and only four papers about the frequency and amplitude modulation of the QPO signal by other timing phenomena. The few results available, however, suggest that some of these endeavours are worthwhile pursuing further.

[3]This is possibly also the case for the upper kHz QPO, and all other QPO signals in these sources.

Most of the models that have been proposed to explain the frequencies of the kHz QPOs assume that one of the QPOs reflects the Keplerian orbital motion at some preferred radius in the accretion disc. For instance, in the sonic-point beat-frequency model [115], the upper kHz QPO is produced at the radius where the radial flow velocity in the disc turns from subsonic to supersonic (the sonic radius), and the lower kHz QPO is a beat between the upper kHz QPO and the spin frequency of the neutron star. Besides the main peaks at $\nu_{\rm low}$ and $\nu_{\rm upp}$, this model predicts some other (weaker) harmonics and sidebands of these QPOs at specific frequencies; for instance, in this model there should be a relatively strong harmonic of the lower QPO at $2\nu_{\rm low}$ (see table 3 of [115], for a list of other harmonic and sideband peaks predicted by the model).

In the relativistic-precession model [148], the upper kHz QPO is also assumed to be Keplerian, and the lower kHz QPO is the periastron precession frequency of a slightly non-circular inner accretion disc. Oscillating disc models (e.g., [133]) in a GR potential yield the same frequencies as those for test particles in the relativistic-precession model, but also predict other frequencies that are linear combinations of the three basic relativistic frequencies in the disc. For instance, in an oscillating disc in which the Keplerian and the periastron precession frequencies are excited, there should be a peak at a frequency equal to $2\nu_{\rm upp} - \nu_{\rm low}$.

Finding one of these other peaks would, on one hand, favour one model over the other, opening up a path to understand the dynamics of the disc and the production of the QPOs and, on the other hand, would confirm that the upper kHz QPO is due to Keplerian orbital motion in the disc. Unfortunately, the only attempt to find any of these peaks using data of Sco X-1 [106] gave negative results. The 95% confidence upper limits of a peak at $2\nu_{\rm low}$ and $2\nu_{\rm upp} - \nu_{\rm low}$ are, respectively, 0.12 and 0.26 times the amplitude of the upper kHz QPO in this source (the amplitude of the upper kHz QPO in Sco X-1 is between 0.6% and 2.5%; see table 1 in [106], for the upper limits at other frequencies). The signal of these other peaks could be attenuated in the corona [29, 84, 112, 115] depending on the frequency of the peak, and the radius and optical depth of the corona. Considering this effect, the upper limits of an unattenuated signal at those two frequencies would be a factor between 0.15 and 0.30 of the unattenuated amplitude of the upper kHz QPO in Sco X-1 (see table 2 in [106], for the unattenuated upper limits at other frequencies). In conclusion, in Sco X-1 none of the secondary QPO peaks predicted by the two main classes of models of the kHz QPOs are detected, with upper limits that imply that these secondary peaks are one to 2 orders of magnitude weaker in power than the upper kHz QPO.

If the amplitude of the signal of the QPOs is modulated, one would expect to see sidebands to the main oscillation, at frequencies that are equal to the frequency of the QPOs plus or minus the frequency of the mechanism that modulates the QPO amplitude. This could happen if, for instance, on very short scales the amplitude of the kHz QPOs depends upon mass accretion rate. Although there is no guarantee that this is the case, given that on long time scales the amplitude of the QPOs depends upon QPO frequency (e.g., Fig. 6.5a in Sect. 6.4) and, in turn, QPO frequency

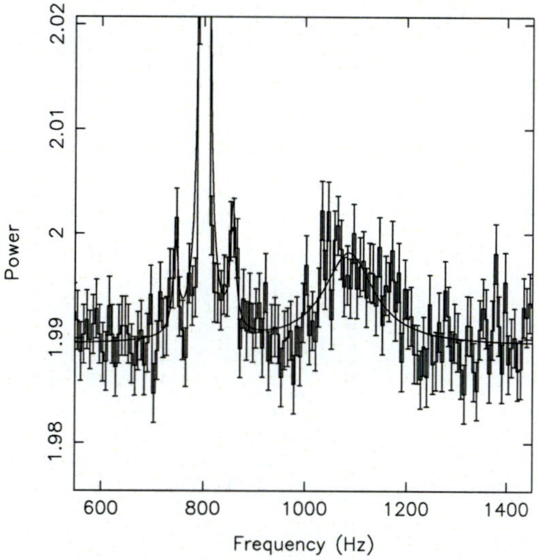

Fig. 6.36 Sideband peaks of the lower kHz QPO in 4U 1636−53. The sideband peaks are visible at a frequency that is ±55 Hz of the frequency of the strong and narrow lower kHz QPOs, which is at ∼800 Hz in this observation. The broad peak at ∼1100 Hz in the Figure is the upper kHz QPO [71]

depends on inferred mass accretion rate (e.g., Fig. 6.3b in Sect. 6.4), it is not unthinkable that this relation could also hold on very short time scales.

Figure 6.36 shows the power spectrum of 4U 1636−53 in the range of frequencies of the kHz QPOs [71]. The strong and narrow peak (off the scale of the plot) at ∼800 Hz and the weak and broad peak at ∼1100 Hz are, respectively, the lower and the upper kHz QPOs. The scale of the y axis in the plot has been chosen to highlight the weak QPO peaks that appear at frequencies that are ∼50 Hz below and above the frequency of the lower kHz QPO. These sidebands peaks are very significantly detected in this observation of 4U 1636−53, and had been observed before in this source and in 4U 1608−52 and 4U 1728−34 [67]. One possible explanation for these phenomena, given in [71], is that the sideband peaks reflect a modulation in the radiation pattern that produces the upper kHz QPO at the Lense-Thirring precession frequency at the inner edge of the accretion rate, and that this modulation would in turn modulate the formation of the lower kilohertz QPO.

Similar to what happens to the amplitude, the frequency of the oscillations that produce the QPOs can also be modulated on the time scale of another variability component in the light curve. Figure 6.37a shows two power spectra of Sco X-1 calculated from very short time intervals to sample the maxima and minima of the light curve of the source on time scales of 0.15 s, corresponding to the frequency range, 6–8-Hz, of the so-called Normal Branch Oscillation (NBO) in this source. The top and bottom panels of this Figure show the power spectrum of, respectively, the maxima and minima of the light curve on those time scales. The frequency of the upper kHz QPO changes significantly, by ∼20 Hz, with the change being anti correlated with the source count rate. This result shows that the frequency of the upper kHz QPO in Sco X-1 is driven by changes that happen on the time scale of the NBO.

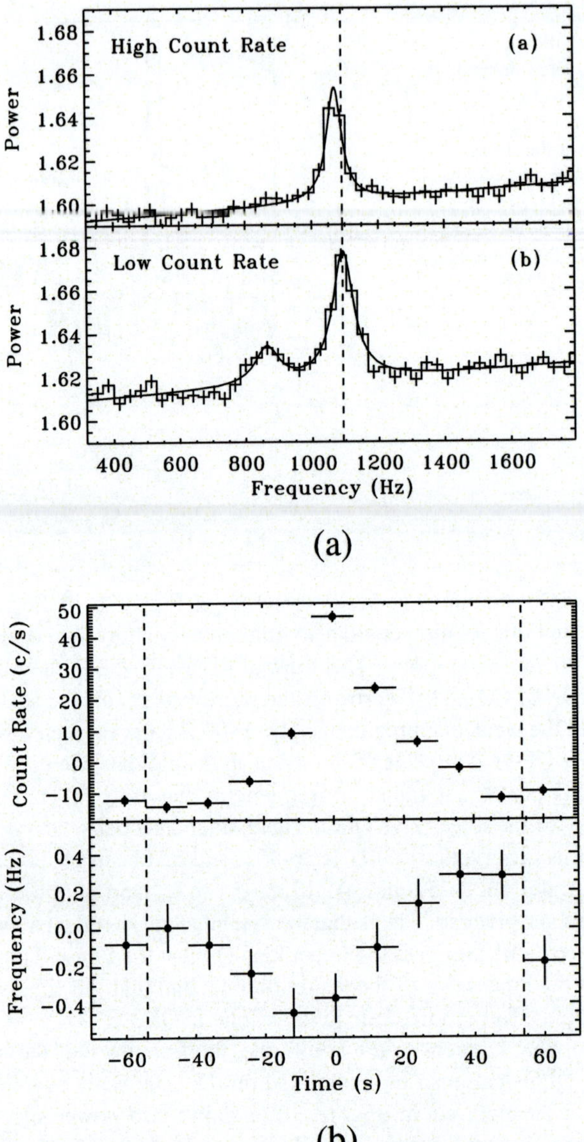

Fig. 6.37 Top (**a**): Power spectra in the range of frequencies of the kHz QPOs in Sco X-1. The power spectrum in the top panel was calculated during intervals of high count rates on the time scale of the normal branch QPO (NBO) at 6–8 Hz in this source. The power spectrum in the bottom panel corresponds to the intervals of low count rates on the time scale of the NBO (originally published as Figure 2 in [186]). Bottom (**b**): For the NS LMXB 4U 1608−52, count rate of the source (top) and frequency of the lower kHz QPO (bottom) over the cycle of the millihertz QPO in this source (originally published as Figure 2 in [185])

Figure 6.37b shows the result of a similar analysis for the lower kHz QPO in 4U 1608−52 on the time scale of the millihertz QPO in this source [137]. In this case, again, the frequency of the QPO changes over the cycle of the millihertz QPO and is anti correlated with the source count rate in that cycle. This result, and the previous one, could be explained if, for instance, the upper kHz QPO in Sco X-1 and the lower kHz QPO in 4U 1608−52 are produced at the inner edge of the accretion disc, and the value of the inner radius of the disc is modulated by radiation coming from the neutron-star surface on the time scale of the NBO signal in the case of Sco X-1, and the time scale of the millihertz QPO in 4U 1608−52.

6.9 Probing Neutron-Star Interiors and GR with kHz QPO

The Keplerian orbital frequency at a radial distance r from the centre of a slowly rotating neutron star (to first order in the specific angular momentum, $j = cJ/GM^2$) is:

$$\nu_{\mathrm{K}} = \frac{1}{2\pi}\left(1 - j\left(\frac{GM}{rc^2}\right)^{3/2}\right)\sqrt{\frac{GM}{r^3}}, \tag{6.2}$$

where M and J are, respectively, the gravitational mass and angular momentum of the star, and G and c are the gravitational constant and the speed of light, respectively [79, 116]. Because the disc must be outside the neutron star, if the frequency of one of the kHz QPOs reflects Keplerian orbital motion at this radial distance in the disc [115, 148], it must hold that $r \geq R_{\mathrm{NS}}$, where R_{NS} is the neutron-star radius. The existence of a radius, R_{ISCO}, inside which no stable circular orbit is possible [10] also implies that $r \geq R_{\mathrm{ISCO}}$. Depending on the equation of state, the radius of the ISCO can be inside or outside the neutron star [114]. Under these conditions, for an observed frequency of a QPO, ν_{QPO}, the maximum allowed mass and radius of the neutron star are, respectively, $M_{\max} = 2.2\mathrm{M}_{\odot}(1 + 0.75j)(1000\ \mathrm{Hz}/\nu_{\mathrm{QPO}})$, and $R_{\max} = 19.5\ \mathrm{km}(1 + 0.2j)(1000\ \mathrm{Hz}/\nu_{\mathrm{QPO}})$. The highest the QPO frequency, the smallest the maximum allowed mass and radius of a neutron star.

The maximum frequency of any of the kHz QPO observed with RXTE from all LMXBs is $\nu_{\mathrm{upp}} = 1288 \pm 8$ Hz in 4U 0614+09, with a 3-σ lower limit of 1267 Hz [168]. This measurement puts an upper limit to the mass of the neutron star in this system of $M < 2.1\mathrm{M}_{\odot}$ that is fairly independent of the assumed neutron-star equation of state. Assuming that this QPO reflects the Keplerian orbital frequency at, or just outside, the ISCO, this frequency leads to a mass of the neutron star in 4U 0614+09 of $2.0 \pm 0.1\mathrm{M}_{\odot}$ [168].

One should keep in mind that the identification of the upper kHz QPO with the Keplerian frequency at the inner edge of the accretion disc is model dependent. For instance, in the model proposed in [128] the lower kHz QPO is the one identified with the orbital frequency at the inner disc radius. On the other hand, as shown for instance in [4, 5, 47], it is possible to have frequencies above the Keplerian frequency at the inner edge of the accretion disc if there is a transition region between the disc and the neutron-star surface where matter in the disc has to slow down to match the rotation speed of the neutron star.

Measurements of QPO frequencies and spectral properties of the source offer another possibility to constrain the mass and radius of a neutron star. A broad iron emission line at 6.5–7 keV has been observed in about a dozen accreting neutron stars [26, 31–33, 43–46, 63, 64, 97, 98, 102, 117, 129, 131, 132, 134, 144, 146, 177, 178]. It has been proposed that, like in accreting black-hole systems [48, 140], the iron line is due to reflection of corona photons off the accretion disc, and that the line profile is driven by special and GR effects. If this is the case, the shape of the line profile would depend upon the inner disc radius. Therefore, detecting kHz QPOs in the power spectrum and a broad iron line in the energy spectrum of the same object would provide separate ways to constrain the neutron-star parameters in that object [25].

There is only one source, 4U 1636−53, in which both a broad iron line in the energy spectrum and kHz QPOs in the power spectrum were observed simultaneously in four separate occasions [145]. Because these four observations sampled different spectral states of the source, and presumably the inner radius of the disc changed between observations [54], this dataset offered the opportunity to test whether the inner disc radius deduced independently from the line profile and from the kHz QPOs changed in a consistent way. The main result of this analysis [145] was that the inner radius of the accretion disc deduced from the frequency of the upper kHz QPO was correlated with the spectral state of the source, whereas the radius deduced from the profile of the iron line was not. Because of this, the combined results from the kHz QPOs and the iron line do not lead to a consistent value of the neutron-star mass. Since those were the only observations available for this test, and no new observations of kHz QPOs are possible with current missions, we have to wait until the eXTP [180] and Athena [122] missions fly to address this question again. For the moment the jury is still out.

We refer the reader to the chapter in this book by Cole Miller for a more detailed discussion on methods of constraining neutron-star masses, radii and the equation of state of the cold dense matter that constitutes these stars.

6.10 Conclusions and Outlook

The kHz QPOs in neutron stars unlocked a new door to study the dynamics of matter, and the time-dependent interplay between matter and radiation, on very short time scales in the violent and turbulent environment that surrounds a neutron star, and under the most extreme conditions in GR. These QPOs also offer a chance to constrain the mass, radius and internal constitution of these extremely compact objects, with the potential to unveil the properties of matter under conditions that are unattainable in laboratories on Earth. The amount and quality of the data collected in almost 15 years with the RXTE satellite are unique and, for the moment, unrivalled by data that have or can be collected by any other satellite in the past or currently in operation. The rich phenomenology of the kHz QPOs discussed in this chapter triggered scores of new ideas about accretion, and will guide this area of research for years to come. The next generation of X-ray satellites with enough time resolution to observe the millisecond time scales in accreting LMXBs with enough collecting area, and the capability to deal with very bright objects, will soon be a reality. The complementary technical capabilities of Athena and eXTP will give a new impulse to this topic in the coming decade. If we were asked to anticipate now what the most fruitful areas of research will be when those missions start providing data, we would dare to say that the most exciting results will come from the correlated study of the properties of the kHz QPOs and the broad iron line (Sect. 6.9) over the shortest possible time scales, and the prospect of finding signatures of the innermost stable circular orbit around a neutron star (Sect. 6.4). The analysis tools that are needed to achieve these goals have to address *simultaneously* the spectral (physical parameters that describe the emission) and timing (rms amplitudes, lags and coherence function) properties of variable signals over the shortest time scales allowed by the data. As we showed in this Chapter, initiatives are already being taken in that direction. This, and the efforts to try and understand the data, will keep us busy for at least two more decades.

Acknowledgments This work is part of the research programme Athena with project number 184.034.002, which is (partly) financed by the Dutch Research Council (NWO). TMB acknowledges financial contribution from the agreement ASI-INAF n.2017-14-H.0. This research has made use of data and/or software provided by the High Energy Astrophysics Science Archive Research Center (HEASARC), which is a service of the Astrophysics Science Division at NASA/GSFC. This research made use of NASA's Astrophysics Data System.

References

1. M.A. Abramowicz, W. Kluźniak, A precise determination of black hole spin in GRO J1655-40. Astron. Astrophys. **374**, L19–L20 (2001)
2. M.A. Abramowicz, T. Bulik, M. Bursa, W. Kluźniak, Evidence for a 2:3 resonance in Sco X-1 kHz QPOs. Astron. Astrophys. **404**, L21–L24 (2003)
3. P.C. Agrawal, A broad spectral band Indian Astronomy satellite 'Astrosat'. Adv. Space Res. **38**(12), 2989–2994 (2006)
4. M. Ali Alpar, QPO frequency derivative-frequency correlation indicates non-Keplerian boundary layer with a maximum in rotation rate. Mon. Not. R. Astron. Soc. **462**(1), L98–L100 (2016)
5. M.A. Alpar, D. Psaltis, The highest dynamical frequency in the inner region of an accretion disc. Mon. Not. R. Astron. Soc. **391**(3), 1472–1476 (2008)
6. M.A. Alpar, J. Shaham, Is GX5 - 1 a millisecond pulsar? Nature **316**(6025), 239–241 (1985)
7. D. Altamirano, M. van der Klis, M. Méndez, S. Migliari, P.G. Jonker, A. Tiengo, W. Zhang, The island state of the atoll source 4U 1820-30. Astrophys. J. **633**(1), 358–366 (2005)

8. D. Altamirano, M. van der Klis, M. Méndez, P.G. Jonker, M. Klein-Wolt, W.H.G. Lewin, X-ray time variability across the atoll source states of 4U 1636-53. Astrophys. J. **685**(1), 436–450 (2008)
9. D. Altamirano, M. van der Klis, M. Méndez, R. Wijnand, C. Markwardt, J. Swank, Discovery of kilohertz quasi-periodic oscillations and state transitions in the low-mass X-ray binary 1E 1724-3045 (Terzan 2). Astrophys. J. **687**(1), 488–504 (2008)
10. J.M. Bardeen, W.H. Press, S.A. Teukolsky, Rotating black holes: locally nonrotating frames, energy extraction, and scalar synchrotron radiation. Astrophys. J. **178**, 347–370 (1972)
11. D. Barret, Soft lags in neutron star kHz quasi-periodic oscillations: evidence for reverberation? Astrophys. J. **770**(1), 9 (2013)
12. D. Barret, G. Vedrenne, Hard X-ray emission from weakly magnetized neutron stars. Astrophys. J. Suppl. **92**, 505 (1994)
13. D. Barret, W. Kluźniak, J.F. Olive, S. Paltani, G.K. Skinner, On the high coherence of kHz quasi-periodic oscillations. Mon. Not. R. Astron. Soc. **357**(4), 1288–1294 (2005)
14. D. Barret, J.-F. Olive, M.C. Miller, An abrupt drop in the coherence of the lower kHz quasi-periodic oscillations in 4U 1636-536. Mon. Not. R. Astron. Soc. **361**(3), 855–860 (2005)
15. D. Barret, J.-F. Olive, M.C. Miller, The coherence of kilohertz quasi-periodic oscillations in the X-rays from accreting neutron stars. Mon. Not. R. Astron. Soc. **370**(3), 1140–1146 (2006)
16. D. Barret, M. Bachetti, M.C. Miller, The drop of the coherence of the lower kilohertz quasi-periodic brightness variations is also observed in XTE J1701-462. Astrophys. J. **728**(1), 9 (2011)
17. T.M. Belloni, D. Altamirano, High-frequency quasi-periodic oscillations from GRS 1915+105. Mon. Not. R. Astron. Soc. **432**(1), 10–18 (2013)
18. T. Belloni, G. Hasinger, An atlas of aperiodic variability in HMXB. Astron. Astrophys. **230**, 103–119 (1990)
19. T. Belloni, M. Méndez, J. Homan, The distribution of kHz QPO frequencies in bright low mass X-ray binaries. Astron. Astrophys. **437**(1), 209–216 (2005)
20. T. Belloni, J. Homan, S. Motta, E. Ratti, M. Méndez, Rossi XTE monitoring of 4U1636-53 - I. Long-term evolution and kHz quasi-periodic oscillations. Mon. Not. R. Astron. Soc. **379**(1), 247–252 (2007)
21. T. Belloni, M. Méndez, J. Homan, On the kHz QPO frequency correlations in bright neutron star X-ray binaries. Mon. Not. R. Astron. Soc. **376**(3), 1133–1138 (2007)
22. T.M. Belloni, A. Sanna, M. Méndez, High-frequency quasi-periodic oscillations in black hole binaries. Mon. Not. R. Astron. Soc. **426**(3), 1701–1709 (2012)
23. J.S. Bendat, A.G. Piersol, *Random Data: Analysis and Measurement Procedures*, 4th edn. (Wiley, Hoboken, 2010)
24. M. Berger, M. van der Klis, J. van Paradijs, W.H.G. Lewin, F. Lamb, B. Vaughan, E. Kuulkers, T. Augusteijn, W. Zhang, F.E. Marshall, J.H. Swank, I. Lapidus, J.C. Lochner, T.E. Strohmayer, Discovery of 800 HZ quasi-periodic oscillations in 4U 1608-52. Astrophys. J. Lett. **469**, L13 (1996)
25. S. Bhattacharyya, Ways to constrain neutron star equation of state models using relativistic disc lines. Mon. Not. R. Astron. Soc. **415**(4), 3247–3252 (2011)
26. S. Bhattacharyya, T.E. Strohmayer, Evidence of a broad relativistic iron line from the neutron star low-mass X-ray binary Serpens X-1. Astrophys. J. Lett. **664**(2), L103–L106 (2007)
27. M. Boutelier, D. Barret, M.C. Miller, kHz quasi-periodic oscillations in the low-mass X-ray binary 4U 0614+09. Mon. Not. R. Astron. Soc. **399**(4), 1901–1906 (2009)
28. S. Boutloukos, M. van der Klis, D. Altamirano, M. Klein-Wolt, R. Wijnands, P.G. Jonker, R.P. Fender, Discovery of twin kHz QPOs in the peculiar X-ray binary circinus X-1. Astrophys. J. **653**(2), 1435–1444 (2006)
29. J. Brainerd, F.K. Lamb, Effect of an electron scattering cloud on X-ray oscillations produced by beaming. Astrophys. J. Lett. **317**, L33 (1987)
30. E.M. Cackett, Are the kHz QPO lags in neutron star 4U 1608-52 due to reverberation? Astrophys. J. **826**(2), 103 (2016)

31. E.M. Cackett, J.M. Miller, S. Bhattacharyya, J.E. Grindlay, J. Homan, M. van der Klis, M.C. Miller, T.E. Strohmayer, R. Wijnands, Relativistic iron emission lines in neutron star low-mass X-ray binaries as probes of neutron star radii. Astrophys. J. **674**(1), 415–420 (2008)
32. E.M. Cackett, D. Altamirano, A. Patruno, J.M. Miller, M. Reynolds, M. Linares, R. Wijnands, Broad relativistic iron emission line observed in SAX J1808.4-3658. Astrophys. J. Lett. **694**(1), L21–L25 (2009)
33. E.M. Cackett, J.M. Miller, D.R. Ballantyne, D. Barret, S. Bhattacharyya, M. Boutelier, M.C. Miller, T.E. Strohmayer, R. Wijnands, Relativistic lines and reflection from the inner accretion disks around neutron stars. Astrophys. J. **720**(1), 205–225 (2010)
34. P. Casella, T. Belloni, L. Stella, The ABC of low-frequency quasi-periodic oscillations in black hole candidates: analogies with Z sources. Astrophys. J. **629**(1), 403–407 (2005)
35. D.J. Christian, J.H. Swank, The survey of low-mass X-ray binaries with the Einstein observatory solid-state spectrometer and monitor proportional counter. Astrophys. J. Suppl. **109**(1), 177–224 (1997)
36. B.M. Coughenour, E.M. Cackett, P. Peille, J.S. Troyer, Modeling the upper kHz QPOs of 4U 1728-34 with X-ray reverberation. Astrophys. J. **889**(2), 136 (2020)
37. M.G.B. de Avellar, M. Méndez, A. Sanna, J.E. Horvath, Time lags of the kilohertz quasi-periodic oscillations in the low-mass X-ray binaries 4U 1608-52 and 4U 1636-53. Mon. Not. R. Astron. Soc. **433**(4), 3453–3463 (2013)
38. M.G.B. de Avellar, M. Méndez, D. Altamirano, A. Sanna, G. Zhang, Phase lags of quasi-periodic oscillations across source states in the low-mass X-ray binary 4U 1636-53. Mon. Not. R. Astron. Soc. **461**(1), 79–92 (2016)
39. B. De Marco, G. Ponti, T. Muñoz-Darias, K. Nand ra, Tracing the reverberation lag in the hard state of black hole X-ray binaries. Astrophys. J. **814**(1), 50 (2015)
40. T. Di Salvo, M. Méndez, M. van der Klis, E. Ford, N.R. Robba, Study of the temporal behavior of 4U 1728-34 as a function of its position in the color-color diagram. Astrophys. J. **546**(2), 1107–1120 (2001)
41. T. Di Salvo, N.R. Robba, R. Iaria, L. Stella, L. Burderi, G.L. Israel, Detection of a hard tail in the X-ray spectrum of the Z source GX 349+2. Astrophys. J. **554**(1), 49–55 (2001)
42. T. Di Salvo, M. Méndez, M. van der Klis, On the correlated spectral and timing properties of 4U 1636-53: an atoll source at high accretion rates. Astron. Astrophys. **406**, 177–192 (2003)
43. T. Di Salvo, R. Iaria, M. Méndez, L. Burderi, G. Lavagetto, N.R. Robba, L. Stella, M. van der Klis, A broad iron line in the Chandra high energy transmission grating spectrum of 4U 1705-44. Astrophys. J. Lett. **623**(2), L121–L124 (2005)
44. T. di Salvo, A. D'Aí, R. Iaria, L. Burderi, M. Dovčiak, V. Karas, G. Matt, A. Papitto, S. Piraino, A. Riggio, N.R. Robba, A. Santangelo, A relativistically smeared spectrum in the neutron star X-ray binary 4U 1705-44: looking at the inner accretion disc with X-ray spectroscopy. Mon. Not. R. Astron. Soc. **398**(4), 2022–2027 (2009)
45. T. Di Salvo, R. Iaria, M. Matranga, L. Burderi, A. D'Aí, E. Egron, A. Papitto, A. Riggio, N.R. Robba, Y. Ueda, Suzaku broad-band spectrum of 4U 1705-44: probing the reflection component in the hard state. Mon. Not. R. Astron. Soc. **449**(3), 2794–2802 (2015)
46. E. Egron, T. Di Salvo, S. Motta, L. Burderi, A. Papitto, R. Duro, A. D'Aì, A. Riggio, T. Belloni, R. Iaria, N.R. Robba, S. Piraino, A. Santangelo, Testing reflection features in 4U 1705-44 with XMM-Newton, BeppoSAX, and RXTE in the hard and soft states. Astron. Astrophys. **550**, A5 (2013)
47. M.H. Erkut, D. Psaltis, M.A. Alpar, Quasi-periodic oscillations as global hydrodynamic modes in the boundary layers of viscous accretion disks. Astrophys. J. **687**(2), 1220–1229 (2008)
48. A.C. Fabian, M.J. Rees, L. Stella, N.E. White, X-ray fluorescence from the inner disc in Cygnus X-1. Mon. Not. R. Astron. Soc. **238**, 729–736 (1989)
49. A.C. Fabian, A. Zoghbi, R.R. Ross, P. Uttley, L.C. Gallo, W.N. Brandt, A.J. Blustin, T. Boller, M.D. Caballero-Garcia, J. Larsson, J.M. Miller, G. Miniutti, G. Ponti, R.C. Reis, C.S. Reynolds, Y. Tanaka, A.J. Young, Broad line emission from iron K- and L-shell transitions in the active galaxy 1H0707-495. Nature **459**(7246), 540–542 (2009)

50. E. Ford, P. Kaaret, M. Tavani, D. Barret, P. Bloser, J. Grindlay, B.A. Harmon, W.S. Paciesas, S.N. Zhang, Evidence from quasi-periodic oscillations for a millisecond pulsar in the low-mass X-ray binary 4U 0614+091. Astrophys. J. Lett. **475**(2), L123–L126 (1997)
51. E.C. Ford, M. van der Klis, J. van Paradijs, M. Méndez, R. Wijnands, P. Kaaret, Discovery of a second kilohertz QPO in the X-ray binary 4U 1735-44. Astrophys. J. Lett. **508**(2), L155–L158 (1998)
52. J. Frank, A. King, D.J. Raine, *Accretion Power in Astrophysics*, 3rd edn. (Cambridge University Press, Cambridge, 2002)
53. K.C. Gendreau, Z. Arzoumanian, T. Okajima, The Neutron star Interior Composition ExploreR (NICER): an explorer mission of opportunity for soft X-ray timing spectroscopy, in *Proceedings of the SPIE*, vol. 8443. Society of Photo-Optical Instrumentation Engineers (SPIE) Conference Series, September 2012, p. 844313
54. M. Gierliński, C. Done, The X-ray spectrum of the atoll source 4U 1608-52. Mon. Not. R. Astron. Soc. **337**(4), 1373–1380 (2002)
55. M. Gierliński, C. Done, K. Page, X-ray irradiation in XTE J1817-330 and the inner radius of the truncated disc in the hard state. Mon. Not. R. Astron. Soc. **388**(2), 753–760 (2008)
56. M. Gilfanov, M. Revnivtsev, S. Molkov, Boundary layer, accretion disk and X-ray variability in the luminous LMXBs. Astron. Astrophys. **410**, 217–230 (2003)
57. J. Homan, M. van der Klis, A 695 HZ quasi-periodic oscillation in the low-mass X-ray binary EXO 0748-676. Astrophys. J. **539**(2), 847–850 (2000)
58. J. Homan, M. van der Klis, P.G. Jonker, R. Wijnands, E. Kuulkers, M. Méndez, W.H.G. Lewin, RXTE observations of the neutron star low-mass X-ray binary GX 17+2: correlated X-ray spectral and timing behavior. Astrophys. J. **568**(2), 878–900 (2002)
59. J. Homan, T. Belloni, M. van der Klis, P. Casella, M. Mendez, W. Lewin, R. Fender, E. Gallo, N. Gehrels, Spectral evolution and kHz QPOs in the transient Z source XTE J1701-462. Astron. Telegram **748**, 1 (2006)
60. J. Homan, M. van der Klis, R. Wijnands, T. Belloni, R. Fender, M. Klein-Wolt, P. Casella, M. Méndez, E. Gallo, W.H.G. Lewin, N. Gehrels, Rossi X-ray timing explorer observations of the first transient Z source XTE J1701-462: shedding new light on mass accretion in luminous neutron star X-ray binaries. Astrophys. J. **656**(1), 420–430 (2007)
61. J. Homan, R. Wijnands, D. Altamirano, T. Belloni, Rapid decay of the neutron star transient XTE J1701-462. Astron. Telegram **1165**, 1 (2007)
62. J. Homan, M. van der Klis, J.K. Fridriksson, R.A. Remillard, R. Wijnands, M. Méndez, D. Lin, D. Altamirano, P. Casella, T.M. Belloni, W.H.G. Lewin, XTE J1701-462 and its implications for the nature of subclasses in low-magnetic-field neutron star low-mass X-ray binaries. Astrophys. J. **719**(1), 201–212 (2010)
63. R. Iaria, A. D'Aí, T. di Salvo, N.R. Robba, A. Riggio, A. Papitto, L. Burderi, A ionized reflecting skin above the accretion disk of GX 349+2. Astron. Astrophys. **505**(3), 1143–1151 (2009)
64. R. Iaria, T. Di Salvo, M. Del Santo, F. Pintore, A. Sanna, A. Papitto, L. Burderi, A. Riggio, A.F. Gambino, M. Matranga, Study of the reflection spectrum of the LMXB 4U 1702-429. Astron. Astrophys. **596**, A21 (2016)
65. A. Ingram, Error formulae for the energy-dependent cross-spectrum. Mon. Not. R. Astron. Soc. **489**(3), 3927–3938 (2019)
66. A. Ingram, M. van der Klis, M. Middleton, D. Altamirano, P. Uttley, Tomographic reflection modelling of quasi-periodic oscillations in the black hole binary H 1743-322. Mon. Not. R. Astron. Soc. **464**(3), 2979–2991 (2017)
67. P.G. Jonker, M. Méndez, M. van der Klis, Discovery of a new, third kilohertz quasi-periodic oscillation in 4U 1608-52, 4U 1728-34, and 4U 1636-53: sidebands to the lower kilohertz quasi-periodic oscillation? Astrophys. J. Lett. **540**(1), L29–L32 (2000)
68. P.G. Jonker, M. van der Klis, R. Wijnands, J. Homan, J. van Paradijs, M. Méndez, E.C. Ford, E. Kuulkers, F.K. Lamb, The power spectral properties of the Z source GX 340+0. Astrophys. J. **537**(1), 374–386 (2000)

69. P.G. Jonker, M. van der Klis, J. Homan, M. Méndez, W.H.G. Lewin, R. Wijnands, W. Zhang, Low- and high-frequency variability as a function of spectral properties in the bright X-ray binary GX 5-1. Mon. Not. R. Astron. Soc. **333**(3), 665–678 (2002)
70. P.G. Jonker, M. Méndez, M. van der Klis, Kilohertz quasi-periodic oscillations difference frequency exceeds inferred spin frequency in 4U 1636-53. Mon. Not. R. Astron. Soc. **336**(1), L1–L5 (2002)
71. P.G. Jonker, M. Méndez, M. van der Klis, Sidebands to the lower kilohertz quasi-periodic oscillation in 4U 1636-53. Mon. Not. R. Astron. Soc. **360**(3), 921–925 (2005)
72. P. Kaaret, W. Yu, E.C. Ford, S.N. Zhang, Correlation between fast quasi-periodic oscillations and X-ray spectral shape in Atoll sources. Astrophys. J. Lett. **497**(2), L93–L96 (1998)
73. P. Kaaret, S. Piraino, E.C. Ford, A. Santangelo, Discovery of microsecond soft lags in the X-ray emission of the Atoll source 4U 1636-536. Astrophys. J. Lett. **514**(1), L31–L33 (1999)
74. E. Kara, J.F. Steiner, A.C. Fabian, E.M. Cackett, P. Uttley, R.A. Remillard, K.C. Gendreau, Z. Arzoumanian, D. Altamirano, S. Eikenberry, T. Enoto, J. Homan, J. Neilsen, A.L. Stevens, The corona contracts in a black-hole transient. Nature **565**(7738), 198–201 (2019)
75. K. Karpouzas, M. Méndez, E. ro M. Ribeiro, D. Altamirano, O. Blaes, F. García, The comptonizing medium of the neutron star in 4U 1636 - 53 through its lower kilohertz quasi-periodic oscillations. Mon. Not. R. Astron. Soc. **492**(1), 1399–1415 (2020)
76. S. Kato, Trapped one-armed corrugation waves and QPO's. Publ. Astron. Soc. Jpn. **42**, 99–113 (1990)
77. S. Kato, J. Fukue, Trapped radial oscillations of gaseous disks around a black hole. Publ. Astron. Soc. Jpn. **32**, 377 (1980)
78. W. Kluźniak, M.A. Abramowicz, Resonant oscillations of accretion flow and Khz QPOS. Astrophys. Space Sci. **300**(1–3), 143–148 (2005)
79. W. Kluzniak, P. Michelson, R.V. Wagoner, Determining the properties of accretion-gap neutron stars. Astrophys. J. **358**, 538 (1990)
80. W. Kluźniak, M.A. Abramowicz, S. Kato, W.H. Lee, N. Stergioulas, Nonlinear resonance in the accretion disk of a millisecond pulsar. Astrophys. J. Lett. **603**(2), L89–L92 (2004)
81. A.S. Kompaneets, The establishment of thermal equilibrium between quanta and electrons. Sov. J. Exp. Theor. Phys. **4**(5), 730–737 (1957)
82. N. Kumar, R. Misra, Energy dependent time delays of kHz oscillations due to thermal Comptonization. Mon. Not. R. Astron. Soc. **445**(3), 2818–2824 (2014)
83. N. Kumar, R. Misra, Constraining the size of the Comptonizing medium by modelling the energy-dependent time lags of kHz QPOs of neutron star system. Mon. Not. R. Astron. Soc. **461**(3), 2580–2588 (2016)
84. N.D. Kylafis, G.S. Klimis, Effects of electron scattering on the oscillations of an X-ray source. Astrophys. J. **323**, 678 (1987)
85. D. Lai, D. Tsang, Corotational instability of inertial-acoustic modes in black hole accretion discs and quasi-periodic oscillations. Mon. Not. R. Astron. Soc. **393**(3), 979–991 (2009)
86. F.K. Lamb, M.C. Miller, Changing frequency separation of kilohertz quasi-periodic oscillations in the sonic-point beat-frequency model. Astrophys. J. **554**(2), 1210–1215 (2001)
87. F.K. Lamb, M.C. Miller, Sonic-point and spin-resonance model of the kilohertz QPO pairs. arXiv e-prints, astro-ph/0308179, August 2003
88. F.K. Lamb, N. Shibazaki, M.A. Alpar, J. Shaham, Quasi-periodic oscillations in bright galactic-bulge X-ray sources. Nature **317**(6039), 681–687 (1985)
89. L.D. Landau, E.M. Lifshitz, *Mechanics, Volume 1*, 3rd edn. Course of Theoretical Physics (Butterworth-Heinemann, Oxford, 1976)
90. D.A. Leahy, W. Darbro, R.F. Elsner, M.C. Weisskopf, P.G. Sutherland, S. Kahn, J.E. Grindlay, On searches for pulsed emission with application to four globular cluster X-ray sources : NGC 1851, 6441, 6624 and 6712. Astrophys. J. **266**, 160–170 (1983)
91. H.C. Lee, G.S. Miller, Comptonization and QPO origins in accreting neutron star systems. Mon. Not. R. Astron. Soc. **299**(2), 479–487 (1998)

92. H.C. Lee, R. Misra, R.E. Taam, A compton upscattering model for soft lags in the lower kilohertz quasi-periodic oscillation in 4U 1608-52. Astrophys. J. Lett. **549**(2), L229–L232 (2001)
93. W.H. Lee, M.A. Abramowicz, W. Kluźniak, Resonance in forced oscillations of an accretion disk and kilohertz quasi-periodic oscillations. Astrophys. J. Lett. **603**(2), L93–L96 (2004)
94. D. Lin, R.A. Remillard, J. Homan, Evaluating spectral models and the X-ray states of neutron star X-ray transients. Astrophys. J. **667**(2), 1073–1086 (2007)
95. D. Lin, R.A. Remillard, J. Homan, Spectral states of XTE J1701 - 462: link between Z and Atoll sources. Astrophys. J. **696**(2), 1257–1277 (2009)
96. M. Linares, M. van der Klis, D. Altamirano, C.B. Markwardt, Discovery of kilohertz quasi-periodic oscillations and shifted frequency correlations in the accreting millisecond pulsar XTE J1807-294. Astrophys. J. **634**(2), 1250–1260 (2005)
97. R.M. Ludlam, J.M. Miller, Z. Arzoumanian, P.M. Bult, E.M. Cackett, D. Chakrabarty, T. Dauser, T. Enoto, A.C. Fabian, J.A. García, K.C. Gendreau, S. Guillot, J. Homan, G.K. Jaisawal, L. Keek, B. La Marr, C. Malacaria, C.B. Markwardt, J.F. Steiner, T.E. Strohmayer, Detection of reflection features in the neutron star low-mass X-ray binary serpens X-1 with NICER. Astrophys. J. Lett. **858**(1), L5 (2018)
98. R.M. Ludlam, J.M. Miller, D. Barret, E.M. Cackett, B.M. Coughenour, T. Dauser, N. Degenaar, J.A. García, F.A. Harrison, F. Paerels, NuSTAR observations of the accreting Atolls GX 3+1, 4U 1702-429, 4U 0614+091, and 4U 1746-371. Astrophys. J. **873**(1), 99 (2019)
99. M. Lyu, M. Méndez, A. Sanna, J. Homan, T. Belloni, B. Hiemstra, Iron-line and continuum variations in the XMM-Newton and Suzaku spectra of the neutron-star low-mass X-ray binary 4U 1636-53. Mon. Not. R. Astron. Soc. **440**(2), 1165–1178 (2014)
100. J. Malzac, A. Merloni, A.C. Fabian, Jet-disc coupling through a common energy reservoir in the black hole XTE J1118+480. Mon. Not. R. Astron. Soc. **351**(1), 253–264 (2004)
101. G. Mastroserio, A. Ingram, M. van der Klis, Multi-time-scale X-ray reverberation mapping of accreting black holes. Mon. Not. R. Astron. Soc. **475**(3), 4027–4042 (2018)
102. S.M. Mazzola, R. Iaria, T. Di Salvo, M. Del Santo, A. Sanna, A.F. Gambino, A. Riggio, A. Segreto, L. Burderi, A. Santangelo, N. D'Amico, A broadband spectral analysis of 4U 1702-429 using XMM-Newton and BeppoSAX data. Astron. Astrophys. **621**, A89 (2019)
103. M. Méndez, On the maximum amplitude and coherence of the kilohertz quasi-periodic oscillations in low-mass X-ray binaries. Mon. Not. R. Astron. Soc. **371**(4), 1925–1938 (2006)
104. M. Méndez, T. Belloni, Is there a link between the neutron-star spin and the frequency of the kilohertz quasi-periodic oscillations? Mon. Not. R. Astron. Soc. **381**(2), 790–796 (2007)
105. M. Méndez, M. van der Klis, Precise measurements of the kilohertz quasi-periodic oscillations in 4U 1728-34. Astrophys. J. Lett. **517**(1), L51–L54 (1999)
106. M. Méndez, M. van der Klis, The harmonic and sideband structure of the kilohertz quasi-periodic oscillations in Sco X-1. Mon. Not. R. Astron. Soc. **318**(3), 938–942 (2000)
107. M. Méndez, M. van der Klis, R. Wijnands, E.C. Ford, J. van Paradijs, B.A. Vaughan, Kilohertz quasi-periodic oscillation peak separation is not constant in the Atoll source 4U 1608-52. Astrophys. J. Lett. **505**(1), L23–L26 (1998)
108. M. Méndez, M. van der Klis, J. van Paradijs, Difference frequency of kilohertz QPOs not equal to half the burst oscillation frequency in 4U 1636-53. Astrophys. J. Lett. **506**(2), L117–L119 (1998)
109. M. Méndez, M. van der Klis, E.C. Ford, R. Wijnand, J. van Paradijs, Dependence of the frequency of the kilohertz quasi-periodic oscillationson X-ray count rate and colors in 4U 1608-52. Astrophys. J. Lett. **511**(1), L49–L52 (1999)
110. M. Méndez, M. van der Klis, E.C. Ford, The amplitude of the kilohertz quasi-periodic oscillations in 4U 1728-34, 4U 1608-52, and Aquila X-1, as a function of X-ray intensity. Astrophys. J. **561**(2), 1016–1026 (2001)
111. A. Merloni, A.C. Fabian, Accretion disc coronae as magnetic reservoirs. Mon. Not. R. Astron. Soc. **321**(3), 549–552 (2001)
112. M.C. Miller, Attenuation of beaming oscillations near neutron stars. Astrophys. J. **537**(1), 342–350 (2000)

113. M.C. Miller, Interpreting QPOs from accreting neutron stars, in *X-ray Timing 2003: Rossi and Beyond*, vol. 714, ed. by P. Kaaret, F.K. Lamb, J.H. Swank. American Institute of Physics Conference Series, July 2004, pp. 365–370
114. M.C. Miller, F.K. Lamb, G.B. Cook, Effects of rapid stellar rotation on equation-of-state constraints derived from quasi-periodic brightness oscillations. Astrophys. J. **509**(2), 793–801 (1998)
115. M.C. Miller, F.K. Lamb, D. Psaltis, Sonic-point model of kilohertz quasi-periodic brightness oscillations in low-mass X-ray binaries. Astrophys. J. **508**(2), 791–830 (1998)
116. M.C. Miller, F.K. Lamb, D. Psaltis, Constraints on the equation of state of neutron star matter from observations of kilohertz QPOs. Nucl. Phys. B Proc. Suppl. **69**(1–3), 123–128 (1999)
117. J.M. Miller, M.L. Parker, F. Fuerst, M. Bachetti, D. Barret, B.W. Grefenstette, S. Tendulkar, F.A. Harrison, S.E. Boggs, D. Chakrabarty, F.E. Christensen, W.W. Craig, A.C. Fabian, C.J. Hailey, L. Natalucci, F. Paerels, V. Rana, D.K. Stern, J.A. Tomsick, W.W. Zhang, Constraints on the neutron star and inner accretion flow in serpens X-1 using NuSTAR. Astrophys. J. Lett. **779**(1), L2 (2013)
118. S. Miyamoto, K. Kimura, S. Kitamoto, T. Dotani, K. Ebisawa, X-ray variability of GX 339-4 in its very high state. Astrophys. J. **383**, 784 (1991)
119. E.H. Morgan, R.A. Remillard, J. Greiner, RXTE observations of QPOs in the black hole candidate GRS 1915+105. Astrophys. J. **482**(2), 993–1010 (1997)
120. A. Mukherjee, S. Bhattacharyya, Highly coherent kilohertz quasi-periodic oscillations from the neutron star X-ray binary EXO 1745-248. Astrophys. J. Lett. **730**(2), L32 (2011)
121. A. Mukherjee, S. Bhattacharyya, Fractional amplitude of kilohertz quasi-periodic oscillation from 4U 1728-34: evidence of decline at higher energies. Astrophys. J. **756**(1), 55 (2012)
122. K. Nandra, D. Barret, X. Barcons, A. Fabian, J.-W. den Herder, L. Piro, M. Watson, C. Adami, J. Aird, J.M. Afonso, D. Alexander, C. Argiroffi, L. Amati, M. Arnaud, J.-L. Atteia, M. Audard, C. Badenes, J. Ballet, L. Ballo, A. Bamba, A. Bhardwaj, E. Stefano Battistelli, W. Becker, M. De Becker, E. Behar, S. Bianchi, V. Biffi, L. Bîrzan, F. Bocchino, S. Bogdanov, L. Boirin, T. Boller, S. Borgani, K. Borm, N. Bouché, H. Bourdin, R. Bower, V. Braito, E. Branchini, G. Branduardi-Raymont, J. Bregman, L. Brenneman, M. Brightman, M. Brüggen, J. Buchner, E. Bulbul, M. Brusa, M. Bursa, A. Caccianiga, E. Cackett, S. Campana, Cappelluti, M. Cappi, F. Carrera, M. Ceballos, F. Christensen, Y.-H. Chu, E. Churazov, N. Clerc, S. Corbel, A. Corral, A. Comastri, E. Costantini, J. Croston, M. Dadina, A. D'Ai, A. Decourchelle, R. Della Ceca, K. Dennerl, K. Dolag, C. Done, M. Dovciak, J. Drake, D. Eckert, A. Edge, S. Ettori, Y. Ezoe, E. Feigelson, R. Fender, C. Feruglio, A. Finoguenov, F. Fiore, M. Galeazzi, S. Gallagher, P. Gandhi, M. Gaspari, F. Gastaldello, A. Georgakakis, I. Georgantopoulos, M. Gilfanov, M. Gitti, R. Gladstone, R. Goosmann, E. Gosset, N. Grosso, M. Guedel, M. Guerrero, F. Haberl, M. Hardcastle, S. Heinz, A. Alonso Herrero, A. Hervé, M. Holmstrom, K. Iwasawa, P. Jonker, J. Kaastra, E. Kara, V. Karas, J. Kastner, A. King, D. Kosenko, D. Koutroumpa, R. Kraft, I. Kreykenbohm, R. Lallement, G. Lanzuisi, J. Lee, M. Lemoine-Goumard, A. Lobban, G. Lodato, L. Lovisari, S. Lotti, I. McCharthy, B. McNamara, A. Maggio, R. Maiolino, B. De Marco, D. de Martino, S. Mateos, G. Matt, B. Maughan, P. Mazzotta, M. Mendez, A. Merloni, G. Micela, M. Miceli, R. Mignani, J. Miller, G. Miniutti, S. Molendi, R. Montez, A. Moretti, C. Motch, Y. Nazé, J. Nevalainen, F. Nicastro, P. Nulsen, T. Ohashi, P. O'Brien, J. Osborne, L. Oskinova, F. Pacaud, F. Paerels, M. Page, I. Papadakis, G. Pareschi, R. Petre, P.-O. Petrucci, E. Piconcelli, I. Pillitteri, C. Pinto, J. de Plaa, E. Pointecouteau, T. Ponman, G. Ponti, D. Porquet, K. Pounds, G. Pratt, P. Predehl, D. Proga, D. Psaltis, D. Rafferty, M. Ramos-Ceja, P. Ranalli, E. Rasia, A. Rau, G. Rauw, N. Rea, A. Read, J. Reeves, T. Reiprich, M. Renaud, C. Reynolds, G. Risaliti, J. Rodriguez, P. Rodriguez Hidalgo, M. Roncarelli, D. Rosario, M. Rossetti, A. Rozanska, E. Rovilos, R. Salvaterra, M. Salvato, T. Di Salvo, J. Sanders, J. Sanz-Forcada, K. Schawinski, J. Schaye, A. Schwope, S. Sciortino, P. Severgnini, F. Shankar, D. Sijacki, S. Sim, C. Schmid, R. Smith, A. Steiner, B. Stelzer, G. Stewart, T. Strohmayer, L. Strüder, M. Sun, Y. Takei, V. Tatischeff, A. Tiengo, F. Tombesi, G. Trinchieri, T.G. Tsuru, A. Ud-Doula, E. Ursino, L. Valencic, E. Vanzella, S. Vaughan, C. Vignali, J. Vink, F. Vito, M. Volonteri, D. Wang, N. Webb, R. Willingale,

J. Wilms, M. Wise, D. Worrall, A. Young, L. Zampieri, J. In't Zand, S. Zane, A. Zezas, Y. Zhang, I. Zhuravleva, The hot and energetic universe: a white paper presenting the science theme motivating the Athena+ mission. arXiv e-prints, arXiv:1306.2307, June 2013
123. P.L. Nolan, D.E. Gruber, J.L. Matteson, L.E. Peterson, R.E. Rothschild, J.P. Doty, A.M. Levine, W.H.G. Lewin, F.A. Primini, Rapid variability of 10-140 KeV X-rays from CYG X-1. Astrophys. J. **246**, 494–501 (1981)
124. M.A. Nowak, R.V. Wagoner, M.C. Begelman, D.E. Lehr, The 67 HZ feature in the black hole candidate GRS 1915+105 as a possible "Diskoseismic" mode. Astrophys. J. Lett. **477**(2), L91–L94 (1997)
125. M.A. Nowak, B.A. Vaughan, J. Wilms, J.B. Dove, M.C. Begelman, Rossi X-ray timing explorer observation of Cygnus X-1. II. Timing analysis. Astrophys. J. **510**(2), 874–891 (1999)
126. J.F. Olive, D. Barret, L. Boirin, J.E. Grindlay, J.H. Swank, A.P. Smale, RXTE observation of the X-ray burster 1E 1724-3045. I. Timing study of the persistent X-ray emission with the PCA. Astron. Astrophys. **333**, 942–951 (1998)
127. P.M. O'Neill, E. Kuulkers, R.K. Sood, M. van der Klis, The X-ray fast-time variability of Sco X-2 (GX 349+2) with RXTE. Mon. Not. R. Astron. Soc. **336**(1), 217–232 (2002)
128. V. Osherovich, L. Titarchuk, Kilohertz quasi-periodic oscillations in neutron star binaries modeled as Keplerian oscillations in a rotating frame of reference. Astrophys. J. Lett. **522**(2), L113–L116 (1999)
129. D. Pandel, P. Kaaret, S. Corbel, Relativistic iron line emission from the neutron star low-mass X-ray binary 4U 1636-536. Astrophys. J. **688**(2), 1288–1294 (2008)
130. P. Peille, D. Barret, P. Uttley, The spectral-timing properties of upper and lower kHz QPOs. Astrophys. J. **811**(2), 109 (2015)
131. F. Pintore, T. Di Salvo, E. Bozzo, A. Sanna, L. Burderi, A. D'Aì, A. Riggio, F. Scarano, R. Iaria, Study of the reflection spectrum of the accreting neutron star GX 3+1 using XMM-Newton and INTEGRAL. Mon. Not. R. Astron. Soc. **450**(2), 2016–2024 (2015)
132. S. Piraino, A. Santangelo, P. Kaaret, B. Mück, A. D'Aì, T. Di Salvo, R. Iaria, N. Robba, L. Burderi, E. Egron, A relativistic iron emission line from the neutron star low-mass X-ray binary GX 3+1. Astron. Astrophys. **542**, L27 (2012)
133. D. Psaltis, C. Norman, On the origin of quasi-periodic oscillations and broad-band noise in accreting neutron stars and black holes. arXiv e-prints, astro-ph/0001391, January 2000
134. R.C. Reis, A.C. Fabian, A.J. Young, Relativistically broadened iron line in the Suzaku observation of the neutron star X-ray binary 4U 1705-44. Mon. Not. R. Astron. Soc. **399**(1), L1–L5 (2009)
135. R.A. Remillard, D. Lin, ASM Team at MIT, NASA/GSFC, New X-ray Transient, XTE J1701-462. Astron. Telegram **696**, 1 (2006)
136. R.A. Remillard, E.H. Morgan, J.E. McClintock, C.D. Bailyn, J.A. Orosz, RXTE observations of 0.1–300 HZ quasi-periodic oscillationsin the microquasar GRO J1655-40. Astrophys. J. **522**(1), 397–412 (1999)
137. M. Revnivtsev, E. Churazov, M. Gilfanov, R. Sunyaev, New class of low frequency QPOs: signature of nuclear burning or accretion disk instabilities? Astron. Astrophys. **372**, 138–144 (2001)
138. E. M. Ribeiro, M. Méndez, G. Zhang, A. Sanna, Relation between the properties of the kilohertz quasi-periodic oscillations and spectral parameters in 4U 1636-53. Mon. Not. R. Astron. Soc. **471**(1), 1208–1214 (2017)
139. E.M. Ribeiro, M. Méndez, M.G.B. de Avellar, G. Zhang, K. Karpouzas, The amplitude of the kilohertz quasi-periodic oscillations in 4U 1636-53 in the frequency-energy space. Mon. Not. R. Astron. Soc. **489**(4), 4980–4991 (2019)
140. R.R. Ross, A.C. Fabian, The spectra of accretion discs in low-mass X-ray binaries. Mon. Not. R. Astron. Soc. **281**, 637 (1996)
141. R.R. Ross, A.C. Fabian, X-ray reflection in accreting stellar-mass black hole systems. Mon. Not. R. Astron. Soc. **381**(4), 1697–1701 (2007)

142. A. Sanna, M. Méndez, D. Altamirano, J. Homan, P. Casella, T. Belloni, D. Lin, M. van der Klis, R. Wijnands, The kilohertz quasi-periodic oscillations during the Z and atoll phases of the unique transient XTE J1701-462. Mon. Not. R. Astron. Soc. **408**(1), 622–630 (2010)
143. A. Sanna, M. Méndez, T. Belloni, D. Altamirano, The time derivative of the kilohertz quasi-periodic oscillations in 4U 1636-53. Mon. Not. R. Astron. Soc. **424**(4), 2936–2942 (2012)
144. A. Sanna, B. Hiemstra, M. Méndez, D. Altamirano, T. Belloni, M. Linares, Broad iron line in the fast spinning neutron-star system 4U 1636-53. Mon. Not. R. Astron. Soc. **432**(2), 1144–1161 (2013)
145. A. Sanna, M. Méndez, D. Altamirano, T. Belloni, B. Hiemstra, M. Linares, Broad iron emission line and kilohertz quasi-periodic oscillations in the neutron star system 4U 1636-53. Mon. Not. R. Astron. Soc. **440**(4), 3275–3284 (2014)
146. N. Shaposhnikov, L. Titarchuk, P. Laurent, Discovery of red-skewed K_{α} iron line in Cyg X-2 with Suzaku. Astrophys. J. **699**(2), 1223–1228 (2009)
147. L. Stella, M. Vietri, Lense-thirring precession and quasi-periodic oscillations in low-mass X-ray binaries. Astrophys. J. Lett. **492**(1), L59–L62 (1998)
148. L. Stella, M. Vietri, S. Morsink, General relativistic effects in Qpos. Astrophys. Lett. Commun. **38**, 57 (1999)
149. T.E. Strohmayer, Discovery of a 450 HZ quasi-periodic oscillation from the microquasar GRO J1655-40 with the Rossi X-ray timing explorer. Astrophys. J. Lett. **552**(1), L49–L53 (2001)
150. T. Strohmayer, W. Zhang, A. Smale, C. Day, J. Swank, L. Titarchuk, U. Lee, U 1728-34. IAU Circulars **6387**, 2 (1996)
151. T. Strohmayer, W. Zhang, J. Swank, U 1728-34. IAU Circulars **6320**, 1 (1996)
152. T.E. Strohmayer, W. Zhang, J.H. Swank, A. Smale, L. Titarchuk, C. Day, U. Lee, Millisecond X-ray variability from an accreting neutron star system. Astrophys. J. Lett. **469**, L9 (1996)
153. R.A. Sunyaev, L.G. Titarchuk, Comptonization of X-rays in plasma clouds. Typical radiation spectra. Astron. Astrophys. **500**, 167–184 (1980)
154. T.M. Tauris, E.P.J. Van Den Heuvel, *Formation and Evolution of Compact Stellar X-ray Sources* (Cambridge University Press, Cambridge, 2010), p. 623
155. Jr. Terrell, N. James, Shot-noise character of Cygnus X-1 pulsations. Astrophys. J. Lett. **174**, L35 (1972)
156. J.S. Troyer, E.M. Cackett, Spectral-timing analysis of the lower kHz QPO in the low-mass X-ray binary Aquila X-1. Astrophys. J. **834**(2), 131 (2017)
157. J.S. Troyer, E.M. Cackett, P. Peille, D. Barret, A systematic spectral-timing analysis of kilohertz quasi-periodic oscillations in the Rossi X-ray timing explorer archive. Astrophys. J. **860**(2), 167 (2018)
158. P. Uttley, I.M. McHardy, The flux-dependent amplitude of broadband noise variability in X-ray binaries and active galaxies. Mon. Not. R. Astron. Soc. **323**(2), L26–L30 (2001)
159. P. Uttley, I.M. McHardy, S. Vaughan, Non-linear X-ray variability in X-ray binaries and active galaxies. Mon. Not. R. Astron. Soc. **359**(1), 345–362 (2005)
160. P. Uttley, T. Wilkinson, P. Cassatella, J. Wilms, K. Pottschmidt, M. Hanke, M. Böck, The causal connection between disc and power-law variability in hard state black hole X-ray binaries. Mon. Not. R. Astron. Soc. **414**(1), L60–L64 (2011)
161. M. van der Klis, Fourier techniques in X-ray timing, in *NATO Advanced Science Institutes (ASI) Series C*, vol. 262, ed. by H. Ögelman, E.P.J. van den Heuvel. NATO Advanced Science Institutes (ASI) Series C (1989), p. 27
162. M. van der Klis, F. Jansen, J. van Paradijs, W.H.G. Lewin, E.P.J. van den Heuvel, J.E. Trumper, M. Szatjno, Intensity-dependent quasi-periodic oscillations in the X-ray flux of GX5-1. Nature **316**(6025), 225–230 (1985)
163. M. van der Klis, J. Swank, W. Zhang, K. Jahoda, E. Morgan, W. Lewin, B. Vaughan, J. van Paradijs, Scorpius X-1. IAU Circulars **6319**, 1 (1996)
164. M. van der Klis, J.H. Swank, W. Zhang, K. Jahoda, E.H. Morgan, W.H.G. Lewin, B. Vaughan, J. van Paradijs, Discovery of submillisecond quasi-periodic oscillations in the X-ray flux of Scorpius X-1. Astrophys. J. Lett. **469**, L1 (1996)

165. M. van der Klis, R.A.D. Wijnands, K. Horne, W. Chen, Kilohertz quasi-periodic oscillation peak separation is not constant in Scorpius X-1. Astrophys. J. Lett. **481**(2), L97–L100 (1997)
166. M. van Doesburgh, M. van der Klis, Testing the relativistic precession model using low-frequency and kHz quasi-periodic oscillations in neutron star low-mass X-ray binaries with known spin. Mon. Not. R. Astron. Soc. **465**(3), 3581–3606 (2017)
167. M. van Doesburgh, M. van der Klis, On the frequency correlations of low-frequency QPOs with kilohertz QPOs in accreting millisecond X-ray pulsars. Mon. Not. R. Astron. Soc. **490**(4), 5270–5284 (2019)
168. M. van Doesburgh, M. van der Klis, S.M. Morsink, The highest frequency kHz QPOs in neutron star low-mass X-ray binaries. Mon. Not. R. Astron. Soc. **479**(1), 426–434 (2018)
169. S. van Straaten, E.C. Ford, M. van der Klis, M. Méndez, P. Kaaret, Relations between timing features and colors in the X-ray binary 4U 0614+09. Astrophys. J. **540**(2), 1049–1061 (2000)
170. S. van Straaten, M. van der Klis, T. di Salvo, T. Belloni, A multi-Lorentzian timing study of the Atoll sources 4U 0614+09 and 4U 1728-34. Astrophys. J. **568**(2), 912–930 (2002)
171. S. van Straaten, M. van der Klis, M. Méndez, The Atoll source states of 4U 1608-52. Astrophys. J. **596**(2), 1155–1176 (2003)
172. S. van Straaten, M. van der Klis, R. Wijnands, Relations between timing features and colors in accreting millisecond pulsars. Astrophys. J. **619**(1), 455–482 (2005)
173. B.A. Vaughan, M.A. Nowak, X-ray variability coherence: how to compute it, what it means, and how it constrains models of GX 339-4 and Cygnus X-1. Astrophys. J. Lett. **474**(1), L43–L46 (1997)
174. B.A. Vaughan, M. van der Klis, M. Méndez, J. van Paradijs, R.A.D. Wijnands, W.H.G. Lewin, F.K. Lamb, D. Psaltis, E. Kuulkers, T. Oosterbroek, Discovery of microsecond time lags in kilohertz QPOs. Astrophys. J. Lett. **483**(2), L115–L118 (1997)
175. B.A. Vaughan, M. van der Klis, M. Méndez, J. van Paradijs, R.A.D. Wijnands, W.H.G. Lewin, F.K. Lamb, D. Psaltis, E. Kuulkers, T. Oosterbroek, Erratum: discovery of microsecond time lags in kilohertz QPOs. Astrophys. J. Lett. **509**(2), L145 (1998)
176. R.V. Wagoner, Relativistic diskoseismology. Phys. Rep. **311**, 259–269 (1999)
177. Y. Wang, M. Méndez, A. Sanna, D. Altamirano, T.M. Belloni, The reflection spectrum of the low-mass X-ray binary 4U 1636-53. Mon. Not. R. Astron. Soc. **468**(2), 2256–2264 (2017)
178. Y. Wang, M. Méndez, D. Altamirano, G. Zhang, T.M. Belloni, E.M. Ribeiro, M. Linares, A. Sanna, S.E. Motta, J.A. Tomsick, Study of the X-ray properties of the neutron star binary 4U 1728-34 from the soft-to-hard state. Mon. Not. R. Astron. Soc. **484**(3), 3004–3016 (2019)
179. A.L. Watts, Thermonuclear burst oscillations. Annu. Rev. Astron. Astrophys. **50**, 609–640 (2012)
180. A.L. Watts, W.F. Yu, J. Poutanen, S. Zhang, S. Bhattacharyya, S. Bogdanov, L. Ji, A. Patruno, T.E. Riley, P. Bakala, A. Baykal, F. Bernardini, I. Bombaci, E. Brown, Y. Cavecchi, D. Chakrabarty, J. Chenevez, N. Degenaar, M. Del Santo, T. Di Salvo, V. Doroshenko, M. Falanga, R.D. Ferdman, M. Feroci, A.F. Gambino, M.Y. Ge, S.K. Greif, S. Guillot, C. Gungor, D.H. Hartmann, K. Hebeler, A. Heger, J. Homan, R. Iaria, J. in't. Zand, O. Kargaltsev, A. Kurkela, X.Y. Lai, A. Li, X.D. Li, Z.S. Li, M. Linares, F.J. Lu, S. Mahmoodifar, M. Méndez, M. Coleman Miller, S. Morsink, J. Nättilä, A. Possenti, C. Prescod-Weinstein, J.L. Qu, A. Riggio, T. Salmi, A. Sanna, A. Santangelo, H. Schatz, A. Schwenk, L.M. Song, E. Šrámková, B. Stappers, H. Stiele, T. Strohmayer, I. Tews, L. Tolos, G. Török, D. Tsang, M. Urbanec, A. Vacchi, R.X. Xu, Y.P. Xu, S. Zane, G.B. Zhang, S.N. Zhang, W.D. Zhang, S.J. Zheng, X. Zhou, Dense matter with eXTP. Sci. China Phys. Mech. Astron. **62**(2), 29503 (2019)
181. N.E. White, L. Stella, A.N. Parmar, The X-ray spectral properties of accretion disks in X-ray binaries. Astrophys. J. **324**, 363 (1988)
182. R.A.D. Wijnands, M. van der Klis, Discovery of two simultaneous kilohertz quasi-periodic oscillations in KS 1731-260. Astrophys. J. Lett. **482**(1), L65–L68 (1997)
183. R.A.D. Wijnands, M. van der Klis, J. van Paradijs, W.H.G. Lewin, F.K. Lamb, B. Vaughan, E. Kuulkers, Discovery in 4U 1636-53 of two simultaneous quasi-periodic oscillations near 900 HZ and 1176 HZ. Astrophys. J. Lett. **479**(2), L141–L144 (1997)

184. R. Wijnands, J. Homan, M. van der Klis, E. Kuulkers, J. van Paradijs, W.H.G. Lewin, F.K. Lamb, D. Psaltis, B. Vaughan, Discovery of kHz quasi-periodic oscillations in the Z source Cygnus X-2. Astrophys. J. Lett. **493**(2), L87–L90 (1998)
185. W. Yu, M. van der Klis, Kilohertz quasi-periodic oscillation frequency anticorrelated with millihertz quasi-periodic oscillation flux in 4U 1608-52. Astrophys. J. Lett. **567**(1), L67–L70 (2002)
186. W. Yu, M. van der Klis, P.G. Jonker, Dependence of kilohertz quasi-periodic oscillation properties on the normal-branch oscillation phase in Scorpius X-1. Astrophys. J. Lett. **559**(1), L29–L32 (2001)
187. W. Zhang, T.E. Strohmayer, J.H. Swank, Discovery of two simultaneous kilohertz quasi-periodic oscillations in the persistent flux of GX 349+2. Astrophys. J. Lett. **500**(2), L167–L169 (1998)
188. G. Zhang, M. Méndez, A. Sanna, E.M. Ribeiro, J.D. Gelfand, Relation between spectral changes and the presence of the lower kHz QPO in the neutron star low-mass X-ray binary 4U 1636-53. Mon. Not. R. Astron. Soc. **465**(4), 5003–5010 (2017)
189. A. Zoghbi, P. Uttley, A.C. Fabian, Understanding reverberation lags in 1H0707-495. Mon. Not. R. Astron. Soc. **412**(1), 59–64 (2011)
190. A. Zoghbi, A.C. Fabian, C.S. Reynolds, E.M. Cackett, Relativistic iron K X-ray reverberation in NGC 4151. Mon. Not. R. Astron. Soc. **422**(1), 129–134 (2012)

Index

© Springer-Verlag GmbH Germany, part of Springer Nature 2021

T. M. Belloni et al. (eds.), *Timing Neutron Stars: Pulsations, Oscillations and Explosions*, Astrophysics and Space Science Library 461,
https://doi.org/10.1007/978-3-662-62110-3